Growth, Development, and Pattern

A Series of Books in Biology

Editors: George W. Beadle, Ralph Emerson, and Douglas M. Whitaker

Growth, Development, and Pattern

N. J. Berrill, *McGill University*

Drawings by Evan Gillespie

 W. H. Freeman and Company

San Francisco and London

for Michael

Preface

As the natural evolution of an individual's interest spanning many years, this descriptive and analytical account of the phenomena of growth and emergent pattern inevitably has a personal quality. This perhaps is as it should be. Thus while the material and topics are drawn from the entire living kingdom, primary interest centers on the processes of development from somatic tissues among the so-called lower forms. And although the property of development is regarded as being inherent in all protoplasm, I believe that the more direct approach to the elucidation of that property lies in the study of the simpler kinds of propagation, reconstitution, and repair exhibited so abundantly by the more vegetative animal forms and by plants in general. This primary problem is the nature of emergent protoplasmic organization at the biological, rather than at the chemical, level.

Some account of the investigations of the shoot meristem of higher plants is included, since there are indisputable connections between the points of view inspiring this work and those pervading the field of animal morphology during the past half-century. However, the indirect effects of hormones, light, and temperature have been excluded—just as they have been omitted from the chapters on animal morphogenesis. Gratitude is here expressed to Dr. W. G. Boll for his helpful discussions on the various problems relating to morphogenesis in plants. It should be borne in mind, however, that this is a book written by a zoologist primarily for zoologists, and the inclusion of plant material represents an excursion rather than an attempt at a balanced survey.

The inherent interrelationships among the rich and diverse material presented in this book are multidimensional; but the written word must follow a linear course. No single organizational plan could possibly convey the complex reality, and the scheme adopted for this work is merely the one—of many alternates—that seemed best suited to a general survey. To overcome in some degree the single-track course of words alone, cross references are supplied throughout the text.

Illustrations, too, reflect a subjective choice. They have been selected to extend as well as to support the narrative. Neither stands alone, and both as a unit are intended to be suggestive—to present enough detail of the various organisms and developmental events so that these may "speak for themselves." Only by the extension of such detail may the gap between the dynamic, living structure and the underlying macromolecular organization be bridged.

Illustrations whose legends carry no indication of source have been taken from my own publications. Personal credit is given to the authors of all other illustrations. Acknowledgment is made here to the publishers of the following journals for permission to make use of the illustrations: *American Journal of Botany, American Naturalist, Annales Société Royale Zoologie Belge, Annals of Botany, Annals South Africa Museum, Année Biologique, Archiv Entwicklungsmechanik der Organismen, Archives d'Anatomie microscopique, Archives de Zoologie expérimentale, Biochemical Cytology, Biological Bulletin, Bulletin Société Zoologie de France, Bulletin Torrey Botanical Club, Experimental Biology Symposium, Hilgardia, Journal of Cellular and Comparative Physiology, Journal of Embryology and Experimental Morphology, Journal of Experimental Zoology, Journal Marine Biology of United Kingdom, Journal of Morphology, Journal of Protozoology, University of California Publications in Zoology, University of Hawaii Press Publications.*

Not least symptomatic of a pervasive personal quality is the selectivity governing the inclusion of individual investigations, past and present. The account concerns both organisms and biologists; and it is hoped that some justice has been done each. It is worth re-emphasizing, however, that the content and scope are limited, and it is obvious that preponderant attention has been given to my own work and to the work of Paul Brien and Yô K. Okada on the phenomena of asexual development in several groups of invertebrates. This I consider to be my privilege. In effect, then, the book reflects a particular point of view of the problems of form—often to the exclusion of other points of view and other biological investigations that are equally important.

There is no final explanation in this book; for with every question answered, several more arise. The tentative conclusions offered, however, derive from a wide range of animal and plant material—much of which reflects the personal bias of fascination and pleasure in particular organisms—and from the amazing diversity of developmental performances—which are of interest to all students of growth and form. Within such admitted limits as the concern more with form than with substance, with morphogenesis more than with histogenesis, both the survey and the bibliographical references are intended as a guide to past work, persistent phenomena, and continuing opportunities for investigation. Thus the book is designed for advanced students and for those of my colleagues who still have time to explore.

March, 1961   *N. J. Berrill*

Contents

Chapter 1

Introduction: Cells, Genes, and Organisms

THE DEVELOPMENT of organisms has presented an unfolding pageant and an intellectual challenge to all who have studied it during the past century. Yet, no general theory of development has emerged, in spite of the mounting mass of observational and experimental information. One reaction to this inability to recognize and formulate laws of development or to produce a satisfying interpretation of the nature and material basis of organization has been to deny the need for them—to aver that there *is* no general explanation of developmental phenomena, but only a multiplicity of more or less independent processes working together in such a way as to produce this or that type of organism. In other words,

> Nuclear division, cell growth, cell division, cell aggregation, movements of cell complexes, differential growth, cytological differentiation, polarity, orientation— these are only a modest selection from the list of component phenomena into which we have learned to decompose development. The revelation of the multiplicity of developmental processes and mechanisms has been a sad disappointment, for it has removed all hope of a general, comprehensive and universal formula of development.[1]

Whether such pessimism is justifiable remains to be seen, although it is not widely shared. On the other hand, the truly remarkable progress in nuclear genetics, for instance, has encouraged faith in an eventual interpretation of development in terms of genetics and chemistry, so that Waddington states,

> . . . it has always been clear that, since genes are the determinants of the characters of an animal, they must also provide a set of terms in which we shall eventually be able to give a causal account of development which is, at its own level of analysis, complete and comprehensive.[2]

[1] Weiss, *Am. Nat.*, **74**:41 (1940).
[2] Waddington, *Fol. Biotheoret.*, **3**:128 (1948).

Even in Waddington's statement, however, the organism is implicitly regarded as the final product of interacting processes originating with the genes; and the question again arises whether any such account can be adequate. Is an organism solely the product of its genes or the sum of its cells? Stern expresses the problem more acutely:

> The same genetic constitution can produce first a caterpillar and then a butterfly, first a tadpole and then a frog, first a polyp and then a jellyfish. There is no difference in the genes of an egg which develops into a large female of the marine worm *Bonellia* or into the microscopic, semi-parasitic male which lives inside the body of the female. The genes in the egg of a termite may lead to the appearance of queens, soldiers, or workers.[3]

Perhaps the difficulty lies in the form of the questions: So long as they are framed in terms of genes or cells, so long will we receive information in those terms and no others. Most organisms consist of many cells of many kinds in various groupings, but each cell is assumed to have the same set of genes as the others; however, some organisms exhibit elaborate form and structure, but do not consist of definitive cells at all. Neither cells nor genes, taken by themselves, sufficiently account for the development and maintenance of pattern.

Weiss recognizes this insufficiency; in his discussion of the cell theory, he states that

> . . . every step in development reveals the cell in a double light: partly as an active worker and partly as a passive subordinate to powers which lie entirely outside of its own competence and control, i.e., supracellular powers. Now it is perfectly true that some of these latter result from interactions of cell individuals and are, therefore, of cellular origin. But it is equally true—and the findings of experimental embryology are one rich store of evidence for our assertion—that many of these are supracellular from the beginning. . . . One frequently refers to these organizing entities under the term of "fields."[4]

Any account of organization and differentiation must recognize this dual nature of cells. Clearly, cells have their individuality, each with its own nucleus, its cytoplasm, and its territorial limits. Furthermore, cell differentiations or specializations are for the most part well-definable types that do not blend into one another, and for any particular kind of organism are of a certain limited number—fewer than 10 in a hydra but many times that number in a mammal. For each organism there seems to be a specific variety of paths of specialization open to the descendants of an unspecialized cell; and knowledge of the factors, internal or external, which set a cell into this or that path, is fundamental to understanding the general phenomena of tissue differentiation. At the same

[3] Stern, *Am. Sci.,* **42**:246 (1954).
[4] Weiss, *op. cit.,* p. 45.

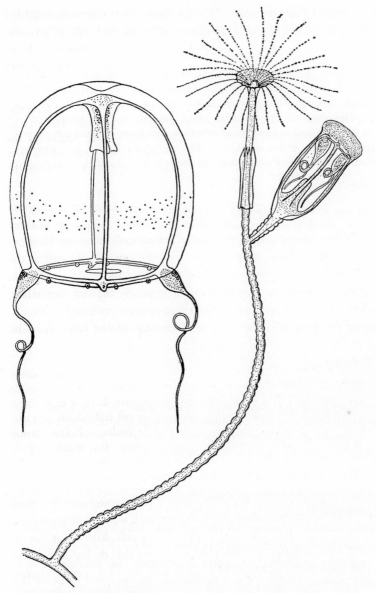

Figure 1-1. Polymorphism in hydroid *Campanulina paracuminata,* showing expanded polyp with attached and free stage of medusa. (After Rees.)

time, such factors seem to be of less importance to an appreciation of pattern (with which this book is more concerned). It is, in fact, doubtful whether a unit of pattern or structure exists. Neither genes nor nucleus, which all cells of a specific organism supposedly share in common, can explain the regional differences in form and finer structure. Nor are shape and pattern necessarily the product of cell proliferation for, as Sinnott said in speaking of the growth of gourds,

> . . . evidence is now available that the organ is equally independent of the next lower unit, the cell. If the growth of a cucurbit fruit from a tiny ovary primordium until several days after fertilization is measured, the rate of increase is found to be constant. . . . The mass of the material being poured into the young fruit is growing at a constant rate regardless of whether this mass is being cut up into new cells rapidly, slowly or not at all. . . . The unity of behavior, and thus presumably the unity of organization, inheres in the whole and not in its elements.[5]

Elsewhere he says that

> . . . before we can intelligently set up experiments to determine the integrating and coordinating growth processes which control development and produce specific forms, we must first obtain precise descriptive information as to *exactly how development proceeds.*—[T]he morphologist . . . *must present his results in quantitative terms.*[6]

The validity of this prerequisite may be brought home by the fact that while experimental data on reconstitution in *Tubularia* have been published in almost every year during the present century, we still have to turn to Louis Agassiz of the nineteenth century for anything approaching an adequate account of the structure of *Tubularia*.

Sinnott concludes, however, with

> . . . this fundamental paradox: that protoplasm, itself liquid, formless and flowing, inevitably builds those formed and coordinated structures of cell, organ and body in which it is housed. If dynamic morphology can come to the center of this problem, it will have brought us close to the ultimate secret of life itself.[7]

The discovery of the endoplasmic reticulum by means of electron microscopy, although showing cytoplasm to be anything but formless, leaves the main problem virtually untouched.

If not the cells as such, what makes the cell state which is the organism? What subordinates them and controls their behavior and developmental destiny? Apart from entelechies and other intangible abstractions, two notable attempts have been made during recent decades to identify the material basis of the organizing force. The older, that of Child, aims at correlating observable axial gradients of structure with gradients in oxidative metabolism and has the merit of coordinating quantities of diverse observations—as his culminative book *Problems of Pattern and Development* fully demonstrates. The other, the

---

[5] Sinnott, *Sci.,* **89**:44 (1939).
[6] *Ibid.,* **85**:64 (1937).
[7] *Ibid.,* p. 65.

electrodynamic theory of Lund, more recently applied by Burr and by Moment, substitutes the pattern of electrical potentials for the metabolic pattern as the over-all control. Both theories appear to suffer from the same weakness, however, in that they assume that protoplasm—to give the substance in question a single name—can be fully described or explained in terms of familiar chemistry and physics.

That such an assumption is almost certainly invalid is indicated by Szent-Györgyi in his discussion of the physiology of muscle contraction:

> The chief property of muscle is that we do not understand it. The more we know about it, the less we understand and it looks as though we would soon know everything and understand nothing. The situation is similar in most other biological processes and pathological conditions, such as the degenerative diseases. This suggests that some very basic information is missing.[8]

His conclusion is that biological phenomena, such as muscular contraction, cannot be described in terms of classical chemistry but belong to the domain of quantum mechanics, to quantum biology:

> . . . we will have to introduce three new factors into our thinking if we want to understand biological reactions: water structures, the electromagnetic field, and triplets or some other unusual form of excitation made possible by water structures.[9]

If organizing forces truly exist, they should have both a material basis and a definable location in a living system. Since the only matter constituting an organism is cells and their products, or nuclear and cytoplasmic material of equivalent nature, and since organizing forces seem to be of a supracellular character, we are virtually compelled to look at the conjoined surfaces of cells and tissues, where cytoplasmic continuities are at least possible for both site and substance. And we may be wise, in the light of Szent-Györgyi's mature reflections, to postpone attempts at chemical or physical analyses until the biological phenomena have been more fully examined and classified at their own *biological* level. Needham (1942) has stated that we must consider the universe as a series of levels of organization and complexity, ranging from the subatomic level, through the atom, the molecule, the colloidal particle, the living nucleus and cell to the organ and the organism, the psychological and sociological entity. It follows that the laws and regularities which we find at one level cannot be expected to appear at lower levels. The conditions for their appearance do not exist there. We must seek to elucidate the regularities which occur at each of these levels without attempting either to force the higher or coarser processes

---

[8] Szent-Györgyi, *Sci.*, **124**:873 (1957).
[9] *Ibid.*, p. 875.

into the framework of the lower or finer processes, or conversely to explain the lower by the higher.

This point of view is dominant throughout the succeeding pages of the present discussion. Several assumptions are provisionally made: (1) that at the cortical cytoplasmic level there is substantial continuity from cell to cell, at least in certain significant tissues—a continuity which provides the material basis for organizational effects; (2) that such cytoplasmic matter exists in states and complexities which at present are for the most part unknown and indescribable in terms of the chemistry and physics of a lower hierarchical level; and (3) that new properties or qualities emerge as the scale of organizational territories increases. Accordingly, it should be possible to describe laws of developmental behavior— for instance, relating the sizes and growth rates of various primordia with their structural expressions—more or less in terms of four-dimensional geometry. By such a description, it may be possible to circumscribe the unknown so as to give it shape and to determine where the well of ignorance truly lies. Such is the aim of this book: to describe the dynamics of development and maintenance of form and structure at a purely biological level of inquiry and in those organisms that best lend themselves to this inquiry.

The study of morphogenesis has naturally been for the most part the analytical study of developing eggs, although in most organisms the egg seems to be a highly specialized developmental unit. The basic phenomena of development are therefore likely to be complicated by secondary processes and events, and certain fundamentals tend to be masked. Child has stated that

> . . . embryonic development is not the only, and perhaps in many respects not the most interesting and significant, material for study of various developmental problems. An adequate general theory of development must be based on the less specialized forms of development; it must recognize and distinguish the factors common to difference forms of development from those characteristic of only a particular form; also, it must attempt to interpret the more highly specialized embryonic form of development as far as possible in terms of the common factors.[10]

There are many kinds of developmental units, all of which exhibit the essentials of development, or evolving form; but some units do so in a simpler manner than do others. Units such as limb buds of vertebrate embryos, regeneration blastemata, small buds and growing terminals produced by many invertebrates (particularly those of colonial status), and the meristems of plants have all been subjected to analytical observation and experiment. Because they are less specifically concerned with reproduction and hereditary mechanisms, all these units are the more likely to express the basic morphogenetic qualities of protoplasm which are general properties of living matter. On the other hand,

[10] Child, *Problems of Pattern and Development,* p. 15.

eggs exhibit an astonishing variation, ranging from those of coelenterates with no discernible trace of preorganization to those of such highly specialized, so-called determinate types as polychaetes and ascidians. The development of all units—egg, embryo, regenerate, or bud—involves the problem of emergent form; while form, being itself dynamic in the case of living substance, is actively maintained after development and growth have brought it into being.

This dualism is also true of the individual cell, whether it be a gamete, a protistan, or a unit in a multicellular organism. Whichever it may be and whatever the relationship between multicellular structure and the cells that contribute to that structure, the cell as such poses a fundamental morphological question: Is the cell the sum of various and numerous self-reproducing components, or is it essentially an indivisible whole? To phrase the question another way: Is cytoplasm the product of interacting genes, or are nucleus and cytoplasm equally autonomous although interacting as a system? Or, in Baitsell's words,

> . . . may it not be possible that the cell is a protoplasmic crystal in which an almost infinite number of protein molecules, beginning with the genes in the chromosomes, are associated in a specific ultramicroscopic pattern characteristic of a particular type of cell? In such a condition independent protein molecules are not present, but all are organized to form the complete cell unit exhibited in the crystalline pattern of a specific type of protoplasm.[11]

Although the fundamental question should not be prejudged, all the newer analytical techniques—X-ray, phase contrast, electron microscopy, and biochemical cytology—indicate both the wholeness of the cell system and the dynamically cystalline quality of its substance.

REFERENCES FOR
Introduction

Agassiz, L. 1862. *Contributions to the Natural History of the United States of America.* Boston, Mass.: Little, Brown and Co.

Baitsell, G. A. 1955. "The Cell as a Structural Unit." *Am. Sci.* **43**: 133-43.

Burr, H. S. 1932. "An Electro-dynamic Theory of Development." *J. Comp. Neurol.* **56**:347-71.

Child, C. M. 1941. *Problems of Pattern and Development.* Chicago, Ill.: Univ. of Chicago Press.

Lund, E. J. 1947. *Bioelectric Fields and Growth.* Austin, Tex.: Univ. of Texas Press.

Moment, G. B. 1952. "A Theory of Growth Limitation." *Am. Nat.* **87**:139-53.

Needham, J. 1942. *Biochemistry and Morphology.* New York: Cambridge Univ. Press.

[11] Baitsell, *Am. Sci.,* **43**:139 (1955).

Sinnott, E. W. 1937. "Morphology as a Dynamic Science." *Sci*. **85**:61-5.
——. 1939. "The Cell and the Problem of Organization." *Ibid*. **89**:416.
Stern, C. 1954. "Two or Three Bristles." *Am. Sci*. **42**:212-47.
Szent-Györgyi, A. 1957. "Bioenergetics." *Sci*. **124**:873-5.
Waddington, C. H. 1948. "The Concept of Equilibrium in Embryology." *Fol. Biotheoret*. **3**:128-38.
Weiss, P. 1940. "The Problem of Cell Individuality in Development." *Am. Nat*. **74**:34-46.

PART I
CELLS AND CELL AGGREGATIONS

Morphogenesis is exhibited at two levels: the progressive development and maintenance of structure within the individual cell, and the formation of structure and pattern, which depend on the aggregation or combination of many cells to constitute organismal units of a higher order. These phenomena are seen in striking form in unicellular and multicellular protistans and in sponges. In each of the following five chapters the various processes and associated problems are discussed approximately in order of increasing complexity, i.e., from the so-called simple unicellular state through the multinuclear unicellular state and the aggregated state to the completely unified multicellular state. New qualities arise at each level of complexity; and it is hoped that the comparative approach, although not explicit, will yield some insight into the nature of emergent organization.

Chapter 2

Cells: Fine Structure of Protoplasm

Intracellular Fiber Systems

One of the earliest discoveries resulting from experimental analysis of the cell was that the eggs of many animals could develop normally after all of their visible inclusions had been stratified by means of the centrifuge, i.e., that the polarity and pattern of organization of an egg are not changed by such dislocation of egg substances. In a classical investigation of the effect of centrifugal force on the structure of *Crepidula* eggs, Conklin (1917) concluded that cell polarity is vested neither in electrical polarity nor in surface-tension forces, but in what he describes as the spongioplasmic framework of the cytoplasm, the strands which may be stretched or bent but are rarely broken:

> The pattern of cleavage is dependent upon the position and direction of the spindles and this is determined, not by inclusions, but by the spongioplasm which holds nuclei, centrospheres and mitotic figures in a definite relation to the cell axes, and which is so elastic that when it is distorted by pressure or centrifugal force, it tends to bring parts back to their normal positions. . . . Protoplasmic flowing and intracellular movements are probably caused by the contractility of the spongioplasm.[1]

Now, several decades later, Conklin's view is supported by findings with the electron microscope and is reflected in current interpretations. Frey-Wyssling (1948) visualizes such an organization of protoplasm as a bonded molecular structure of threadlike molecules, with linear proteins connected in such a way as to form an irregular three-dimensional mesh with side chains as points of attachment, the mesh strands corresponding to individual polypeptide chains and forming a very fine molecular lattice. In this concept, residual end groups remain free to interact with water, hydrophilic groups, and lipids, while the side

[1] Conklin, *J. Exp. Zool.*, **22**:366 (1917).

chains are not fixed bridges but temporary bonds continually being broken, re-formed, and rearranged. Cell substance, in other words, has the properties of liquid crystals. An organized molecular structure, of which cell polarity is but one outstanding feature, is automatically produced by the components and re-forms after being disturbed.

Cell polarity in fact is basic to much of a cell's character and activity. In animal eggs it is fixed while the cell is still a part of the ovarian tissue, but in the eggs of algae such as *Fucus* the primary polarity is established after the eggs have been shed, by one of several agents. Whitaker (1936) has shown that gradients of auxin, hydrogen ion, or carbon dioxide or the incidence of white light are clearly of great importance. Yet, according to Bünning (1952), the ultimate cause of such gradients, i.e., the polarity induced within the young egg cell or spore, cannot be a chemical gradient but only a specific protoplasmic structure. Only by the assumption that a polar structure is the ultimate cause of polarity, can it be explained why polarity does not disappear during a plant's period of dormancy, when all metabolic energy-yielding processes necessary for the maintenance of a chemical or electrical gradient ceases. Apparently, this polar structure need not be present throughout the entire protoplasm; it seems to be confined to the peripheral parts of the cell closely adjoining the cell wall, since polarizing influences such as light are effective only if they strike the peripheral strata of the protoplasm. Moreover, the predominant importance of the cortex in the determination of animal eggs is indicated by experiments of Hörstadius (1949) and his co-workers, in which 50 per cent of the interior cytoplasm of a sea-urchin egg was removed before fertilization without causing any deviation, except size, from normal larval development.

By revealing the fine structure of protoplasm, electron microscopy has added much to the portrait of the cell, which thereby appears to be far more complex than was previously thought. Electron microscopy is, of course, mainly a study of dead and dehydrated material; yet, whereas many investigators speak of lamellar and tubular structures, or the "ergastoplasm" of secretory cells, Bretschneider (1952) describes the basic cytoplasm as fine protein filaments of 80–200 millimicrons (mμ) in diameter. He concludes that if the approximate diameter of a linear protein molecule is 20 Å (the length of the longest amino acid in the chain), then a cross section contains not more than 4–10 protein chains accommodated side by side—and probably less because of spiraling and separation. The filaments form a very regular, three-dimensional, quasi-hexagonal lattice with definite transverse and longitudinal axes. Structure of this kind appears even in the living cell (as shown by the surface-cell film technique of electron microscopy), for all of the plasm of neuroglia cells, before fixation, consists of a regular meshwork of filaments 150 Å thick and a mesh length of 400 Å.

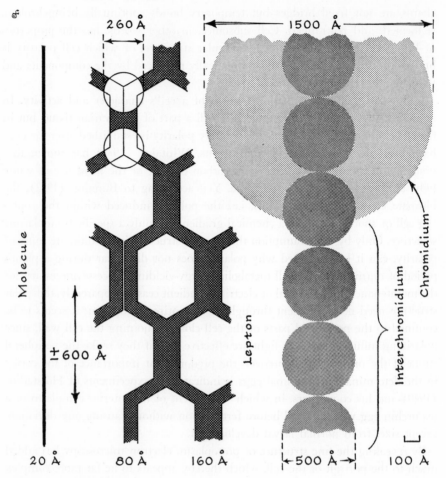

Figure 2-1. Comparative scale of protein molecule, lepton lattice, and chromidium. (After Bretschneider.)

Bretschneider calls the mesh a "lepton lattice," the units of the lattice constituting the "leptons" of the cytoplasm. As he points out, it is possible that a true molecular lattice of the type envisioned by Frey-Wyssling may underlie it, and leptons may represent structure of a second order of magnitude and characteristic of the hyaloplasm generally. On this basis, too, the so-called ergastoplasm of secretory cells, rich in RNA, is formed of agglutinated bundles of leptons 500-600 Å thick, which are joined by bridges of leptons of normal thickness; whereas, the axoplasm of neurons, instead of containing definitive neurofibrils, has a basic structure of leptons 75-200 Å thick, parallel to one another and coalescing to form thicker fibrils which are capable of taking up metal ions and dyes.

The lepton lattice seems to be essentially the same as the endoplasmic reticulum described by Pallade (1956) and others, the reticulum being a continuous network of membrane-bound cavities permeating the entire cytoplasm—from cell membrane to nucleus, actually contiguous with the cell membrane, at least at various points. With particular reference to the intensive membrane activity known as pinocytosis (exhibited by macrophages in tissue culture), i.e., the continuous movement of "ruffles" which results in the incorporation into the cytoplasm of droplets of fluid from the surrounding medium, Pallade interprets electron micrographs as evidence that the incorporated cell membrane is fed into the smooth surfaced parts of the endoplasmic reticulum. He believes that, at a submicroscopic level, this phenomenon is common among animal cells. If so, animal cells, in general, may be capable of incorporating matter in bulk, not merely as molecules or ions, from the surrounding medium—a possibility of importance in relation to induction phenomena.

Polarized light microscopy has been applied most intensively to studies of the cell membrane of erythrocytes and of the structure of myofibrils. By measuring the diffraction pattern produced at the surface of the spherical red-cell ghost and comparing that pattern with the theoretical diffraction patterns calculated for spherical birefringent membranes of varying thickness, Mitchison (1953) determined that the thickness of the dehydrated membrane is about 0.5 μ. The amount of lipid in the ghost is known to be enough to form a double monolayer about 40 Å thick at or very near the surface. It is also known that the form birefringence of the red-cell membrane is positive with respect to the tangent, while the intrinsic birefringence is negative in the same respect, indicating that a predominantly radial molecular orientation occurs throughout the thickness of the dehydrated-protein part of the membrane and therefore of the protein itself. In other words, the cell cortex contains protein chains that are looped or coiled so as to lie radially, but at the same time arranged in elongate bundles lying tangentially to the surface. There is also much indirect evidence that the cortex of other cells is very similar to that of the red cell; i.e., the cortex consists of lipid and protein, with the lipid as the thin outer layer. The same combination of positive and negative birefringence has been found in nerve cells, *Amoeba,* and sea-urchin eggs.

On the other hand, X-ray diffraction and optical polarization studies, confirmed by electron microscopy, of muscle fibrils indicate that the myosin filament, which forms the basis of a fibril, is built of myosin rods (micellae) 45 Å in diameter and 500 Å in length placed about 70 Å apart. Myosin fila-

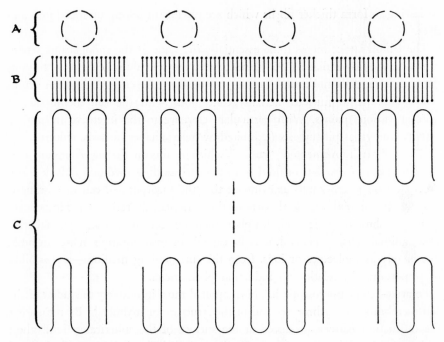

Figure 2-2. Diagram of lipid and protein layers in membrane of red-cell ghost. **A.** Layer of antigenic and amphoteric protein (20-50 Å). **B.** Permeability layer of lipid (40 Å). **C.** Structural layer of protein (5,000 Å). (After Mitchison.)

ments in vertebrate cross-striated muscle were originally thought to be arranged only at the surface of the fibril, as a cylindrical casing with a liquid center; however, the electron microscope shows them to be entirely filled and, according to Bretschneider, forming a beautifully ordered crystalline structure, the whole muscle fiber, including the I-bands, being doubly refracting in the direction of its longitudinal axis. Although the myofibrillar proteins are sharply contrasted with the proteins of the sarcoplasm of the muscle fiber (so far as physical properties are concerned), in the cell the functions of both types of protein are closely integrated, according to Perry (1955).

Bretschneider views such an ordered structure of myofibrillar mitochondria, etc., as a hierarchical extension of the lepton basis, i.e., a regular development of the lattice as a periodically repeating structural pattern of a system of elementary filaments not arranged merely according to the statistically favored orientation, but organized in a very definite way: molecular periods, determined by valence bonds, are overlaid by micellar periods, which include all those cross-striations that appear in the ultrastructure as macro- or microperiods such as nodular structures or lattices; and upon this periodic structural pattern are

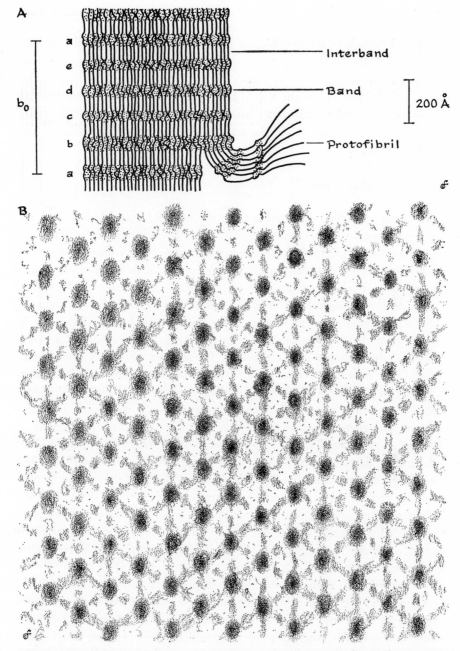

Figure 2-3. Features of fibrils. **A.** Structural features of collagen fibrils accessible to electron optics and small-angle diffraction. Segment **b** is to be repeated many times vertically; and details, such as number and location of bands **a-e**, are only roughly indicated. (After Bear.) **B.** Cross section of portion of insect flight muscle, showing crystal-lattice arrangement of fine and coarse fibrils. (×200,000.) (After Huxley and Hanson.)

superposed periods of the next-highest order, adding new structural and functional units to the filamentary structures already formed, and so on, extending to higher levels such as the sarcomeres of cross-striated muscle. It has become increasingly evident that in the cells of both animals and plants, the polymerization of proteins and carbohydrates involves a strict regularity not only in the end-to-end linkages, but also in the transverse direction, to form filaments of constant thickness.

Cilium and Flagellum

The fibrous proteins, which seem to be the main structural basis of cell and tissue organization, are classified by Astbury and his co-workers (1955) into two groups, according to their large-angle X-ray diagrams: collagen group, and keratin-myosin-epidermin-fibrinogen group (k-m-e-f, for short). The two groups were so named to express or imply this concept in molecular biology: that underlying the phylogeny of visible structures, there must be a corresponding— and in a sense more clear-cut—lineage of the fundamental biological macromolecules, particularly of the proteins. The k-m-e-f group produces long-range elastic structures; the collagen group, relatively inextensible structures. The former can undergo intramolecular transformations from the normal, folded α configuration to the extended β configuration or to the supercontracted state— processes presumed to underlie biological elasticity and movement generally, i.e., as a comprehensive system of polypeptides folded in definite configurations for definite purposes. Thus the gap between the extremes of muscles fibers and bacterial flagella seems to have been bridged by the discovery that their X-ray diagrams are so closely related that not only do they belong to the k-m-e-f group but appear to have the same axial macroperiod of about 410 Å. Further, the bacterial flagella contain two configurations, α and supercontracted, of polypeptide chains simultaneously—suggesting that rhythmic bending processes may be produced by propagation of waves of folding along selected chains. Astbury calls such flagella "monomolecular muscles."

It is remarkable, whatever the precise explanation, that flagella and cilia throughout the plant and animal kingdoms have one and the same basic constitution. In as diverse forms as the spermatozoa of sea-urchins, mollusks, insects, and fowl; the spermatozoids of green and brown algae, fungi, bryophytes, and pteridophytes; the flagella of flagellates; the cilia of ciliate protozoa and of molluscan and amphibian ciliated epithelium; and the combs of ctenophores, the construction is the same (consult References, p. 70). In all of these the flagellar or ciliary matrix is sheathed in a membrane which is continuous with that of the cell surface, and the matrix invariably contains a pair of single axial filaments surrounded by 9 double strands of contractile protein. The central pair remains separate and confers bilaterality on the structure as a whole, the plane of which is oriented normal to the plane of pendular motion. The tail of mammalian spermatozoa is fundamentally similar to the above, but possesses an external additional girdle of 9 fibers, with bilateral symmetry, also of a mitochondrial nature. Even in the rods of vertebrate retinal cells, which develop from ciliumlike structures, there are 9 double fibers, although the 2 axial filaments are missing. In fact, Fauré-Fremiet (1958) has pointed out the great resemblance of the central rod-cell structure with its associated rod sacs to the organization and fine structure of the flagellum base and eyespot chamber of a green flagellate.

In commenting on the constancy of flagellar and ciliary fiber pattern, Bradfield (1955) considers it impossible at present, because of lack of knowledge of mechanics of fibrogenesis, to show that the 9 plus 2 arrangement is inevitable (although this is no more than an admission of ignorance concerning the nature of the macromolecular basis). Consequently, he inclines to the view that the particular pattern observed has arisen by chance and has become a constant cytological feature, though it is just one of several possible patterns.

The formation of the 9 primary double fibers of the flagellum, seen most clearly during spermatogenesis in insects and mollusks, has been studied with

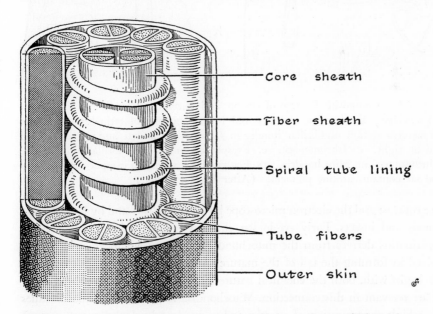

Core sheath

Fiber sheath

Spiral tube lining

Tube fibers

Outer skin

Figure 2-4. Diagrammatic reconstruction of plant cilium. (After Manton and Clarke.)

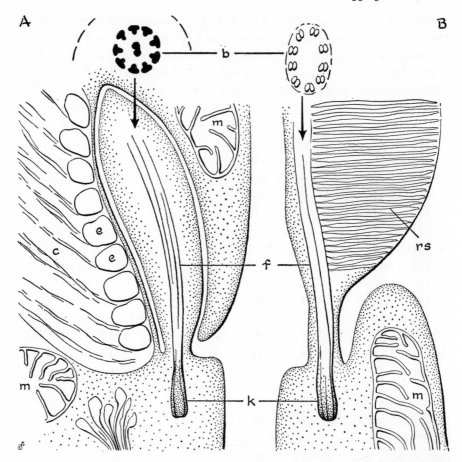

Figure 2-5. Composite diagram of eyespot and of retinal rod cell. **A.** Eyespot in *Chromulina.* **B.** Junction of outer and inner segments of retinal rod cell. **b.** Transverse section of fibrillar bundle in *Chromulina* at left and of retinal rod cell at right. **c.** Chromoplast. **e.** Eyespot chamber. **f.** Fibrillar bundle. **k.** Kinetosome. **m.** Mitochondria. **rs.** Flattened rod sac. Both eyespot chamber and rod sac contain carotine pigment. (After Fauré-Fremiet.)

phase contrast and the electron microscope by Grassé (1956) and by Kaye (1958) in snails and insects. Early cytologists clearly established that the nebenkern of spermatids derives from the mitochondria of the spermatocyte and later is involved in forming the tail of the mature spermatozoan.

To begin with, both the chemical nature and the fine structure of mitochondria are relevant in this connection. Mitochondria isolated from somatic tissues have a high concentration of protein and of lipid and a low concentration of nucleic acid. Electron-microscope studies have shown that, structurally, mitochondria in both animals and plants have the same basic organization, which consists of a continuous double membrane bounding an inner structure which, in turn, consists of double membranes called cristae. During the meiotic division of spermatogenesis in both snail and insect, small mitochondria fuse to form several long rods, which are distributed among the spermatide in an orderly way so that each gets 4 or 5. These round up to form the nebenkerm spheres

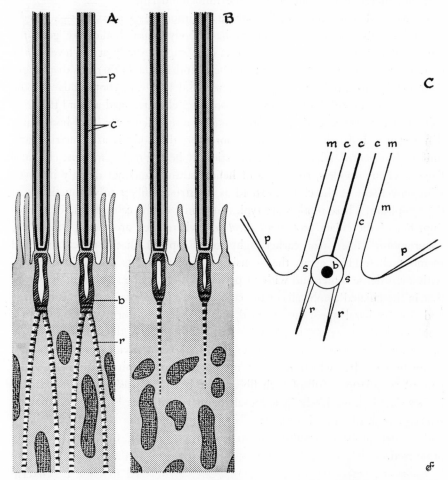

Figure 2-6. Diagrams of ciliary apparatus, showing cilium, basal corpuscle, and fiber rootlets. **A.** Intestinal epithelium of bivalve mollusk. (After Fawcett and Porter.) **B.** Frog's pharynx. (After Fawcett and Porter.) **C.** Cilium of ciliate *Spirostomum*. **m.** Membrane of cilium continuous with pellicle **p.** **c.** Longitudinal section through inner and outer ciliary filaments. **s.** Septum or membrane surrounding basal granule **b** or kinetosome. **r.** Rootlet. (After Randall.)

which surround the axial filament. During the late stages of sperm-lengthening, the mitochondria elongate and twist around the axial filament to form an envelope, which becomes the mature mitochondrial sheath of helical strands enclosing the greater part of the central filament. At the same time the mitochondrial fine structure undergoes a reorientation (although it does not break down at any time): the cristae are grouped into blocks which then line up perpendicularly to the long axis of the mitochondrian.

Virtually all cilia and flagella have a common basal apparatus—a bundle of straight fibers plus some kind of basal body. Basal bodies are generally separated from the cilium by an opaque diaphragm exactly at the level of the body surface, although according to Fawcett and Porter (1954) this diaphragm is absent in mammals, where cilium and basal body are continuous and the basal body is not solid, but a ring of 9 small circles arranged around the cross section of the central cavity. In mollusks and amphibians the basal body has a dense center and a less-dense shell; moreover, the body is asymmetrical and uniformly oriented relative to the direction of beat. Finally, in most cilia and flagella, ciliary rootlets, consisting of helical strands and not directly involved with movement of the cilium, extend as nonstriated fibers to various levels of the cytoplasm. These rootlets are typical not only of the cells of ciliated epithelium but of the ciliate protozoan *Euplotes,* where they are thought to be the "neuromotor" fibers (of earlier light microscopists) that connect cirri and membranelles. Here again they consist of fine filaments 120 Å in diameter with a less-dense center and without periodicity. It may be significant, however, that in the ciliated comb cells of the ctenophore *Pleurobrachia,* there is no large and distinct basal apparatus. Instead, the 9 fibers transform gradually into a single root, described by Bradfield as being about 400 Å thick and showing a regular, major periodicity of 500 Å; the single root divides proximally into many fine rootlets which fade out in the nuclear region of the cell, whose general cytoplasm is full of such fibers. The basal body as a whole appears to be associated—most likely in a causal manner—with the growth and maintenance of both distal and proximal fiber systems, the difference between cilium and rootlet probably resulting from differences among the materials being organized.

Centriole and Mitotic Apparatus

In a discussion of fiber system, the question of the centriole inevitably arises, since centrioles or their nonstainable equivalents seem to be centers for the initiation and growth of the internal fiber systems which constitute the asters and spindles responsible for mitosis.

The mitotic apparatus of sea-urchin eggs has been made available for study in isolation by Mazia and Dan, who used digitonin to remove the rest of the cell by a process of selective solubilization. In a review of his work, Mazia (1955) states that the fiber system accounts for about 2 per cent of all egg protein and consists of a simple protein together with small amounts of nucleoprotein. Electron-microscope studies of the isolated mitotic apparatus show a gel with regions where fibrils are condensed and oriented to form larger fibers. (Astral and chromosomal rays seem to be the result of such condensation and orientation.) A mass of sulfur-bonded protein chains would give an unoriented gel that is chemically homogeneous; and the hypothesis is put forward that the secondary bonding, and hence the geometric pattern of fibers in the mitotic apparatus, is determined by diffusion of an agent from the mitotic centers (i.e., the centriole, centrosome, or centrosphere of various authors) and from the chromomeres of chromosomes. Mazia raises the question of whether the protein used to form the mitotic apparatus pre-exists as such before division, derives from a pre-existing protein, or is synthesized anew, emphasizing the fact that the fiber system is broken down after use and re-established each time.

In animal cells, at least, the centriole is a stainable center or body whose self-duplication Huettner (1933) has observed through many mitotic generations in early stages of developing egg of *Drosophila*. More recently, de Haarven and Bernhard (1956) have used the electron microscope to observe the centriole in amphibian, mammalian, and cancerous tissues. They describe it as a cylinder about 150 Å wide and two or three times as long, consisting of about 9 centriolar tubules whose average diameter is 20 Å. During autoduplication each tubule doubles in size in the course of a preparatory phase for complete division. De Haarven and Bernhard conclude that the centriole is a highly differentiated organelle capable of synthesizing fibrous proteins that are indispensable to the diverse stages of cytogenesis. That this is so is shown by the dual role of the central body of spermatozoans in producing the flagellum and later in forming the fertilization astral system, and also by the part it plays in various unicellular flagellates (p. 52).

Extracellular Fiber Systems

Organized fibrous systems appear, as a result of polymerization processes, adjacent to, but outside of, the living cell or tissue in both animals and plants. In animals, collagenous material, consisting of the unique fibrous protein collagen together with a varying quantity of mucoid, is found mainly internally; the polysaccharide chitin, mainly externally on the body surface. In plants, cellulose is the principal extracellular material, although chitin also occurs (e.g., in *Phycomyces,* p. 27); cellulose is also produced by some animals, notably ascidians.

The main interest in animal tissues concerns collagen, which is capable of developing striking organization external to the cells that produce it. Thus Weiss and Ferris (1956) describe the basement lamella of the larval amphibian skin as consisting of about 20 layers of ground substance in which cylindrical fibers about 500 Å in diameter are embedded parallel to one another, but alternate 90° from layer to layer. During repair, after epidermal cells have covered a wound, rather uniform fibers of small size, less than 200 Å in diameter, appear at random in the space between the epidermal underside and the subjacent fibroblasts. Then, proceeding from the epidermal surface downward, a wave of organization spreads over this primitive tangle, resulting in straightening, orientation, packing into characteristic layers, and finally building of fibers 500 Å in diameter. The crossbands of all fibers of a given ply are in register; even the small (200 Å) fibers are spaced at rather regular intervals about 500 Å apart. The significance of the spacing seems to be that it corresponds to the segmental period of the fibers, to the diameter of the mature fibers, and to the distance between the granules in the intermediary film. In the formative stages the nucleus of most epidermal cells lies flat and closely applied to the basal surface where the lamella arises, with no more than 0.3 μ of space between the nuclear membrane and the cell surface. In fact, Porter (1954) describes the collagenous basement membrane of vertebrates as actually originating within the proximal parts of the epidermal cells. Accordingly, both the collagenous material and its progressive organization are closely associated with the basal surface of the epidermal epithelium, and possibly originate within the cortex of the epithelial cells themselves. This last possibility raises the question of the symplasmic state in tissues (see p. 107).

The development of patterned structure in collagenous material, however, can occur in vitro, in the absence of cells altogether. The work of Gross and his colleagues (1956) is summarized as follows: Under the influence of physico-chemical forces, an apparently homogeneous population of macromolecules can be precipitated, from solution, in the form of highly ordered structures, even when a variety of other substances of high molecular weight is present. At the same time, such precipitates may vary so radically in their morphological characteristics that they seem to be entirely unrelated to one another, with the result that, depending on readily controlled conditions in solution, they can be readily dissolved and transformed into one another. Identical structural types can be precipitated from solutions of greatly varying properties and

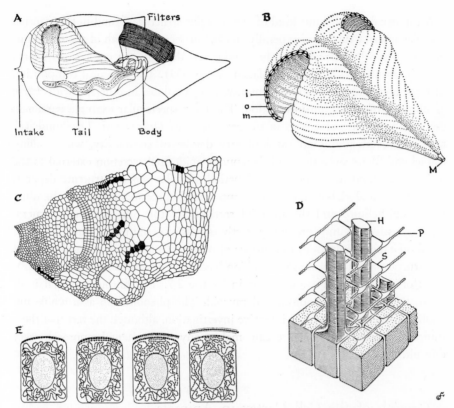

Figure 2-7. External cuticular structures secreted by epidermis in larval tunicate *Oikopleura dioica*. **A.** Animal, consisting of trunk and tail within secreted "house," showing external food intake and dorsal filters. **B.** Structure of food funnel. **M.** Attachment to mouth. **i, o,** and **m.** Inner, outer, and middle membrane of funnel substance. **C.** Surface view of oikoplastic epithelium (epidermis) of one side of trunk, with mouth at left, showing pattern of specialized secretory cells. **D.** Scheme, showing secretion of extracellular supporting fibers **H** and layers of primary fibers **P** cross-connected by secondary fibers **S.** **E.** Secretion and lifting off of fiber from cortical region of epidermal cell. (After Körner.)

can be produced by way of different intermediate stages in structure formation. In Gross' words,

> . . . it is conceivable that structures and organelles such as cell membranes, chromosomes, mitotic spindles, endoplasmic reticulum, mitochondria, etc., are formed, dissolved, and reformed by the spontaneous association of high polymer building blocks under the influence of physical chemical forces, the collagen systems being a simplified example of such a process.[2]

[2] Gross, *J. Biophys. and Biochem. Cytol.,* **2**(Supp.):272 (1956).

An unusual form of this kind of extracellular structure, found in the external coat (or test) of tunicates, generally consists of secretions both of the epidermis and of mesenchymal cells that have migrated through the epidermis to form what is essentially an external collagen (see p. 359). The chemical composition in some ascidians seems to be primarily pure cellulose; in others, a wide range of fibrous differentiation in addition. The most spectacular example is seen in the so-called house of larvacean tunicates such as *Oikopleura.* The house (or test), however, attains an extraordinary degree of complexity, with feeding funnel and filters, even though it is a purely epidermal secretion external to the body surface and has to be renewed periodically. Its general structure depends on a precise and elaborate pattern of specialized epidermis known as the oiko-plastic epithelium, and the material secreted by the cells is of two general types: massive columns growing directly out from certain cells, and fine fibers lifted off tangentially from the surface of other cells. The pattern and mode of construction and the process of secretion have been described by Körner (1952), some of whose findings are shown in Figure 2-7. The fibrous material itself seems to be mainly chondroid and mucoid. The phenomenon as a whole un-doubtedly merits much more intensive investigation, although the fact that these tunicates inhabit the open ocean makes any experimental approach very difficult.

Growth of the Cell Cortex in Animals

Growth of the cortical layer of animal cells has been studied mainly in eggs, particularly those of sea urchins. The ectoplasmic layer of such eggs is relatively thick, and the growth of new cortical substance during cleavage has received much attention.

Electron-microscope studies by Afzelius (1956) of the cortex of the eggs of several genera of sea urchins show that the cortical granules characteristic of this region first appear as spiral or concentric lamellae in the endoplasm and then, during maturation of the egg, move to the cortex, where the spiral order may be lost and peripheral spaces appear. Both the outer layer of the cortex and the cortical granules at this time are rich in sulfhydril groups. There is strong evidence that the so-called hyaline layer is formed from or by the cortical gran-ules which do not merge with the fertilization membrane and that the fine structure is either fibrous or amorphous. Most of the new cell surface of the fertilized egg comes from the cortical granules of the outer membrane.

During cleavage of the sea-urchin egg, according to Swann and Mitchison (1957), decrease in birefringence starts at the very beginning of the anaphase in the region around the chromosomes which are moving outward at 5-6 μ per minute, and is probably the result of disorientation of the protein structure of the spindle. The essential function of the wave of change spreading outwardly seems to be the initiation of disorientation at the cell surface. Optical changes in the cortex, i.e., decrease in birefringence implying disorientation, proceed until cleavage results. During cleavage the entire cell surface, and not merely the equatorial ring, becomes about twelve times more rigid than before, while the cell has little or no internal pressure. Dan and Ono (1954) have shown that, in this type of egg, no area expansion occurs in the cleavage furrow itself. From this fact, Swann and Mitchison conclude that the furrow wall is passive and is pushed inward by active expansion of the rest of the egg surface; furthermore, if this scheme is to fit in with the diffusion pattern of the disorienting substance, it must be assumed that high concentrations of this substance will so disorient the protein structure of the surface that it becomes plastic and loses its power of expansion. Surface expansion alone, however, cannot go on indefinitely without undue thinning of the cortex; if cortical expansion is responsible for cleavage, re-formation and growth of the cortex is necessary at some point in the cycle.

On the other hand, during division of the newt egg, according to Selman and Waddington (1955), the pigmented cortex moves outward from the furrow region; and new, unpigmented cortex grows within the furrow. Simple expansion of the surface of the original cortex does not take place, for if it did, movement of pigment would be obvious. The cortical gel is first synthesized before cleavage, as an additional layer beneath the existent pigmented cortex, and is then transferred to the required region during cleavage. Cortical growth in the amphibian egg and cortical expansion in the echinoderm are probably two aspects of the same fundamental process. Dan's study (1954) on sea-urchin eggs with pigment granules uniformly distributed around the surface of the egg before cleavage indicates that the net result in the two types is very similar, for here also the pigmented surface remains on the outside, while new, unpigmented cortex forms beneath it. In the newt the new, unpigmented cortex, by which the daughter blastomeres remain in contact after cleavage, is first formed as a sheet of gel, later seen as a double layer, which grows downward by a process involving gelation at its lower edge. Thus cleavage in animal cells must involve both expansion by molecular reorientation and growth of the gelated surface cortex or structural membrane; in all probability, both processes are initiated in some way by the daughter groups of anaphase chromosomes.

Cytoplasmic growth, however, appears to be both a cortical and an endoplasmic phenomenon, varying with cell type and cell phase. Perhaps the clearest evidence concerning the site of greatest cytoplasmic synthesis comes from the study of the growth of axons from neurons, where the linear extension offers exceptional opportunities for observation and experiment. Weiss and Hiscoe (1948) employed arterial cuffs to produce local narrowing and constriction, causing displacement of axonal substance and dividing the nerve into two parts of unequal growth: an unimpaired proximal region; and a distal region with a diameter limited to the narrowest part of the constriction, the local narrowing acting as a bottleneck or dam. Since the narrowed distal region, however, cannot be fully accounted for by a reduced supply of material passing through the barrier, the question becomes: What actually moves along the axon—fully synthesized axoplasm or merely growth accessories?

The general picture resulting from damming is one of telescoping, ballooning, coiling, and beading at the proximal side of the dam, with fibrillar material

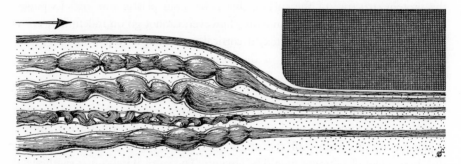

Figure 2-8. Damming of axoplasms of nerve fiber proximal to a constriction. (After Weiss and Hiscoe.)

piled up and contorted throughout the swollen part and clearly not of a fluid character. After release of the dam, during the first two days, the advancing front of the axoplasmic "flood wave," which is recognizable by its lumpy form, progresses at the measurable rate of 1–2 mm per day—which may well be the normal rate at which axoplasm is propelled along a nerve fiber. The conclusion is that, in mature fibers of stationary size, for every quantity of substance received from the perikaryon (or the body of a nerve cell exclusive of the nucleus), an equivalent quantity is lost at the distal end, possibly by metabolic degradation. In other words, the youngest cytoplasm is close to the nucleus, while the oldest is far from it, at least with respect to axons and dendrons and, presumably, the oldest cytoplasm is also peripheral for the cell body proper. There is, now, ample evidence that nucleic acid plays a necessary role in protein synthesis; but cytochemical studies on RNA of nucleoli and chromosomes have failed to demonstrate any simple relation to RNA of cytoplasm, and evidence by Swift and his co-workers (1956) strongly suggests that cytoplasmic RNA is synthesized in the cytoplasm.

Growth of the Cell Wall in Plants

Growth of the cortical region of plant cells has been studied mainly in the single-celled fungus *Phycomyces* and in such higher plant cells as cotton hairs and onion-root tips.

Phycomyces

The tubular cell *Phycomyces* is favored for the study of growth primarily because of its size: the full-grown cell may measure 100 μ in diameter and several centimeters long. According to Grehn (1932), *Phycomyces* growth is spiral: the cell extends and twists itself by active growth in area of the cell wall along an axis which winds helically through the localized growing zone, much like the wrappings of a cardboard mailing tube.

According to Castle (1936), within a few days after germination of a spore on a suitable (preferably solid) medium, a luxuriant mycelium develops, which subsequently gives rise to numerous aerial sporangiophores. These are un-branched, tubular cells, without cross walls, which grow up from primordia formed by the mycelium close to the substrate. For a day or two, each sporangio-phore grows at its tip, which is sensitive to gravity, light, and other environ-mental agents or gradients. When growth in length ceases, the tip apex balloons into a nearly spherical swelling that is full of yellow pigment and densely granular protoplasm, which is the spore-forming mass. A cross wall, the columella, is then laid down, which eventually cuts off the tip apex from the fluid-filled cavity of the sporangiophore. The spore mass then cleaves into a large number of spores; and the mature sporangium thus formed, darkened to brown and then black, is henceforth borne as an isolated load at the top end of the sporangiophore.

The sporangiophore again commences to elongate by growth occurring ex-clusively in a zone 1–2 mm below the sporangium and proceeding at a more rapid rate than earlier. This major period of growth, which has received particu-lar experimental attention, may proceed for hours and even days. At 25° C the elongation may be as fast as 3 mm per hour, and the terminal spore mass, if suitably marked with attached particles or glass fibers, can be seen to rotate about the long axis of the sporangiophore; i.e., the cell not only elongates but twists at its terminal end during the process of growth. The steeply spiral

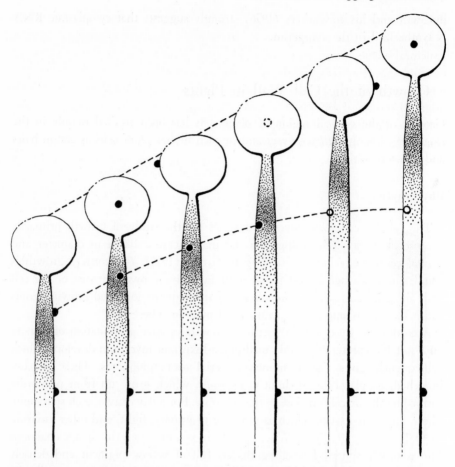

Figure 2-9. Spiral growth of sporangiophore of *Phycomyces.* (Zone of growth stippled; broken lines trace relative rotation.) (After Castle.)

orientation of the axis of growth is confirmed by the pattern of helical striation in the mature cell wall and in the growing zone itself. The spiral course is generally, but not invariably, left-handed, i.e., clockwise when viewed from above. It concerns the primary cell wall and is most likely a radically different process from the thickening and spiral layering commonly seen in the secondary walls of many plants. In any case, the nature of the primary spiral growth continues to be the main problem, which breaks down to the nature of spiral growth itself, the origin of spiral growth, and the question of clockwise versus counterclockwise progression.

Electron microscopy of the primary wall of *Phycomyces* reveals a rather dense texture of chitin microfibrils, which, however, do not exhibit a main direction that could be recognized as the pitch of a spiral. Frey-Wyssling (1952) believes that enlargement of the wall area and intussusception of new structural elements can occur only by the loosening of the microfibrillar mesh in spots, that it must be the living protoplast—not the texture of the cell wall— which causes spiraling, and that the cell wall is the consequence and not the origin of the spiral growth. Castle (1953) finds this interpretation incompatible with the distribution of intensity of growth over the zone in which it takes place, which he has shown rises to a maximum about 0.3 mm below the point where the sporangium is attached and then declines to zero as the lower end

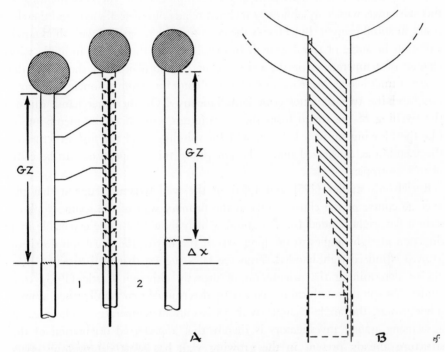

Figure 2-10. Diagrams of growing zone of *Phycomyces,* showing kinematics of growth. **A.** Steady state. Growing zone **GZ** shown at left and again at right displaced by amount Δx. Middle diagram decomposes this continuous process into two partial processes: in process **1**, growing zone is stretched, separating micro-fibrils in wall; in process **2**, gaps are filled by interpolation of new fibrils, and bottom section of length Δx is converted into secondary wall. **B.** Spiral growth. Position of row of markers, vertical at time zero, is shown again after 10 minutes, during which interval growing zone has moved up by $\frac{1}{6}$ of its length (shown by broken-line box). Top marker has made half a revolution. Since intermediate markers have rotated less, their vertical displacement is correspondingly smaller. Therefore, tilts of directions of motion of markers are alike. (After Delbrück and Reichhardt.)

of the zone is reached; i.e., growth does not occur in local spots, whether spiraling or not. By attaching a single, small particle to the wall of the growing zone and following its relative displacement longitudinally and rotationally during growth, Castle demonstrated that there is a definite axis throughout the length of the growth zone and that helix has a constant pitch. This pitch, observationally confirmed by striation in the wall deposited by the process of growth, shows that growth is oriented along, and takes place parallel to, such an axis throughout the whole growing zone of a cell.

Heyn (1939) suggested that structural peculiarities of crystaline chitin in the wall might lead to enlargement along slip-planes in the crystal-lattice of this substance, which are obliquely inclined relative to the cell's axis and thereby result in spiral growth. But Heyn's theory cannot be reconciled with the known variation in angle of spiral growth from cell to cell, the change in angle with change of temperature, or the existence of the fibrillar mesh shown by the electron microscope. Moreover, rotation is a constant accompaniment of elongation, and the two activities cease simultaneously. On the other hand, during the swelling of the tip to form the sporangium, growth of the membrane is like the blowing up of a balloon, without spiraling; and the fine structure of the membrane, as viewed under the polarizing microscope, seems to be completely isotropic.

Roelefson's studies (1951 and 1953) of the wall structure suggest that the average course of the chitin fibrils in the primary wall of the sporangiophore is in a flat, right-handed (or Z-) spiral, which should be more extensible in a direction at right angles to the long axis of the fibrils; the usual spiral axis of growth is indeed right-handed. Thus the facts suggest that wall structure does aid in determining the orientation of growth, although Castle (1953) concludes that spiral growth cannot be strictly determined structurally since, among other things, the angle changes with variations in temperature. The area of agreement among investigators is merely that a preferred orientation of the structure already present in the growing wall has some determining effect on the axis of growth.

Sporangium formation involves growth which appears to be truly unoriented, so that some influence is able to abolish the usual mode of spiral tubular enlargement. Castle emphasizes that it has never been possible to attach even one marker as small as a *Lycopodium* spore in order to measure satisfactorily the distribution of growth at the pointed end of a stage-1 cell while measurement is so easy with the stage-4 cell—a circumstance pointing to the ultra-sensitivity of the stage-1 cell. Furthermore, there seems to be no clue to why growth of an adult sporangiophore should be almost always along a left-handed, rather than a right-handed, spiral, or to why the direction of spiraling should be less

than an absolute preference—whether it is an outcome of chitin-protein com-
plexes, molecular asymmetry, or some more-familiar developmental occurrence.

Since Castle's findings, Delbrück and Reichhardt (1956) have stated that
the growing zone stretches without elongating, since for every amount it grows
in length, a corresponding amount at its bottom is converted into secondary
wall, which ceases to grow in length. Similarly, the wall of the growing zone
does not actually become thinner as it stretches, since for every amount it thins,
a corresponding amount of new wall material is formed. The principal layer
in the wall consists of chitin microfibrils 200 Å in diameter and arranged approximately horizontally; as the wall stretches, the microfibrils separate from
each other and new fibrils are laid down to fill the gaps.

Delbrück and Reichhardt's analysis of the light-growth response refers to
experiments in which specimens are symmetrically illuminated on two or more
sides. At a constant intensity of illumination, the rate of growth is also constant,
whatever the particular intensity; but if the intensity is changed, a striking and
transient change occurs in the rate of growth only of the base of the growing
zone. After a short period of higher illumination, growth is maintained at the
normal rate for 2.5 minutes; then at double the normal rate for a few minutes,
followed by a decrease to below normal; until, about 15 minutes after the initial
stimulus, the rate returns to normal. The net growth is zero, and the effect is
an altered distribution of growth rates in time.

Cotton Hairs

The basic structural material of the primary or secondary cell wall of higher
plants is cellulose, rather than chitin. At present the fundamental unit appears
to be the microfibril, which varies in size between 50 and 250 Å, is made up of
polymolecular chains organized in a very orderly manner, and has a high
intrinsic birefringence. The first electron-microscope work indicated a network
(or reticulum) of microfibrils, but more detailed studies suggest that there
are different layers within the primary wall. Roelefson and Houwink (1951
and 1953), in examining staminal hairs and young cotton hairs, which lack
secondary walls, showed that on the outside of the primary wall the microfibrils
are longitudinally arranged in the form of a network, while on the inside there
is a larger and more transverse layer of microfibrils. These investigators postulated a flat, helical structure of microfibrils for the primary wall as a whole,
which they envisioned as a sheaf of nets whose outer, older layers were pushed
outward by new layers from the inside. Thus the general orientation of the
microfibrils making up the nets on the outside changes from a transverse to a

more longitudinal direction as the cell increases in length, the nets being extensively interwoven and not disparate.

Onion-Root Tips

In a detailed electron-microscope study of the meristematic tissue of onion root, Scott and his co-workers (1956) concluded that the primary wall may be developed in four continuous stages:

Stage 1. The reticulate stage of a thin, loosely interwoven mesh, during which the cell increases in diameter and volume.

Stage 2. The stage most apparent in the bulk of the meristematic tissue, during which the microfibrils of the lateral walls appear with a preponderant parallel orientation transverse to the longitudinal axis, which is now bipolar. In this stage the cell attains its maximum girth and doubles or triples in length.

Stage 3. Active elongation of the cell—a stage characterized by the appearance of parallel, horizontal microfibrillar orientation which continues until growth slows down and eventually stops.

Stage 4. Secondary-wall formation with characteristic helical orientation of the microfibrils.

According to Northcote (1958), growth of the cell wall in area is generally supposed to be an expansion of the microfibrillar network of the primary wall by intussusception, either at the tip or in mosiac fashion. The more elaborate microfibrillar organization known to occur in the cell wall indicates, however, that growth can be thought of as a process of apposition of microfibrillar nets, i.e., the multinet theory. It can also be conceived of as an extension of a continuous matrix in which microfibrils are successively laid down.

Chapter 3

Cells: Cell Cortex in Protista

Cortex and Nucleus in Acetabularia and Stentor

Interaction of nucleus and cytoplasm relative to the control of pattern is seen in striking form in the unicellular Mermaid's-Cap, comprising species of *Acetabularia* and related genera (all members of the Siphonales algae), and in the ciliate *Stentor*. Because both types are remarkable for large size, they have been subjected to experimental procedures hardly possible in smaller forms.

Acetabularia

Normal Development

An individual in *Acetabularia mediterranea* is the sole product of one zygote and is essentially a giant cell with a single nucleus. The plant consists of a stalk about 5 cm long with a rhizoidal base, in one branch of which the nucleus lies, and with somewhat everted umbrellalike cap.

The zygote itself results from the conjugation of a pair of very small, biflagellate mononuclear cells. During development, the embryo elongates to form a rhizoidal part, which clings to the substratum, and an ascending stem. When the stem reaches a certain height, a whorl of branching hairs forms beneath the apex; and after further growth, other whorls form above the first whorl. The whorls are short-lived, however: all but two of them drop off, so that only the uppermost one of two whorls is present at any given time.

When mature, i.e., when the upward growth of the stem has reached its maximum, the individual develops a fertile whorl of gametangia, which grow together to become segments of a cap or umbrella. Above this cap a short-lived second whorl forms, the hairs of which grow together to form the corona.

Throughout growth of the zygote, the nucleus remains in one branch of the

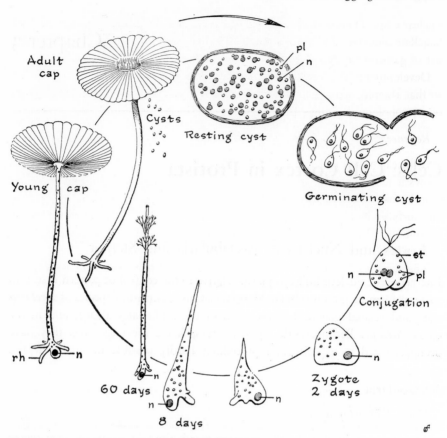

Figure 3-1. Life cycle of *Acetabularia mediterranea*. **pl.** Plastids. **n.** Nucleus. **st.** Stigma. **rh.** Rhizoid. (After Brachet.)

rhizoid base and grows to an enormous size, with an equally large nucleolus. Despite the nucleus size, the nucleocytoplasmic ratio steadily shifts to the disadvantage of the cytoplasm; i.e., nuclear material becomes relatively less. The extent of this disadvantage may possibly determine the maximum size of the organism.

When the cell or organism reaches full size, the giant nucleus in the rhizoid undergoes a series of rapid divisions to form thousands of small, secondary nuclei which fill the rhizoid processes and then begin to move up into the stem together with the green chromatophores. Both nuclei and chromatophores are carried up from the rhizoids by plasma circulation, and the previously bright-green rhizoids consequently become a lighter green.

Almost all the cytoplasm containing the secondary nuclei and the chromatophores wanders into the chambers of the cap, where it subdivides into mononuclear cysts. Then within each cyst repeated divisions produce as many as 1,800 flagellate gametes; and since a single cap may have 15,000 cysts, the total number of gametes formed from the original cell material is extremely high.

Development of *Acetabularia mediterranea* always proceeds in this order, so that rhizoid, stem, hair whorls, cap, and cysts comprise a definite sequence of morphological differentiations within a single individual.

Regeneration

If the cap—even full size—is cut off anywhere above the rhizoid, the basal end grows a new apex and goes through the normal successive formation of hair whorls and ultimately cap. Differentiation of a cap can also occur at the posterior cut surface of an upper, anucleate part—in this part giving rise to hetero-morphosis. Thus, parts of the anucleate stem, also, can regenerate and may live and grow for several months.

The capacity for apical differentiation decreases from the apex toward the base; for rhizoid differentiation, from the base toward the apex, as judged from relative frequencies of perfect and imperfect regenerations. In either type of differentiation, the longer the piece of stem, the greater its regenerative capacity. Hämmerling (1934) postulates two kinds of form-producing substances: one for the cap and another for the rhizoid, with two opposing concentration gradients.

Whatever the significance of the "form-producing substances," there seems to be no doubt that the capacity for apical differentiation depends on something produced by the nucleus. For example, 5-mm piece of anucleate stem cut from the basal end does not regenerate apical structure. But if a slightly more-distal piece of stem is removed first, leaving the basal piece attached to that rhizoid which contains the nucleus, and after a while the rhizoid together with the base of the nucleus is cut off, then the basal piece possesses full regenerative capacity. Clearly, something is supplied; but whether it is growth material, form-producing agents, or some sort of catalyst is not yet known. Nevertheless, the production of new cytoplasm close to, or by, the nucleus of nerve cells and the subsequent migration of that cytoplasm along the axon seems to be comparable to the agent diffusing from the vicinity of the nucleus of *Acetabularia*.

Grafting Experiments

The two species *Acetabularia mediterranea* and *A. wettsteinii* differ in size, in number of chambers constituting the cap, in type of chromatophores, and in the presence in *A. mediterranea* of an upper and lower corona while *A. wettsteinii* contains an upper corona only.

A nonnucleated piece of *A. mediterranea* stem can be grafted to a short piece

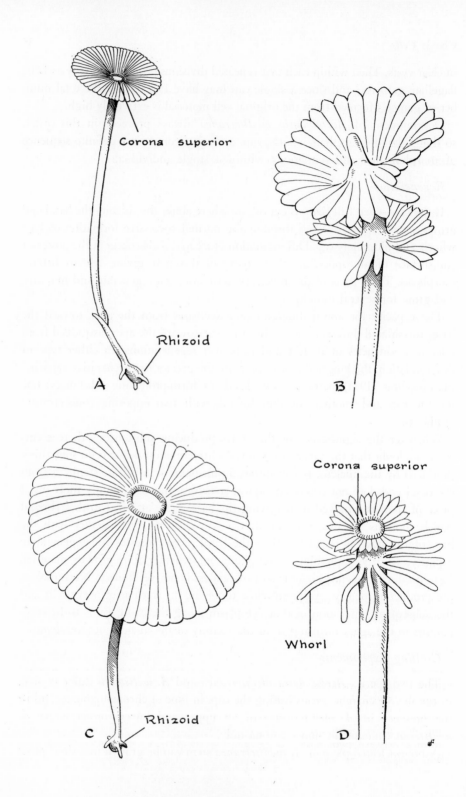

of the nucleus-containing rhizoid of *A. wettsteinii* if, as soon as the cellulose stem is cut through, the two pieces are pushed together so that their naked cytoplasms come into contact.

In one series of such experiments by Hämmerling (1946), the apical structures to result from the grafts were all in conformity with the nucleated *A. wettsteinii* piece. In another set of experiments, the formations were of the *A. mediterranea* type at first, but later transformed to the *A. wettsteinii* kind. Since the chromatophores are very different in the two species, it was clear in both series of experiments that the *A. wettsteinii* pattern was exhibited by the *A. mediterranea* cytoplasm, and Hämmerling (1946) concluded that the determining influence spread from the *A. wettsteinii* nucleus. Although this conclusion is probably right, no definitive statement can be made until the effect of grafting nonnucleated pieces of one species into the system of the other has been investigated.

Similar results were obtained from grafts between *A. mediterranea* and *A. crenulata* and between *A. mediterranea* and *Acicularia schenkii*. In both *A. crenulata* and *Acicularia schenkii* the cup chamber splits open and a spur sticks out, and in the latter organism many caps are formed in series. Stems of *Acicularia schenkii* grafted to nucleus-containing bases of *A. mediterranea* produce caps of an intermediate character, indicating, according to Hämmerling, that the "cap-forming substance" is stored in the *Acicularia* cytoplasm.

The development of the *Acetabularia* cell comes to an end with cyst formation. If a mature cap is cut off before a certain time, so that regeneration ensues, division of the primary nucleus is delayed until a new cap is formed—a procedure that can be repeated many times. In grafts, division of the primary nucleus is delayed also when the stem of an immature plant is grafted onto the rhizoid of an older one; conversely, a stem with mature cap grafted onto the rhizoid of a younger individual induces premature division of the nucleus—a result comparable to that of grafted stentors of differing ages (p. 50).

Once the secondary nuclei have begun to migrate, regeneration can no longer take place. On the other hand, when a young cap is grafted to a decapitated plant with migratory nuclei, the migration ceases until the cap is full grown. When almost all the secondary nuclei have migrated from the stem to the cap, and the cap with its nuclei is cut off and replaced with a younger cap which is allowed to attain maximum size, the few nuclei remaining in the stem will

---

Figure 3-2. Regeneration of cap from basal end. **A** and **C.** Young and adult caps of *Acetabularia mediterranea* regenerated from nucleated posterior parts; hence, stalks shorter than normal. **B.** Two caps of *A. crenulata*, with stalk growing out above upper cap and beginning whorl formation (whorl is generally formed between two caps). **D.** Young cap of *A. crenulata*, with pointed rays and remains of old whorl. (After Hämmerling.)

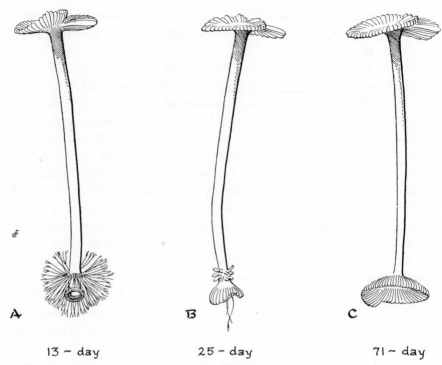

13 ~ day   25 ~ day   71 ~ day

Figure 3-3. Regeneration of cap in *Acetabularia mediterranea.* **A.** Anucleate portion with original cap. **B.** Posteriorly whorl and cap are formed, but only cap grows. **C.** Anterior cap also increases in size. (After Hämmerling.)

undergo division. The resulting nuclei then migrate to the cap chambers, where they form cysts in the usual way.

In a summary of his work, Hämmerling (1953) interprets his results as demonstrating the existence of nucleus-controlled morphogenetic substances which differ from species to species. Accordingly, in binucleate grafts a balanced ratio between substances controlled by the nuclei is established and maintained —a situation explaining the formation of intermediate caps which do not transform to one or the other species type. In trinucleate grafts containing one *A. mediterranea* and two *A. crenulata* nuclei, a morphological shift occurs in the direction of *crenulata* caps—an understandable situation because of the imbalance between nuclear substances.

Hämmerling postulates that the morphogenetic substances originate from gene action and are comparable to those substances which function between gene and character; i.e., the morphogenetic substances do not merely induce a predirected development, but determine the direction itself, directing the metabolism in such a way that from protoplasm, as the substrate of morphogenesis,

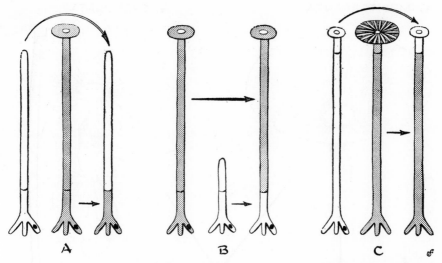

Figure 3-4. Transplantation experiments of *Acetabularia mediterranea* individuals of different ages. **A.** Young stem on old rhizoid. **B.** Old stem on young rhizoid. **C.** Young cap on end of stem whose secondary nuclei are already wandering into cap region. (After Hämmerling.)

a cap of definite shape is eventually formed. Essential to this interpretation is the assumption that these compounds react effectively with the cytoplasm of other species, as, for example, the morphogenetic substance of *A. mediterranea* causes the formation of a *mediterranea* cap from *A. crenulata* cytoplasm.

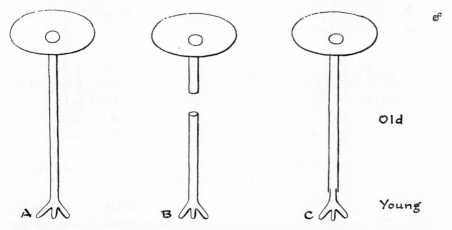

Figure 3-5. Suppression and induction of division of primary nucleus. **A.** Normal division, i.e., at stage when cap attains maximum diameter. **B.** No division when cap is cut off. **C.** Division when grafted to anucleate cap of maximum diameter. (After Hämmerling.)

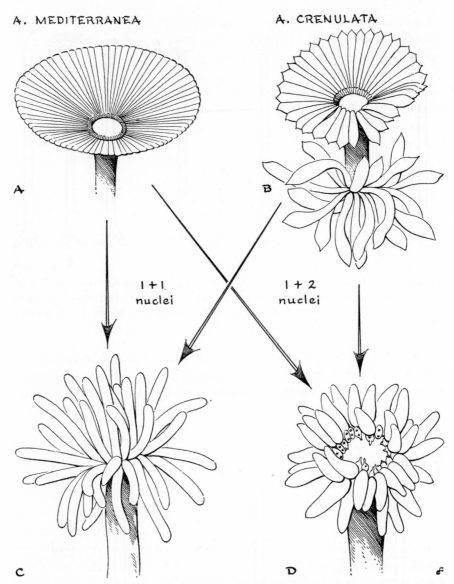

A. MEDITERRANEA   A. CRENULATA

1 + 1
nuclei

1 + 2
nuclei

Figure 3-6. Formation of intermediate caps in binucleate grafts of *Acetabularia mediterranea* and *A. crenulata*. **A.** *med* cap with 72 rays, united, without spurs. **B.** Two *cren* caps: the lower with isolated rays; the upper with rays still united; both with spurs. **C.** *cren-med* intermediate cap. 33 rays, isolated, slim, without spurs. **D.** *cren*-like intermediate cap of a cren-med transplant with 34 rays, partly united, some pointed. (After Hämmerling.)

Experiments with interchanging morphogenetic substances have been conducted by Werz (1955), who examined cytoplasmic systems containing nuclear material from *A. crenulata* and *A. mediterranea* in ratios varying from 4 *cren* : 0 *med* to 0 *cren* : 4 *med* in relation to, e.g., initiation and form of caps and form and number of chambers. Varying the number of nuclei from the same species had no effect, but adding the various mixtures again produced mixed results—though not proportionate to the mixture, for *mediterranea* structures were more numerous. Experiments with anucleate stalks showed a basipetal gradient of species-specific substances, i.e., a higher concentration the greater the distance from the nuclei that determine the cytoplasmic constituents.

Environmental factors also play a part in influencing growth. For example, unusually intense illumination produces individuals with abnormally short stalks and disproportionately large caps; whereas too-weak light induces abnormally long stalks and correspondingly small caps, as though the cap were formed out of whatever material remains after stalk growth is complete. Anucleate pieces of stalk retain their regenerative capacity even after 2 weeks in the dark, so that, when light is restored, they develop as well as those in the light; and the pieces in the light slowly lose the capacity at about the same rate as those in the dark. Hämmerling (1946), however, concludes that changes in morphogenetic regenerative capacity of anucleate pieces are not due to changes in the reactivity of the protoplasmic substrate, but are consequences of changes in the activity of the nucleus during development. He considers such changes to be not autonomous but dependent on the state of the cell system. Thus the nucleus is less active in old cells than in younger ones, but becomes fully active if the cap is cut off. The conditions established by the morphogenetic activity of the nucleus in turn react upon the nucleus to control its activity.

This theory is somewhat in keeping with the conclusions derived from experiments on frog embryos. Experiments by Hertwig and by Spemann on cleaving frog and newt eggs and by Seidel (1932) on the migration of nuclei in the dragonfly egg demonstrated that the nuclei of the segmenting eggs were qualitatively equivalent and interchangeable. Almost thirty years after Spemann's work, King and Briggs (1955a and b) found that this interchangeability holds for both the blastular and young gastrular stages in the frog, but not for later stages. When introduced into enucleated eggs that were activated by pricking, nuclei of the late gastrular chordomesoblast allowed development to proceed only as far as the blastular stage in most cases and as far as an abnormal neurular stage in a very few cases. The conclusion from these and similar experiments was that nuclei undergo some very important changes during differentiation and that nuclear function undergoes progressive specialization during cell differentiation. Few cells, however, are more specialized, structurally at least, than spermatozoa; yet the sperm nucleus is obviously fully potent once it has entered the cytoplasm of an egg of the same species—a change of environment comparable perhaps to that when the *Acetabularia* cap is cut off.

Stentor

Normal Development

With the exception of *Paramecium, Stentor* has been studied experimentally more than has any other protozoan ciliate, mainly because of its relatively large size (1-4 mm) and complex structure (*Stentor coeruleus* especially). Fission,

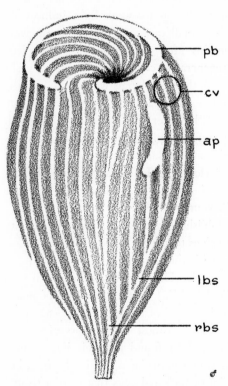

regeneration, and reorganization phenomena all demonstrate the dynamic aspect of *Stentor* form and pattern. Since 1744, when Trembley published a remarkably clear and accurate account of its form and behavior, and 1893, when Johnson described its general biology and morphology, the animal has continued to arouse interest.

The somewhat cone-shaped *Stentor* body is attached at various points to a substratum at its narrow end, which terminates in a disklike foot. The wide anterior, or adoral, end has a peristomial band of membranelles and cilia which carries food in a clockwise path to the buccal pouch, which in turn spirals down to the cytosome and gullet. (For convenience the peristomial band and adoral ciliary field in general are called the head; the buccal pouch, cytosome, and gullet may be called the mouth.) A single, contractile vacuole lies within the cell body, to the right and in front of the cytosome; and a beaded macronucleus, with 15-20

Figure 3-7. Typical *Stentor.* **pb.** Peristome. **cv.** Contractile vacuole. **lbs** and **rbs.** Left and right boundary stripe of ramifying zone of pigment stripes. **ap.** Prospective adoral zone. (After Weisz.)

lobes, lies along the left side, the cytosome being considered ventral. Micronuclei lie close to the macronuclear lobes.

Apart from the structure described above, the cortex of the body shows about 100 pigmented stripes passing from head to foot, with a row of cilia, an infraciliature, and a contractile myoneme lying between each pair of stripes. The cortical pattern of pigmented stripes shows a left-to-right gradient in the width of stripes and in the spaces between them. Commencing at the oral meridian and passing from left to right around the body, the stripes increase progressively in width and in the size of spaces between them and, at the start, in length as well. The finest and most closely packed lines are in the region of the oral meridian, extending somewhat to the right of it. This region, the fine-line zone, abuts on its right with the widest stripe. Along the line of abutment—known as the ramifying zone—the stripes become progressively shorter toward the right and eventually each splits off from the anterior part of the adjoining wide stripe. The ramifying zone seems to be the normal source of new stripes and may be regarded as the primary growth region. Each new stripe thus produced also grows in length in conformity with the body as a whole. By inference, it is probable that each new ciliary row, infraciliature, and myoneme is produced in the same region in a conformable manner.

Regeneration

The following description of some preliminary experiments will clear the way for an analysis of the various morphogenetic events in *Stentor*.

1. By literally shaking stentors to pieces, Lillie (1896) and Morgan (1901) showed that most fragments at least $\frac{1}{64}$ the size of the original individual can reconstitute into whole individuals, and that regeneration or reconstitution was an all-or-none event. Weisz (1948), however, has shown that pieces, even if large, lacking certain essential organelles are incapable of regeneration, while pieces as small as $\frac{1}{100}$ the size of the original individual can reconstitute if they contain such organelles. He also found (1948 and 1951) that no matter where a stentor is transected, a foot regenerates from the cut surface of the anterior piece and a head from the posterior piece, that endoplasm from any region can support regeneration, and that there is no gradient in regenerative capacity: the rate of regeneration is the same for all sizes and parts for fragments exceeding 70 μ in diameter.

2. Schwartz (1935) has shown that micronuclei are not necessary for regeneration, although they are for fission. On the other hand, a single lobe of the macronucleus can support regeneration, and at the same rate whether 1 lobe or 7 are present. Weisz (1949) found that only the anterior lobes are competent

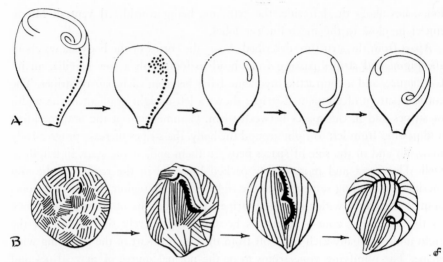

Figure 3-8. Regeneration and reconstruction *Stentor*. **A.** Intact individual at left with peristomial band, gullet, and row of kinetosomes associated with left boundary stripe. From left to right, development of anarchic field of kinetosomes from cut terminal after decapitation and subsequent formation and development of membranelle primordium. (After Weisz.) **B.** All original structures in minced *Stentor* removed except one short piece of adoral band, whose pattern is cut into many disarranged patches. Subsequently, patches align and rejoin, new primordium appears at abnormal locus of stripe-width contrast, and normal stentor reconstitutes. (After Tartar.)

during prefission and prereorganizational stages, but that all lobes are competent at other times.

3. The surface ectoplasm of stentors can be cut into tiny patches with random orientation and still recover normal arrangement. The patches slowly reorient themselves—not all at once as by a magnet, but in piecemeal fashion—so that, at first, areas of parallel striping appear and then the stripes join together. Tartar (1941 and 1956c) has shown that there is no extensive migration of patches, but only torsion *in situ*.

Whatever their importance, the pigmented stripes clearly mark distinctive elongated areas of the cortical ectoplasm—as though the ectoplasm had been experimentally stained for the purpose of enabling the investigator to recognize and follow the growth and development of the various regions.

The left boundary stripe, from which new stripes arise, merits close attention. From a series of excision experiments, Weisz (1951) found that adoral differentiation and contractile-vacuole formation always occur at a free anterior terminal of the left boundary stripe of the ramifying zone, that differentiation of a foot always occurs at the free posterior terminal of the same stripe, and that every point along the stripe is capable of mediating either oral or foot differentiation—this capability indicating an inherent polarity of the stripe. It is therefore not surprising that almost any transection of a stentor can regenerate since it inevitably includes a section of the left boundary stripe and probably at least 1 lobe of the macronucleus. However, if the stripe is excised altogether, another ectoplasmic stripe may take over its specific functions, and normal regeneration follows after considerable delay.

This region, where narrow and wide pigment stripes meet and a new head normally appears, is called the anarchic zone or reorganization field by Weisz (1949) and the primordium site by Tartar (1956b). It can be excised and implanted elsewhere in the same or another individual. When the region is grafted into the back of another individual, the host develops a second set of feeding organelles and a second primordium site, and upon transection regenerates a doublet head and two primordia. Such doublet stentors can divide, reorganize, and regenerate as doublets for weeks. According to Tartar (1956a), removal of one of the primordium sites causes the doublet to revert to the single normal type.

Excision of the primordium site from a stentor does not, however, even

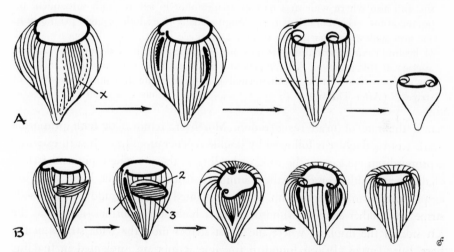

Figure 3-9. Grafts of primordium sites in *Stentor*. **A.** Implantation of extra primordium site *x* between two wide stripes of decapitated host leads to double oral regeneration: one from graft and one from host primordium. Also shown is double regeneration from posterior fragment of such a doublet. **B.** Effect of grafting primordium site at right angles into wide-stripe area, showing primordium formation at all loci of stripe contrast: at host primordium site 1, at normal site in graft 2, and at site where fine lines of graft meet wide stripes of host at right angles 3, followed by stripe reorientation and final formation of symmetrical doublet. (After Tartar.)

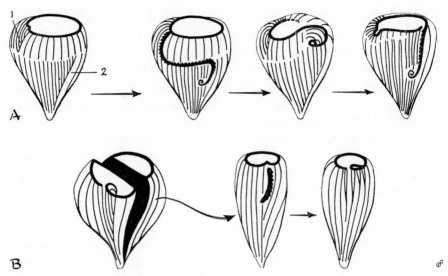

Figure 3-10. Rotation and excision of halves in *Stentor.* **A.** Anterior half is
  rotated 180° on posterior half, so that original primordium site is divided into
  two displaced sections 1 and 2. Primordium formation occurs at both parts of
  site and also where wide and narrow stripes abut in suture, with subsequent reorganization. **B.** Aboral half is cut longitudinally to exclude original primordium
  site and widest and narrowest stripes. Difference in stripe width appears at line
  of healing because remaining stripes are of graded width, and new primordium
  appears at this junction. After regeneration, wide stripes split anteriorly into narrow stripes, reconstituting both normal primordium site and normal number of
  stripes. (After Tartar.)

retard the rate of head regeneration. Moreover, removal of both primordia
heads from a doublet is followed by double regeneration. In each such excision,
a primordium appears at the time of healing at the site where pigment stripes
of different widths come together. Similarly, if a sector of fine lines excluding
any of the normal primordium site is implanted into the middle of the wide
stripes of another stentor, primordia appear both at the normal site and at the
left side of the implant, where the finest stripes meet the wide stripes of the
host. But, Tartar (1956a) found, if fine-line sectors are implanted in fine-line
zones, or wide-line sectors in wide-line zones, no primordium regenerates.
Therefore, the apposition of wide- and narrow-stripe areas is crucial for oral
differentiation.

Tartar has further explored this relationship by making a circumferential cut
around the middle of a stentor and rotating the anterior half 180° on the
posterior, thus dividing the primordium site so that its two parts lie on opposite
sides of the cell. Primordium formation then occurs not only at both parts of

the primordium site but throughout the long suture between, which joins the one part to the other. Primordium structure accordingly appears wherever wide- and fine-line material come together, whether side by side or end to end.

Two other aspects of stripe behavior are notable: Wherever posterior ends of stripes come together, a foot or holdfast is produced. If a stentor is split longitudinally so as to exclude the original primordium site and the widest and narrowest stripes, it nevertheless heals and regenerates a new primordium, where a difference in stripe width still exists; moreover, each of the widest stripes of the residual individual split at their anterior ends to form the normal number of stripes. Despite the general coincidence of oral-primordium forma- tion with contrasting pigment-stripe widths, it is very improbable that the stripes themselves are intimately involved in the elaboration of oral structures, although they do serve as an index of cortical developmental events. In fact, in *Stentor coeruleus,* Johnson (1893) saw the primordium "break through" the pigment striping as a definite, bordered "lesion" in the latter.

The nature of the stripes and clear bands has been studied by both light and electron microscopy. The dark bands in the ectoplasmic zone consist of re-

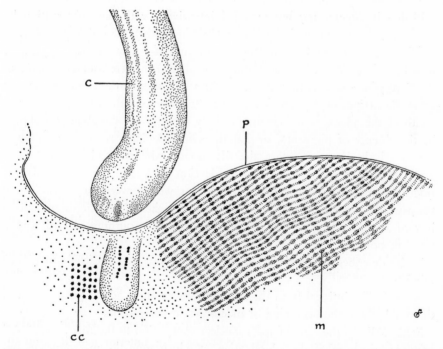

Figure 3-11. Section through myoneme of *Stentor.* (×34,000.) **p.** Pellicle. **c.** Cilium. **cc.** Ciliary corpuscle. **m.** Fibrillar plates of myoneme. (After Fauré-Fremiet.)

fringent granules which are uncolored protrichocysts in *Stentor polymorphus* and *S. roeseli,* blue in *S. coeruleus,* and black in *S. niger.* Each of the clear bands is occupied by a myociliary ensemble which Villeneuve-Brachon (1940) found to consist of (1) a longitudinal range of vibratile cilia with a corresponding series of kinetosomes; (2) a longitudinal fibril, the kinetodesma, immediately to the right of the kinetosomes; and (3) a large fiber, the myoneme, situated to the right of the kinetodesma, with plates joined to the epiplasmic pellicle. These observations have been confirmed and extended with the electron microscope by Fauré-Fremiet and his co-workers (1956), who found that the cilia, kinetodesma, and myoneme constitute a definite topographical ensemble: the myociliary complex, which occupies the clear zone of the ectoplasm, the myoneme being in close contact with the amorphous pellicle and organized in the form of parallel plates at right angles to the surface of the cell. Within the endoplasm, however, are other myoid fibers, whose fibrils have a reticular aspect, are denser than the environmental cytoplasm, and show an irregular, often ill-defined trabecular structure.

Fission

Fission in *Stentor* has been studied intensively, with close observation by Johnson and experimentally by Weisz (1951 and 1956). According to Johnson, the first sign of fission is a rift (the primordium of a new adoral zone) in the pellicle and ectoplasm near and almost parallel to the left boundary stripe of the ramifying zone. At first, the rift is in a longitudinal direction; yet, it cuts across the stripes of the ramifying zone. At the outset, the rift spans the full width of the adoral zone, opening entirely through the ectoplasm (as shown by the absence of pigment), so that the soft endoplasm is exposed and protrudes a little. From the start, the rift surface exhibits a ciliary motion and gradually, as the young membranelles develop from the exposed endoplasmic ridge, the motion is restricted to undulations which sweep lengthwise along the new adoral zone. As soon as the new zone reaches its full length and curves inward at the ends, a new gullet begins to develop, pushing inward with a spiral motion.

Weisz describes the whole division process as four stages, each lasting 45-60 minutes.

Stage 1. The interval from the first appearance of an adoral primordium at the mid-body to the beginning of macronuclear contraction.

Stage 2. From the start of macronuclear contraction to the first pronounced leftward curvature of the adoral zone, at which time the macronucleus has contracted to 5 fusion masses from the original 8.

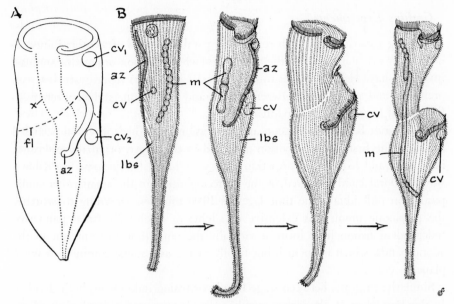

Figure 3-12. Fission in *Stentor*. **A.** Diagram. **fl.** Course of fission line. **x.** Path of ectoplasmic stripes in mid-ventral region. **az.** Stage in adoral-zone development during fission. **cv₁** and **cv₂**. Old and new contractile vacuole, respectively. (After Weisz.) **B.** Four stages of fission, showing reconstitution of typical form in each new individual. **m.** Macronucleus undergoing condensation and subsequent extension. **cv.** New contractile vacuole. **az.** New adoral zone. **lbs.** Left boundary stripe. (After Johnson.)

Stage 3. From the end of stage 2 to the appearance of a visible fission line, during which micronuclear division takes place and the macronucleus further contracts until it is a single fusion mass.

Stage 4. From the end of stage 3 to complete separation, during which the macronucleus splits and renodulates and a new gullet forms posteriorly at the adoral zone.

The path of the fission line is precisely determined in both space and time, and any decrease of cell volume on either side of the rift yields daughters of unequal size. Weisz (1956) considers that micronuclear mitosis, macronuclear splitting, and fission-line formation, which occur almost simultaneously near the end of stage 3, are interrelated causally and that micronuclear karyokinesis may well be the trigger. He also believes that the fission line is the visible expression of a local resorption or solvation affecting all the organelles in the ectoplasm, for the line does not become visible simultaneously throughout the cell and cannot be a fibrous contraction, as suggested by Johnson.

Grafting Experiments

Grafting experiments which resulted in parabiotic pairs of stentors joined in homopolar orientation along the right side (where division organelles are least involved) have yielded further information. Weisz (1956) made grafts between prefission and postfission, between stage-1 and postfission, and between post-fission and later-division individuals.

In the graft between prefission and postfission individuals, division occurred in both. The first sign of division appeared in the larger, originally prefission individual and led to the formation of a new adoral band. Sometime later a similar adoral band appeared in the correct position in the smaller, originally postfission individual. The time lag was 10-90 minutes, according to whether the parabiotic union was extensive or slight: in extensively fused pairs two independent fission lines form at virtually the same time just anterior to the adoral bands, which is normal, but finally merge as a single common cleavage plane.

Similarly, in grafts between stage-1 and postfission individuals, both divided. In this type of graft, the larger member already possessed a developing adoral band at the time of union.

In grafts between stage-2 or stage-3 and postfission individuals, the postfission individuals did not undergo division.

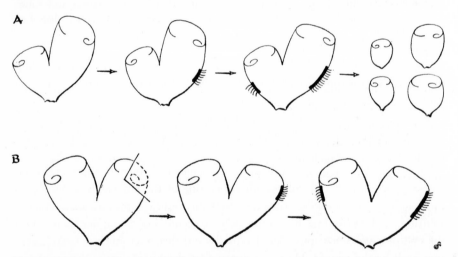

Figure 3-13. Graft union of whole individuals. **A.** Result of union between pre-fission and postfission individuals: division occurs simultaneously in both. **B.** Regeneration induction: at zero hour gullet of one partner is excised; then it regenerates. Later, new primordium appears in intact member, and original oral structure is resorbed. (After Weisz.)

Somewhat similarly, if the oral region of only one of a parabiotic pair of equivalent individuals is excised, a new adoral primordium regenerates in the usual manner about 4 hours later. Sometime later a new adoral zone also forms in the intact individual, but is accompanied by dedifferentiation of the original oral structures, just as in the physiological reorganization of normal stentors.

The induction effect, in both fission and regeneration, disappears with the end of stage 1. Furthermore, if grafted individuals are separated before adoral membranelles are visibly differentiated in either, induction again fails to occur; this result suggests that the inducing system requires a certain amount of time to take effect.

Excision experiments with parabiotic pairs also demonstrate that neither the oral apparatus nor the left boundary stripe is the source of the inducing stimuli. The question remains: What kind of transmission is responsible for the induction effect? Weisz (1956) points out that there is fairly good correlation between amount of time lag and the protoplasmic distance involved and states that the possibility of diffusible trigger substances is not ruled out. At the same time, he favors an as-yet indefinable, initially reversible physiological reorganization of the endoplasm, during early stages of division, which would spread into a parabiotically connected individual to induce division. Undisturbed endoplasm and a large body volume seem to be the prerequisites for division. In regeneration there is no early reversible phase; nor does macronuclear splitting, micronuclear division, or fission-line formation occur.

Fiber Systems in Flagellates and Ciliates

Flagellates and ciliates in general have complex cortical patterns of flagella and cilia, respectively. The flagellar systems are directly or indirectly controlled by a pair of centrioles; in ciliates, the centrioles are apparently involved only in mitosis, and the ciliary complex, with its basal granules, is a self-sustaining system. Together, flagellates and ciliates display cortical pattern and behavior more vividly than does any other material.

Flagellates

Centrioles and Cortex

The flagellates most profitably studied have been those commensal in the gut of termites and certain wood-feeding cockroaches—from which flagellates in abundance can always be obtained—species of the flagellate genera *Trichonympha, Barbulanympha,* and *Thynchonympha.* According to Cleveland, no

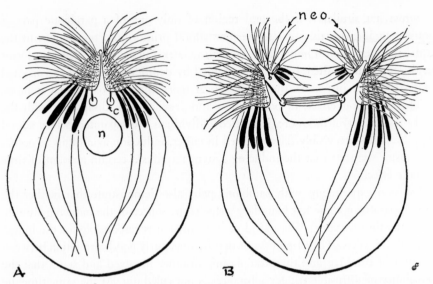

Figure 3-14.   Resting and dividing cells in *Barbulanympha*.   **A.** Resting cell, showing centrosome **c** on distal (inner) ends of elongate centrioles, flagellated areas,
axostyles (long lines), and nucleus **n.**   **B.** Dividing cell, showing old extranuclear
organelles and new ones **neo.** Each daughter cell after cytokinesis has two sets of
these organelles.   (After Cleveland.)

other organism shows the centrioles and the manner in which they function in
producing the achromatic figure as clearly as do these flagellates, in which far
more of centriole morphology and function can be seen than in all other organisms combined and in which centriole and centrosome can be readily
distinguished from each other.

In the resting cell of *Barbulanympha,* the two centrioles are elongate and
equal in length, and each has a centrosome enveloping its distal end. The distal
end of each centriole terminates between the two flagellated areas and the
nucleus, extending anteriorly underneath a flagellated area to the end of the
cell where the two centrioles join together by an interconnection. At the onset
of the prophase preparatory to division, the connection is lost, and immediately
the proximal end of each elongate centriole produces a new centriole by outgrowth. At the same time, the distal end of the old, elongate centriole begins
to form astral rays, which grow from that region of each centriole that is surrounded by a centrosome.

This process is clearly visible in living material by means of phase contrast:
the astral rays are seen to pass through the centrosome to the centriole. The
rays from both centrioles soon become long enough to meet and join, thereby
forming the central spindle of the achromatic figure. As the astral rays continue

to grow, the spindle elongates, pushing the distal ends of the centrioles apart. At the same time, the new centrioles formed from the proximal ends of the old ones grow in volume and increase in length until a centrosome begins to form at the distal end of each. Meanwhile, the two parent flagellated areas and

Figure 3-15. Telophase stage of division in cell of *Pseudotrichonympha*. Cytoplasm has divided for short distance, and two daughter rostra have been produced. From each rostrum, a long, coiled centriole extends to end of large central spindle. Distal ends of centrioles are surrounded by centrosome, through which astral rays, chromosomal fibers, and central spindle pass. These fibers come from distal ends of centrioles. Daughter chromosomes are connected via centrosomes and chromosomal fibers to ends of centrioles.

the two groups of parabasal bodies and axostyles, which persist throughout cell division, are pushed apart by the central spindle's increase in length. As the parent areas separate, a new set of organelles is formed for each daughter cell, each set consisting of a flagellated area, a group of parabasal bodies, and a group of axostyles—all formed from the anterior, slightly enlarged proximal end of each new centriole. This enlarged proximal end, the growing region of a centriole, has the ability to reproduce not only a new centriole but directly or indirectly all the other extranuclear organelles as well. Thus the centrosome is produced directly from the distal end of a centriole, which in turn was produced by the enlarged proximal end. After the astral rays and spindle complete their function, they slowly disappear. Each new centriole grows and elongates and takes its place alongside the old centriole in each daughter cell.

During cell reproduction each centriole usually duplicates itself once, but occasionally there are two or more duplications, or none, giving rise to daughter cells with many centrioles, as well as with only one. Sometimes, for instance, 6 centrioles are grouped in, and anchored to, one flagellated area, and 2 to the other. Whatever the number of centrioles in such large, multinucleate forms, which have many widely separated sets of extranuclear organelles, the cytoplasm is still capable of synchronizing both the activities of all the centrioles (all are active or inactive at the same time) and the entire period of growth of the many achromatic figures. Cleveland (1957b) suggests that each cell, no matter how large it is or how many nuclei and centrioles it possesses, manufactures a substance which freely circulates in the cytoplasm and controls and synchronizes both types of centriole activity (i.e., formation of achromatic figure and of organelle) with chromosome activity. When more centrioles than the usual one are duplicated, each attains full growth and has at its distal end a centrosome of normal size and activity. Astral rays are never produced without centrioles, and micromanipulation and other experimental procedures have shown that the rays are fastened securely to the distal end of a centriole. Centrosomes, on the other hand, apparently play but a minor role in cell reproduction, for they are absent altogether in about half the genera of both hypermastigote and hypomastigote flagellates; thus the centriole must play the vital role. Furthermore, the centrioles, rather than the chromosomes, seem to determine the nature of the extranuclear organelles (which are the main basis of classification in flagellates); and chromosomes, because they have directed the course of flagellate evolution, must have done so indirectly through the centrioles.

Not all hypermastigote flagellates are alike in centriole behavior. Cleveland has described several types. In the resting cell of *Trichonympha,* there are one long centriole and one short centriole connected together, the short one having been formed from the end of the long one during the preceding cell division.

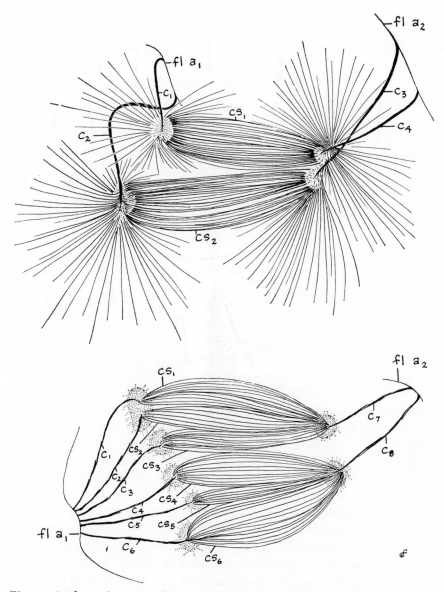

Figure 3-16. Achromatic figure formation by multiple centrioles in *Barbula-nympha*. **fl a$_{1-2}$**. Flagellated area. **cs$_{1-6}$**. Central spindle. **c$_{1-8}$**. Centriole. (After Cleveland.)

With the onset of the prophase, the short one elongates, and only when the two are of equal length do astral rays form from their distal ends. As soon as the connection between their proximal ends is broken at about the time the spindle begins to form, each proximal end at once forms a new short one, with

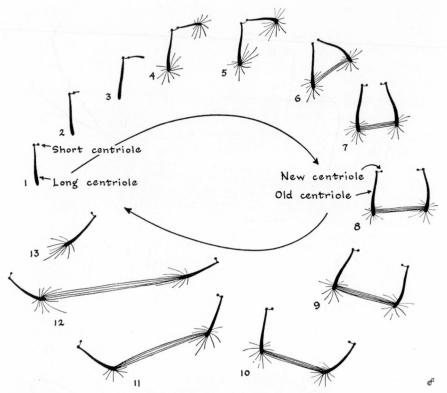

Figure 3-17. Life cycle of centriole in *Trichonympha,* showing reproduction of new centriole at one end and production of astral rays at other. (After Cleveland.)

which it remains connected. The new centriole of each daughter cell begins to form new organelles early in the prophase and continues to do so until the time of actual cell division, so that in each daughter half the flagella, half the parabasal bodies, and half the axostyles are new and half are old. Thus each centriole in *Trichonympha,* as in many other flagellates, acts in one capacity in one generation and in another in the next: producing organelles from its proximal end in the first generation and astral rays from its distal end in the second, and so on. The distal end may form astral rays repeatedly, but the proximal (or anterior) end forms new flagella and parabasal bodies only once— with the exception described below.

It seems clear from this account that as long as two centrioles remain connected, centriole duplication does not occur, but that as soon as the connection is broken, duplication takes place at once—and only from the end closest to the cell cortex. The inner, or so-called distal, end functions only to produce astral rays of simple protein, comparable perhaps to the ciliary roots of basal granules; whereas, the proximal end, in addition to possessing self-duplicating properties, has the capacity to organize cortical cytoplasm into characteristic organelles.

Dependence of organelles upon centriole activity is seen negatively in many hypermastigote flagellates. In *Joenia,* for example, there is only one set—not two as in *Trichonympha* and *Barbulanympha*—of extranuclear organelles, all produced by a single centriole. During the prophase, nucleus and centrioles become dissociated from the cortex and move to the posterior end of the cell, leaving the old flagella, parabasal bodies, and axostyles to degenerate. None of this is carried over during cell division except nucleus and centrioles, and although each daughter cell receives one old and one new centriole, only the new centriole can initiate organelle production; thus only one set of organelles is formed. Evidently, therefore, the differentiated organelles not only are produced under the influence of a new centriole but are dependent directly or indirectly upon connection with, or proximity to, the centriole. In *Pyrsonympha,* a related genus, Grassé and his co-workers (1956) observed that 4-8 flagella, the parabasal bodies, and the axostyles all originate in a homogeneous, dense, ellipsoid centriole.

In *Trichonympha* and other genera whose flagella are numerous and arranged in orderly rows, Cleveland (1949) found that numerous small basal granules grew from the centriole and arranged themselves in rows in the cortical cyto-plasm, each granule subsequently giving rise to a flagellum. In electron-micro-scope studies of *Trichonympha,* Pitelka and Schooley (1958) found that the flagellar roots, each consisting of a tube of 9 double hollow fibers, either stand in contact with the outer striated tube of the rostral tube or with the parabasal bodies posterior to the tube or, below this level, lie free in the cytoplasm in what seems to be peripheral endoplasm, but are never in direct contact with the centriole. These investigators suggest that once a few basal granules are pro-duced by, or organized in, the region of the centriole, they may migrate and serve as models for the production of additional basal bodies to complete the required number of flagellar rows.

At this point, interrelationships become even more subtle. If the flagellar rows depend on the propagation and migration of granules from the vicinity of a centriole, what determines their number and arrangement? Centrioles produce different fiber structures in different places and reproduce themselves in different ways at different times. Are they self-determining, or are they in turn responsive to their cytoplasmic environment? In other words, are they key inductive cen-

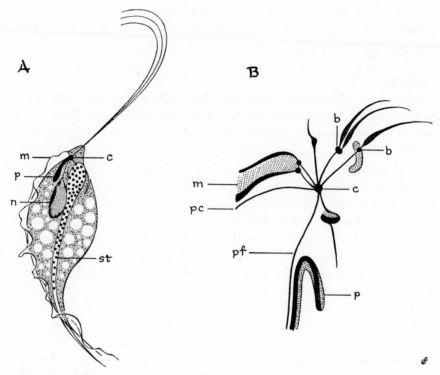

Figure 3-18. Relationship of organelles to centriole. **A.** *Trichomonas lacertae.*
m. Costal membrane. **c.** Centriole. **p.** Parabasal body. **n.** Nucleus. **st.** Axostyle. **B.** *Macrotrichons hirsuta.* **b.** Basal granule of flagellum. **pc.** Paracostal
filament. **pf.** Parabasal filament. (After Grassé.)

ters, of a self-propagating kind, which perform according to the nature of, and
changes in, a fine cytoplasmic structure of the lowest hierarchical level? And,
as seems to be indicated by the phenomena of artificial parthenogenesis, can
they arise from cytoplasm without having precursors? It may be significant that
the *Trichonympha* ectoplasm, particularly in the flagellated area, consists of
several layers: a double membrane external to two parallel sheets of parallel
fibers, with a deeper third layer of fibers at an angle to the others—an arrangement similar to the organization of collagen in basement membranes. In
Trichonympha the organization lies within the cell boundary and it is conceivable that organizational properties at the lepton-lattice level of this region
may underlie and control all visible pattern and the activity of bodies such as
centrioles and kinetosomes. Much depends on the extent and nature of structural
continuities in depth, other than that already seen as centriolar extension in,
for example, *Barbulanympha.*

Ciliates

Differentiation of Cortex

Ciliates in general are outstanding because they have a cortical fibrillar system in which no cytoplasmic differentiations stand alone, but are all interconnected to form an integrated pattern. Tartar (1941) suggests that the fibrillar system may be the one through which the shape of the cell and the type and pattern of differentiation of the outer layer are to be interpreted; in Tartar's words, "Is there some as yet invisible groundwork which determines the shape of the cell following which the fiber system simply spreads over this contour and adapts itself accordingly, or is the fiber system itself the active, immediate agent in determining pattern and form?" [1]

The following facts support the second theory:

1. Balamuth (1940) found that when the neck region separating the head and base of the heterotrich *Lichnophora macfarlandi* is transected, the elements of the fibrillar system frequently grow out beyond the edge of the body, the individual fibrils extending free into the medium. Later, outgrowth of the body proper encloses the fibrils, so that this particular fiber system initiates growth and does not first invade a pre-existing structure. Only cuts which injure fibrils of the adoral zone result in the regeneration of that structure.

2. Transformations of the fiber system are epigenetic in character. Klein (1932) gives many examples of an early generalized narrow-net system in which at certain intersections the granular primordia of cilia and trichocysts later appear.

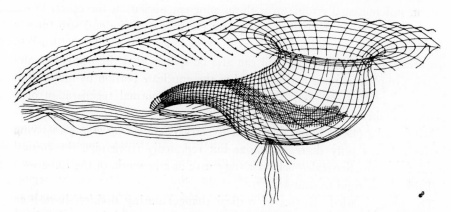

Figure 3-19. Diagram of basal bodies (blepharoplasts or kinetosomes) and pharyngoesophageal network of *Paramecium*. (After Lund.)

[1] Tartar, *Growth* Supp. (Symp. 3), **5**:22 (1941).

3. In *Paramecium caudatum* the shape of the daughter cells resulting from division is closely bound to growth of the fiber system: Tartar (1941) found that when multiplication of the fibers and their derivatives does not take place, the resulting cells are incomplete and abnormal in shape. In this species each cilium with its basal granule lies in the center of a hexagonal pit in the outer surface; the hexagons are formed by meridional lines which alternate with the ciliary rows and by cross-connections which pull the meridians into zigzag form, thereby creating the hexagons. In the center of each cross-connection lies an opening for a trichocyst. During division the major events are: the cross-connections in the outer polygonal system break down, and the prospective daughter cells actually grow apart. Otherwise there is no significant dedifferentiation of cortical structure; and all the original body cilia, trichocysts, and fibers are passed on—all new structures other than the contractile vacuoles arising genetically from the old.

In support of the first possibility that a "groundwork . . . determines the shape of the cell," is *Condylostoma magnum.* One end of this large, elongate marine protozoan bears a hooded peristome with a membranelle and a row of ciliary plates. During division there is no change in the meridional system of body cilia; new oral structures arise entirely separate from pre-existing ones; the old peristome partly dedifferentiates and is remodeled on a smaller scale; and the two daughter cells are shaped from the original, without growth of new ends as in *Paramecium.* Furthermore, the greater part of *C. magnum* endoplasm can be eliminated without precluding regeneration: When the ciliate is placed in fresh water, it swells to the limit of extensibility of the outer layer; and the endoplasm aggregates in the form of a ball at one end, leaving a very transparent, differentiated ectoplasm with an adherent nucleus at the other. When cut so as to remove the ball of endoplasm together with the oral structures, a new peristome regenerates as soon as the remainder of the cell is returned to sea water. In this species the cell can also be split so that the nucleus resides in one cytoplasmic mass and is united to the anucleate portion by only a thin bridge of protoplasm; yet, Tartar (1941) found, normal regeneration of a peristome takes place.

Dembowska (1938) found that in the hypotrich *Stylonychia mytilus,* starving individuals not only decrease in size but repeatedly reorganize the cortical pattern on a smaller scale—behavior suggestive of a network, of the full-grown individual, too rigid to contract.

The most detailed analyses of cortical changes during division have been made by Chatton and Seguéla (1940) and Bonner (1954) of the closely related hypotrich genus *Euplotes* and by Fauré-Fremiet (1954) of the holotrich *Urocentrum.* The time interval between two divisions of *Euplotes* is 94 hours, but

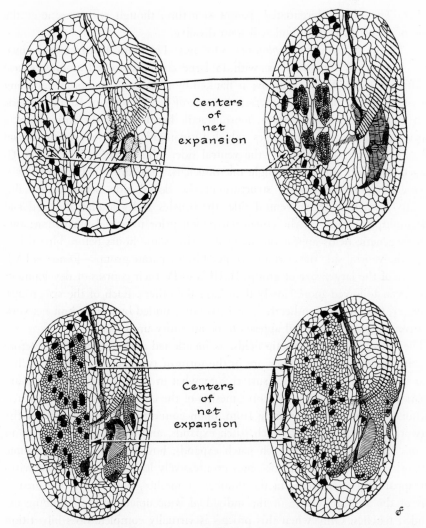

Figure 3-20. Ventral surface of predividing individuals in *Euplotes,* showing progressive expansion of new network rudiments with contained cirri bases. (After Chatton and Seguéla.)

the morphogenetic processes involved start 8 hours before the cytoplasmic cleavage. As Lwoff (1950) emphasizes, the important phenomenon—the great mystery in this division—is that a long time before nuclear division two "morphogenetic fields" are formed; for it is quite obvious that it is not a "differentiated" ciliate which "divides," but that cleavage separates two already highly differentiated ciliates and that the real division has taken place 8 hours earlier. The old, differentiated, parent structure, though still present at the time of cleavage, is dead and will soon dissolve.

Briefly, the structure of *Euplotes* is as follows: The ventral surface of the oval body is typically hypotrichous, with 18 large cirri distributed among several specific locations. The oral groove is flanked by a series of membranelles along one side; on the dorsal side there is a series of 8 or 9 rows of basal granules, each granule supporting a short noncontractile bristle. Over the entire body is a network of silver-staining lines, which form regularly meshes on the dorsal side and an irregular mesh on the ventral side. Within the cell is a large **C**-shaped macronucleus and a single micronucleus. According to Bonner (1954), during division the cortical structure of the dorsal side behaves very differently from that of the ventral side: the bristles and granules on the dorsal side multiply rapidly in the equatorial region prior to actual bipartition; two morphogenetic fields appear on the ventral side some hours before bipartition.

On the ventral side the cirri are arranged in 6 separate groups—locations I-VI. Because of the larger size of groups II, III, and IV their courses of development have been followed more closely than have the others. Each of the six groups, however, first arises as either 1, 2, or 3 cirri surrounded by a tight-knit network of silver-staining fibers, and appears to be an entity unto itself.

The so-called morphogenetic fields, as mentioned above, appear long before the time of cell division. They are in the form of two sets of elongate patches near the middle of the ventral surface somewhat in front of the mouth and are separated from each other by 2 or 3 meshes of the cortical network, each patch within a set apparently forming from, or in connection with, a strand of the network. Subsequently each such patch expands, and 3 cirri grow outward in groups II, III, and IV. As each patch expands, however, the almost invisible network of the primordium becomes progressively loosened or stretched without apparently an increase in the number of meshes, until finally the greater part of the ventral surface of the individual is occupied by the adjoining expanded patches. Only when this process is virtually complete does bipartition begin: the cleavage plane then divides the cell between the anterior and posterior network areas. The original mouth and membranelle, which lie within the forward area of the cell, remain and become part of the anterior daughter cell. A new mouth and membranelle form within the posterior daughter cell; and, along one strand of one mesh close to the old mouth, the primordium of these structures is discernible as early as are the two morphogenetic areas of cirri. During the expansion of the new networks, the old network does not seem to be pushed aside, so much as to dissolve as the new encroaches.

On the dorsal side the final process of cleavage follows a striking elongation of the individual as a whole. The elongation, clearly the result of growth in the middle region of the body, is expressed both as the extensive stretching or expansion of the network meshes of this region and as the multiplication of granules along the longitudinal strands. Granule multiplication provides each prospective daughter cell with a full complement of the linearly arranged granules; cleavage itself is the furrowing of the cortical layer in the middle zone of the growth region—the zone where growth has in all probability been most extensive or rapid. Most of the dorsal expansion takes place after the two daughter cells have separated, so that the new, crowded, basal granules— originally in the middle region of the parental cortex—become as separated from one another as are the surviving granules from the anterior or posterior ends.

The cirri, which attain full size on the ventral surface before much significant expansion of the basal network takes place, are uniform in size irrespective of species or size of an individual as a whole. This phenomenon led Tartar to suggest that there are constant-size, or atomistic, organelles that are independent of the varying sizes of individuals. The network, on the other hand, expands in proportion to the distance between the cirri, and mesh size is an index to the degree of expansion. In this connection it is noteworthy that at first the cirri in each group are equidistant from one another irrespective of the proportions of a particular group's future expansion. Both Fauré-Fremiet (1948) and Bonner (1954) have suggested that the network with its granules may be the expression of a crystal structure, whose nature is as yet unknown. Significantly, injury to as little as a single hypotrich cirrus or its basal granule leads to the same profound reorganization as that initiated by the bisection of a whole animal.

Kinetosomes

The so-called silver-line system has been known in ciliates since Klein (1928), but there is a fibrillar complex, at least in many ciliates, which can be seen in vivo and, according to Chatton and Lwoff (1935), is distinct from the silver-staining fibers. This infraciliature consists of superficial fibrils adhering to the pellicle as rectilinear (never sinuous) threads. Other granules as well as the ciliary basal granules are similarly attached to the pellicle. These represent successive stages of the multiplication of the basal granules. The basal granules are called kinetosomes, and a row of kinetosomes has on its right a parallel and associated fibril called the kinetodesma. Electron-microscope studies by Metz, Pitelka, and Westfall (1953) of *Paramecium* show that a kinetodesma consists of a bundle of parallel fibrils each of which is connected to a kinetosome at one end, but terminates freely within the kinetodesma bundle at the other.

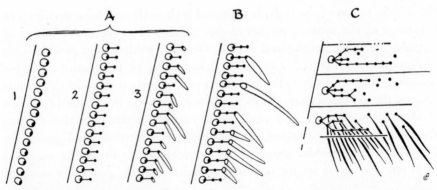

Figure 3-21. Multiplication of basal granules to form trichocysts. **A.** Gymnodini-
idae. **B.** *Polyspira.* **C.** *Foetteringii.* 1. Normal ciliary row, the row constituting
a kinety, showing small kinetosomes within satellite corpuscles, with a parallel
fiber or kinetodesma at one side. 2. Division of kinetosomes with production of
trichocystomes. 3. Formation of trichocysts. (After Chatton and Lwoff.)

The combined kinetodesma and row of associated kinetosomes form a kinety,
and all the kineties together constitute the infraciliature.

The origin of a new mouth and gullet has been described in greater detail
in genera other than *Paramecium.* During *Glaucoma* and *Chromodina* stoma-
togenesis, for example, according to Lwoff (1950), one of the ventral somatic
kineties gives birth to a granular field by localized multiplication of its kinetosomes; this granular field forms the rudiment of the new mouth, which becomes
secondarily organized by the regrouping of kinetosomes into oblique rows to
form the vibratile membranes. This regrouping must be the result either of
interactions among the newly formed kinetosomes or of the kinetosomes' response to a more subtle pattern in their cytoplasmic surroundings. Such somatic
origin of new buccal ciliature seems to be characteristic of the ciliates.

It has been followed closely in *Urocentrum* by Fauré-Fremiet (1954), who
describes it as an autonomous system with its own morphology, in which a
completely somatic morphogenesis is achieved by kinetosomes following a
specific and definite plan of multiplication and rearrangement. In this genus,
bipartition as a whole is preceded by an equatorial cell expansion that is associated with multiplication of kinetosomes and extension of kineties similar
to those processes of the dorsal side of *Euplotes*—though even more striking.
As the middle zone expands, the kinetosomes follow a specifically ordered,
complete plan, comparable to a process of regeneration in that there is reconstitution of a posterior half-ciliate from the anterior part and an anterior half-
ciliate from the posterior part. The complex mechanism of bipartition results
in the organization of two equivalent homogeneous networks of polarized

kineties, in a geometrical and crystallographical sense, and the surface of the ciliate is alone responsible for these characters and mechanisms. The bipartition mechanism is reminiscent of the concept of a cortical morphogenetic field, and, according to Fauré-Fremiet (1954), the network of characteristic kineties presents a sort of objective image of such a cortical field. Whether the network is primary or merely responsive, however, is still in doubt.

Besides being capable of self-duplication, the kinetosomes may contribute to, or participate in, any one of the following functions: (1) perpetuation of the generating kinety; (2) lengthwise growth of the generating kinety; (3) forma-

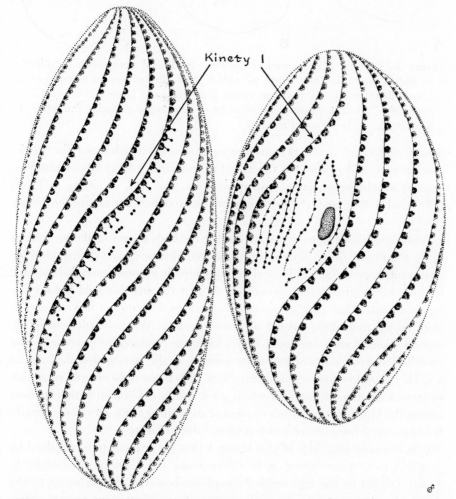

Figure 3-22. Formation of oral ciliature in *Chromodina* by division of kinetosomes of kinety 1. (After Chatton and Lwoff.)

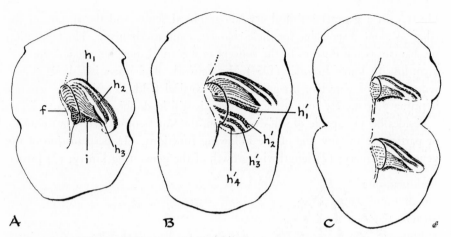

Figure 3-23. Oral ciliature in *Urocentrum.* **A.** Buccal cavity with polykineties h_1 and h_2, and composite kinety h_3 with 8 fibers directed toward buccal opening f, and ciliary field l. **B.** Division: ciliary field grows and subdivides into 4 oblique zones h'_{1-4}, which constitute oral ciliature of prospective posterior daughter. **C.** Separation. (After Fauré-Fremiet.)

tion of new kineties; and (4) surface morphogenesis, which may influence the development, and control the functional maintenance, of (a) cilia, trichocysts, flagella, membranelles, undulating membranes, and cirri, according to Lwoff (1949 and 1950); and (b) gullets, vacuoles, and holdfast organelles, Weisz (1951).

When specialized kineties are cut so that the ends cannot or do not fuse, as in the cutting of fission lines or in merotomy, kinetosomes at the cut ends become morphogenetically active and multiply rapidly to form a so-called anarchic field, in which granules are numerous. At first the anarchic field lacks a discernible pattern, but later it becomes arranged in double rows of kinetosomes, from which more-differentiated structures develop. When kineties are intact, anarchic fields form from one side of a kinety to produce, for example, a new mouth. The process is essentially the same as in the cutting described above: a single row of kinetosomes yields successive rows to the right in rapid succession to form a small anarchic field, which gives rise to the oral structure. In most ciliates the kineties extend from one end of the animal to the other in longitudinal or in spiral lines, one of which is identifiable as kinety 1.

It is from the right side of this kinety 1 that the oral ciliature is derived by propagation of kinetosomes, as in *Chromodina* and *Glaucoma,* according to Lwoff (1950). In the heterotrich *Licnophora,* however, the peristome of the daughter ciliate originates from the escape of one kinetosome or a very few kinetosomes from the original peristomial system of the parent; the "escapee(s)" then multiply to form an anarchic and homogeneous field of closely packed kinetosomes. The new kinetosomes later become aligned and organized as if controlled by, as-yet obscure, orienting forces; the basal disk divides simply through elongation of its constituent kineties, according to Villeneuve-Brachon (1940). Balamuth (1942) found that, significantly, the peristomial structure is readily regenerated, but that the basal disk is not.

According to Lwoff (1950), the kinetodesma is the visible agent by or through which the somewhat mysterious morphogenetic forces exert their orienting action; moreover, the local disappearance of even part of a kinetodesma causes the kinetosomes to scatter and initiate a new anarchic field. During the growth and division phases of the life cycle of certain parasitic ciliates, such as *Foetteringii,* the kinetodesmas and their associated rows of kinetosomes exhibit striking alterations of torsion and detorsion, ranging from perfectly bipolar structures to tight whorls. It is evident, in *Foetteringii* at least, that elongations and shortenings of the kinetodesmas and the torsions of the cytoplasm are independent phenomena. In most highly evolved mature ciliates, however, the cortical network is a rigid structure; whereas kinetosomes may multiply, as do other particles endowed with genetic continuity.

Yet, the kinetosomes and kinetodesmas are oriented by some property of the cortical structure—some property which can only be some kind of electronic or intermolecular force, possibly at the macromolecular level of the lepton lattice.

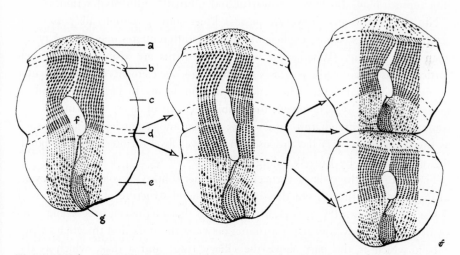

Figure 3-24. Ventral body ciliature in *Urocentrum* during fission. Kineties and kinetosomes are portrayed only in median band—sufficient to show characteristic disposition. **a.** Apical cap. **b.** Anterior transverse groove. **c.** Fascia. **d.** Cingula. **e.** Posterior hemisphere. **f.** Oral aperture. **g.** Scopula. (After Fauré-Fremiet.)

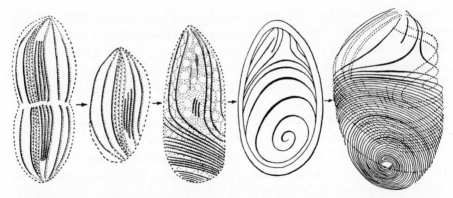

Figure 3-25. Postfissional growth of anterior daughter cell in *Foetteringii,* showing torsion and formation of invaginated spiral coil at posterior end. (After Chatton and Lwoff.)

If the kinetosomes continue to increase in number after the cortical meshwork has attained its maximal limit, they will have no place for attachment. Their position no longer controlled, the free blocks of kinetosomes then crystallize according to their structure and environment and form new systems or new organelles. Since a single kinetosome such as a centriole can exert an attracting and orienting influence in flagellates and in metazoan cell divisions, groups of kinetosomes may also have a powerful and mutually influential action.

Not all protozoans, however, possess kinetosomes, though all possess an ectoplasmic cortex, whose explicit morphology differentiates one kind of protozoan from another. Whether or not such elements are visibly distinct, each cortical locus undoubtedly possesses specific diagnostic elements on the molecular or macromolecular level. The crucial question seems to be: What kind of molecular hierarchy is involved: the visible network of fibers and kinetosomes comprising the infraciliature as a whole; a macromolecular lepton lattice, such as Bretschneider (1952) has proposed; or a lower level of intermolecular bonding, as suggested by Frey-Wyssling (1953)?

The nature of the stalk of vorticellids, whether solitary or colonial, is relevant to this question, though not decisive. The general structure of this stalk, well described by Entz (1893), is clearly related to that of the cell as a whole. The vorticellid cell body consists of a funnel between the stalk and the ciliary ring; a bell, which is the part above the ciliary ring; and a disk, which is the peristomial border together with the adoral zone and the cytostome. The fibrillar system, which lies immediately below a sculptured pellicle, consists of an outer and inner complex of fibrils or myonemes, each complex in turn consisting of an outer circular, spiral layer and an inner longitudinal layer—

totaling 4 layers in all. The outer fibrillar complex continues uninterruptedly into the protoplasmic lining of the stalk sheath. As shown by electron-microscope studies by Romillar and his co-workers (1956), the main structure of the stalk is a flexible bundle of scleroprotein fibrils which result either from the ciliary membrane (in *Campanella*) or from the 9 ciliary fibrils (in *Opercularia* and *Zoothamnion*). Since the stalk seems to be fundamentally an extension of the fibrillar complex, it is difficult to suppose that its organization depends on anything more subtle.

Whatever its degree of autonomy, the fibrillar complex of the cortex exists in relation to the endoplasmic cytoplasm and is at least influenced by the nucleus, even though the nucleus does not directly determine cortical pattern. Although the evidence supports the concept that the fibrillar system as a whole is the actual functional and essential organization of the differentiated organism,

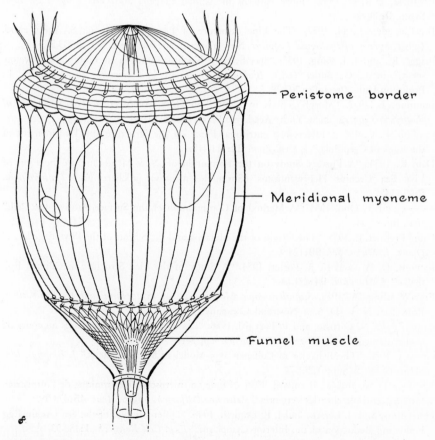

Figure 3-26. Pattern of second complex of body fibrils in *Epistylis*. (After Entz.)

the system is epigenetic and can develop from cortical substance of submicroscopic structure.

REFERENCES FOR
Cells

GENERAL

Afzelius, B. A. 1955. "The Fine Structure of the Sea Urchin Spermatozoa As Revealed by the Electron Microscope." *Zeitschr. Zellforsch. mikro. Anat.* **42**:134-48.

———. 1956. "The Ultrastructure of the Cortical Granules and Their Products in the Sea Urchin Egg As Studied in the Electron Microscope." *Exp. Cell Research.* **10**:257-85.

Astbury, W. T., E. Breighton, and C. Weibull. 1955. "The Structure of Bacterial Flagella." *Soc. Exp. Biol.* Symp. **9**:282-305.

Bradfield, J. R. G. 1955. "Fibre Patterns in Animal Flagella and Cilia." *Soc. Exp. Biol.* Symp. **9**:306-34.

Bretschneider, L. H. 1952. "The Fine Structure of Protoplasm," in G. S. Avery *et al.* (eds.). *Survey of Biological Progress.* New York: Academic Press, Inc. **2**:233-57.

Briggs, R., and J. J. King. 1955. "Specificity of Nuclear Function in Embryonic Development," in E. G. Butler (ed.). *Biological Specificity and Growth.* Princeton, N. J.: Princeton Univ. Press. Pp. 207-28.

Bünning, E. 1952. "Morphogenesis in Plants," in G. S. Avery *et al.* (eds.). *Survey of Biological Progress.* New York: Academic Press, Inc. **2**:105-40.

Conklin, E. G. 1917. "Effects of Centrifugal Force on the Structure and Development of the Eggs of Crepidula." *J. Exp. Zool.* **22**:311-419.

Dan, K. 1954. "A Further Study on the Formation of 'New Membrane' in the Eggs of Live Sea Urchins Henricentrotus (Strongylocentrotus) pulcherrimus." *Embryologia.* **2**:99-114.

———, and T. Ono. 1954. "A Method of Computation of the Surface Area of the Cell." *Ibid.* Pp. 87-98.

Fauré-Fremiet, E. 1958. "The Origin of the Metazoa and the Stigma of the Phytoflagellates." *Quart. J. Micro. Sci.* **99**:123-9.

Fawcett, D. W., and K. R. Porter. 1954. "A Study of the Fine Structure of Ciliated Epithelia." *J. Morphol.* **94**:221-64.

Frey-Wyssling, A. 1953. *Submicroscopic Morphology of Protoplasm and Its Derivatives.* Princeton, N. J.: D. Van Nostrand Company, Inc.

Grassé, P.-P., N. Carusso, and P. Farvard. 1956. "Les ultrastructures cellulaires au cours de la spermatogénèse de l'Escargot." *Ann. sci. nat.* Série zool. **18**:339-80.

Gross, J. 1956. "The Behavior of Collagen as a Model in Morphogenesis." *J. Biophys. Biochem. Cytol.* **2**(Supp.):261-74.

Haarven, E. de, and W. Bernhard. 1956. "Étude au microscope électronique de l'ultrastructure du centriole chez les vertèbres." *Zeitschr. Zellforsch. mikro. Anat.* **45**:378-98.

Hörstadius, S., I. J. Lorsch, and J. F. Danielli. 1949. "Differentiation of the Sea Urchin Egg Following Reduction of the Interior Cytoplasm." *Exp. Cell Research.* **1**:188-93.

Houwink, A. L., and P. A. Roelefson. 1954. "Fibrillar Architecture of Growing Plant Cell Walls." *Acta bot. néerl.* **3**:385-409.

Huettner, A. F. 1933. "Continuity of the Centriole in Drosophila melanogaster." *Zeitschr. Zellforsch. mikro. Anat.* **19**:119-34.

Kaye, J. S. 1958. "Changes in the Fine Structure of Mitochondria during Spermatogenesis." *J. Morphol.* **102**:347-99.

King, J. J., and R. W. Briggs. 1955. "Changes in the Nuclei of Differentiating Gastrula Cells, As Demonstrated by Nuclear Transplantation." *Nat. Acad. Sci. U.S. Proc.* **41**:321-5.

Körner, W. F. 1952. "Untersuchungen über die Gehäusbildung bei Appendicularien (Oikopleura dioica)." *Zeitschr. Morphol. Okol. Tiere.* **41**:1-53.

Manton, I. 1956. "Plant Cilia and Associated Organelles," in D. Rudnick (ed.). *Cellular Mechanisms in Differentiation and Growth.* Princeton, N. J.: Princeton Univ. Press. Pp. 61-71.

Mazia, D. 1955. "The Organization of the Mitotic Apparatus." *Soc. Exp. Biol. Symp.* **9**:335-57.

Mitchison, J. M. 1953. "A Polarized Light Analysis of the Human Red Cell Ghost." *J. Exp. Biol.* **30**:397-432.

Northcote, D. H. 1958. "The Cell Walls of Higher Plants: Their Composition, Structure and Growth." *Biol. Rev.* **33**:52-102.

Pallade, G. E. 1956. "The Endoplasmic Reticulum." *J. Biophys. Biochem. Cytol.* **2**(Supp.):85-97.

Perry, S. V. 1955. "The Components of Myofibrils and Their Relation to Structure and Function." *Soc. Exp. Biol. Symp.* **9**:203-27.

Porter, K. R. 1954. Report in International Conference on Electron Microscopy. London.

Rudall, K. M. 1955. "The Distribution of Collagen and Chitin." *Soc. Exp. Biol.* Symp. **9**:49-71.

Scott, F. M., K. C. Hamner, E. Baker, and E. Bowler. 1956. "Electron Microscope Studies of Cell Wall Growth in the Onion Root." *Am. J. Bot.* **43**:313.

Seidel, F. 1932. "Die Potenzen der Furchungskerne im Libellenei und ihre Rolle bei der Aktivierung des Bildungszentrums." *Arch. entwick. Org.* **126**:213-76.

Selman, G. G., and C. H. Waddington. 1955. "The Mechanism of Cell Division in the Cleavage of the Newt's Egg." *J. Exp. Biol.* **32**:700-33.

Swann, M. M., and J. M. Mitchison. 1957. "The Mechanism of Cleavage in Animal Eggs." *Biol. Rev.* **33**:103-35.

Swift, H., L. Rebhun, E. Rasch, and J. Woodward. 1956. "The Cytology of Nuclear RNA," in D. Rudnick (ed.). *Cellular Mechanisms in Differentiation and Growth.* Princeton, N. J.: Princeton Univ. Press. Pp. 45-60.

Weiss, P. 1956. "The Compounding of Macromolecular and Cellular Units into Tissue Fabrics. *Nat. Acad. Sci. U.S. Proc.* **42**:819-30.

————, and W. Ferris. 1954. "Electron Microscopy Study of the Texture of the Basement Membrane of the Larval Amphibian Skin." *Ibid.* **40**:528-40.

————. 1956. "The Basement Lamella of Amphibian Skin." *J. Biophys. Biochem. Cytol.* **2**(Supp.):275-82.

Weiss, P., and H. Hiscoe. 1948. "Experiments on the Mechanism of Nerve Growth." *J. Exp. Zool.* **107**:315-96.

Whitaker, D. M. 1936. "The Effect of White Light upon the Rate of Development of the

Rhizoid Protuberance and the First Cell Division in Fucus furcatus." *Biol. Bull.* **70**:100-8.

PHYCOMYCES

Castle, E. S. 1936. "The Origin of Spiral Growth in Phycomyces." *J. Cell. Comp. Physiol.* **8**:493-502.

——. 1937. "The Distribution of Velocities of Elongation and Twist in the Growth Zone of Phycomyces in Relation to Spiral Growth." *Ibid.* **9**:477-89.

——. 1953. "Problems of Oriented Growth and Structure in Phycomyces." *Quart. Rev. Biol.* **28**:364-72.

Delbrück, M., and W. Reichhardt. 1956. "I. System Analysis for the Light Growth Reactions of Phycomyces," in D. Rudnick (ed.). *Cellular Mechanisms in Differentiation and Growth.* Princeton, N. J.: Princeton Univ. Press. Pp. 3-45.

Frey-Wyssling, A. 1952. *Deformation and Flow in Biological Systems.* Amsterdam: North Holland Pub. Co. Pp. 236-40.

Grehn, J. 1932. "Untersuchungen über Gestalt und Funktion der Sporangienträger bei den Mucorineen. I. Entwicklungsfeschichte der Sporangienträger." *Jahrb. wiss. bot.* **76**:93-165.

Heyn, A. N. J. 1939. "Some Remarks on the Mechanism of Spiral Growth of the Sporangiophore of Phycomyces and a Suggestion for Its Further Explanation." *Ned. akad. weten. Proc.* Series C. **42**:431-47.

Preston, R. D. 1948. "Spiral Growth and Spiral Structure. I. Spiral Growth in Sporangiophores of Phycomyces." *Biochim. biophys. acta.* **2**:155-66.

——. 1952. *The Molecular Architecture of Plant Cell Walls.* New York: John Wiley & Sons, Inc. Pp. 182-201.

Roelefson, P. A. 1950. "The Origin of Spiral Growth in Phycomyces Sporangiophores." *Rec. Trav. bot. néerl.* **42**:72-110.

——. 1951. "Cell-Wall Structure in the Growth Zone of Phycomyces Sporangiophores. II. Double Refraction and Electron Microscopy." *Bochim. biophys. acta.* **6**:357-73.

——, and A. L. Houwink. 1953. "Architecture and Growth of the Primary Cell Wall in Some Plant Hairs and in Phycomyces Sporangiophores." *Acta bot. néerl.* **2**:218-38.

ACETABULARIA

Hämmerling, J. 1934. "Über formbildende Substanzen bei Acetabularia mediterranea, ihre räumliche und zeitliche Verteilung und ihre Herkunft." *Arch. entwick. Org.* **132**:1-72.

——. 1946. "Neue Untersuchungen über die physiologischen und genetischen Grundlagen der Formbildung." *Naturwiss.* **33**:377-42; 361-5.

——. 1953. "Nucleo-cytoplasmic Relationships in the Development of Acetabularia." *Internat. Rev. Cytol.* **2**:475-98.

Schultze, K. L. 1939. "Cytologische Untersuchungen an Acetabularia mediterranea und Acetabularia wettsteinii." *Arch. Protistenk.* **92**:179-225.

Werz, G. 1955. "Kernphysiologische Untersuchungen an Acetabularia." *Planta.* **46**:113-53.

FLAGELLATES

Cleveland, L. R. 1949. "Hormone-induced Sexual Cycles of Flagellates. I. Gametogenesis, Fertilization, and Meiosis in Trichonympha." *J. Morphol.* **85**:197-296.

———. 1955. "Hormone-induced Sexual Cycles of Flagellates. XIII. Unusual Behavior of Gametes and Centrioles of Barbulanympha." *Ibid.* **97**:511-42.

———. 1957a. "Types and Life Cycles of Centrioles of Flagellates." *J. Protozool.* **4**:230-41.

———. 1957b. "Achromatic Figure Formation by Multiple Centrioles of Barbulanympha." *Ibid.* Pp. 242-8.

———, S. R. Hall, E. P. Sanders, and J. Collier. 1934. "The Wood-feeding Roach Cryptocercus, Its Protozoa and the Symbiosis between Protozoa and Roach." *Mem. Am. Acad. Arts Sci.* **17**:185-342.

Grassé, P.-P. 1952a. *Traité de Zoologie.* Paris: Masson & Cie. **1**(*Ciliés*):862-915.

———. 1956. "L'ultrastructure de Pyrosonympha vertens: Les flagelles et leur coaptation aver le corps, l'axostyle contractile, le paraxostyle, le cytoplasme." *Arch. biol.* **67**:595-611.

Pitelka, D. R., and C. N. Schooley. 1958. "The Fine Structure of the Flagellar Apparatus in Trichonympha." *J. Morphol.* **102**:199-230.

CILIATES: GENERAL

Balamuth. W. 1940. "Regeneration in Protozoa; A Problem of Morphogenesis." *Quart. Rev. Biol.* **15**:290-337.

———. 1942. "Studies on the Organization of Ciliates. II. Reorganization Processes in Licnophora macfarlandi during Binary Fission and Regeneration." *J. Exp. Zool.* **91**:15-43.

Bonner, J. T. 1954. "The Development of Cirri and Bristles during Binary Fission in the Ciliate *Euplotes*." *J. Morphol.* **95**:95-108.

Chatton, E., and A. Lwoff. 1935. "Les Ciliés apostomes. I." *Arch. zool. exp. gén.* **77**:1-453.

Chatton, E., and J. Seguéla. 1940. "La continuité génétique des formations ciliaires chez les Ciliés hypotriches. Le cinétome et l'argyrome au cours de la division." *Bull. Biol. France-Belgique.* **74**:349-442.

Dembowska, W. S. 1938. "Körperreorganisation von *Stylonychia mytilus* beim Hungern." *Arch. Protistenk.* **91**:89-105.

Entz, G. 1893. "Die elastischen und kontraktile Elemente der Vorticellen." *Math. naturwiss. ber. ung.* **10**:1-48.

Fauré-Fremiet, E. 1948. "Les mécanismes de la morphogénèse chez les ciliés." *Fol. Biotheoret.* Series B. **3**:25-58.

———. 1954. "Morphogénèse de bipartition chez Urocentrum Turbo (cilié Holotriche)." *J. Embryol. Exp. Morphol.* **2**:227-38.

Grassé, P.-P. 1952b. *Traité de Zoologie.* Paris: Masson & Cie. **2**(*Ciliés*).

Klein, B. 1928. "Die Silberlinien system der Ciliaten." *Arch. Protistenk.* **62**:177-260.

———. 1932. "Das Ciliensystem in seinen Bedeutung für Lokomotion, Koordination, und Formbildung mit besonderer Berücksichtigung der Ciliaten." *Ergebn. Biol.* **8**:75-179.

Lwoff, A. 1949. "Kinetosomes and the Development of Ciliates." *Growth* Supp. (Symp. 9) **13**:61-91.

———. 1950. *Problems of Morphogenesis in Ciliates.* New York: John Wiley & Sons, Inc.

Metz, C. B., D. R. Pitelka, and J. A. Westfall. 1953. "The Fibrillar System of Ciliates As Revealed by the Electron Microscope." *Biol. Bull.* **104**:408-25.

Randall, J. T. 1957. "The Fine Structure of the Protozoan *Spirostomum ambignum.*" *Soc. Exp. Biol.* Symp. **10**:185-98.

Romillar, C., E. Fauré-Fremiet, and M. Gauchery. 1956. "Origine ciliaire des fibrilles Scléroprotéiques pedunculaires chez les ciliés peritriches." *Exp. Cell Research.* **11**:527-41.

Tartar, V. 1941. "Intracellular Patterns: Facts and Principles concerning Patterns Exhibited in the Morphogenesis and Regeneration of Ciliate Protozoa." *Growth* Supp. (Symp. 3) **5**:21-40.

Villeneuve-Brachon, S. 1940. "Recherches sur les ciliés hétérotriches." *Arch. zool. exp. gén.* **83**:1-180.

Weisz, P. B. 1954. "Morphogenesis in Protozoa." *Quart. Rev. Biol.* **29**:207-29.

CILIATES: STENTOR

Fauré-Fremiet, E., C. Rouillen, and M. Gauchery. 1956. "Les structures myoides chez les ciliés. Étude au microscope électronique." *Arch. d'anat. micro. morphol. exp.* **45**:138-61.

Johnson, H. P. 1893. "A Contribution to the Morphology and Biology of the Stentors." *J. Morphol.* **8**:467-562.

Lillie, F. R. 1896. "On the Smallest Parts of Stentor Capable of Regeneration; A Contribution on the Limits of Divisibility of Living Matter." *J. Morphol.* **12**:239-49.

Morgan, T. H. 1901. "Regeneration of Proportionate Structures in Stentor." *Biol. Bull.* **2**:311-28.

Schwartz, V. 1935. "Versuche über Regeneration und Kerndimorphismus bei *Stentor coeruleus.*" *Arch. Protistenk.* **85**:100-39.

Tartar, V. 1954. "Reactions of Stentor to Homoplastic Grafting." *J. Exp. Zool.* **127**:511-76.

———. 1956a. "Grafting Experiments concerning Primordium Formation in Stentor coeruleus." *Ibid.* **131**:75-122.

———. 1956b. "Further Experiments Correlating Primordium Sites with Cytoplasmic Pattern in Stentor coeruleus." *Ibid.* **132**:269-98.

———. 1956c. "Pattern and Substance in Stentor," in D. Rudnick (ed.). *Cellular Mechanisms in Differentiation and Growth.* Princeton, N. J.: Princeton Univ. Press. Pp. 73-100.

Weisz, P. B. 1948. "Time, Polarity, Size and Nuclear Content in the Regeneration of Stentor coeruleus." *J. Exp. Zool.* **107**:269-87.

———. 1949. "The Role of Specific Macronuclear Nodes in the Differentiation and the Maintenance of the Oral Area in Stentor." *Ibid.* **111**:141-56.

———. 1951. "An Experimental Analysis of Morphogenesis in Stentor coeruleus." *Ibid.* **116**:231-58.

———. 1956. "Experiments on the Initiation of Division in Stentor coeruleus." *Ibid.* **131**:137-62.

Chapter 4

Cell Combinations: Slime Molds

In the majority of organisms, cells are united to form tissues with varying degrees of unity, and under particular circumstances tissues may become disaggregated and subsequently reaggregated or reunited. Organizational phenomena, including both histogenesis and morphogenesis, are involved throughout such aggregations and disaggregations, and in many ways are different in kind from those seen in individual cells. Total mass, volume, or surface area (whichever aspect is the most significant) is obviously of paramount importance in determining organizational possibilities and limitations—as is clearly seen even in a uninucleate organism such as *Acetabularia,* in which the large, vegetative individual exhibits an organization and a structural pattern strikingly different from those of its relatively minute biflagellate gamete. In similar fashion, the minimal structure of the individual amoebae of the vegetative stage of a cellular slime mold contrasts sharply with the multicellular structure of the fruiting stages which the individuals combine to form.

The general relationship of cells to the tissues they constitute and to the organism as a whole has been in debate for the better part of a century—the history of which has been reviewed at length by Baker (1948-1952). Many questions arise; for example: What are the means by which separated cells are drawn together? What is the degree of fusion between cells comprising aggregates or tissues? What is the basis of cell segregation within tissue masses? What are the factors determining the path of cell specialization? What is the basis of tissues and morphogenetic movement as distinct from the movement of individual cells. These and other problems may be profitably approached through the study of various multicellular organisms of comparatively low degree of organization, such as the slime molds, sponges, and protistan colonies, and through disaggregation-reaggregation experiments with the tissues of more highly organized organisms.

Cellular Slime Molds

The cellular slime molds, comprising the single order Acrasiales (Olive, 1902), are characterized by a unique fruiting phase in which great numbers of individual myxamoebae enter communal groups, known as pseudoplasmodia, which later transform into organized units, called sorocarps, of a determinant pattern. They are different from the myxomycetes proper, such as *Physarum* (p. 92), in that the cells do not fuse to form true plasmodia and no flagellated swarm cells appear at any time. Whatever the relationship of cell to tissue in cellular slime molds may be, the activities and transformations of these very simple organisms raise most of the problems of morphogenesis. The genus most studied, *Dictyostelium,* has been cultured under a wide variety of conditions, and the analysis of developmental and related phenomena in *Dictyostelium discoideum,* the species primarily employed, has been carried out mainly by Raper and Bonner with their respective co-workers.

Life Cycle

The life cycle of the slime mold lasts about 4 days and has been divided arbitrarily into four stages.

Stage 1 (Vegetation). The interval from the appearance of a single amoeba until the food supply is exhausted. The single, unicellular, nonciliated amoeba hatches out of a single spore, feeds on bacteria, grows and divides, and glides about independently of other amoebae in the culture, repeating growth and division until the food supply is exhausted.

Stage 2 (Aggregation). A process of aggregation, beginning when the culture reaches its maximum density, i.e., when growth and division are inevitably slowing down as the food supply becomes depleted; during this stage groups of amoebae move toward central collection points and form a single large mass at each such center.

Stage 3 (Migration). From aggregation to the end of migration. Each large mass, composed of about 1,000-200,000 or more single amoebae, becomes elongate, or sausage-shaped, and covered by a sheath secreted by the constituent cells. Before they are fully formed, these masses are erect, compact, and cylindrical. Once formed, they bend over, touching the substrate, and migrate in the horizontal position for varying periods of time. The migratory path of a mass is indicated by a slime streak (composed of the sheath) which has been secreted and left behind posteriorly. This pseudoplasmodium exhibits both heliotropism and sensitivity to minute temperature gradients, moving toward sunlight and warmer regions. During migration the constituent amoebae are in constant pseudopodial motion; the sheath does not move, but the cells move forward within it.

Stage 4 (Differentiation). At the end of migration, the pseudoplasmodium again assumes a vertical position to form a stalk that supports a smooth, round, spore mass at its apex; the whole is known as a sorocarp. The cells which constitute the anterior region of the pseudoplasmodium form the base of the stalk structure; they become large, vacuolated, and enclosed in a cellulose sheath. Cells from the posterior end flow up the outside of the basal stalk cells, increase the length of the stalk, and by differentiating into spores, finally form the spore mass on the top of the stalk. The resultant multicellular sorocarp may well be compared with the unicellular sporangiophore of *Phycomyces*.

Each of the four stages—vegetation, aggregation, migration, and differentiation, or sorocarp formation—presents problems of fundamental importance.

Stage 1 (Vegetation)

The problems relating to the vegetative stage are general, concerning the nature of amoeboid movement, growth, the growth curve, cell division, etc., and are in no way confined to myxamoebae; for mitotic activity in myxamoebae is thought to continue to some extent into the aggregative stage.

Stage 2 (Aggregation)

The aggregation phenomenon has, understandably, attracted much attention. The beginning of the aggregation of amoebae in a given culture toward a common center coincides with decrease in food supply and marks the end of growth and cell division. Whether or not the amoebae are haploid and conjugate in pairs at this time to produce diploid cells—a subject of debate—does not seem particularly relevant to an understanding of the succeeding events.

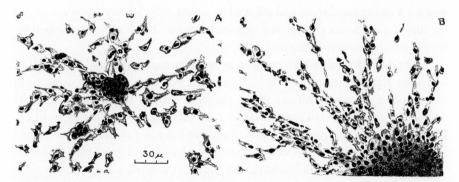

Figure 4-1. Myxamoebae in *Dictyostelium*. **A.** Aggregation center. **B.** Streaming. (After Bonner.)

At the commencement of aggregation, the shape of each individual changes from a nondescript, pseudopodial outline to a decidedly longer, thinner one. All the individuals move in groups toward common centers in a given field, joining larger and larger tributaries until, when the streaming is finished, each group forms a united mass of about 1,000 to about 200,000 individuals. Although the amoebae take on definite orientations, not all of them move radially: those outside an aggregation center migrate indirectly toward the center, for, apart from the aggregation centers, larger streams are themselves attraction loci. According to Shaffer (1956), the streaming of a group of amoebae toward a center also exhibits the curious phenomenon of sudden pulses of increased movement at roughly 5-minute intervals. Toward the end of the aggregation stage, the central area mounds up as more and more myxamoebae enter it, and in *Dictyostelium* the mound grows upward and flops over to start the migrating stage.

Natural questions that arise at this point are: What underlies or initiates the change from the vegetative to the aggregative stage? What causes the amoebae to move toward a common center? What is the local nature of an attraction center? What is the significance of the rhythmic pulses of movement?

Bonner (1947) has shown that amoebae on the undersurface of a thin glass shelf in a tank of water so orient themselves as to stream toward a center on the top surface of the shelf; that if two shelves are placed side by side with a gap between, the amoebae on one shelf attempt to reach a center on the other, if not dominated by a center of their own; and that if the amoebae are in a gently flowing stream of water, those downstream are attracted toward the center, but those upstream are not. Chemotaxis seems indicated, and Bonner (1947) has named the diffusible chemical substance acrasin. This conclusion is in line with that of earlier workers and is supported experimentally by the fact that when some myxamoebae are placed on nonnutrient agar and underneath a Visking membrane and allowed to aggregate, other myxamoebae placed on top of the membrane become oriented toward the already established aggregation centers and streams.

Shaffer (1956) sandwiched amoebae between an agar block and a glass slide, and then rapidly and continuously transferred washings, drop by drop, from living centers of other cultures to the meniscus around the block, so that the acrasin of the other cultures diffused into the interior of this agar block. Chemotaxis then occurred, the sensitive cells moving in parallel fashion toward the nearest edge instead of converging toward a central point as in normal aggregation. When washings were left on the meniscus for 1 minute, their attraction was strong; when left for 5 minutes, attraction was weak; and after 15 minutes, there was no attraction. Thus an unstable chemical seems to be the source of the attraction. Since Shaffer's experiment, it has been demonstrated that such instability is due to the presence of another slime-mold product, possibly an enzyme, of greater molecular size than acrasin.

The pulses of movement may be the result of pulses in acrasin secretion, although, as mentioned above, the significance of such pulses remains obscure. According to Shaffer (1957), since the acrasin gradient is produced by diffusion, the direct range of a center must be limited to the size and maximum sensitivity of the responding amoebae, the amount of chemical produced, and the time needed for orientation to occur after secretion has begun. A center's range is, however, rapidly and economically extended by a relay system, for acrasin not only orients an amoeba but induces it to secrete acrasin in turn. This relay system further reduces the importance of any influence the chemical secreted at the center can have at the periphery. Amoebae move toward the center, not because of a sustained gradient of secretion outward from it, but because of the time sequence in which the relay operates. If the timing of the relay is seriously disturbed, aggregation becomes chaotic. Since secretion at a center may fluctuate, a periodic re-excitation of the relay system may result; thus the resulting pulses which spread out over an aggregation may reinforce the guidance of centrifugal flow in its streams and may repetitively reassert the authority of a center even in an aggregation whose oriented amoebae are not in contact with one another.

Some questions yet to be answered satisfactorily are: What determines the focus of a center? How is it that large streams are attraction centers and yet themselves flow toward a center? Why do amoebae start to aggregate?

Shaffer (1958) suggests that primary centers are formed by the first cells to integrate and set up acrasin gradients adequate for orientation over even short distances, thereby being the first to activate the acrasin relay; and these cells subsequently control the movements of surrounding cells. Whether the initiator is a single cell or a small group of cells has not been decided. Serological investigations by Gregg (1956) of cell adhesion in several species of slime molds indicate that new surface antigens appear as the time of aggregation approaches; and since aggregation coincides with the cessation of the maximal rate of growth, cortical differentiation—incompatible with such growth—then becomes possible. In other words, the few cells that, inevitably, cease growing and dividing before the majority do may be the first to secrete acrasin and become aggregation centers.

Stage 3 (*Migration*)

Interest in migration centers around (1) the puzzling nature of the mechanics of migration of the multicellular mass and (2) the link between migration and some form of differentiation within the pseudoplasmodium.

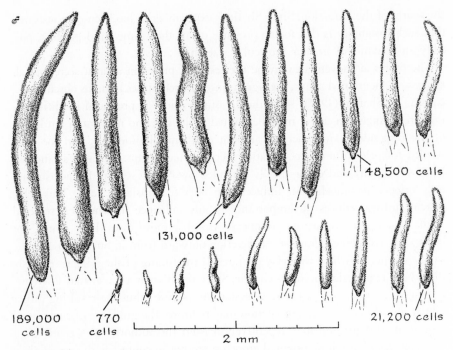

Figure 4-2. Migrating pseudoplasmodia of *Dictyostelium discoideum,* showing essentially constant pattern and extreme variation in size. (After Raper.)

MECHANICS OF MIGRATION

Structurally the migrating pseudoplasmodium is very simple: it consists of undifferentiated myxamoebae surrounded by a noncellular sheath, or envelope, apparently secreted by the whole mass. The size of the mass as a whole is most variable, and, according to Bonner (1959), the particular rate of migration is in direct proportion to the volume. At all times. whether the pseudoplasmodium is migrating in response to light, to increased temperature, or at random, the pointed end is to the fore, where, according to Raper (1941), there is apparently a "receptive center." As the pseudoplasmodium migrates, it becomes smaller and the rate of migration slows down with the decrease in size. At the culmination, when the pseudoplasmodium rises into the air, the rate of upward rise is faster, the larger the mass.

Individuals among the actively pseudopodial internal amoebae may travel at somewhat different rates. Bonner suggests that all the amoebae progress forward, somehow gaining traction both from the secreted slime sheath and from one another beneath the surface of the cell mask; he contends that the concept of forward progression as the achievement solely of the outside superficial layer of cells is not supported by the evidence. Conceivably, each amoeba gives off slime, which has some rigidity for traction and is ultimately deposited inside the slime track.

Raper (1940) found that when, in the presence of bacteria, a pseudoplasmodium is completely disintegrated so that its constituent myxamoebae are separated, the separated individuals will immediately return to the vegetative stage until the available bacteria have been consumed. In the total absence of bacteria, similarly dissociated myxamoebae reaggregate into minute, but typical pseudoplasmodia and give rise to diminutive sorocarps. Apparently, growth is incompatible with aggregation.

DIFFERENTIATION WITHIN THE PSEUDOPLASMODIUM

It is generally agreed that the cells comprising an aggregation center become the anterior part of the pseudoplasmodium and eventually stalk cells and that the latecomers to the central region become the posterior part of the pseudoplasmodium and eventually spore cells. Experiments reported by Raper (1940 and 1941) indicate, however, that the fate of the cells in the mass is not determined irreversibly at the time the mass is formed; for when the anterior tip of the migrating body is removed, it produces a normal fruiting body, and

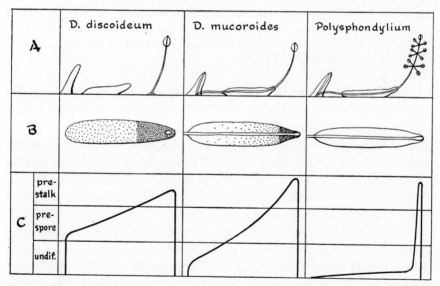

Figure 4-3. Species of cellular slime molds. **A.** Morphology. **B.** Alkaline phosphatase staining characteristics during approach to culmination stage. **C.** Hypothetical level of differentiation along axis of cell mass at stage shown in **B.** (After Bonner, Chiquoine, and Kolderie.)

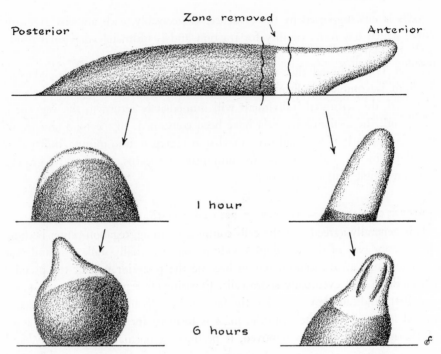

Figure 4-4. Partly differentiated migrating cell mass, stained with Nile blue sulphate for polysaccharides, is bisected. Reconstruction occurs in both parts, and anterior end of each fragment has reversed its staining properties: one from light prestalk condition to dark prespore condition; vice versa in the other. (After Bonner, Chiquoine, and Kolderie.)

an isolated posterior part will do the same. It appears that throughout its life span any single amoeba is capable of differentiating into either a spore or a stalk cell and that the determination of which it will be (if not interrupted experimentally) is governed by its position in the mass. Moreover, when the anterior tip of a migrating mass is cut off and grafted to the side of another migrating pseudoplasmodium, the graft and the mass posterior to it go off in a direction different from the original tip—an event suggesting that some sort of organizing power resides in the tip.

In some of Raper's experiments, when the anterior 10-20 per cent of pseudoplasmodia was excised and placed in a considerably distant location, the anterior segment usually resumed migration and subsequently fruited; whereas the unexcised posterior segment merely crowded forward a short distance to the original position of the apical fragment and fruited immediately. In other experiments, removal of as little as 1 per cent of the anterior end destroyed the migratory capacity of the posterior part.

Stage 4 (Differentiation)

At the end of migration the pseudoplasmodium again assumes a vertical position and proceeds to build a sorocarp by the progressive and orderly differentiation of its constituent cells into either spore cells, stalk cells, or basal disk cells. Just before differentiation gets under way, the pseudoplasmodium rounds up and the anterior tip surmounts the whole to become an apical papilla. At this time the sheath initial is a delicate hyaline membrane which appears

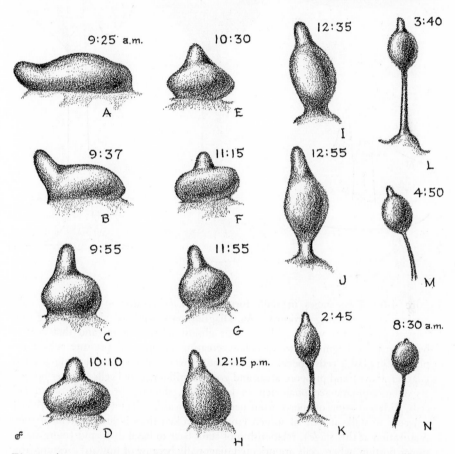

Figure 4-5. Chronology of single developing sorocarp in *Dictyostelium discoideum*. **A.** Migrating pseudoplasmodium. **B.** End of migration. **C–D.** Assumption of vertical orientation. **E–F.** Initiation of stalk formation. **G–H.** Beginning of sporophore elongation coincident with basal-disk formation. **I–J.** Initiation of sorogen ascent. **K.** Beginning of spore formation in peripheral area. **L–M.** Inward progression during successive stages. **N.** Mature sorus culmination completed previous evening. (After Raper and Fennell.)

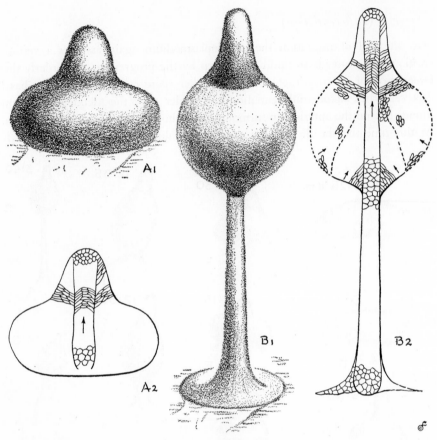

Figure 4-6. Two stages in stalk formation in *Dictyostelium discoideum*. **A₁**. Surface view of young sorocarp. **A₂**. Diagrammatic sectional view of young sorocarp, showing that young stalk extends almost to substratum and that cells in basal area are beginning to vacuolate, creating an upward pressure which displaces still-plastic cells above. **B₁**. Surface view of later stage, showing prestalk (darker above) and prespore areas and zone of still-amoeboid cells (darker below). **B₂**. Diagrammatic sectional view of later stage, showing orientation of prestalk cells, differentiation of spores from periphery inward, orientation of prespore cells adjacent to stalk at base of sorgen (internal broken lines indicate extent of spore maturation at this stage), relationship of stalk base to basal disk, and young stalk's upper portion, where cells are oriented diagonally because of upward pressure from vacuolating cells below. (After Raper and Fennell.)

beneath the apex. According to Raper and Fennell (1952), the sheath is of the greatest significance in sorocarp development: It is the product of the whole community of cells acting as an organized unit, its proportions bear a definite relationship to the mass of cooperating cells, and it determines the dimensions which the supporting tissues will assume. The sheath is apparently formed extracellularly and then deposited in the intercellular spaces that lie in a critical area within the mass of amoeboid cells. According to Gezelius and Rånby (1957), the sheath seems to consist of partly mercerized cellulose of a very low crystalline order.

After the pseudoplasmodium rounds up, it flattens into a crownlike shape; as a result, the sheath initial comes into contact with the substratum. As the mass of myxamoebae both above and outside the sheath begin to ascend the stalk—as a result both of amoeboid movement of constituent cells and of the lifting effect of the differentiating stalk—a creasing of the slime envelope constricts the bottommost amoebae and separates them from the rest to produce a basal disk. By this time the area of cells which will form the spore mass (prespore cells) and that which will further extend the stalk (prestalk cells) are well defined; the former represent the main body of the mass, and the latter a smaller body of cells surrounding the apical papilla and extending down into the shoulders of the mass. The cells in the upper and outer regions of the rising mass are transformed into spores, the transformation spreading inward and upward until all the cells that have not found their way into the stalk become spores.

Both Raper (1940) and Bonner (1944) have used vital dyes to follow the fate of various parts of the pseudoplasmodium. Raper also made grafts between pigmented and colorless fragments: When a pigmented anterior portion was joined to a colorless posterior one, the pigmented cells maintained their apical position during migration and early differentiation; then they invaginate through their colorless neighbors to produce a pigmented stalk, while the spore mass, basal disk, and upper stalk remained colorless. After marking various regions of the aggregative stage, Bonner found that those myxamoebae which arrived first at an aggregation center formed the apex of the pseudoplasmodium and that the apical cells became part of the lower stalk; intermediate cells, of the upper stalk and spores; and the hindmost cells, of the basal disk. On the other hand, Raper's experiments, mentioned earlier, demonstrate that cells ordinarily destined to become spores can become stalk cells, and vice versa, when their position in the mass is altered. Thus Bonner and Slifkin (1949) report that myxamoebae grown in the light at 23°C, incubated in the dark at 17°C but during the differentiation stage, produced structures containing 14 per cent stalk cells; incubated in the light at 17°C, 24 per cent stalk cells; whereas in the dark at 27°C, only 2 per cent stalk cells, the units consisting of large spore masses and tiny stems.

Raper had already observed that when a pseudoplasmodium is transected, the posterior part stops migrating and forms a normal fruiting body of somewhat reduced over-all size. The anterior part continues to migrate; but if the

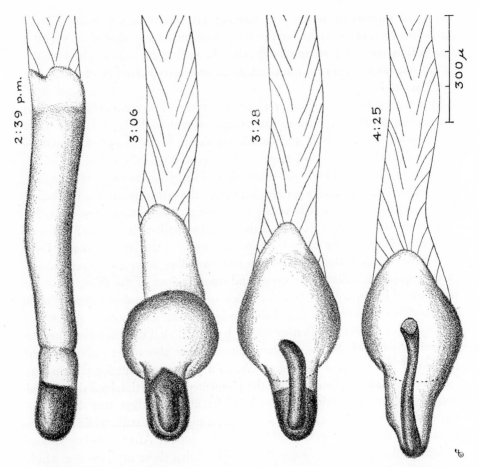

Figure 4-7.   Surface views of *Dictyostelium discoideum,* showing stalk forming at tip and being pushed downward through prespore cells to substratum. Dark tip was obtained by grafting tip of colored migrating cell mass onto decapitated colorless one. (After Bonner.)

migration is for only a very short distance, almost all the cells differentiate as stalk cells, forming virtually no spore mass; i.e., most of them differentiate according to their original destiny. If, however, migration of the anterior cells is extensive, the fruiting structure becomes progressively more normal with more and more cells differentiating into spores.

Various questions arise which have yet to be answered satisfactorily; for example: What factors are responsible for the initiation of sorocarp formation? What factors determine the cessation of migration? Are these two questions the same? If not, is there any connection between them; and if so, what is it?

What is the precise nature of the difference between the cells in the tip and those in the remainder of the body? Is the stalk sheath secreted by those cells in the stalk, or does the sheath form from cells that differentiate partly because they are cut off from the interior? What are the mechanisms controlling the orientation of the cell mass upon the substratum? What is the mechanism that keeps the spore cell mass at the stalk's growing tip, and what prevents the tip from growing all the way through the body?

At this point statements by the several leading investigators may be salutatory.

When one looks over these facts as I have presented them here it is alarming to realize that even though we know some details of the development of Dictyostelium, and a great many details concerning the regulatory development of other animals, we understand no better than did Driesch the explanation of a harmonious equipotential system and yet a half century of embryology has elapsed. It may be

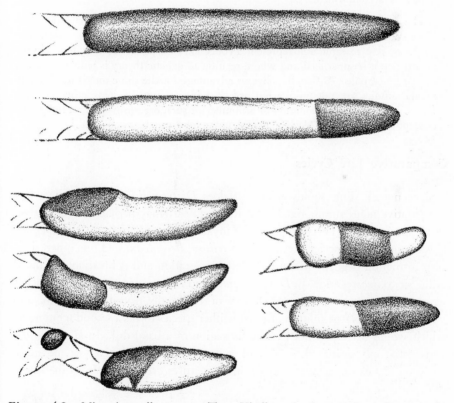

Figure 4-8. Migrating cell masses. **Top.** Vitally stained migrating cell mass of uniform coloration alters to one possessing dark tip and light posterior portion. **Bottom.** Rapid forward movement of colored anterior portion which has been grafted into posterior region of intact migrating cell mass. (After Bonner.)

that we no longer find entelechies a temptation, but our substitutes are poor indeed.[1]

Perhaps, then, the important fact in sorocarp building is not that cellulose begins to be formed at this time, but that it then first begins to be deposited in an orderly manner and in a strategic position. . . . At no stage in its developmental cycle does the slime mold Dictyostelium represent a true metaphyte—yet in its fruiting phase it exhibits many of the characteristics of multicellular forms. By the close integration and coordination of thousands of individual cells it produces a functioning organization capable of exhibiting unit responses to various external stimuli. By the continued intimate and regulated association of its constituent cells, this organization succeeds in building a unit fruiting structure of specific pattern and proportions. This it does by a sharp division of labor within its population to yield both reproductive cells and supportive "tissue," the latter arising as a result of the extracellular formation of cellulose and its deposition within a critical plane, thus fashioning a form or mold, within which final cellular differentiation is accomplished.[2]

The most general conclusion we can reach at this stage of investigation into slime mold development is that no general conclusions are as yet possible. The results of past research still represent a collection of separate stories whose physiological depth is superficial and whose relation to genetic theory is tenuous. . . . It should be emphasized that the obvious advantages of the slime molds for developmental studies are not unique but are held in common by some other protistal forms. It is only by exploitation of a number of such organisms that an integrated concept of comparative developmental physiology can be established.[3]

Comparative Life Cycles

Since not all slime molds behave as does *Dictyostelium discoideum,* a few comparative notes summarized from Raper (1940) may perhaps be suggestive: In all species of the Dictyosteliacea the myxamoebae collect in upright, peglike masses, and in all species except *Dictyostelium discoideum* and one other, sorocarp formation begins without delay after migration and is invariably initiated at the site of aggregation. In *D. mucoroides* and *D. purpureum* the first evidence of sorocarp formation is the appearance of a centrally placed hyaline tube and within it a vacuolated column of cells; neither species forms a second membrane, which would be responsible for the formation of a basal disk. In *D. minutum,* apparently only a limited number of myxamoebae can effectively cooperate in sorocarp formation, so that the fruiting structures are smaller from the outset than those of other species.

In *Polyspondylium violaceum,* as the differentiation stage develops and the

[1] Bonner, *Am. Nat.,* **86:**88 (1952).
[2] Raper and Fennell, *Torrey Bot. Club Bull.,* **79:**47, 49 (1952).
[3] Sussman, in Hutner and Lwoff (eds.), **2:**220 (1955).

body of aggregated myxamoebae is raised above the substratum, masses of myxamoebae are periodically left behind while the ever-decreasing body ascends the lengthening stalk. Characteristically, these masses are deposited at comparatively regular intervals, and in general each mass is smaller than the one deposited before it. As the stalk is extended and additional masses of myxamoebae are left behind, the mass deposited earliest forms vertical segments, each of which develops a side branch perpendicular to the main axis. Progressing upward, subsequent masses segment and develop side branches. An individual side branch duplicates the main stem, differing only in its smaller dimensions, its more prominently tapered stalk, and its anchorage to the main axis rather than the substratum (for a full account, see Harper, 1929 and 1932).

Such differences in the morphogenesis and gross morphology of the multicellular structures must presumably be traced to differences in the constituent cells—although not necessarily to differences of a recognizably similar kind. It is surely significant, however, that mixtures of cells from different species undergo species-specific segregation at least to some degree. Working with two cell populations thoroughly mixed at the migration stage and following experiments of Raper and Thom (1941), Bonner and Adams (1958) found either that the different strains underwent complete separation; that they partly merged to form two fruiting bodies, one of which stood on the sorus of the other; or that they formed a single fruiting body, in which the prespores from the two strains regrouped to form two cohesive blocks of cells in one sorus— an event suggesting an intermediate degree of surface compatibility.

It is also evident that form is independent of cell number, for the proportional size of stalk to fruiting body and the same over-all shape of each are the same despite varying numbers of cells involved in construction. Moreover, the individual cells which comprise the stalk assume a shape in conformity with the shape and dimensions of that stalk. This phenomenon is comparable to that in heteroploid salamander larvae, described by Fankhauser (1945): both the renal tubules and the lens epithelium have a constant epithelial thinkness despite a wide range in cell size, and in the number of cells, which adjust in shape and number to supracellular requirements.

Amoeboid Movement and Protoplasmic Streaming

The amoeboid movement of isolated *Dictyostelium* cells during the vegetative stage and the mass movements of cell aggregates during the migratory and differentiation stages seem to be related phenomena. Whether or not these movements are akin to morphogenetic movements in general, some under-

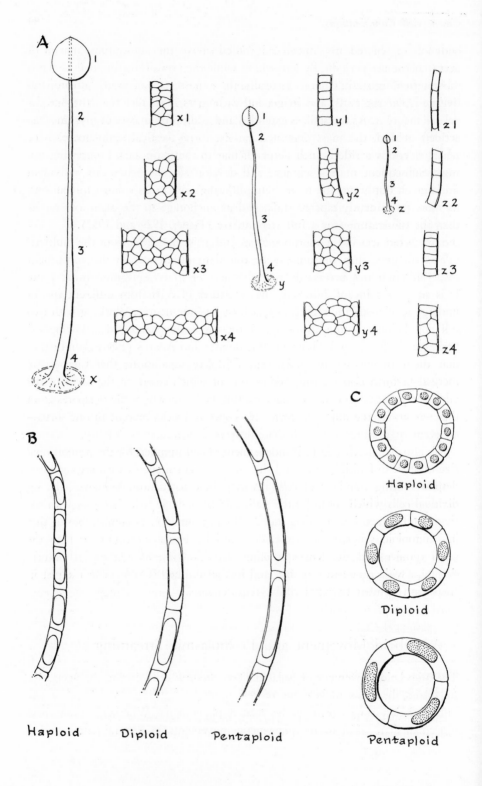

A

x1
x2
x3
x4
x

y1
y2
y3
y4
y

z1
z2
z3
z4
z

B

Haploid   Diploid   Pentaploid

C

Haploid

Diploid

Pentaploid

standing of the nature of amoeboid movement and its associated activities seems relevant.

Amoeboid movement and protoplasmic streaming have been reviewed by Seifriz (1943 and 1952), De Bruyn (1947), and Noland (1957). The reversible sol-gel theory, proposed by Pantin (1923) and elaborated by Mast (1926), described an amoeba as a contracting tube of gelated ectoplasm and closed at the posterior end, with the anterior end occupied by fluid ectoplasm of the advancing pseudopodium. The fluid endoplasm streams forward through the tube from a center of liquefaction near the posterior end; on reaching the anterior end, the streaming endoplasm forms the pseudopodial fluid ectoplasm, which is continuously added to the contracting tube by gelation at the sides. In studying the amoeboid cells of disaggregated amphibian embryos, however, Holtfreter (1948) observed that while the streaming endoplasm was enclosed by a plasmagel layer, the outer membrane, or plasmalemma, was separated from the plasmagel by a liquid, hyaline layer of considerable thickness, particularly in the region of the advancing pseudopodium; he therefore assigned the motive force to the outer membrane itself, and not to a contracting gel.

Earlier, Seifriz (1943) stated his opinion that any protoplasmic contractility is due to the shortening of protein fibers by molecular folding, with oxidative processes supplying the energy; and subsequently Goldacre (1952) put forward a more detailed theory along this line. Briefly, the theory of these investigators is as follows: In the fluid endoplasm the individual protein molecules are folded in relatively compact form; thus they move readily on one another, giving the endoplasm its liquid character. At the tip of the advancing pseudopodium, they unfold, interdigitating as they straighten out, and become attached to one another to form a felted layer of elongated molecules, i.e., an ectoplasmic gel. As the amoeba advances so that the plasmagel comes to lie near the posterior end, the protein molecules, with the aid of energy from ATP, begin to fold again; this folding activity contracts the felted layer, squeezing out liquid; with further folding, the molecules become loosened from one another and the gel reverts to a sol. This theory does not distinguish between plasmalemma and plasmagel, but most other investigators report that the plasmalemma actually moves forward as fast as, or even faster than, the amoeba as a whole—that it slides over the plasmagel. Bairati and Lehmann (1953), using polarized light, observed that plasmasol streaming through a small break in the outer membrane exhibits

---

Figure 4-9. Adaptation of cells to pattern. **A.** Pattern of sorocarp remains constant despite size. Individual myxamoebae constituting stalk assume whatever shape is necessary to pattern. (After Raper.) **B–C.** Lens epithelium and pronephric tubules, respectively, of haploid, diploid, and pentaploid salamander larvae, showing adaptation of cells to constant structure. (After Fankhauser.)

distinct birefringence of flow—a phenomenon indicating that the endoplasmic molecules have at least some degree of linear form.

The nature of the plasmalemma, or pellicle, itself is ambiguous. It is certainly semisolid, since it can be lifted from the underlying plasmagel by means of microdissection or injection. But it is also a semifluid, as observed by McClendon (1909), who found that a needle thrust completely through an amoeba does not interrupt motion: both pellicle and cytoplasm part in front of the needle and fuse behind it, leaving the needle behind. From electron-microscope studies, Lehmann, Manni, and Bairati (1956) conclude that the outer layer of an amoeba consists of a thin sheet of mucoprotein; beneath it, is a felted, fibrous protein layer, which is in contact with the plasmagel or ectoplasm beneath. Noland (1957) also reports that the external layer has fluid properties, so that it acts like a thin layer of slime moving over the visible membrane. He suggests that theories of amoeboid movement are inadequate because they do not take into account some underlying molecular mechanism which can produce streaming in the absence of hydrostatic pressure and can explain more kinds of streaming than can the gel tube theory. What is needed is an explanation of how molecules can orient themselves in one direction and crawl forward on any solid surface, while others of the same sort crawl forward on the backs of the molecules beneath them.

Noncellular Slime Molds

For many reasons the plasmodium stage of multinuclear, noncellular slime molds are more rewarding for experimental analysis than are individual amoebae, *Physarum polycephalum* having been most widely studied to date. For example, this species is commonly found within the tissues of decaying wood, where it feeds on bacteria, fungal hyphae, and possibly protozoa. Furthermore, its substance is organized much like that of an amoeba, though on a much more massive scale. Early in its development the plasmodium shows a characteristic tendency to become subdivided into veinlike branches, which divide in turn and then reunite to form a fanlike network with a continuous layer of protoplasm at the advancing margin of the plasmodium. An outer, hyaline layer surrounds the inner, granular region which contains the nuclei.

The interior material, the flowing endoplasm, exhibits a series of pulsating movements, at first in one direction and then in the reverse direction, though always a little stronger in the direction which the plasmodium as a whole is pursuing. Growth and migration continue as long as food is abundant; but, according to Camp (1937), often within a few hours after food is exhausted, the plasmodium passes into the fruiting stage—an occurrence strikingly reminiscent of the phase reaction in *Dictyostelium*.

A. R. Moore (1935) discovered that the plasmodium of its own accord readily passes through silk-gauze or filter-paper pores as narrow as 0.001 mm in diameter. Within a few hours after inserting a filter pocket containing a piece of plasmodium in nutrient agar, Moore observed that the plasmodium had flowed through the filter onto the agar surface. If the filter was too fine, the plasmodium continued to move about within the pocket for days. On the other hand, *forcing* the plasmodium through gauze with meshes as large as 0.2 mm in diameter caused death. Moore concluded that there were fibrillar elements present which were at least 0.2 mm long, but not more than 0.0005 mm wide. Kamiya and

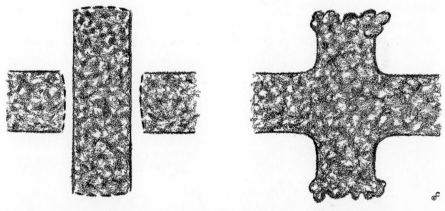

Figure 4-10. Fragment cut from any part of *Physarum* and placed in gap cut in vessel fuses. (After Winer and Moore.)

his co-workers (1957) have identified actomyosinlike, ATP-sensitive proteins in the outer membrane, or plasmalemma, which seem to be similar to those detected in amoeba, although Loewy (1952) believes that the motive force lies within the endoplasm at the interface with the plasmagel.

The reversible streaming movement of *Physarum* endoplasm is certainly comparable in many ways to the forward movement of amoeboid endoplasm. According to Winer and Moore (1941), the streaming is locally controlled, and the walls of the protoplasmic vessels are elastic, although they expand and contract (channels 60 μ wide expand and contract by 5-10 μ) without any observable change in the thickness of the ectoplasmic walls and hence without significant change in the plasmasol-plasmagel relationship. Using time-lapse cinemaphotography, Seifriz (1943) observed that the plasmodium as a whole goes through rhythmic contractions and expansions which are synchronized with the outward and inward flows of the protoplasm, each pulsation requiring about 95 seconds altogether. All the simultaneous lines of flow approach one another or radiate from a central point—a situation which can be partly accounted for by the assumption that there are many scattered centers of contraction and expansion or that the motive force is in every stream. Winer and Moore found that the mechanism for streaming is present in fragments as small as 1 mm wide, although when the fragment is within a large channel, it seems to assume the rhythm of the contiguous endoplasm. Thus when two ends of an isolated fragment are placed close to each other, they fuse and the streaming becomes circular. When a fragment is placed in the gap between two cut ends, or when one fragment is placed on top of another, fusion also occurs. It is clear, therefore, that when the walls of two channels touch, the two walls coalesce.

Two different species of plasmodia, not unexpectedly, do not fuse; but, surprisingly, when two cultures from the same stock are grown separately on oat

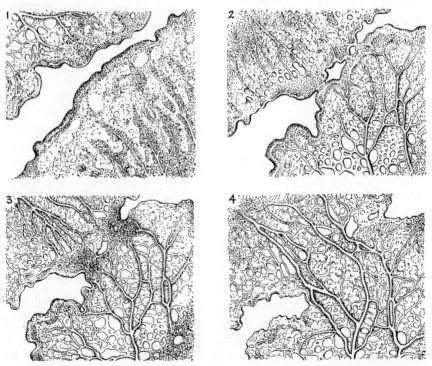

Figure 4-11.   Fusion of approaching edges of plasmodium with establishment of vessel continuity.   (After Winer and Moore.)

agar and rice agar, respectively, and then put together on plain agar, they also do not fuse, often withdrawing from each other as if they were of two different species. Yet pieces from the same culture fuse at once, with channels forming a continuum. The nature of the protoplasmic surface, therefore, varies not only with the genetic constitution but also in some manner with nutritive intake.

Chapter 5

Cell Combinations: Cell Specificity and Aggregation

Studies of the reaggregation process in sponges of various kinds have been made ever since Wilson (1907) first devised the technique of subjecting sponge tissue to cellular dissociation. To this day, however, there is considerable difference of opinion concerning the status of cells as individuals in the reconstructed organisms—a conflict which may well reflect the lack of uniformity among different types of sponges.

Dissociation and Reaggregation

Cells in sponge tissue are dissociated by pressing the sponge through bolting silk to form a suspension of separate cells and cell clusters, which soon settle on the bottom of the containing vessel. About a day later, the cells aggregate. Galtsoff (1925) found that in aggregates of about 2,000 cells, or aggregates 1.0-1.5 mm in diameter, reorganization to form new sponge organisms occurred after 5-6 days; whereas in large aggregates (3-4 mm in diameter) no new sponges developed. Moreover, Wilson (1910) and Galstoff (1925) found that cell mixtures of the red encrusted sponge *Microciona prolifera* and those of the yellow sulfur sponge *Cliona cellata* produce aggregates which are species-specific, i.e., not composed of mixtures. In dissociation-reaggregation studies on *Ficulina ficus,* Fauré-Fremiet (1932) facilitated the recognition of original cell types by staining the sponges with vital dyes before filtering them through bolting silk.

Upon settling, both the archeocytes of the interior and the pinacocytes of the original, outer sponge surface flatten and exhibit amoeboid movement. Accord-

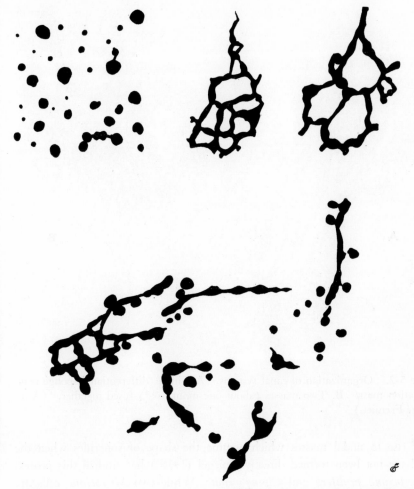

Figure 5-1. Sponge aggregation masses, showing direct formation of spherules and indirect formation from aggregation network. (After Fauré-Fremiet.)

ing to Brien (1937), who worked with the fresh-water sponge *Ephydatia fluviatilis,* aggregation is due to the emission of long pseudopodia, which, in *Ficulina,* Fauré-Fremiet called "the hyaloplasmic veil." Both archeocytes and amoebocytes are highly thigmotactic; this protoplasmic adhesiveness is responsible for some sort of coalescence that causes cells to make contact with the greatest possible mutual surface. Amoeboid movement seems to be of small importance in the actual coalescence.

After concentration by coalescence, cell adhesiveness leads to the formation, on the vessel bottom, of networks whose meshes become increasingly enlarged,

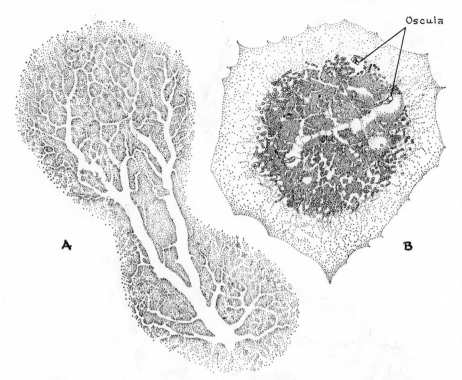

Figure 5-2. Organization of canal systems. **A.** Single, differentiated sponge reaggregation mass. **B.** Two masses (about one month old) fused together. (After Fauré-Fremiet.)

giving rise to nodal masses which assume the shape of spherules when the latter have not been formed directly. Spiegel (1954b) has studied this process in *Microciona prolifera* and *Cliona cellata*. While two *Microciona* cells are coalescing, their granules undergo a series of intense movements and shifts in position, which result in a striking reorganization of the granuloplasm.

Upon coming into contact, the cells coalesce. The plasmic membrane of opposed cells join together in what seems to be a zipperlike manner, proceeding from the area of initial contact. After 5-10 minutes the surface of union between two cells is indistinct or has vanished altogether; and there is little or no doubt that the clear, outer hyaloplasm has formed a common matrix for the two granuloplasms, which do remain distinct and separate from each other.

Spiegel also noted that an aggregate of 2 cells extruded pseudopodia about 15 μ long (twice the length produced by a single cell), an aggregate of 3 cells formed pseudopodia 30-35 μ long, and aggregates of 100-200 cells produced pseudopodia 200-300 μ long. In general the pseudopodial size varied according to the 1-2 power of the cell number. The hyaloplasm of all the cells in a coalesced group acts as a unit, becoming freely available to the aggregate as a single pseudopodium.

Within a few hours, a mixed suspension of red and yellow sponges aggregates as a mixture consisting of islands of one type inside the other, but shortly thereafter the two kinds separate from each other. Spiegel found that adding antisera, made in rabbits, to cell suspensions of each species and of a mixture reversibly inhibited aggregation in homologous antisera and led to complete segregation in normal serum. In the presence of antiserum heterologous to both species, large, mixed aggregates formed at random.

Reconstitution

Brien (1937) has studied the process of reconstitution in the greatest detail in *Ephydatia fluviatilis*. The progressive reorganization of the initially massive spherule or aggregate formed by the coalescence of filtered cells is briefly as follows: At the commencement of aggregation the pinacocytes, derived from the original epidermis, form a peripheral epithelium thereby isolating a mass of mesenchyme from the external medium and enclosing the archocytes, amoebocytes, collencytes, all other mesenchymal cells together with the collar cells,

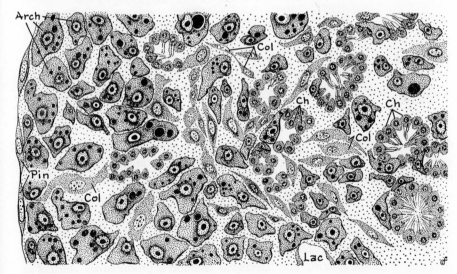

Figure 5-3. Part of section through peripheral region of *Ephydatia* spherule 17 hours after filtration. **Pin.** Fine peripheral epithelium of pinacocytes. **Arch.** Enclosing archeocyte. **Col.** Collencyte. **Ch.** Choanocyte. **Lac.** Lacuna. (After Brien.)

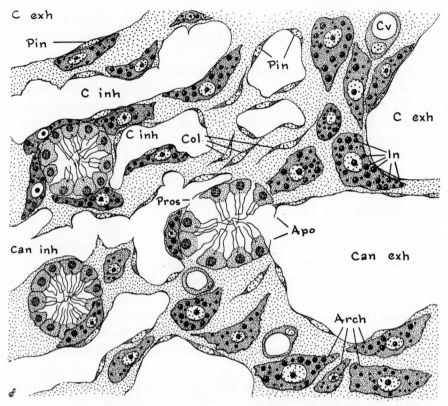

Figure 5-4. Later stage of differentiation of *Ephydatia* spherule. **C inh.** Inhalant canal system. **C exh.** Exhalant canal system. **Pin.** Pinacocyte. **Arch.** Archeocyte. **Col.** Collencyte. **Pros.** Prosopyle. **Apo.** Apopyle. **Cv.** Contractile vacuole.

or choanocytes. Only in the peripheral zone, where respiratory exchange is adequate, does reorganization proceed directly. In contrast, the central region undergoes histolysis and is later invaded by lacunae, which represent the exhalant water system. Archeocytes in the central region ingest the debris of necrotic cells.

In the peripheral region the collencytes line the entire epithelium and give rise to the hypodermal spaces of the inhalant water system, which is thus formed after the exhalant system. The flagellated chambers, produced by clusters of choanocytes, remain normal and active, taking a position between the two water systems so that the systems can communicate only through the intermediation of the chambers. All cells undergo division during the reorganization; their sustenance is probably derived from the histolysis of the more central cells. By the fourth day after the commencement of aggregation, the epithelium has attached the aggregate to the substratum and has acquired a marginal film.

Involution in Large Aggregations

In large aggregates (3-4 mm rather than 1.0-1.5 mm in diameter) reconstitution does not take place. Involution occurs after 10-20 days, the aggregate is deprived of choanocytes, and the lacunae and ostia disappear, leaving the epithelium, which surrounds a dense mass of, primarily, enormous archeocytes and of collencytes. In general, involution in aggregates is manifested by the absorption of choanocytes, the arrested construction of water systems, the absence of scleroblasts, and the preponderance of archeocytes, which absorb and digest the choanocytes. It seems clear enough that choanocytes are indispensable to the reorganization of the young sponge.

But what are the factors responsible? Size alone is not sufficient cause, since involution occurs within the normal size range of sponges which do undergo reorganization when all conditions of culture are identical. According to Brien, reorganization seems to be correlated with the presence of especially small initial quantities of choanocytes. He suggests that the necessary ratio of choanocytes to archeocytes is not realized, as it is in the normally reconstituting sponge, because the archeocytes absorb choanocytes and precipitate involution. In other words, certain tissues weaken, which provokes phagocytosis, which in turn prevents reorganization from taking place.

As mentioned above, aggregates in involution lose their choanocytes, so that they become a mesothelial, archeocytic culture under a pinacocytic, epithelial envelope. The aggregates expand to attain a new equilibrium that is incompatible with the continued existence of choanocytes, but conducive to the growth and multiplication of archeocytes and collencytes. Such aggregates form true organisms in the restricted sense of Ephrussi in speaking of tissue cultures, but organisms different from the normal sponge, whose structure responds to the coexistence and correlation among the three categories of sponge cells: archeocytes, choanocytes, and collencytes.

The presence, on the same substratum, of aggregates in involution and of aggregates undergoing normal reconstitution has led to a striking observation: At the beginning of aggregation, all the spherules formed are apparently capable of at least confluence—of uniting as a single large sphere. Once the aggregates become attached to their support, however, this capacity is lost: two categories of aggregates appear and any possibility of confluence between them disappears.

Evolving aggregates, which are immediately distinguishable by their thickness, transparency, small lacunae, and abundant flagellated chambers, are capable of confluence. In fact, as with sponges formed from gemmules (p. 106), confluence between adjoining aggregates is inevitable. Similarly, aggregates which commence involution at the onset of fixation can join to form a large involuting mass. But between the one kind and the other, confluence seems to be incompatible: they do not establish contact; the two marginal zones advancing toward one another stop moving when they are less than 1 mm apart. Although the two categories of aggregates at first contain the same kinds of cells, differing only in the presence or absorption of choanocytes, the new equilibrium somehow establishes incompatible physiological states.

Brien concludes from such comparisons that structure or form is not an entity—that it is not something imposed upon substance, as in the Aristotelian duality of substance and form—but that form is the expression of a physiological equilibrium which confers unity upon an organism. On the other hand, the different cellular elements, having attained an equilibrium which gives the organization its normal structure and harmonious physiological unity, can regroup after dissociation. The pattern of regrouping is not haphazard: the structure of the elements dictates that regrouping will be the same as in the normal aggregative process. There are fixed possibilities of regrouping: the elements can subsist only in associations that they necessarily adopt, although within the compatible associations any quantitative variation inevitably leads to different states of equilibrium. Structure is thus seen as the expression of cellular correlations and their reciprocal physicochemical activities. Such a concept gains meaning if the physicochemical basis includes the interaction of high-polymer molecular systems.

Reaggregation in Other Tissues

The ratio of cell types in an aggregation is also significant in aggregations other than sponge. In aggregations of dissociated embryonic chick cells, Moscona (1956) found that the role of cell types varies with the relative number in the total population. When less than a certain minimal proportion, the cells of a given kind did not reconstitute their typical tissue fabric, although the same number of cells, when cultured separately, did so.

Fell and Gruneberg (1939) found that mouse and chick tissues can be cultured together in heterologous sera and remain identifiable. In experiments by Moscona (1957), presumptive chondroblasts (as stellate mesenchymal cells) from mouse and chick limb buds, after being cultured together for 6 days, formed typical cartilage, the matrix of mouse cells merging with that of chick cells. Similarly, typical liver tissue was produced by a mixture of presumptive mouse- and chick-liver cells. Random mixtures of the *same* histogenetic type, therefore, combine to form uniform chimeric tissues. On the other hand, chick and mouse cells of different histogenetic types do not combine: each differentiates according to its nature.

The evidence strongly suggests that the process of tissue formation is preceded or accompanied by a reshuffling of the aggregated cells. This theory is in keeping with the dissociation-reaggregation experiments on amphibian embryos made by Spiegel (1954a) and Townes and Holtfreter (1955). According to the latter investigators, different cell types in a composite aggregate, in consequence of directed movements, are sorted into distinct, homogeneous layers whose stratification corresponds to the normal germ layer arrangement. Tissue segregation becomes complete because selectivity of cell adhesion emerges, so

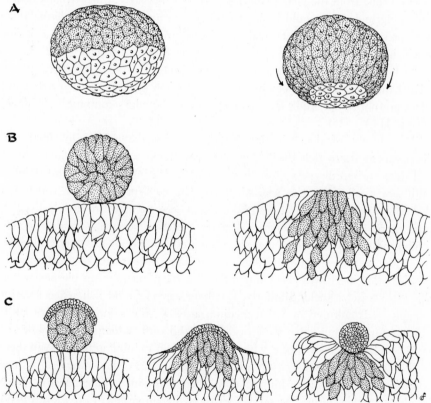

Figure 5-5. Spreading and segregation of tissues in reaggregated cell masses of amphibian embryos. **A.** Coated endodermal cells spread over uncoated area. **B.** Aggregate of uncoated endoderm becomes incorporated into endodermal substratum. **C.** Endodermal graft covered by ectoderm invaginates into endoderm; ectoderm spreads at first and then becomes isolated. (After Townes and Holtfreter.)

that when homologous cells meet, they remain permanently united to form functional tissues; whereas a cleft develops between certain nonhomologous tissues (e.g., between neural or endodermal and adjacent tissues; whereas mesodermal elements, as in normal embryogenesis, take an intermediate position connecting the inner with the outer epithelium). Feldman (1955), working with the comparable material *Triturus alpestris*—though he dissociated cells of neurulae rather than gastrulae—came to similar conclusions. Thus in aggregates of ectoderm and endoderm, ectodermal cells at first invaginated into the mesoderm, but were later expelled to the surface, where they formed a separate epidermal layer; in other words, an initial affinity was later lost. Prospective medullary plate and prospective epidermis separated, respectively forming a neural tube from the plate tissue and an epidermal layer. Furthermore, neural and epidermal tissue could be kept from separating if some mesodermal cells were present to form an intermediate layer adhesive to both. There is a similar specificity in the healing of epithelial wounds. Thus Chiakulas (1952) found that when a hole is made in a sheet of epithelium, the tissue extends until the cells of the free edges fuse again, each with its own kind. If the two advancing edges are not of the same species or character, the edges continue to migrate. (See p. 114.)

The concept of determination associated with the analysis of metazoan development states that during embryonic differentiation the various organs and tissue primordia pass through stages of increasing specificity and stability until they finally form characteristic types. This concept leads to the conclusion that the various tissues and their constituent cells become irreversibly fixed in type; therefore, in culture media, cells either remain irreversibly fixed or merely exhibit modulation, in which there is an apparent—not real—loss of type identity. Thus pigmented epithelium depigments during 8-10 passages in culture media which is rich in growth-promoting embryo extract, but re-forms pigment when this extract is withheld. Similarly, Kasahara (1935) found that when anterior hypophysis is cultured in a rich medium at 37°C, no eosinophile or basophile cells are formed; but when nutrients are reduced and the temperature is lowered to 28-30°C, many cells appear with abundant eosin-staining granules. And there is the older pioneering observation of Drew (1923) that sheets of kidney epithelium redifferentiate into tubules when a fragment of connective tissue is added to the culture. Mass also seems to be important. Lopaschov (1935) noted that explants of presumptive head mesoderm from a single, early gastrula of the newt *Triton taeniatus* forms only muscle tissue in vitro; whereas, explants from large masses consisting of identical cells and derived from the fusion of several of these same primordia produce muscle, chorda, brain, sensory primordia, and ectoderm.

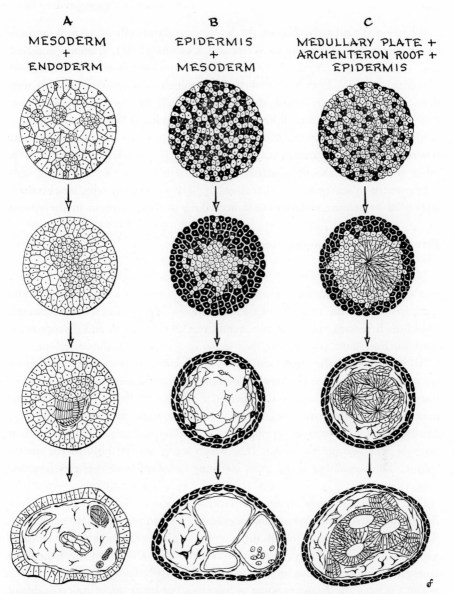

Figure 5-6. Combination and segregation of various combinations of dissociated tissues of amphibian embryos. **A.** Mesoderm and endoderm mixtures result in centripetal migration of mesodermal cells. **B.** Epidermal cells, when combined with mesodermal cells, move outward, and mesodermal cells inward. **C.** Axial mesoderm, medullary plate, and epidermis cell mixtures all sort out and re-form respective tissue types. (After Townes and Holtfreter.)

Whatever the causative factors, the type-specificity of cells seems to be attained progressively and, according to Weiss and Moscona (1957), relates to constructive cell behavior as much as to cellular specialization. Precartilaginous blastemata of chick limb buds (3-4-day-old) and chick sclera (6-7-day-old) were dissociated; and after mixed suspensions settled, the new aggregates were cultivated on plasma clots. Both types of cells produced cartilage, but each according to its specific pattern of origin: the first formed lumpy nodules with whorl-shaped cell arrangements; the second, flat, pseudostratified plates. In other words, the rules of assembly were ingrained even in such young cells.

In contrast to segregation and reassembly of differentiated cells, unspecialized cells tend to acquire differentiation according to their position in the system.

Development of Sponge Gemmules

Many sponges, including *Ephydatia,* reproduce through the natural formation of gemmules consisting entirely of archeocytes; thus sponges afford direct comparison with the more histologically complex aggregation masses, although, unlike such masses, the gemmule archeocytes contain lipids and glycoproteins, are binucleate, and are encased, as a mass, in a triple chitinous envelope.

At germination the archeocytes, having become polynucleate, undergo a complete segmentation, so that they become mononucleate histoblasts. The first histoblasts to emerge through the micropyle are mesothelial cells. Extending out over the support to cover the gemmule envelope, these mesothelial cells are a mesenchymal mass whose peripheral cells become the pinacocytes and whose internal cells become the collencytes, which make up the bulk of the mesenchyme. The remaining archeocytes continue to escape from the envelope and subdivide.

First central and then hypodermal lacunae appear in the mesenchyme whose delimiting collencytes become pinacocytes after forming the lacunary epithelium. The central and hypodermal lacunae represent the beginnings of the exhalant and inhalant water systems, respectively. At this time the last of the archeocytes emerge, all mononucleate, to become distributed in the mesenchymal matrix which encloses the lacunae. Some archeocytes persist as such; others continue to divide into small cells which become the choanocytes that make up the flagellated chambers.

Thus all the cells of the sponge derive from archeocytes; and cellular differentiation progressively yields pinacocytes and collencytes, both of which comprise the epithelial envelope, the canals, and the mesenchymal framework; then archeocytes; and finally choanocytes. The destiny of the histoblasts derived from the initial archeocytes is linked to the organogenetic stage; i.e., it is a function of the morphogenetic moment. On the other hand, it seems clear that the archeocytes act on one another in some way so that the histoblasts from these initial archeocytes will not all have the same character. At the outset, all the histoblasts are collencytes; later in the newly formed mesothelial tissue, in which lacunae appear, the histoblasts produced by multiple division become choanocytes, while other archeocytes persist as such or differentiate as sclero-blasts, vacuolate spherical cells, and other specialized cells. Evidently, the first cells produced and differentiated exert some sort of influence on subsequent cells, so that differentiation of the former progressively brings into being the typical structure of the sponge and its corresponding physiological unity. This is the well-known principle of Driesch: that the differentiation of cells is a function of their position—that organogenesis and morphogenesis determine histogenesis.

In the reorganization of a sponge after dissociation, the inverse of Driesch's principle holds: There is no histogenesis, but only the morphogenesis and organogenesis implicit in the nature of the culture cells. As Huxley (1911) has stated: The position of the cells is a function of their nature. However, it is in-teresting that in gemmules the archeocytes are totipotent and give rise to all categories of cells; whereas regenerating aggregates must consist of three types of cells. The question remains: Is the greater potentiality of the gemmule archeocyte linked to its enormous size and/or rich content of lipids and glyco-proteins?

The Symplasmic State

Sponges

More acutely than other organisms, sponges raise the question of the sym-plasmic state: Do the constituent cells retain their full individuality while contributing to, and maintaining, the unity of the whole; are there true syncitia, which are of fundamental significance to the basic organization; or does an intermediate symplasmic condition prevail? Perhaps symplasmic states vary so greatly among sponges that what is true of one is not true of another. Certainly, there is a wide divergence of opinion among investigators.

According to Ijima (1901), the sponge *Euplectella marshalli* (Venus's-flower-basket) in place of a comparatively compact parenchyma and canals lined with epithelial and epitheloid membranes, has exceedingly cavernous tissue so penetrated by intercommunicating spaces that it looks like a cobweb formed of delicate, threadlike, sometimes filmlike trabeculae. Ijima describes the whole

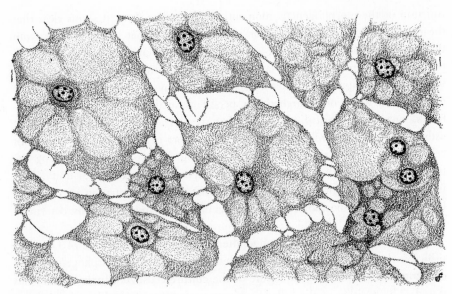

Figure 5-7. Symplasmic state of superficial epithelium of young sponge, showing individualized pinacocytes united by intercellular bridges. (After Brien.)

system of trabeculae as a syncitial network of nucleated protoplasm, although along with the basic syncitial structure, there are free cells, consisting of archeocytes and their derivatives. Okada (1928) found a similar syncitial, reticular structure in *Farrea*. Ijima found that there are no layers of flattened epithelial cells and that the development of pinacocytes both as constituents of the exterior and as a lining of the internal cavities and passages is entirely suppressed, all membranes being formed as adaptations of the general trabecular system.

In the relatively simple sponges *Leucosolenia folcata* and *Grantia compressa* and some others, Rio-Hortega and Ferrer (1917) describe membranes that seem to be intermediate between true syncitia and typical epithelia. On the contrary, in *Ficulina ficus* reconstituted following dissociation, Fauré-Fremiet (1931 and 1932), found that separate and distinct cells make up not only the epidermis and canal membranes, the cells being appressed together, but also the whole mesenchyme as well, the mesenchymal cells being interlaced by lamellate processes. Fauré-Fremiet's description was based on the fact that during treatment with an excellent macerating fluid, the interior of the sponge breaks down into separate cells, but overlooked the fact that macerating fluids characteristically break down intercellular connections in tissues generally.

Similarly, Brien (1937) is convinced that in young spongillids reared from gemmules of *Spongilla lacustris* and *Ephydatia fluviatilis,* pinacocytes are of the conventional cell type and appressed to one another. Bronstedt (1936), however, describes the spongillid epidermis as a continuous sheet of cytoplasm in which cells are not individually delimited—that silver-nitrate lines are not cell boundaries. Studying the effects of narcotics and so-called aquarium degeneration on the epidermis of *Microciona prolifera,* Wilson (1938) reaffirms that in the marginal films of reconstituting sponges the epidermal membrane is normally syncitial, but that under the influence of narcotics the syncitial epidermis breaks up into cell-like fragments by means of vacuolar degeneration.

There seems no reason to doubt the accuracy of the various descriptions of these investigators, for the records and observations in every case are obviously meticulous. The conclusions seem unavoidable that, on the one hand, sponge tissues are variable and, on the other, that histological processes do entail significant tissue changes from the living staate. In a study of young sponges developed from *Ephydatia fluviatilis* larvae, Brien and Meewis (1938) reaffirm that the epidermis, superior or basal, is not a syncitium and that the constituent pinacocytes retain their individuality. At the same time, they describe, even in fixed histological preparations, intercellular hyaline spaces traversed by numerous fine cytoplasmic strands that connect the adjacent cells; these spaces seem to be invariably present. Brien admits that in the living condition a syncitial state is apparent, but that contraction resulting from fixation demonstrates the reality of cellular individuality and that a true syncitium does not exist. In the strict definition of "syncitium," this conclusion is undoubtedly true, but if the term "symplasmic state" as employed by Studnička (1934) is substituted for all cases in which the cytoplasm of tissues is continuous in any form, the difficulty and disagreement are largely resolved.

Coalescence of sponge cells, as already noted, consists primarily of fusion of the hyaline layer, leaving the endoplasmic territories, each with its contained nucleus, intact and separate from one another. Depending on the distinctness and distribution of the common hyaline layer, this condition could appear as syncitial, as a typical epithelial sheet, or as cells actually separated for the most part by intercellular collagen though still connected by fine ectoplasmic strands. Continuity, in other words, may be either all-embracing or tenuous; but whichever it is, the cytoplasmic continuity remains a fact of the utmost importance.

General

The unity of tissues and the individuality of cells are seemingly incompatible concepts that together express a paradoxical reality. Cells do possess true individuality, and tissues true unity. The unity of the latter to a considerable degree overrides the individuality of the former and exhibits characteristics of an emergent sort, i.e., not predicated on recognizable qualities of individual cells; on the other hand, the individual cell contributes to that supracellular unity which subordinates the cell. Since this dilemma is not a metaphysical question, the unity of tissues must be rooted in substance and structure as concretely as the individuality of a cell is attested by its morphology.

The old cell theory—that the organism is essentially a cellular state made up of cooperating cells—was forcefully criticized long ago by both Wilson and Sedgewick as being clearly inadequate. On the other hand, the organism as a true syncitium, in which cells merge completely and only nuclei retain their integrity, is demonstrably untrue. The question is: Where between these two extremes does the truth lie; i.e., to what extent can cells merge and still retain their essential individuality and integrity? It is to such mergers that Studnička (1934) applies the term "symplasmic state." In his opinion most cells, as seen in preparations, consist of nuclei and endoplasmic territories only and that various degrees of union exist at the ectoplasmic level, at least in those tissues that can be regarded as unified structures. Beyond the ectoplasm is the exoplasm; i.e., material, to some extent organized, secreted by or through the ectoplasm, which forms the ground substance of various connective tissues and the basement lamellae of epithelia.

The question may to a great extent be: Where does a cell end and what lies between cells? Collagenous material may develop typical organized structures in the absence of cells. On the other hand, the fibrillar structure of basement membranes in vertebrates has been seen to originate in material that is within the limits of the basal ectoplasm of the epidermal cells, although outside of the ectoplasm the fibrils are clearly continuous in the membrane far beyond the confines of any one cell. Is this continuity attained before or after final secretion from the cells, and is there a very definite point at which a cell ends and exoplasm begins? Is the glycoprotein film on the outside of an amoeba part of the living system or not? Studnička says that the metazoan body is in a symplasmic state, throughout which the living substance is continuous, and that the fibrillar ground substance in general—whatever else may be added to it— is a true exoplasmic extension of the ectoplasm. Although this may be a valid and valuable point of view, there are examples of a true symplasmic state which does not involve any deliberation of the role of exoplasm and is limited to ectoplasmic structure.

Muscle Bands

The tail of the ascidian larva develops precociously during embryonic development, and the band of muscle tissue formed along each side of the notochord consists of a comparatively small number of relatively very large cells. During

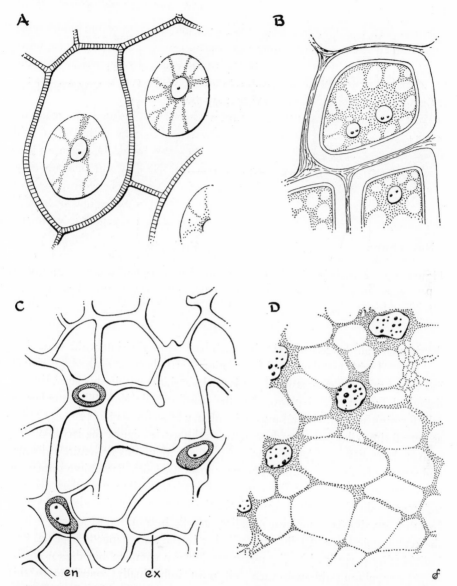

Figure 5-8. Symplasmic states. **A.** Epidermoid cells from notochord of *Cyprinus auratus*, showing broad, homogeneous exoplasm. **B.** Parenchymatous cartilage from tail fin of *Petromyzon*, showing capsular structures. **C.** Reticular epithelium from dorsal fin of foetal *Spinax niger*, showing differentiated exoplasm **ex** and endoplasm **en**—the latter present only in neighborhood of nuclei. **D.** Reticular epithelium of soft cells from horny teeth of *Petromyzon planeri*. (After Studnička.)

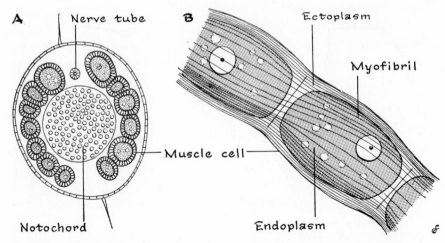

Figure 5-9. Myofibrils of tail muscle of ascidian tadpole *Distomus aggregatum* passing across cell boundaries in spiral, longitudinal course through continuous ectoplasmic territory. **A.** Cross section through tail. **B.** Two muscle cells from muscle band.

development all the cells constituting each band cease division at the same time and proceed with histological differentiation. Partly because of the large size of individual cells, the ectoplasmic layer is thick; and it is within this layer, sharply delimited from the endoplasm, that the myofibrils develop. The number of cells forming an entire lateral muscle band may be as few as 18 or 20 or as many as several times that number, depending on the species size of the larva. Within each such band the nuclei and their respective endoplasms are entirely distinct and separate from one another. The myofibrils, however, extend in a somewhat spiral course continuously from one end of the band to the other, passing uninterruptedly from cell to cell within the ectoplasmic layer (Berrill, 1948). In other words, the ectoplasm of adjacent cells is continuous, at least where the ectoplasmic layer of adjoining cells lies in the same tangential plane. Cell individuality is retained by the nucleus-endoplasm complex, but does not exist in the ectoplasm, although each cell must individually contribute to this substance. Yet, during larval metamorphosis, when the tail undergoes resorption, the muscle bands break up into separate cells, each with its own ectoplasmic territory, though the cells are no longer functional.

Accordingly, ectoplasm can form layers continuous from cell to cell—layers in which structure can differentiate without disrupting cell individuality and without necessarily preventing a cell from withdrawing from the system. It may be significant, however, that in such ectoplasmic layers the symplasmic state is developed by a group of cells which are at the same stage of histological maturity. It is noteworthy that these observations, made independently by several investigators, result in the main from the remarkably large dimensions of the individual muscle cells, especially of the ectoplasmic layer; in more normal-sized tissues the processes and their effects would be too easily overlooked.

Epithelial Contraction Waves

A second example, functional rather than histological, is also found in ascidians. Epidermal ampullae occur as outgrowths of the body or of the vascular epidermis of various ascidians, either in colonies of adult individuals as in *Botryllus* or as long, slender structures which anchor metamorphosed larvae of styelid and molgulid ascidians (Berrill, 1929). Especially in the latter type the long, ampullary outgrowths undergo a rhythmic wave of contraction, commencing at the base and progressing toward the wider tip. Two successive waves may be visible at one time, the first occurring about 2 minutes before the second. The significant fact in the present context is that the tube exhibiting the contraction wave consists of a simple epidermal epithelium, one cell thick, and contains only haemolymph with a few floating blood cells—no muscle cells. As the wave progresses from the proximal to the distal end of an ampulla, the constituent epithelial cells undergo a regular change in shape, from subcubical to columnar and back to subcubical. Two possibilities are apparent: Either there is a basement membrane—as yet undetected but nonetheless likely—which undergoes a rhythmic myoid contraction, not at all in keeping with the general

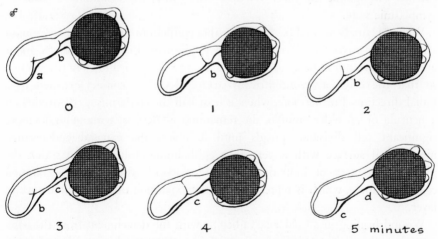

Figure 5-10. Contraction waves in ascidian epidermal ampullae. Contractile epithelium consists of single layer of cuboid cells. **a–d.** Successively forming waves arising near base of ampulla.

character of the collagenous proteins that are typical of such structures; or the basal ectoplasmic layer of the epithelium is symplasmic and contains continuous microfibrils of the contractile protein group. If valid, the latter, more plausible possibility suggests that the symplasmic state of cells in unified tissues is generally overlooked, because in the living state the ectoplasmic layer of such small cells is extremely difficult to see and in the preserved state cellular contraction is most likely to interrupt the original continuity—an eventuality similar to the initial separation of the larval muscle cells during metamorphosis of the ascidian tadpole.

Epithelial Expansion

The healing of wounds by epithelial expansion has been studied mainly in amphibian embryos by Holtfreter (1943) and in amphibian larvae by Chiakulas (1952) and Lash (1955).

In amphibian eggs the cell surface layer is not hyaline, but contains small yolk granules; it is characterized by a coating of black pigment granules, which do not constitute a distinct line of demarcation between the coating and the inner cytoplasm. During division, however, the cells of the "surface coat" (Holtfreter's term) divide only superficially, and the surface membrane amalgamates the peripheral cells into a plastic supracellular unit which is an effective force "that makes for a morphogenetic integration of the dynamic functions of the single cells into a co-operative unity."[1] Holtfreter suggests that the properties of the coat result from a feltlike, submicroscopic fibrillar structure which extends throughout the cells. Whatever the cause, the condition is clearly a symplasmic state.

When a small wound is made in the epithelium of the archenteron of *Ambystoma,* leaving the deeper layer of unpigmented cells exposed, the tissue around the wound at first retracts and then spreads inward to close the wound, so that the originally concentrically arranged cells become reoriented in a radial direction. Furthermore, when at least half the epidermis is removed from a neurula or an older embryo, the remaining epidermis, without undergoing significant cell division, spreads until it covers the exposed endodermal-mesodermal surface with a very thin epithelium. Holtfreter states that the gliding movement of individual cells are evidently directed by a common centripetal force, which is represented by the expansion of the syncitial surface layer.

Wound healing in an older larva begins with the detachment from the basement lamella of the epidermal cells bordering the wound, proceeds to a wave of mobilization by these cells to more-distant areas, and results in the active

[1] Holtfreter, *J. Exp. Zool.,* **93**:266 (1943).

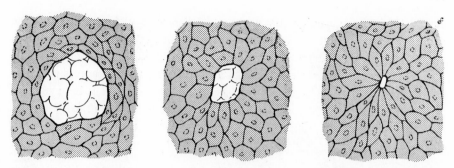

Figure 5-11. Contraction of epithelial wound in *Ambystoma*. (After Holtfreter.)

migration of the epidermal sheet toward the center of the wound. The rate of movement is relatively constant for all cells within the sheet, and the migration can be considered mainly, though not exclusively, a mass movement which involves all types of epidermal cells from all epidermal layers. Movement of all the cells is arrested simultaneously, as soon as closure or juncture is effected, at which time fusion with the homologous tissue takes place, forming a continuous epithelium with a smooth outer contour. Although this phenomenon does not in itself demonstrate the existence of a symplasmic state, it is difficult to account for this type of multilayered expansion except in terms of the extension of a fibrillar structure that controls and carries along the individual cells as in the surface coat of the embryo.

Intercellular Connections

Protoplasmic intercellular connections, consisting of ectoplasmic substance alone or together with an endoplasmic core, have been described for many organisms. For example, cytodesms or plasmodesms, i.e., fine protoplasmic strands which penetrate the secondary cellulose wall of plant cells join cytoplast to cytoplast. Similarly, in vertebrates, tonofibrils link together the cytoplasm of adjacent epidermal cells. Cell bridges are also known and relatively accessible for study in the early cleavage stages of invertebrate eggs, those of echinoderms and mollusks having been well described. According to M. M. Moore (1932), working with sea-urchin eggs, and Costello (1939), with nudibranch eggs, in addition to the enveloping hyaline layer (corresponding to Holtfreter's surface coat of the amphibian egg)—which plays an important part in binding together the blastomeres—there arise, independently of that layer, protoplasmic bridges which draw together adjacent blastomeres. Coherence of cells, in Moore's opinion, is due to structures within the living cellular wall, as was suggested by Conklin (1917), who centrifuged *Crepidula* eggs.

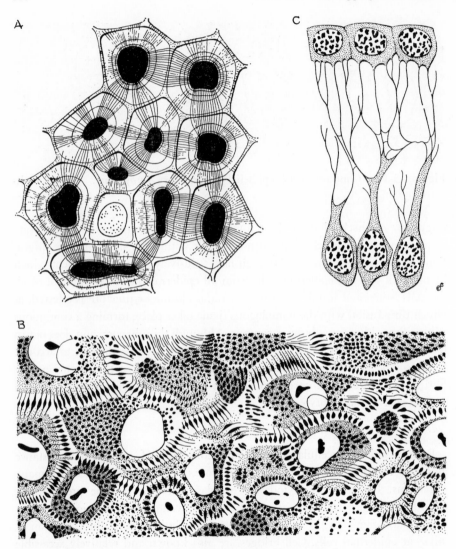

Figure 5-12. Intercellular protoplasmic strands. **A.** Endosperm of *Strychnos nux-vomica*. (After Tangl.) **B.** Malpighian layer of epidermis of human palm. (After Maximov.) **C.** As primary mesostroma between ectoderm and mesoderm of human embryo. (After Studnička.)

The behavior of the living intercellular connections between blastomeres of echinoderm eggs has been observed by Andrews (1897) and Whong (1931), who describe two kinds of connections: primary cell bridges, which are comparatively large, hyaline structures between blastomeres and which result from cell division; and secondary bridges, which are formed by direct protoplasmic contact between blastomeres that previously may have been separated. Andrews observed such secondary bridges grow as fine, actively spinning filaments reaching toward the adjacent cells; these filaments may be comparable to the long, active pseudopodial extensions at the tip of the invaginating archenteron in the sea urchin, observed cinematographically by Gustafson and Kinnander (1956), and may also be comparable to the fertilization cone thrown out by the cortex of various invertebrate eggs to meet an advancing spermatozoan. Whong found that within a few minutes after isolated blastomeres of various stages had been brought together, connections formed; even when two cell plates derived from two different eggs after the seventh division, they united by the same means. Such secondary bridges are contractile and extensible, as shown both by manipulation and in cell division. With decreasing cell size, these bridges become proportionately more effective in uniting cells—and correspondingly more difficult to observe.

It is clear, therefore, that in an organism whose cells constituting some sort

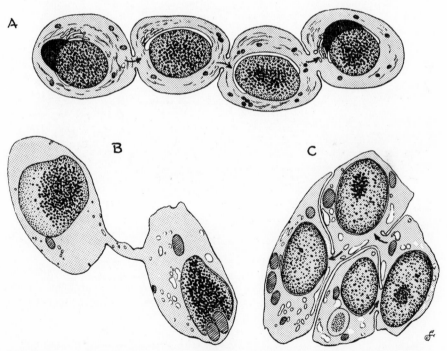

Figure 5-13. Differentiating cells joined by intercellular cortical and endoplasmic bridges. **A.** Four conjoined guinea-pig spermatids. Two central cells are in same stage as two end cells, but their acrosome is out of place of section. **B.** Pair of spermatids of *Hydra* testis. **C.** Battery of early cnidoblasts from ectoderm of *Hydra*. (After Fawcett.)

of epithelial tissue are comparatively large, there are usually intercellular connections, tangential or radial. Furthermore, it is reasonable to suppose that such connections also exist in organisms whose constituent cells are relatively small, so that ectoplasmic continuities have not yet been detected. The nature and function of such intercellular connections, however, are little known. It is noteworthy, however, that Fawcett (1959) has shown, by electron microscopy, that in groups of differentiating mammalian spermatids and in groups of differentiating *Hydra* cnidoblasts the protoplasm of adjoining cells is continuous not only cortically but endoplasmically—a state also reported by Whaley (1959) for meristematic tissue in the onion-root tip.

Chapter 6

Cell Combinations:
Growth and Differentiation

The Cell State in Colonial Protista

The ciliate *Zoothamnion* and the green flagellate *Volvox*, both of which are colonial Protista, present the basic problems of individual cell differentiation, intercellular connection and coordination, growth pattern and general organization, and regional differentiation or specialization of constituent cells.

Zoothamnion

Zoothamnion alternans, most intensively investigated by Sumners (1938 and 1941), forms regularly branching colonies which are attached to the base and other sites of marine hydroids. A colony consists of a main axial stem which terminates in a single, large apical cell and has branches alternating to left and right; that is, a two-dimensional branching system. An individual zooid is subconical, with a ring of peristomial cilia around the oval disk. The basal end tapers to a stalk which, when viewed in section, resembles a sheathed nerve fiber with an elastic surface membrane, a thick cortical region of hyaline, gelatinous material, and a neuromuscular core which is continuous from branch to branch and from cell to cell.

A new colony originates from a ciliospore, which is a free-swimming zooid that has broken away from its position in an established colony. Auxilliary branch zooids at various levels of a colony metamorphose into ciliospores by resorbing their peristomial cilia and oral disk, growing a heavy equatorial girdle of cilia, and changing from the subconical to a biconvex shape. After breaking away and swimming for a few hours, the ciliospore becomes attached to a subtractum, its ciliary girdle becomes inactive and disappears, a stalk grows out (is secreted, according to Sumners) at the point of attachment, and the body reverts to its original subconical shape.

The first part of the stalk, or peduncle, to appear is a heavy, noncontractile cylinder consisting of a thickened cuticle and a homogeneous hyaline medullary layer. When the cylinder is 200-300 μ long, its formation is interrupted, the basal end of the cell increases in diameter, and outgrowth of the neuromuscular cord begins.

Fifteen hours after attachment, the ciliospore divides unequally, the plane of division becoming the plane of the colony. Successive small daughter cells are formed from the larger terminal cell (the macrozooid), first on one side of the macrozooid and then on the other. Dichotomous branching occurs when the basal part of a cell constricts during division, after which stalk formation begins and the daughter cells move apart from one another. Several types of zooids are differentiated: the large, terminal macrozooid; branch terminal zooids similar to the macrozooid except smaller; and the nutritive microzooids, which constitute the rest of the colony and from each of which are derived ciliospores.

The unity of a colony, as expressed by its over-all pattern, general contractility, and regional differentiation, is clearly based on the continuity of the core material or fibers of the stalk. The fact that a colony is truly symplasmic without being syncitial is shown even more definitely in experiments by Sumners (1938). A break made in the neuromuscular cord creates a regional independence: the apical cell of the upper part persists normally, but a new apical cell for the lower part forms from a cell there. Similarly, when the apical cell alone is removed, the microzooid immediately below it usually replaces it as the new terminal macrozooid. How coordination is effected through the fibrous core of the main stalk and its branches is an unanswered question, except that Sumners rejects transfer of nutritive substances as an explanation.

At certain times asexual propagation by ciliospores is replaced by sexual conjugation, in which a ciliospore assumes the role of a male gamont and fuses with an apical cell, or female gamont. When a free-swimming microgamont fuses with a sexually differentiated apical cell, axial growth and development is arrested for several days, until a new apical cell originates from one of the exconjugants. Meanwhile, all the cells on the 3 or 4 uppermost and youngest branches divide precociously, and common branch cells are also activated (cf. induced division in *Stentor,* p. 50). Yet this response to conjugation never happens when the apical cell is removed and the proliferative activity seems to be initiated by qualitative changes in the coordinating mechanism. Knowing the nature of such changes and the relationship between the cord and regional differentiation could well yield great insight into the general problem of organization.

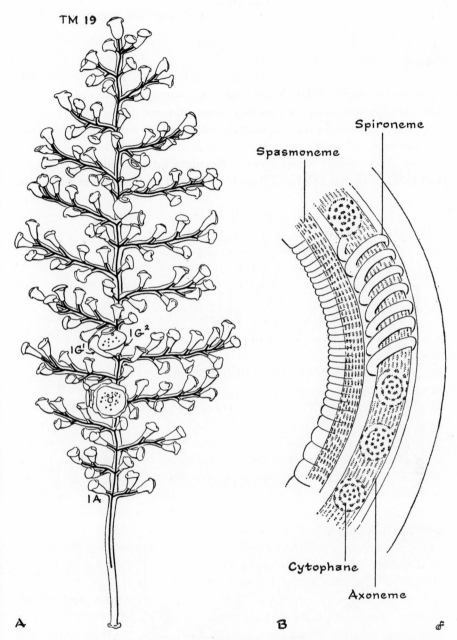

Figure 6-1. Colony and stalk in *Zoothamnion*. **A.** Mature colony in *Z. alternans*, showing alternate arrangement of branches and cells. Apical cell of nineteenth generation **TM 19** represents growing point of primary axis. Each branch has its own terminal cell. Lateral cell of first branch generation **1A** or its two immediate descendants **1G^1** and **1G^2** represent potential ciliospores. (After Sumners.) **B.** Stalk in *Z. arbuscula*. (After Entz.)

Volvox

The green, multicellular, hollow, mucilaginous spheres of *Volvox*, rarely more than 1 mm in diameter were first observed by and described by Leeuwenhoek in 1719. Since that time an extensive literature has come into being, from both botanists and zoologists. Studies of *Volvox* in vivo, however, have been few; the only accounts which approach monographic proportions are those of Janet (1912, 1922, and 1923) and Pocock (1933a and b). The information in the

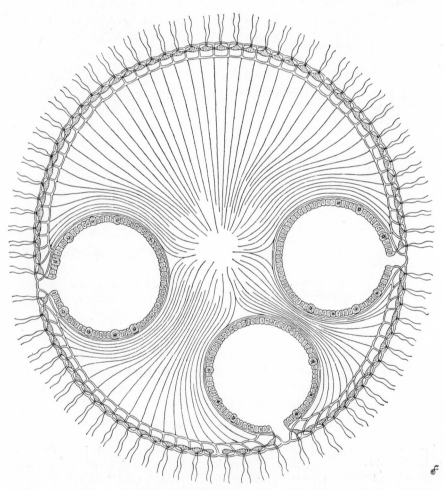

Figure 6-2. Diagrammatic section through asexual individual in *Volvox globator*, showing daughter colonies nearing term of their development, with gonidia forming in their walls. (After Janet.)

present chapter is drawn mainly from these two sources and from a taxonomic account by Smith (1944).

The colonies, as the spheres are usually called, consist of as many as 50,000 cells (or protoplasts) embedded in the outer, transparent mucilaginous layer. In some, but not all, species the cells are closely connected to their neighbors by stout strands of protoplasm. In all species, if the colony is suitably stained, its surface is seen as a pattern of hexagonal or somewhat circular areas with a cell at the center of each.

Each protoplast contains a nucleus, a large chromatophore containing the chlorophyll, an eyespot sensitive to red light, two or more contractile vacuoles, and one or more proteinaceous pyrenoids. Passing from the chloroplast to the outside of the sphere are two flagella (cf. flagella-eyespot-complex, p. 18), whose combined action causes the colony to rotate in the water with one part of it always foremost. That a polar axis exists is shown by the fact that the protoplasts near the anterior pole are better developed and more widely spaced than those near the opposite pole. The rotation of a colony around its axis is generally counterclockwise, although continuous observation shows that a single colony pauses every so often and reverses direction.

Volvox thus raises the two basic problems posed by most multicellular organisms: the nature of the integration of cells to form the whole, and the basis for the regional variation in cell character.

Cells and Protoplasts

A vegetative, or somatic, cell consists of a flagellated protoplast enclosed within somewhat gelatinous walls that are bounded by limiting membranes. As a rule, neighboring cells are connected to one another by strands of protoplasm. Even in those species in which the cells of grown colonies are unconnected, the protoplasmic strands have been seen in the younger stages of development; in other words, such connections exist at the outset, but are broken later. However, connections may disappear in unhealthy and fixed colonies. As a colony grows, the space within the cell walls continually increases in size; whereas the protoplasts, which also grow at first, soon cease except for the progressive extension of the interconnecting strands.

Both the size and the shape of the protoplast depend on the cell's age and position in the colony. At the time of birth, it is oblong, viewed from the side, and polygonal, viewed from above, owing to initial pressure from other cells; then, as the colony matures, its shape changes. The protoplast continues to increase in size until about the time the reproductive cells begin to divide, when individual protoplast growth ceases. As the colony begins to mature, protoplasts diminish in size, particularly in the posterior region of the colony, where they

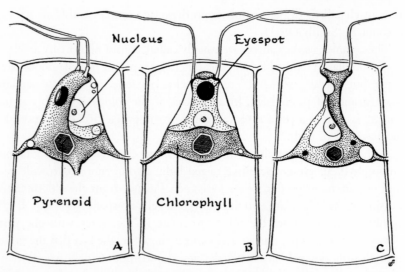

Figure 6-3. Cells with protoplasts from *Volvox rousseletti.* **A–B.** Side and front views, respectively, of cells from anterior pole. **C.** Front view of cell from posterior pole, showing reduction in size of distal end and absence of eyespot. (After Pocock.)

are more closely crowded together and where some are comparatively less well formed than others, usually lacking the eyespot and other distal substance. As the protoplasts become smaller with age, the connecting strands appear as a fine network with the protoplasts forming the nodes.

Protoplasmic Connections

It is tempting, therefore, to regard the continuity of behavior and of the organization of the colony as a whole as the consequence of protoplasmic continuity between cell and cell (the term "cell" is used here as equivalent to the protoplast without its extraterritorial productions). Support for this conclusion derives from the observation of protoplasts which have been separated from one another without injury, apart from the broken interconnections, severed by slight pressure against a cover-slip: the protoplasts assume the shape of an elongated pear and continue moving, aimlessly, for a while. However, the ability of a separated cell to carry on the basic life processes is very short-lived, so that the cell soon dies. Consequently, the nature and origin of the connections are of primary interest.

As a rule, the strands connect two cells directly, although, in older colonies especially, they may show considerable anastomosing. The 5 or 6 strands in each cell, observed under large magnifications, are tapered elongations of the broad, basal part of the protoplast: a strand is broadest at its base (the point of origin), often containing contractile vacuoles, and narrowest at the membrane which separates one cell from another; at that point, a strand merges with the corresponding strand of the adjacent cell. In a young colony the green color of the chloroplast extends far into the strand; but in an old colony, whose strands become attenuated, the strand loses its green color and a contractile vacuole is often seen near the junction node of adjacent strands. The conjoined protoplast substance which is drawn out as neighboring protoplasts separate from one another is of a very intimate nature, embracing much more than merely a surface coat.

An understanding of the origin of the interconnecting strands may result from a discussion both asexual and sexual reproduction of daughter colonies within the parent—processes which introduce a number of associated phenomena.

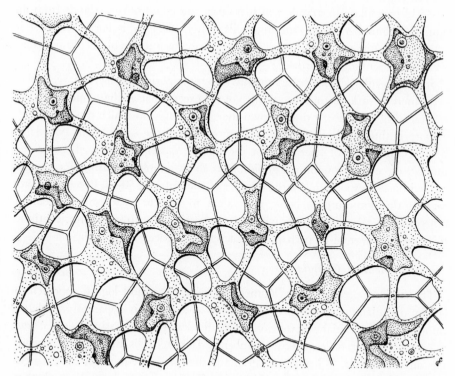

Figure 6-4. Subsurface view of *Volvox globator,* showing cell walls, protoplasts, and intercellular strands. (After Janet.)

Asexual Reproduction

Volvox asexual reproduction, which occurs whenever conditions are favorable, is remarkable in a number of ways. It demonstrates, as in the development of eggs, that the organizational unity of the full-grown organism descends directly from the original unity of an initial cell. Accordingly, this cell, known as the gonidium, is one point of departure.

The number of gonidia in a single colony is usually 8, although among the various species as few as 1 and as many as 20 have been seen. The gonidia are usually confined to the posterior half of a colony and are arranged alternately in two planes, one approximately equatorial and the other midway between the equator and the posterior pole.

At first a gonidium is distinguished from its neighboring vegetative cells by its larger size, darker color, two or more pyrenoids, and large nucleus. Like the vegetative cells, it has two flagella and several contractile vacuoles, and, viewed from above, it is more or less star-shaped; unlike the vegetative cells, it has 5-12 protoplasmic strands which connect the gonidium to strands from an equivalent number of surrounding cells. After a colony is born, i.e., after it escapes from the parental interior, the gonidia continue to enlarge, losing their flagella and sinking inward.

Since this enlargement occurs mainly below the peripheral layer of vegetative cells, there is little disturbance to the colony except that as the gonidium sinks from its original place, it draws down the circle of cells to which it is connected. A pore is thus made in the original outer wall of the colony; this pore later becomes the point of liberation for the mature daughter colony. Because the number of cells surrounding a gonidium is about double that around a somatic cell, with a correspondingly large number of connecting bridges, Pocock (1933a) regards the gonidium as being equivalent to at least two somatic cells. In other words, gonidia are cells which prematurely cease to divide but continue to grow at a normal or perhaps even accelerated pace.

Development of Gonidia

The study of gonidium development is perhaps the most logical approach to colony organization. Growth of the colony as a whole is apparently predetermined, since the initial cell, the gonidium, divides at a progressively slower rate, to produce a more or less specific number of daughter cells, the number depending on the species. Whether the divisions themselves require an increasingly longer time, whether the intervals between divisions are prolonged, or whether both are progressively protracted has not been determined. The final number of cells produced seems to be that of the colony at birth. Thereafter the constitu-

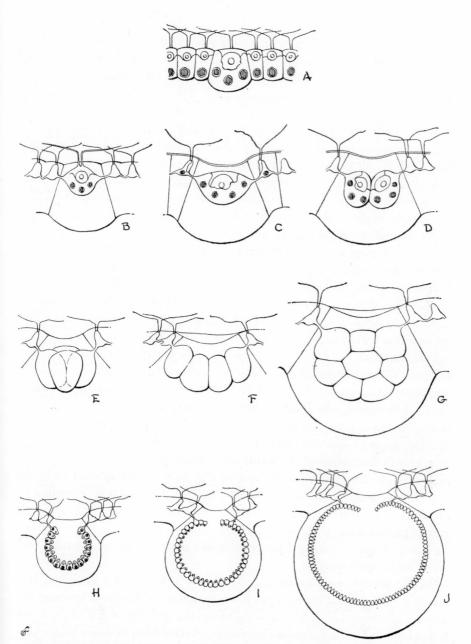

Figure 6-5. Development of gonidium in *Volvox globator*. **A.** Gonidium of relative original size in wall of young colony. **B–C.** Enlarging gonidia before and after, respectively, loss of flagella. **D–G.** 2- and 16-cell stages, respectively. **H–J.** Sections of embryos, showing development of free lip of phialopore. (After Pocock.)

ent cells grow somewhat larger, their flagella lengthen proportionately, and the amount of mucilage external to the protoplast increases, progressively isolating the protoplasts from one another and causing the intercellular protoplasmic bridges to become more and more attenuated. The two phases of growth, that of cell proliferation and of cell enlargement, are thus rather sharply separated.

In the several *Volvox* species of the Euvolvox section (see p. 137), the gonidium begins to divide some time after the birth of a colony, but while the colony is still young; i.e., the gonidium begins to divide a relatively short time after cell division in nonreproductive cells has terminated. The first division is longitudinal, or radial, in relation to the parent; the second is also radial, but at right angles to the first; and in the third the cleavage planes are oblique to the preceding two divisions, forming an 8-celled bowl-shaped stage. The process of cleavage is repeated in regular sequence for a number of times; the number varies with the species and to some extent even within the species: in *Volvox aureus* the final cell number may be as low as 200 and as high as 4,000; in *Volvox globator* it may be as high as 50,000. However, the total range in number of successive cleavages is but tenfold among all species of *Volvox,* and the variation within each species is rarely more than 1 or 2 divisions.

Two questions emerge: What determines the growth potential of the gonidium? What underlies the more-limited, but known variability in final cell number within a species? Growth potential of the gonidium seems to be a fairly specific inheritance, which in turn almost certainly determines the number of cleavages. Since protoplasmic synthesis depends on photosynthesis with its variable supply of salts and other materials in the environment, there may well be a growth fluctuation of the whole cell within rather narrow limits during the period of development—a fluctuation that may add or subtract from the over-all expansion of the initial gonidial surface. Expansion or extension of the original cell surface is a valid verbal expression, for the cleavage planes remain perpendicular to the surface throughout the process of division, although curvature of the expanding layer is inevitable since neighboring somatic cells do not yield much space.

On the possible assumption that the growth potential of a gonidium is a function of storage of, for example, RNA within the cell, what determines the location or selection of those 8 or so cells from the majority of cells of the lower hemisphere, and what determines location or selection from the lower hemisphere only? Despite the fact that the problem of local specialization arises in gonidia in possibly its simplest form, the answer is as obscure as ever; for it is also possible to regard the gonidial cells not as local specializations, but as centers where growth-inhibiting forces that operate generally are locally inoperative. Whether local specializations or growth-inhibited centers, the question remains:

Why do the cells of a newborn colony cease to divide upon termination of initial expansion and cleavage? The pattern of events is repetitive: every generation of daughter colonies is born with about the same number and location of gonidia, each of which repeats the space-time pattern of its predecessor. In other words, within a rather restricted area, an initial large cell undergoes a specific expansion, expressed as a certain number of divisions, within whose expanding network there appear certain loci that possess the properties of the initial cell.

Does the initial cell grow and divide in this manner because it has, in a sense, escaped from the basic organization of the new daughter colony; or is the growth potential of this cell the first distinguishable quality, and the subsequent growth and escape from the parental organization the consequence? It is significant that throughout the cleavage phase the chloroplast is dark green and has several pyrenoids—characteristics indicative of great metabolic activity in the dividing cells. Also, as development proceeds, cells in the daughter colony increase in size but not in proportion to the rate of cell division, so that at each division the cells are smaller than those in the preceding division. The initial relatively large size of the gonidium, the intense metabolism particularly of early division stages, the progressive diminution in cell size, and the final arrest of cell division followed by local renewals of the cycle are all part of a unified picture. These aspects are, of course, essential characteristics of most developing units, such as eggs, buds, and meristems, and are of great and general interest.

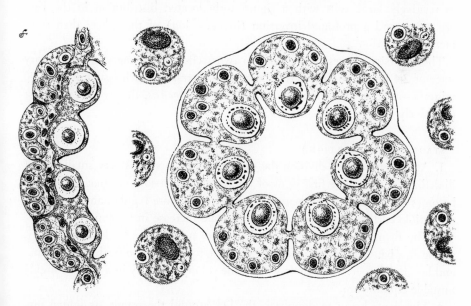

Figure 6-6. Developing gonidia in *Volvox globator,* showing cell bridges (between dividing cells), nuclei, and pyrenoids. (After Janet.)

Origin of Protoplasmic Connections

The protoplasmic unity represented by the intercellular strands that unite the protoplasts of the adult colony persists from the beginning; i.e., the integrity of the original surface, or cortex, of the undivided gonidium is never lost as the result of subsequent cell divisions, and disappears in a few species as a consequence of unusual expansion of a colony only after cell division has ended. Janet (1922) has shown very clearly that from the first division of the gonidium onward cell separation is never complete and that the cortical substance of the developing colony is a continuum from start to finish.

Inversion during Development

The termination of cell division in the developing gonidium leaves the daughter colony as a hollow sphere within the parent and attached to the parental network of cells by the original protoplasmic strands that connected the gonidium to its somatic neighbors. The sphere is open to the outside, the opening being known as the phialopore or ostiole. As mentioned earlier, when the growing and subdividing gonidium sinks below the parental surface, it leaves a pore (called the parental pore) at its original location. The terminal point of cleavage is critical in several ways: the final cell number of the new colony has been attained; the gonidia of the next generation are recognizable as relatively large cells with a significantly located distribution in the new organism; and a profound inversion (invagination) of the sphere then commences.

The inversion process takes place in two well-marked stages: a preparatory stage of 2-4 hours, and the actual process of inversion which lasts for about 40 minutes—at least in *Volvox rousseletti* and *V. capensis,* the two species most intensively studied, by Pocock (1933a and b).

STAGE 1 (PREPARATION)

1. At the end of cell division the ostiole, or phialopore, is very small and of ill-defined outline. The first signs of the coming inversions are that the outline becomes more definite, the marginal cells become more regular, and the lips of the ostiole curl inward. Such changes indicate increase in tension.

2. The firm, rounded contour of the sphere now begins to dent irregularly, giving it a battered appearance: dents appear first at one point then at another and become deeper. The whole wall of the colony is clearly in a state of highly unstable equilibrium.

3. The dents begin to disappear, the lips around the ostiole straighten out, and the sphere again becomes firm and round. Its diameter, however, is some-

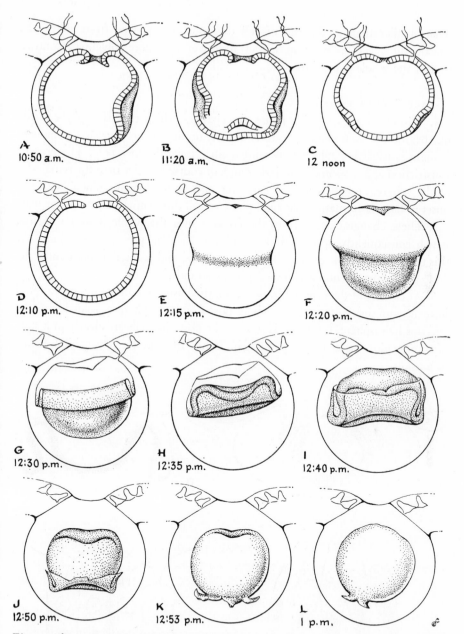

Figure 6-7. Inversion of asexual daughter colony in *Volvox capensis*. **A–D.** Preparation for inversion: depression, denting, and smoothing out. **E.** Beginning of actual inversion: hourglass stage. **F.** Contraction spreading over posterior half. **G–H.** Infolding of posterior half. **I.** Emergence of posterior half (bowler-hat stage). **J–L.** Flash ending and closure of phialopore. (After Pocock.)

what smaller than at the end of cell division, and, viewed from above, the constituent cells of its walls are longer, narrower, and more polygonal. Contraction of the whole wall of the colony is evident.

STAGE 2 (INVERSION)

4. Without any perceptible pause, the changes which turn the sphere inside out begin. The first change, scarcely visible, occurs in the equatorial region as a slight constriction which may shift a little toward either pole. The area of constriction appears somewhat darker and, of course, smaller than the remainder of the sphere, and the cells are narrower and more elongate. Then, with the onset of flagella formation, the final phase of cell differentiation gets under way.

5. These changes spread from the equator over the whole posterior region, which consequently appears darker and smaller than the anterior region. At this time the posterior cells are narrow, elongate, and pointed at both ends, the end commonly extending as mucilaginous strands.

6. At the site of the original constriction, the posterior half commences to fold into the anterior half until the whole posterior hemisphere is folded into the anterior hemisphere. During this process the posterior cells are clearly in a state of compression. Later they expand again, while the anterior cells exhibit a state of tension.

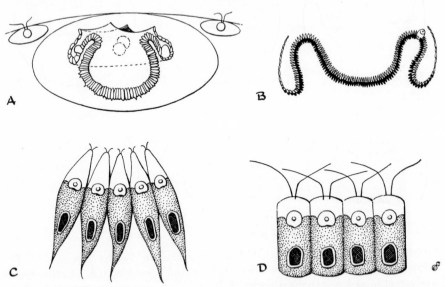

Figure 6-8. Inversion detail. **A–B.** Inversion in *Volvox africanus* and *V. capensis,* showing gonidia. **C–D.** Cells from posterior pole of inverting and recently inverted stage. (After Pocock.)

7. The pore or ostiole enlarges greatly as the infolding approaches completion, until it is as wide as the colony. Since stress is extreme at this stage, the protoplasmic strands connecting the colony to its parent break, so that the daughter colony is free to move within its enclosing vesicle. As the pore enlarges, the cells around it are distorted under the stress, and the formerly obscure protoplasmic connections between cells can be plainly seen.

8. The so-called bowler-hat stage ensues: the posterior half, with its pole still dented downward, begins to protrude through the widely enlarged pore, and the flagella are seen to be active in living material. As the dense, dark-green posterior half protrudes the paler, stretched, anterior half slips down, forming the brim of a hat-shaped structure whose crown is the posterior part.

9. As more and more of the colony protrudes, the cells of the rim contract, so that the rim folds inward, enclosing the contracting pore; the dent remaining at the anterior end straightens out; and the colony begins to move forward with a steady, rotary movement. The cells, by this time closely packed, no longer are elongate with pointed ends. According to Pocock (1933b), the earlier elongation and pointing of the cells are directly related to flagella formation, while the continuity from protoplasmic connections between cells enables the whole colony to act as a single entity.

DEVELOPMENT AFTER INVERSION

After inversion the daughter colony increases in size: both somatic and reproductive cells become larger and the flagella lengthen proportionately. Escape from the parent is usually effected by boring through the wall. The parent colony itself may continue to be active for some time after the escape of the daughter colonies, but eventually it disintegrates. As a daughter colony escapes, its rate of rotary progression becomes exceedingly rapid. At first the colony appears very dark green because its constituent cells are closely packed. However, as development proceeds, the colony appears paler, since the protoplasts become further and further apart from one another because the intervening walls become gelatinous. Furthermore, the protoplasmic connections can be more easily seen, the eyespots in the anterior half become well developed, and ultimately the gonidia begin to divide.

Development of Sexual Colonies

In the dioecious *Volvox rousseletti* several hundred male initial cells are scattered thickly over most of the surface of the colony, leaving only a comparatively small area free at the anterior pole. In other species the number of male cells may be considerably less, although in one, *V. spermatosphaera,* every cell in the male colony develops into a sperm pocket. Although the dioecious type

Figure 6-9. Surface view of hermaphrodite colony in *Volvox aureus,* showing cell boundaries and protoplasmic connections from protoplasts to one another and to large male initial cell and larger female initial cell. (After Janet.)

predominates, many species are protandrous hermaphrodites whose male pockets develop precociously within the embryonic colony. The main questions—for the most part unanswered—concern the nature of the sexual reproductive cells, their initiation, their distribution within the colony, and their development.

MALE INITIAL CELL

Young male initial cells, though similar to gonidia, are much more numerous, remain undivided until a colony enlarges during maturity, and do not broaden before division. As the cell enlarges, the pyrenoids multiply, the flagella disappear, and the cell begins to sink inward. Then the whole cell, with all its contained constituents, swings around until its anterior pole points sideward or inward; the protoplasmic strands connecting the cell to its neighbors remain intact even in this position. This reorientation does not occur in the gonidium and is at least not visible in the egg cell. The male cell, then about 15-20 μ in diameter, begins to divide and progresses until it forms a hollow spheroid about 40 μ in diameter and consists of about 512 cells. The number of divisions is usually 9, though occasionally 8. Throughout this process of blastulation, the process of cleavage is accompanied by active growth of the cells.

At the end of division the cell undergoes a process of inversion similar to that of the developing gonidium except that no denting has been observed—possibly because the process occurs on a smaller scale. Nevertheless, the cells

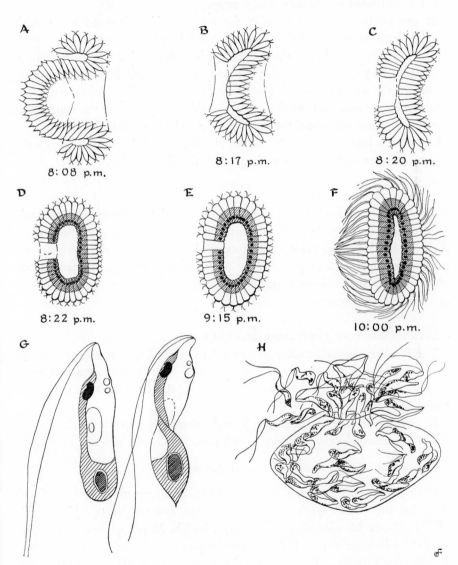

Figure 6-10. Development of sperm in *Volvox capensis,* showing inversion of sperm globoid, mature sperm, and escape of sperm from sperm pocket. (After Pocock.)

change in shape as does the gonidium, acquiring pointed inner ends; and a typical inversion proceeds to form a "thimble" stage. As inversion nears completion, the pointed ends of the cell become flagella which are much longer than those in a daughter colony. The sperm globoid then secretes a delicate external membrane within which the individual cells become free from one another and from which they finally escape.

In general the period preparatory for inversion is at least 2-3 hours, with another hour required for the assumption of the globoid shape. Thus, compared with the inversion of the daughter sphere developing from a gonidium, the preparatory stage is of similar duration, but the actual inversion process is slower in the smaller male globoid. At an early stage of inversion, tension is evident on the outer side of the made globoid, especially near the rim, where inversion begins; otherwise, the process seems to be essentially the same as in a gonidium. In both the male cell and the gonidium, actual inversion follows the end of cleavage and coincides with cytological differentiation.

FEMALE INITIAL CELL

The very young female initial cell is similar to the gonidium and the male initial cell at the same level of maturity, but soon becomes distinguished from them by its deepening color, to an intense dark green and the simultaneous profuse multiplication of pyrenoids. Flagella disappear when the female cell attains a diameter of about 30 μ; and as it matures, the green color is replaced or masked by red haematochrome.

DEVELOPMENT OF FERTILIZED EGG CELL

The fertilized egg cell undergoes a sequence of 7-9 divisions, usually in about 6 hours. Then after a pause of 1-3 hours, the egg cell undergoes a process of inversion essentially similar to that of developing daughter colonies and sperm globoids.

In general the conversion to the sexual phase is a response of developing colonies to changing environmental conditions, usually of a seasonal nature.

Comparative Development of Volvox Species

With the exception of two or three species of intermediate or exceptional character, the various species of *Volvox* fall into two categories, which Smith (1944) calls the Merrillosphaera and Euvolvox. The Merillosphaera section contains the following species: *Volvox spermatosphaera, V. carteri, V. weismannia, V. obversus, V. tertius, V. gigas,* and *V. africanus.* None has protoplasmic connections between protoplasts in the adult colony, although in two of these species such connections have been seen in the embryonic colonies. Gonidia undergo only 10-12 cleavages; the number may be one more or one less within a species. Gonidia are 30-70 μ in diameter; about half this range occurs within a single species. The gonidia appear precociously during the embryonic development of a daughter colony and undergo extensive growth before they start to divide.

The Euvolvox section contains *Volvox globator, V. capensis, V. merrilli, V. barberi, V. amboensis, V. perglobator, V. prolifica,* and *V. rousseletti.* In contrast to Merillosphaera, all Euvolvox have coarse intercellular protoplasmic strands. Gonidia divide 13-15 times; the range, again, within a single species being only one cleavage. Gonidia at their maximum size are 12-15 μ in diameter. They appear relatively late in colony development, and gonidial divisions begin only after cell division in the parental colony has ceased, and usually after its birth.

The correlations between the categories are significant. In Merillosphaera, gonidia differentiate relatively precociously in the embryonic wall. In both categories they cease to divide before somatic cells do, although the Merillosphaera gonidia cease several division earlier than do the Euvolvox; accordingly, the former are correspondingly larger, and during their own development undergo three cleaves fewer. The growth potential, as expressed by the number of cleavages, is precise; but it is noteworthy that the growth potential of the smaller, Euvolvox gonidium is the greater. Still, the nature of either category's potential—so striking, yet so specifically limited—is an unanswered question.

Differentiation

Though they acquire striking substantiation from the foregoing accounts, the two main conclusions are by no means new. (1) In aggregations of diverse types of cells, the cells assume positions in the system according to their special differentiation. (2) In developing systems of unspecialized cells, the cells acquire special differentiation according to their position in the system. The first conclusion is most evident in the reconstitution of sponge reaggregations; the second, in the development of *Volvox* colonies; and both seem to be operative in the cellular slime molds.

Specialized cortical protoplasms of a particular kind fit together, uniting in a meaningful, oriented manner—cell to like cell in sponge and amphibian; cortical fragment to like fragment in the disarranged cortex of *Stentor;* and, as will be seen, tissue fragment to like fragment in the disarranged tissue of *Hydra.* The mechanics of such unions and the primary nature of regional specializations, however, remain obscure.

The developing *Volvox* exhibits a graded differentiation which is obviously significant. The over-all organization which evolves determines or is determined by graded differences in certain features of the constituent cells. The mature organism has a so-called sensitivity pole, which is also the advancing pole of the axis rotation. The pole can be traced through inversion and cleavage to the center of the gonidium's outer surface, corresponding to the eyespot region of the neighboring cells. Similarly, when the gonidium is viewed as a disk united peripherally to a ring of adjacent cells, not only is the center of the disk the presumptive pole of evaginative emergence and organismal activity but there is a gradient in potentiality from the center to the periphery. This graded potential is exhibited in two ways: in the progressive relative reduction of the anterior, or stigmatic, region of cells in successive generations resulting from cleavage; and in the regional location of gonidial differentiation.

The basis for the reduction of the anterior region must somehow be related to, or a property of, the initial difference between the apical point and the periphery of the undivided gonidial cell. Apical material for the sheet of daughter cells is drawn from cytoplasm which is both progressively remote from the apical center and progressively closer to the peripheral connections with neighboring cells. A metabolic gradient, as conceived by Child (see References, pp. 6, 286), might well constitute that difference, although such is pure speculation and the progressively lower potential of increasing peripheral territory is probably related to much more subtle, yet more material characteristics. The problem remains.

Gonidia of a developing colony are located mainly in the posterior, more vegetative hemisphere, within certain broad limits of latitude, and are quite regularly spaced from one another. Given the innate tendency to form gonidia, the positioning of new gonidia is another example of innate response to a certain location in the system. At present it is perhaps unwarranted to assume that the *initiating* agents or organizational clues are derived from the same relative locations in the cortex of the parental gonidium, but undoubtedly something derives from those equivalent locations which induces gonidial differentiation in the developing system.

A comparably striking case of correlation between cell peculiarities and location in a coordinated system is that of the retina in fish and birds. In these two classes of vertebrates—possibly in others, but not in mammals—both single and double cones occur in the retina in regular configurations. A lattice pattern is usual in fish; as a rule, the unit of pattern is a square of 9 cones in which a single cone occupies each corner and the center (5 single cones) and a double cone separates each pair of corners (4 double cones). In some species the single cones at the corners are missing. Lyall (1957) found only single cones in young

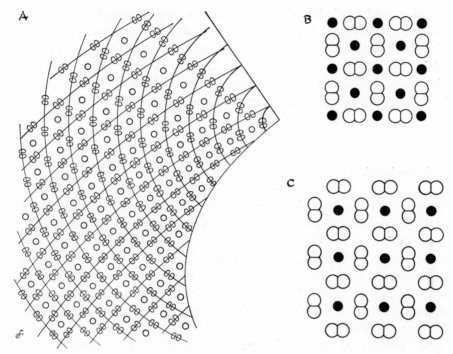

Figure 6-11. Distribution pattern of double and single cones in retina. **A.** Black bass. (After Shafer.) **B.** Young trout. (After Lyall.) **C.** Adult trout and great titmouse *Parus major*. (After Engström.)

trout and only double cones in the adult; Engström (1948) found only double cones in the great titmouse, *Parus major*. The general pattern of the photosensory layer, viewed from above, strongly resembles the crystal-lattice structure of striated muscle, viewed in section.

The retina of the fish eye grows from a peripheral zone in the anterior region, with growth centers in the dorsal and ventral positions of the retina. Shafer (1900) mapped the positions of double and single cones in the anterior region of the large-mouth black bass, *Micropterus salmonoides,* and found an amazing correlation between the distribution and orientation of single and double cones on the one hand and a system of intercrossing coordinates on the other: Single cones occurred in the middle of each lattice space, double cones midway on each coordinate section between crossovers, even when the general picture seemed to lack arrangement or to be in rows of doubles. Furthermore, the separation line in each double cone is at right angles to the coordinate line. It is difficult to avoid postulating the existence of some sort of symplasmic state, or at least comparable substantial continuity, in which a periodicity arises in two directions, and concluding that the peculiar pattern of cones is a response to periodic qualities emerging in the buildup of the two basic systems. From an extensive investigation of the development of visual cells in amphibians, Saxén (1954) concludes that double cells, which arise after the differentiation of other visual cells, are formed by the fusion of a rod and a cone. Whether or not Saxén's conclusion is valid, the pattern of fusions calls for an integrated material basis.

Invagination

Inversion in *Volvox* is a phenomenon comparable to invaginative gastrulation in the development of many invertebrates. In the simpler invertebrates gastrulation is an inversion of the posterior portion of an epithelial vesicle into the anterior portion. The outcome, of course, is different from that in *Volvox,* but, fundamentally, the inversion process is almost certainly similar. In both types of organisms, the epithelium is not in contact with other tissues, so that the inversion force or action is inherent in the epithelium since it cannot result from interaction of adjoining tissues or surfaces. The process has been studied mainly in sea urchins; but because the presence and activity of micromeres in sea urchins introduce complexities, gastrulation studies in starfish afford better comparison.

Moore and Burt (1939) found that when various portions of the invaginating larval sphere in starfish are removed without injury to the ventral or gastral plate, invagination proceeds. Moreover, in both *Dendraster excentricus* and *Patiria miniata,* even when the vegetal plate is isolated by cutting away everything else, the rim of the plate begins to roll up and eventually closes to form a spherical larva which proceeds with invagination. However, when the vegetal plate is cut into radially, it flattens and springs apart, leaving an opening in the line of cut. Moore and Burt concluded that in the vegetal plate there are forces which constitute a radial gradient from the center outward, in intensity and direction; and these investigators suggest that differential cohesion is an important function in this gradient. At present it seems more likely that the tensile and extensive properties of symplasmic fiber systems are involved.

Whatever the cause, it is obvious that the outer hyaline membrane is not a significant factor, for Moore (1952) found that normal invagination occurs when this layer is removed by administration of trypsin, and once more showed that infolding results from properties inherent in the plate.

If intercellular fiber systems are concerned, whether truly symplasmic or not, the two most obvious possible sites are the inner and outer surfaces of the plate epithelium; either or both may be involved since there is contiguity of the surfaces and there may very well be ectoplasmic continuity. On the other hand, both the primary and secondary cell bridges lie about midway between blastomers; i.e., the blastomers are pulled together by strands at their interfaces. In *Volvox,* from the first cleavage through the inversion process, the intercellular strands of each cell are approximately equatorial, and both before and during inversion the inner and outer ends of the cells separately project. Since it is difficult to conceive how either the inner or outer surface of the sphere could develop tension and retain such individual cell configuration, the logical site is where the intercellular strands have been observed: between the surfaces. A continuous fibrillar system at this site throughout the cellular sheet seems reasonable, for just such a structural continuity is seen in the ectoplasmic myo-

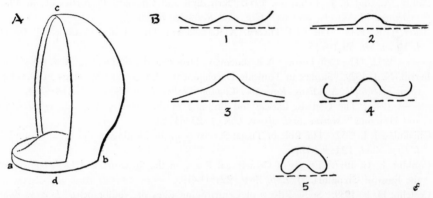

Figure 6-12. Diagrams, showing gastral plates. **A.** Relations of two planes considered in theories of gastrulation. **ab.** Plane of gastral plate. **cd.** Plane through animal and vegetal poles. **B.** Horizontal views of successive positions assumed by gastral plate in starfish *Patiria miniata* during hour following excision. (After Moore.)

fibrils of the tail muscle of the ascidian larva. It should also be noted that fibrillar continuity in the ectoplasmic territory by no means excludes the possibility of a fibrillar system extending far into endoplasmic territory. However this may be, it is significant that although the entire vegetal plate in vivo infolds, commencing at the center, the isolated plate rolls up at the margin to reconstruct a sphere so that the invaginating area is proportionately reduced. Evidently, more is involved than simple contraction, by whatever means, of an epithelial sheet.

In a study of gastrulation in the sea-urchin egg, Dan and Okazaki (1956) hold that the following must be operating: (1) a factor responsible for the initiation of the invagination of the endodermal plate; (2) a process that changes the shape of the constituent cells of the endodermal plate; and (3) the traction exerted by the pseudospodia at the advancing surface. With respect to the second factor, i.e., cells becoming columnar as the plate itself differentiates and then flattening laterally as invagination proceeds, Gustafson and Lenique (1952) suggest that a structural protein is responsible for the stretching.

R E F E R E N C E S   F O R
Cell Combinations

GENERAL

Andrews, G. F. 1897. *The Living Substance*. Boston, Mass.: Ginn and Co.
Bairati, A., and F. E. Lehmann. 1953. "Structural and Chemical Properties of the Plasmalemma of Amoeba proteus." *Exp. Cell Research*. **5**:220-32.
Baker, J. R. 1948. "The Cell Theory: A Restatement, History and Critique. I." *Quart. J. Micro. Sci.* **89**:103-25.
———. 1952. "The Cell Theory: A Restatement, History and Critique. III." *Ibid*. **90**:157-90.
Berrill, N. J. 1929. "Studies in Tunicate Development. I. General Physiology of Development of Simple Ascidians. *Roy. Soc. London Philos. Trans.* Series B. **218**:37-78.
———. 1948. "The Gonads, Larvae and Budding of the Polystyelid Ascidians Stolonica and Distomus." *Marine Biol. Assoc. U.K. J.* **27**:633-50.
Chiakulas. J. J. 1952. "The Role of Tissue Speciality in the Healing of Epithelial Wounds." *J. Exp. Zool.* **121**:383-418.
Conklin, E. G. 1917. "Effects of Centrifugal Force on the Structure and Development of the Eggs of Crepidula." *J. Exp. Zool.* **22**:311-419.
Costello, D. P. 1939. "Some Effects of Centrifuging Eggs of Nudibranchs." *J. Exp. Zool.* **80**:473-99.
Dan, K., and K. Okazaki. 1956. "Cyto-embryological Studies of Sea Urchins. III. Role of the Secondary Mesenchymal Cells in the Formation of the Primitive Gut in Sea Urchin Larvae." *Biol. Bull.* **110**:29-43.
De Bruyn, P. P. H. 1947. "Theories of Amoeboid Movement." *Quart. Rev. Biol.* **22**:1-24.
Drew, A. H. 1923. "Growth and Differentiation in Tissue Cultures." *Brit. J. Exp. Pathol.* **4**:46-52.
Engström, K. 1958. "On the Cone Mosaic in the Retina of Parus major." *Acta Zool.* **39**:65-70.
Fankhauser, G. 1945. "Maintenance of Normal Structure in Heteroploid Salamander Larvae, through Compensation of Changes in Cell Size by Adjustment of Cell Number and Cell Shape." *J. Exp. Zool.* **100**:445-555.
Fawcett, D. W. 1959. "Changes in Fine Structure of the Cytoplasmic Organelles during Differentiation," in D. Rudnick (ed.). *Developmental Cytology*. New York: Ronald Press Co. **2**:161-90.
Feldman, M. 1955. "Dissociation and Reaggregation of Embryonic Cells of the Triturus alpestris." *J. Embryol. Exp. Morphol.* **3**:251-5.
Fell, H. B., and H. Grüneberg. 1939. "The Histology and Self-differentiating Capacity of

the Abnormal Cartilage in a New Lethal Mutation in the Rat (*Rattus norvegicus*)." *Roy. Soc. London Proc.* Series B. **127**:257-77.

Goldacre, R. J. 1952. "The Folding and Unfolding of Protein Molecules as a Basis for Osmotic Work." *Internat. Rev. Cytol.* **1**:135-64.

Gustafson, T., and H. Kinnander. 1956. "Microaquaria for Time-Lapse Cinematographic Studies of Morphogenesis in Swimming Larvae and Observations on Sea Urchin Gastrulation." *Exp. Cell Research.* **11**:36-51.

Gustafson, T., and P. Lenicque. 1952. "Studies on Mitochondria in the Developing Sea Urchin Egg." *Ibid.* **3**:251-74.

Holtfreter, J. 1943. "Properties and Functions of the Surface Coat in Amphibian Embryos." *J. Exp. Zool.* **93**:251-323.

———. 1948. "Significance of the Cell Membrane in Embryonic Processes." *N. Y. Acad. Sci. Trans.* **49**:709-60.

Kasahara, S. 1935. "On the Cultivation in Vitro of the Hypophysis." *Arch. Exp. Zellforsch.* **18**:42-76.

Lash, J. W. 1955. "Studies on Wound Closure in Urodeles." *J. Exp. Zool.* **128**:13-28.

Lehmann, F. E., E. Manni, and A. Bairati. 1956. "Der Feinbau von Plasmalemma und kontraktiler Vacuole bei Amoeba proteus in Schmitt- und Fragment-prapäraten." *Rev. suisse zool.* **63**:246-55.

Lopaschov, G. 1935. "Die Entwicklungsleistungen des Gastrulamesoderms." *Biol. Zentralbl.* **55**:606-15.

Lyall, A. H. 1957. "Cone Arrangements in Teleost Retinae." *Quart. J. Micro. Sci.* **98**:189-202.

Mast, S. O. 1926. "Structure, Movement, Locomotion and Stimulation in Amoeba." *J. Morphol.* **41**:347-425.

Medawar, P. 1947. "Cellular Inheritance and Transformation." *Biol. Rev.* **22**:360-89.

Moore, A. R. 1941. "On the Mechanism of Gastrulation in Dendraster excentricus." *J. Exp. Zool.* **87**:101-11.

———. 1952. "The Process of Gastrulation in Trypsin Embryos of Dendraster excentricus." *Ibid.* **119**:37-46.

———, and A. S. Burt. 1939. "On the Locus and Nature of the Forces Causing Gastrulation in Embryos of Dendraster excentricus." *Ibid.* **82**:159-71.

Moore, M. M. 1932. "On the Coherence of the Blastomeres of Sea Urchin Eggs." *Arch. entwick. Org.* **125**:487-94.

Moscona, A. 1956. "Development of Heterotypic Combinations of Dissociated Embryonic Chick Cells." *Soc. Exp. Biol. Med. Proc.* **92**:410-6.

———. 1957. "The Development in Vitro of Chimeric Aggregates of Dissociated Embryonic Chick Cells." *Nat. Acad. Sci. U. S. Proc.* **43**:184-94.

Noland, L. E. 1957. "Protoplasmic Streaming: A Perennial Puzzle." *J. Protozool.* **4**:1-6.

Pantin, C. F. A. 1923. "On the Physiology of Amoeboid Movement." *Marine Biol. Assoc. U.K. J.* **13**:24-69.

Saxén, L. 1954. "The Development of the Visual Cells: Embryological and Physiological Investigations on Amphibia." *Suomal. Tiedakat.* (*Helsingfors*) *Toim. Annal.* Series 4. **23**:5-93.

Seifriz, W. 1943. "Protoplasmic Streaming." *Bot. Rev.* **9**:49-123.

Shafer, G. D. 1900. "The Mosaic of the Single and Twin Cones in the Retina of Micropterus salmonoides." *Arch. entwick. Org.* **10**:685-91.

Spiegel, M. 1954a. "The Role of Specific Surface Antigens in Cell Adhesion. II. Studies on Embryonic Amphibian Cells." *Biol. Bull.* **107**:149-55.

Studnička, F. K. 1934. "The Symplasmic State of the Tissues of the Animal Body." *Biol. Rev.* **9**:263-98.

Townes, P. L., and J. Holtfreter. 1955. "Directed Movements and Selective Adhesion of Embryonic Amphibian Cells." *J. Exp. Zool.* **128**:53-118.

Weiss, P., and A. Moscona. 1957. "Type-specific Morphogenesis of Cartilages Developed from Dissociated Limb and Scleral Mesenchyme in Vitro." *J. Embryol. Exp. Morphol.* **6**:238-46.

Whaley, W. G. 1959. "Dynamics of Cell Ultrastructure in Development and Growth." *Sci.* **130**:1425-6.

Whong, S. H. 1931. "On the Formation of Cell Bridges in the Development of the Egg of Dendraster excentricus." *Protoplasma.* **12**:123-8.

SLIME MOLDS: DICTYOSTELIUM AND POLYSPONDYLIUM

Arndt, A. 1937. "Untersuchungen über Dictyostelium mucoroides Brefeld." *Arch. entwick. Org.* **136**:681-747.

Bonner, J. T. 1944. "A Descriptive Study of the Development of the Slime Mold Dictyostelium discoideum." *Am. J. Bot.* **31**:175-82.

——. 1947. "Evidence for the Formation of Cell Aggregates by Chemotaxis in the Development of the Slime Mold Dictyostelium discoideum." *J. Exp. Zool.* **106**:1-26.

——. 1957. "A Theory of the Control of Differentiation in the Cellular Slime Mold." *Quart. Rev. Biol.* **32**:232-46.

——. 1959. *The Cellular Slime Molds.* Princeton, N. J.: Princeton Univ. Press.

——, and M. S. Adams. 1958. "Cell Mixtures of Different Species and Strains of Cellular Slime Molds." *J. Embryol. Exp. Morphol.* **6**:346-56.

Bonner, J. T., and M. K. Slifkin. 1949. "A Study of the Control of Differentiation: The Proportions of Stalk and Spore Cells in the Slime Mold Dictyostelium discoideum." *Am. J. Bot.* **36**:727-34.

Gezelius, K., and B. J. Rånby. 1957. "Morphology and Fine Structure of the Slime Mold Dictyostelium discoideum." *Exp. Cell Research.* **12**:265-89.

Gregg, J. H. 1956. "Serological Investigations of Cell Adhesion in the Slime Molds, Dictyostelium discoideum, Dictyostelium purpureum and Polyspondylium violaceum." *J. Gen. Physiol.* **39**:813-20.

Harper, R. A. 1929. "Morphogenesis in Polyspondylium." *Torrey Bot. Club Bull.* **56**:227-58.

——. 1932. "Organization and Light Relations in Polyspondylium." *Ibid.* **59**:49-84.

Olive, E. W. 1902. "Monograph of the Acrasieae." *Boston Soc. Nat. Hist. Proc.* **30**:451-513.

Raper, K. B. 1940. "Pseudoplasmodium Formation and Organization in Dictyostelium discoideum." *Elisha Mitchell Sci. Soc. J.* **56**:241-82.

——. 1941. "Developmental Patterns with Slime Molds." *Growth* Supp. (Symp. 3) **5**:41-76.

——. 1951. "The Isolation, Cultivation and Conservation of Simple Slime Molds." *Quart. Rev. Biol.* **26**:169-90.

——, and D. I. Fennell. 1952. "Stalk Formation in Dictyostelium." *Torrey Bot. Club Bull.* **79**:25-51.

Raper, K. B., and C. Thom. 1941. "Interspecific Mixtures in the Dictyosteliaceae." *Am. J. Bot.* **28**:69-78.

Shaffer, B. M. 1956. "Acrasin, the Chemotactic Agent in Cellular Slime Moulds." *J. Exp. Biol.* **33**:645-57.

——. 1957. "The Properties of Slime-Mould Amoebae of Significance for Aggregation Integration in Aggregating Slime Moulds." *Quart. J. Micro. Sci.* **98**:377-406.

——. 1958. "Integration in Aggregating Slime Moulds." *Ibid.* **99**:103-21.

Sussman, M. 1955. "Developmental Physiology in the Ameboid Slime Mold," in S. H. Hutner and A. Lwoff (eds.). *Biochemistry and Physiology of the Protozoa.* New York: Academic Press, Inc. **2**:201-23.

SLIME MOLDS: PHYSARUM

Camp, W. G. 1936. "A Method of Cultivating Myxomycoete Plasmodia." *Torrey Bot. Club Bull.* **63**:205-10.

——. 1937. "The Fruiting of Physarum polycephalum in Relation to Nutrition." *Am. J. Bot.* **24**:300-3.

Goldacre, R. J., and I. J. Lorsch. 1950. "Folding and Unfolding of Protein Molecules in Relation to Cytoplasmic Streaming, Amoeboid Movement and Osmotic Work." *Nature* (London). **166**:497-500.

Kamiya, N., H. Nakajima, and S. Abe. 1957. "Physiology of the Motive Force of Protoplasmic Streaming." *Protoplasma.* **48**:94-112.

Loewy, A. G. 1952. "An Actomysin-like substance from the Plasmodium of a Myxomycoete." *J. Cell. Comp. Physiol.* **40**:127-56.

Moore, A. R. 1935. "On the Significance of Cytoplasmic Structure in Plasmodium." *J. Cell. Comp. Physiol.* **7**:113-29.

Seifriz, W. 1952. "The Rheological Properties of Protoplasm," in A. Frey-Wyssling (ed.). *Deformation and Flow in Biological Systems.* New York: Interscience Pubs., Inc. **1**:3-156.

Winer, B. J., and A. R. Moore. 1941. "Reactions of the Plasmodium Physarum polycephalum to Physico-chemical Changes in the Environment." *Biodynamica.* **3**:323-45.

SPONGES

Brien, P. 1932. "Contribution à l'étude de la régénération naturelle chez les Spongillidae." *Arch. zool. exp. gén.* **74**:461-506.

——. 1937. "La réorganisation de l'Éponge après dissociation par infiltration et phénomènes d'involution chez Ephydatia fluviatilis." *Arch. biol.* **48**:185-268.

——, and H. Meewis. 1938. "Contribution à l'étude de l'embryogénèse des Spongillidae." *Ibid.* **49**:177-250.

Brønstedt, H. V. 1936. "Entwicklungsphysiologische Studien über Spongilla lacustris (L.)." *Acta Zool.* **17**:75-172.

Fauré-Fremiet, E. 1931. "Étude histologique de Ficulina ficus L. (Demospongia)." *Arch. d'anat. micro. morphol. exp.* **27**:421-48.

——. 1932a. "I. Morphogénèse expérimentale (reconstitution) chez Ficulina ficus L." *Ibid.* **28**:1-80.

Fauré-Fremiet, E. 1932b. "II. Involution expérimentale et tension de structure dans les cultures de Ficulina ficus." *Ibid*. Pp. 121-57.

Galtsoff, P. S. 1925. "Regeneration after Dissociation (An Experimental Study of Sponges). II. Histogenesis of Microciona prolifera." *J. Exp. Zool.* **42**:223-55.

Huxley, J. S. 1912. Some phenomena of regeneration in Sycon. *Phil. Trans. Roy. Soc. London*. Ser. B. **202**:165-189.

Ijima, I. 1901. "Studies on the Hexactinellida. Contribution I (Euplectellidae)." *Tokyo Univ. Fac. Sci. J*. Series 4. **15**:1-299.

Okada, Y. K. 1928. "On the Development of a Hexactinellid Sponge, Farrea sollasii." *Tokyo Univ. Fac. Sci. J*. Series 5. **2**:1-27.

Rio-Hortega, P., and F. Ferrer. 1917. "Contribución al conocimiento histólogica le las Esponjas." *Soc. españ. hist. nat. bol.* **17**:1-41.

Spiegel, M. 1954b. "The Role of Specific Surface Antigens in Cell Adhesion. I. The Reaggregation of Sponge Cells." *Biol. Bull.* **107**:130-48.

Wilson, H. V. 1894. "Observations on the Gemmule and Egg Development of Marine Sponges." *J. Morphol.* **9**:277-406.

————. 1907. "On Some Phenomena of Coalescence and Regeneration in Sponges." *J. Exp. Zool.* **5**:245-258.

————. 1910. "A Study of Some Epithelioid Membranes in Monaxid Sponges." *Ibid*. **9**:537-77.

————. 1935. "Some Critical Points of the Halichondrine Sponge Larva." *J. Morphol.* **58**:287-345.

————. 1938. "Behavior of the Epidermis of Sponges (Microciona) When Treated with Narcotics or Attacked by Aquarium Degeneration." *J. Exp. Zool.* **79**:243-69.

————, and J. T. Penney. 1930. "The Regeneration of Sponges (Microciona) from Dissociated Cells." *Ibid*. **56**:73-148.

ZOOTHAMNION

Sumners, F. M. 1938a. "I. Some Aspects of Normal Development in the Colonial Ciliate Zoothamnion alternans." *Biol. Bull.* **74**:117-29.

————. 1938b. "II. Form Regulation in Zoothamnion alternans." *Ibid*. Pp. 130-54.

————. 1941. "The Protozoa in Connection with Morphogenetic Problems," in G. N. Calkins and F. M. Sumners (eds.). *Protozoa in Biological Research*. New York: Columbia Univ. Press. Pp. 772-817.

VOLVOX

Dobell, C. 1932. *Antony Van Leeuwenhoek and His "Little Animals."* New York: Harcourt, Brace and Co., Inc.

Iyengar, M. O. P. 1933. "Contributions to Our Knowledge of the Colonial Volvocales of South India." *Linnean Soc. London J*. **49**:323-73.

Janet, C. 1912. *Le Volvox*. Limoges. 151 pp.

————. 1922. *Le Volvox*. Mém. 2. Paris. 66 pp.

————. 1923. *Le Volvox*. Mém. 3. Paris. 179 pp.

Pocock, M. A. 1933a. "Volvox and Associated Algae from Kimberley." *S. Afr. Mus. Ann.* **16**:473-521.

――――. 1933b. "Volvox in South Africa." *Ibid*. Pp. 523-45.

――――. 1938. "Volvox tertius Meyer." *Quekett Micro. Club J*. Series 4. **1**:1-25.

Rich, F., and M. A. Pocock. 1933. "Observations on the Genus Volvox in South Africa." *S. Afr. Mus. Ann*. **16**:427-71.

Smith, G. M. 1944. "A Comparative Study of the Species of Volvox." *Am. Micro. Soc. Trans*. **63**:265-310.

Zimmerman, W. 1921. "Zur Entwicklungsgesichte und Zytologie von Volvox." *Jahrb. wiss. bot*. **60**:256-94.

PART II
GROWTH AND FORM
IN METAZOA

The phenomena and problems of growth and form in animals are presented in this part almost exclusively in relation to asexual and reconstitutive development in invertebrates. The purpose of the brief introductory chapter, contrasting sexual and asexual reproduction in bryozoans, is to emphasize that the same morphological outcome can be reached by very different paths depending upon initial circumstances; in these examples the outcome is total development from comparably simple beginnings in each example. Following the introduction is a discussion of the processes of growth and development in nonsexual tissues of hydroid, turbellarian, annelid, and tunicate forms. This discussion culminates in various observations relating to growth, development, and pattern; the conclusions, which are stated in biological terms, are essentially observational rather than interpretative— that is, they represent observed rules of developmental behavior which require explanation.

Chapter 7

Introduction: Sexual and Asexual Development

METAZOAN organisms in many forms develop either from eggs or from somatic buds: the individuals resulting from eggs are free; those from buds are organically joined, constituting a colonial organism. Thus the development from eggs and that from buds are two decidedly different paths leading to essentially the same end product, although colonial organisms exhibit patterns of growth not characteristic of nonbudding organisms.

Because the Bryozoa offer perhaps more diverse developmental and growth phenomena than do other budding forms, they are an introduction to many of the problems of growth and development in other organisms. As a group, however, Bryozoa is generally considered to be diphyletic, as Brien and Papyn (1954) maintain, so that the Endoprocta and Ectoprocta should not be discussed synonymously although the two groups, despite separate lines of descent, exhibit similar developmental features.

Both Endoprocta and Ectoprocta produce small eggs which cleave in a highly determinate manner to form specialized and in certain ways precociously differentiated larvae which later metamorphose into adults. The specialized character of the larva in each group calls for analysis of the egg development that leads to its formation, and the specialized character of the egg itself leads to a subsequent metamorphic transformation. These eggs therefore raise virtually all the problems of classical embryology (discussed briefly at the end of this book). On the other hand, at least one genus of the Ectoprocta, namely, *Crisia*, forms eggs that undergo polyembryonic fission, each egg and primary embryo producing numerous secondary embryos that, as Harmer (1893 and 1929) found, develop relatively simply and directly, i.e., without passing through a specialized larval stage. Development of this sort raises the question of the nature of eggs in

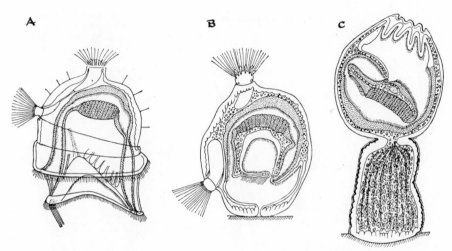

Figure 7-1. Larval stages in *Pedicellina cernua*. **A.** Swimming larva. **B.** Fixed larva. **C.** Metamorphosed larva. (After Brien and Papyn.)

general, of the significance of their specialized features, if any, and of the developmental processes common to all (see pp. 468-482).

The immediate interest lies in the type of development described by Harmer (1893) in *Crisia,* in which unspecialized cell clusters undergo direct development to form the complete zooid. In this genus and probably in others of the same order, for example, *Lichenophora,* the single ovum of an individual (polypide or zooid) divides into several blastomeres within a mass of follicle cells which provide nourishment and also separate the blastomeres from one another. The blastomeres continue to grow as well as divide, and subsequently recombine as a single mass embedded and growing in follicular tissue. This mass, the so-called primary embryo, becomes progressively more irregular in outline as the embryo grows. The irregular protrusions separate from the parental mass and become so-called secondary embryos, some of which may produce separate masses of embryonic cells. Altogether the original ovum can produce as many as 100 secondary embryos, and it is evident that such embryos cannot have inherited a specific organization, even if the original egg had one. Clearly, these embryos—primary or secondary—are clusters of unspecialized, rapidly growing, and dividing cells, whose developmental course is determined by the innate potentiality of the unspecialized cell itself.

Accordingly, it is significant that secondary embryonic masses as they form from the primary mass are almost uniform in size and during growth attain approximately the same extent. Each small mass acquires an outer and inner cell layer, as does a gastrula, by a means not yet clearly observed. The later

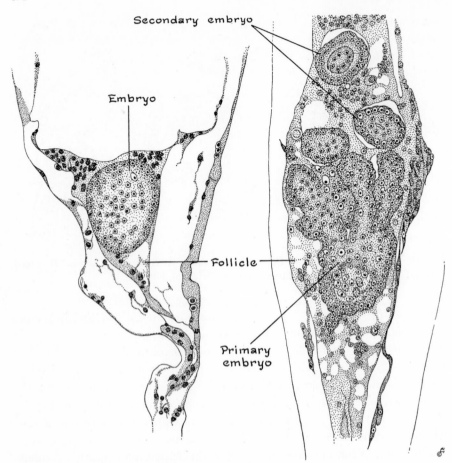

Figure 7-2. Primary and secondary embryos of *Crisia,* showing irregular growth of primary mass and process of budding. (After Harmer.)

development is virtually unknown, except that proliferating egg fragments first form a didermic vesicle, which is capable of developing into the whole organism in its multitudinous entirety, and that this form of development is essentially the same as the formation and development of the somatic buds. The somatic buds of the ectoprocts resemble not only the secondary embryos of *Crisia,* except for the former's incorporation in the parental body wall, but essentially the secondary embryos of the endoprocts, although the patterns of bud production and the detailed structure of the emergent zooids are not distinctive.

The direct and total development of complex organisms from somatic tissue may involve initially so few cells, themselves of so simple a histological character, that the phenomenon is regarded as a complete development of nongametic origin, for the process seems to be far more comprehensive than what is generally thought of as regeneration or reorganization.

Total Development in Somatic Buds

Endoprocta

The origin and development of buds in endoproct bryozoans have been described with the greatest precision by Brien (1956-57) for *Pedicellina cernua*. His account confirms in all essentials the older one of Seeliger (1890b).

In this species the stolon, bearing the individual zooids, is not continuous, but segmented by septal diaphragms which are equally spaced from one another. Each zooid belongs to a so-called fertile segment, which is separated from the next zooid-bearing segment by an infertile one, without a zooid. The tip of the stolon is the point of growth for the successive formation of young blastozooids, which are accordingly arranged in single file in order of size. The distal extremity of the stolon actually constitutes the youngest zooid. On the oral side

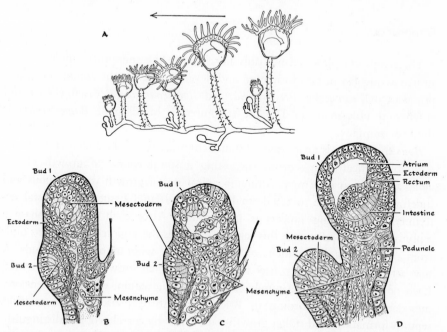

Figure 7-3. Colony in *Pedicellina cernua*. **A.** Growth terminals and developing zooids. **B–D.** Development of terminal buds from mesectoderm. (After Brien.)

of the zooid are two growth centers: a basal area, which grows out and enlarges as the stolon; and an area immediately below the developing calyx, or chalice, which gives rise to the next new zooid.

At their inception the two growth regions of the blastogenic outgrowth are histologically similar. The ectoderm thickens: the cells become columnar and more basophile, proliferate, and become pressed against one another, forming a lenticular thickening. At the blastogenic tip the ectodermal layer thickens anteriorly to produce an ectoblastic nodule whose cells, high and proliferating, soon constitute a mass which becomes surrounded by subjacent mesenchymal cells. This ectoblastic mass segregates from the ectodermal epithelium which now covers it distally and gives rise directly, by cavitation, expansion, and folding, to the organs of the calyx. The basal growth region repeats the entire growth process; i.e., it gives rise both to a new apical zooid and to new basal growth of the stolon.

The morphogenetic process itself is remarkable, first because the essential developmental mass is solely of ectodermal origin, as in the case of the more-typical entocodal medusa buds (p. 170), and second because zooid structure develops very directly. The unity of the initial ectodermal growth and the subsequent expansive differentiation of the ectodermal mass are striking.

Ectoprocta

The ectoprocts, although probably unrelated to the endoprocts, undergo very similar sequences in bud formation and development. Of the accounts of these processes in *Plumatella, Fredericella,* and *Bowerbankia* by Brien (1936) and Brien and Huysmans (1937), the stoloniferous budding of *Bowerbankia* is the least complex.

Bowerbankia buds arise some distance from the tip of the stolon and assume a double helical arrangement, suggesting a spiral course of growth of the elongating stolon extremity. Although apical stolon growth is continuous and indefinite, giving rise to ectodermal and mesodermal tissue at the distal extremity, the budding process is intermittent. After each budding phase a septum forms in the stolon; this septum separates the region that contains buds from the distal stolon, which continues to grow without producing buds. A new set of buds arises only after the distal stolon has grown a certain amount. Each new bud is at first a purely ectodermal evagination, which later encloses mesodermal cells. According to Brien, these three phases—bud formation, septum formation, and stolon growth—are evidently correlated, but distinguishable from one another. He also emphasizes that these morphogenetic processes

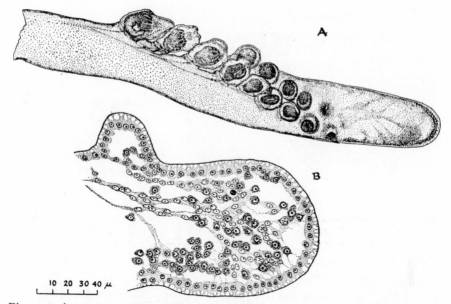

Figure 7-4. Stolon in *Bowerbankia*. **A.** Pustule with double helical row of buds. **B.** Tip of stolon and youngest bud, showing epidermal (ectodermal) initiation. (After Brien and Huysmans.)

should be susceptible to experimental studies in vivo because live material is particularly abundant, resistant, and transparent.

The budding process is essentially the same in *Bowerbankia, Plumatella,* and *Fredericella,* although the region of formation varies; and in all three, the bud in its initial phase is a didermic ectomesodermal mass. This mass results from the proliferation of ectodermal cells at the blastogenic site, cells which are markedly columnar and basophile and have a large nucleus and nucleolus. The ectoblastic mass becomes covered internally by large mesodermal cells, and subsequently a cavity develops within the ectoblastic mass, though not by invagination: as the bud enlarges and becomes ovoid, the cells of the ectoblastic mass become arranged irregularly to form a double internal epithelium, whose two layers are separated by a slit which progressively becomes a cavity. The bud thus attains the closed didermic vesicle stage. In later development the inner ectodermal layer gives rise to all the organs—primarily the digestive canal—of the new individual; the outer mesodermal layer constitutes only the mesodermal tissue.

Plumatella and *Fredericella* (i.e., phylactolematous genera as distinct from the stoloniferous genus *Bowerbankia*), buds arise as a set of three so-called adventitious polypodial buds from the anterior body wall and as a series of

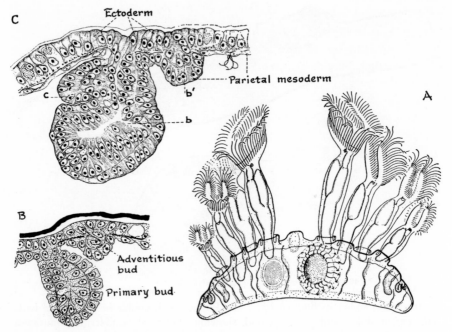

Figure 7-5. Colony in *Plumatella*. **A.** Transverse section, showing common cystid cavity with two statoblasts and polypide (zooid) in regression at center. On each side new zooids form in sequence from marginal epidermis. **B.** Two early stages in formation of bud from epidermis. **C.** Later stage, showing doubling of primary bud **b–c** and growth of adventitious bud **b¹**. (After Brien.)

cystid buds from the funicular strand. In both types of genera, however, the inner vesicle layer is of ectodermal origin. The second and third adventitious buds are formed, by a doubling process, from the neck of the first bud formed.

The morphogenetic phenomena, essentially similar to those in endoprocts, present the same problems: the nature of bud initiation at a prescribed site, the basis of a bud's unity of growth, and the progressive elaboration of pattern as expansion proceeds.

Brien (1936) also made a few cutting experiments on cystid zooids of *Fredericella*. When the lophophore is amputated, the polyp (polypide) grows a new one. When the lophophore is cut off and the digestive canal detached, a new polyp regenerates within the cicatrized cystid from the region corresponding to the site of attachment of the original esophagus. Then in the cicatrized cystid with its newly constituted polyp, an adventitious bud appears at the normal place.

In other words, not only can the whole zooid be reconstituted from a minute cluster of epidermal cells (as in the development of, for example, the adventitious bud) and a lost part regenerated by the remaining part of the organ system (as in lophophore amputation) but the greater part of the whole, i.e., the polypide, can be regrown within the body chamber from a small local region. Further, reconstitution of the entire anterior end, after transection, includes the precisely determined distal budding territory, which is therefore evidently as definite and integrated a part of the total pattern as is any other structural feature. When the diverse types of development, the variable patterns of budding, the capacity for regeneration, the many types of colonies produced, the differing life cycles, and—not least—the general accessibility of both marine and fresh-water bryozoans are considered, it is surprising that bryozoans have not been used as experimental material to a greater extent. On the other hand, the brief survey of this introduction to growth and form in metazoa barely does justice to Brien's meticulous descriptions of the histological processes basic to any understanding of morphogenesis.

REFERENCES FOR
Introduction

Brien, P. 1936. "Contribution à l'étude de la reproduction asexuée des Phylactolemates." *Mél. Pelseneer. Mem. Mus. Roy. Hist. Nat. Belgique.* Series 2. **2**:569-625.

———. 1954. "A propos des Bryozoaires Phylactolemates." *Bull. Soc. Zool. France.* **79**:203-39.

———. 1956-1957. "Le bourgeonnement des Endoproctes et leur phylogénèse." *Ann. Soc. Roy. Zool. Belgium.* **87**:27-43.

———, and G. Huysmans. 1937. "La croissance et le bourgeonnement du stolon chez les Stolonifera (Bowerbankia)." *Ibid.* **68**:13-40.

Brien, P., and L. Papyn. 1954. "Les Endoproctes et la classe des Bryozoaires." *Ibid.* **85**:59-87.

Harmer, S. F. 1893. "On the Occurrence of Embryonic Fission in Cyclostomatous Polyzoa." *Quart. J. Micro. Sci.* **39**:199-242.

———. 1929. "Polyzoa." *Linnean Soc. London Proc. 1928-29.* Pp. 69-118.

Seeliger, O. 1890a. "Bemerkungen zur knospenentwicklung der Bryozoen." *Zeitschr. wiss. Zool.* **50**:560-99.

———. 1890b. "Die ungeschlechtliche Vermehrung der Endoprokten Bryozoen." *Ibid.* **49**:168-208.

Chapter 8

Pattern and Growth: Development of Pattern

THE PATTERN OF development of a multicellular organism is generally one of three types: (1) primarily radial; (2) predominantly bilaterally symmetrical; or (3) essentially repetitive, as in segmented animals, colonial organisms, and herbaceous plants. In some ways the radial organization offers the most direct analytical approach, particularly in coelenterates, whose histology is comparatively simple. In the present discussion the medusa is considered to be the adult archetype, with other types either juvenile or secondary. As such, it appears in two groups very different from each other, the Scyphomedusae and Hydromedusae. On the other hand, polymorphism is striking in hydromedusan life cycles, so that many developmental variables can be studied in living material, without resort to experimental interference.

Developmental Analysis of Medusae

Scyphomedusae

The large, oceanic jellyfish have long attracted attention, but Dalyell (1834) and Sars (1841) were the first to observe the origin of the free-swimming medusae from a small attached polyp, or scyphistoma; a somewhat later (1862) account of this type of origin, by Agassiz, is still one of the most outstanding. The eggs of certain scyphomedusans develop directly into medusae; those of others never do so, and in some the developmental course is dependent on the size of the egg, which is variable. Eggs of the sessile medusae *Haliclystus* and *Lucernaria* are only 0.03 mm in diameter; of the permanently pelagic *Pelagia*, 0.30; of *Cyanea*, 0.12-0.15; and of *Aurelia*, 0.15-0.23 (Berrill, 1949a). To a great extent, the size of the egg determines the manner of gastrulation, which in turn sets the later course of development. In general the smallest eggs gastrulate by unipolar ingression of cells, filling the blastocoele; the largest eggs, by invagination alone; and the intermediate eggs, by a combination of invagination and ingression in varying degree according to species. In light of the fact that there has been considerable disagreement about the mode of gastrulation in *Aurelia*, it is significant that its egg is the most variable in size.

The variable nature of the gastrulation process, in view of its correlation with egg size, may arise from the varying ratio not only of the diameter of the blastula to the thickness of the blastular wall but also of the extent of the presumptive endoderm to the presumptive ectoderm. Whatever the cause, there is a graded series of gastrulae, ranging from the peculiar *Haliclystus* embryo, with its single column of ingressed endodermal cells, derived from an exceptionally minute egg; to the somewhat larger eggs of *Mastigias, Cotylorhiza, Cyanea,* and *Chrysaora,* whose relatively solid ingressed endoderm later develops a coelenteron by splitting; to the relatively large eggs of *Nausithoë, Linuche,* and *Aurelia,* which invaginate in the grand manner; and finally to the large eggs of *Pelagia* and the largest eggs of *Aurelia,* which gastrulate not only by invagination but in such a way that the invaginated endoderm fills no more than half of the segmentation cavity. In the species with the largest eggs, the egg develops directly into a young medusa (ephyra) and in *Haliclystus,* the smallest-egg species, into a larva too small to form even a polyp directly. How this minute larva is able to grow into a relatively enormous polyembryonic mass is an as-yet unanswered question.

The main question is whether the manner of gastrulation in any way determines the subsequent course of development, for there is no doubt that there is a correlation between them; and the problem can be approached from both the development of the polyp and the direct development of the medusa.

In the development of all but the largest eggs, whether gastrulation is by unipolar ingression or invagination, the endodermal tissue fills the blastocoele, and the inner end of the endodermal ingrowth presses against the ectoderm at the opposite pole; the almost spherical gastrula then elongates in that direction. The process of elongation along the polar axis seems to be caused either by the mechanical extension of the gastrula because of the rapidly growing endodermal ingrowth or possibly by an actual contagion of growth between the advancing region of the endoderm and the part of the ectoderm with which it comes in contact. Once such polar growth, or extension, becomes established, the original blastoporal end inevitably remains narrow, and direct development into the medusoid ephyra is apparently no longer possible. Whether this interpretation is valid or not, it is a fact that an elongated embryo is formed and a

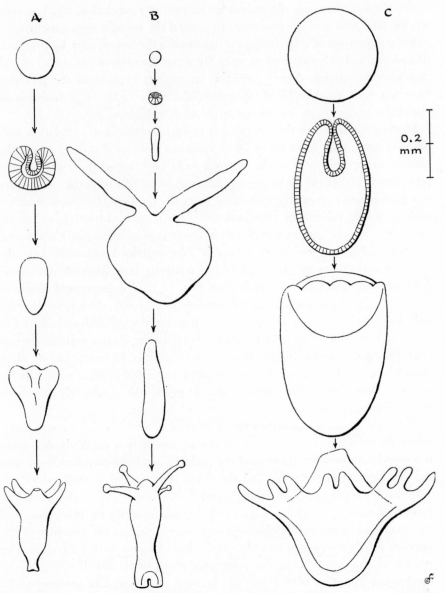

Figure 8-1. Correlation of egg size and development in Scyphomedusae. **A.** Egg, gastrula, planula, and polyp in *Aurelia*. (After Hyde.) **B.** Egg, gastrula, planula, polyembryonic stage, secondary planula, and polyp in *Haliclystus*. (After Wietrzy-skowi.) **C.** Egg, gastrula, and developing medusa-ephyra in *Pelagia*. (After Delap.)

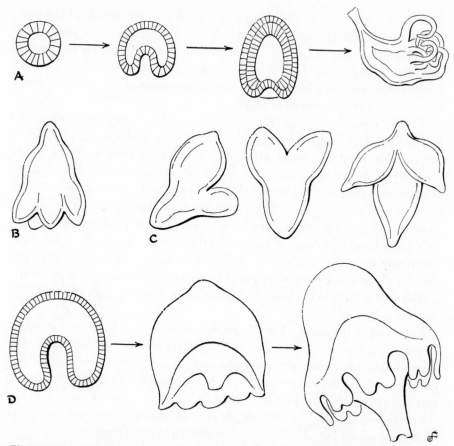

Figure 8-2. Variability of development in *Aurelia*. **A.** Egg, gastrula, planula, and polyp in small-egg type. **B.** Free-swimming actinula with 4 tentacle lobes. **C.** Large planular larva producing buds. **D.** Direct transformation of gastrula into medusa in large-egg type. (After Haeckel.)

polyp develops whenever the invaginated, or ingressed, endoderm fills the blastocoele and comes into close contact with the adjoining ectoderm.

This embryo and polyp formation does not occur when a gastrula develops directly into the medusa stage. In the development of all *Pelagia* eggs and of the largest *Aurelia* eggs, gastrulation is by invagination, although by itself the manner of gastrulation is not distinctive. In these eggs and apparently in the comparatively large eggs of the trachymedusan genera *Geryonia* and *Aglaura,* which develops directly into medusae, the invaginating endodermal area does not increase in proportion to the size of the blastula as a whole. Consequently, the endoderm not only fails to make contact with the inner surface of the ectoderm but remains separated from it by an extensive blastocoele. Under these circumstances the embryo does not elongate to any great extent along its polar axis; instead, the blastoporal region progressively increases in diameter, becoming significantly wider than the egg or blastula. The 8 lappets of the ephyra then develop as evenly spaced extensions of the ectoderm where circumference of the embryo is widest.

Whether these developmental circumstances and outcomes can be described in terms of cause and effect is possible, but at present debatable. Whether or not the elongation in a small blastula to the planular form is caused by the relatively extensive invagination or ingression of endodermal ingrowth, the result, when invagination occurs, is that the blastoporal region narrows and the consequent inadequate quantity of available tissue inhibits lappet formation, so that a polyp with simple tentacles is all that can be produced. Conversely, in a large blastula, failure of the endoderm to grow as rapidly as, or more extensively than, the ectoderm precludes the elongation that might have produced a planula and subsequent polyp formation; furthermore, the alternative endodermal growth in the horizontal plane so increases the girth of the embryo that the ephyra is formed.

Regardless of the course of development—resulting in either a pelagic ephyra or a sessile polyp—growth continues: The ephyra becomes the large and more complex jellyfish, retaining the original pattern of 8 sensory lappets but progressively adding filiform tentacles in the expanding bell margin between each pair of lappets—tentacles that are essentially similar to those of the polyp; and the polyp becomes the scyphistoma, not only increasing in size and in number of distal tentacles but also becoming progressively more elongate.

As a rule, the polyp eventually transforms into ephyrae by strobilation, but at this early stage in development it is decidedly a polyp, by function as well as appearance. When the trunk of the polyp is sectioned at various horizontal levels or in vertical strips, typical regeneration occurs, with the familiar gradation in capacity of different horizontal levels to form distal or pedal structure and with the occurrence of bipolar forms from narrow vertical strips; but, as Gilchrist (1937) confirmed, reconstitution of either type of section is of the polyp. Furthermore, polyps of various scyphomedusans produce lateral, initially small and slender outgrowths from the wall of the lower part of the trunk; each outgrowth usually gives rise to a new individual, which invariably becomes a polyp. The polyp pattern or organization, once it exists, is accordingly "built in," as if it were permanent, and is established in a dynamic, self-perpetuating, and self-restoring form.

A scyphistoma eventually transforms into one or more ephyrae; i.e., it may be monodisk or polydisk. The scyphistomae of *Cassiopea* and *Cotylorhiza* are

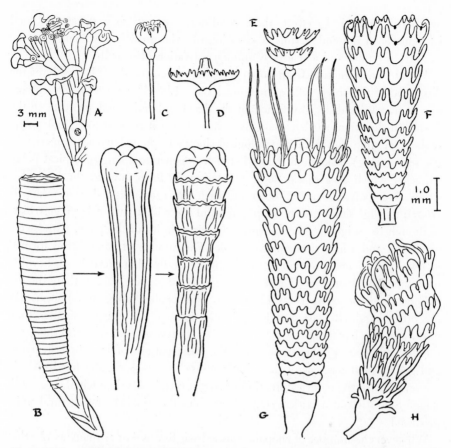

Figure 8-3. Monodisk and polydisk strobilation. (All but **A** drawn to same scale.)
A. Scyphistomae with budding ring. **B.** Annulated theca, prestrobila, and strobila in *Nausithoë*. **C.** Monodisk strobila in *Cotylorhiza*. **D.** Monodisk strobila in *Cassiopeia*. **E–F.** Strobilae in *Cyanea*. **G–H.** Strobilae in *Aurelia*.

usually monodisk: of *Nausithoë, Aurelia,* and *Chrysaora,* usually polydisk; and of *Cyanea,* either monodisk or polydisk. Interest centers around the conditions initiating strobilation, the relation of monodisk to polydisk organisms, and the morphology and histology of the metamorphosis.

All investigators agree that a long nutritive preparation, similar to that which usually precedes gonadogenesis in *Hydra,* is necessary for strobilation, although in *Hydra* the preparatory nutrition leads to a form of asexual reproduction. As a rule, the process depends on the following conditions in sequence: (1) a heavy diet of suitable food for usually several months; (2) a definite, low critical temperature, usually the winter minimum; and (3) a good supply of food and oxygen at the critical period (for detailed discussion and references, see Berrill, 1949a). In effect, these conditions amount to a storage of food particularly in the endodermis, a cessation of axial growth, a widening of the column or trunk, and a relatively sudden activation of growth.

Actual transformation from polyp to ephra structure takes place only at the distal end of the polyp, where the tentacles are. Elsewhere a process of segregation of transverse disks commences at the distal end and slowly proceeds in the distobasal direction, so that each disk ceases to be part of the original polyp. If there were no significant physiological change in the polyp, each constricting (or segregating) disk might be expected to regenerate a ring of typical polyp tentacles, which as a matter of fact occasionally form, toward the base of a strobilating polyp. Usually, however, each emergent disk develops lappets around its edge similar to the lappets around the widest part of large gastrulae, as though the original, inherent protoplasmic polarity of the polyp tissue were wiped out and a new start made possible. The number of lappets thus formed is usually, but not invariably, the eight generally characteristic of Scyphomedusae. The first 1 or 2 disks segregated are usually larger than the remainder, so that as many as 12 lappets, or rhopalial lobes, may appear. On the other hand, the last and lowest disks isolated are usually smaller than most, and the lappets formed may number as few as 4. Evidently, then, the number of primary lobes, which sets the basic pattern of marginal structure and gastrovascular canals of the adult medusa, is not genetically fixed directly but only indirectly, by determination of size at certain developmental or growth stages. Size itself at a critical stage determines the number of structural units produced.

Whatever the number of rhopalia formed—as few as 4 or as many as 12—that number persists throughout subsequent development and determines the structural pattern of the full-grown medusa. Together with topographically associated canals, the rhopalia reflect the primary radial organization. Uchida (1926) found that the medusoid disk as a whole continues to grow interradially, so that the marginal growth centers are situated midway between each pair of rhopalia, with new tissue and structure forming continuously from either side of each center.

The pattern, once established, is fixed; for such medusae readily regenerate missing territory, including rhopalia, always to reconstruct the original form. Child (1933) witnessed this kind of regeneration in the sessile medusa *Haliclystus,* which regenerates the whole set of marginal tentacles and rhopalia. On the other hand, as already described, a scyphistoma transected in various planes always reconstitutes a scyphistoma. Thus the specific character of regeneration is determined by the structural characteristics of the surviving tissue at the time, not by potential qualities.

Hydromedusae

Development of Medusae

The asexual development of medusae in Hydromedusae is in sharp contrast to that in Scyphomedusae, particularly in the three hydromedusan orders Limnomedusae, Anthomedusae, and Leptomedusae. In these orders medusae never develop directly from eggs, but from outgrowths of the manubrium, radial canals, bell margin, and tentacle bases of other medusae or from the stolon, stem, gonangial, or hydranth wall in hydroids; the particular nature of the local growth process determines whether a medusa or some other structure is produced.

As an introduction to hydromedusan development in general, the process in *Limnocnida tanganyicae,* the common fresh-water medusa of southern Africa—fully described by Bouillon (1955a and b)—is discussed here. In this species, medusae form from the wall of the manubrium of the medusa, under certain conditions, and as an outgrowth from the minute, sessile hydroid. The following description of the histology of development applies to the former mode, and there is no evidence that the histology of the latter is significantly different.

The first step in the formation of a medusa bud is the evagination of the manubrial wall at the margin of a budding zone, including both epidermis and endodermis (the evagination closely resembles the initial stage in a *Hydra* bud). The evagination occurs in a region of the manubrium that is very deficient in interstitial cells, and the cells which take part in blastogenesis result from dedifferentiation of the epithelial epidermal and endodermal cells—if dedifferentiation is possible in such generalized cell types.

Within this evaginated unit, only a sharply defined region, limited to the tip, undergoes blastogenesis; only in the tip do intracellular changes occur: the cells of both layers lose their vacuoles and acquire condensed and highly basophile cytoplasm.

The apical epidermis then thickens, becomes many-layered, and proliferates to form a massive epidermal nodule, or button. By contrast, the endodermal cells multiply less prolifically and become a plate under the epidermal mass. The epidermal mass thus interposed between apical peripheral epidermis of the bud and the adjoining endodermis is the entocodon, which seems to be the essential and primary structure required for medusa development in Hydromedusae.

The entocodon invaginates to form a cavity that eventually becomes the subumbrellar cavity; and the endodermal cup now enveloping the epidermal mass grows distally to form 4 (rarely, 5) two-layered strands, arranged as a

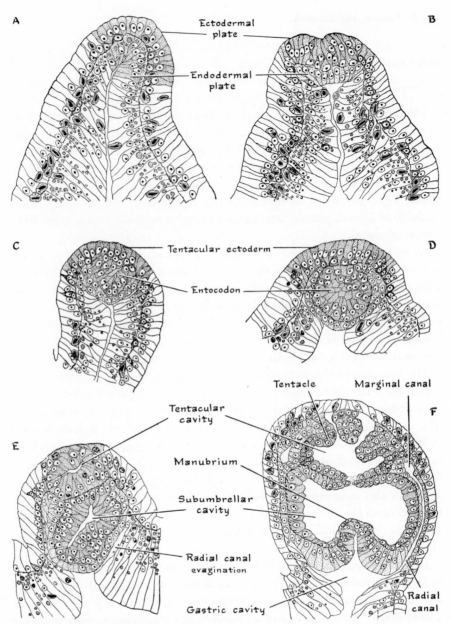

Figure 8-4. Organogenesis of medusa bud in *Limnocnida*. **A.** Apical region of bud evagination with highly basophile ectoderm and endoderm. **B.** Ectodermal cells proliferating to form thick plate. **C.** Origin of ectodermal nodule, or ectocodon. **D.** Entocodon with apical tentacular ectoderm, and origin of radial canals from endodermal cup. **E.** Appearance of subumbrellar cavity within entocodon and of tentacular cavity within apical ectoderm, and radial canal evaginations. **F.** Medusa bud almost fully formed, showing manubrium, marginal canal, gastric cavity, tentacle, and radial canal. (After Bouillon.)

cross, which are the rudiments of the 4 radial canals. At the same time, the apical epidermis proliferates to form a new ectoblastic mass, which presses against the subumbrellar entocodon and subsequently develops its own cavity, the tentacle chamber. The distal ends of the 4 radial canals then fuse to form the circular canal.

Finally the endodermal floor of the bud forms a median diverticulum, the spadix, which everts together with the adjoining central region of the entocodal, or subumbrellar, epithelium. The two everted layers together constitute the manubrium, whose extremity becomes perforated to form a mouth and whose central region perforates to form the velum from the adjacent layers of distal entocodon and proximal tentacle-chamber lining.

Budding to form medusae is limited to a particular band of the manubrium and to the early months of the year, ceasing in the spring, when the same region of the manubrium forms male or female gonadal tissue, depending on the individual; subsequently, eggs are set free to be fertilized and develop into planula larvae. The life cycle of *Limnocnida tanganyicae* from the larval stage onward is very similar to that of *Craspedacusta sowerbii*, except that the medusae of the latter and even of some other *Limnocnida* species do not bud new medusae directly; furthermore, the gonads of *Craspedacusta* are formed from the walls of the radial canals and not from the manubrium. In both genera, however, the planula settles and transforms into a typical microhydra, i.e., a solitary polyp about $\frac{1}{3}$ mm long and $\frac{1}{10}$ mm wide, and supplied with a distal ring of sting-cell batteries. The absence of tentacles may well be a consequence of the minute size of the microhydra, for it is difficult to imagine how even one or two minimum-sized tentacles could form at such a scale.

In both genera the polyps present three types of budding: (1) simple budding resulting from lateral growths similar to the budding process in *Hydra*, except that *Limnocnida* and *Craspedacusta* individuals remain attached together by their base; (2) frustulation, or the production of planuloids, also called frustules; and (3) medusa budding. In frustulation by these two genera and by the allied genus *Gonionemus,* an elongated fragment (frustule, or planuloid) forms from the body wall as a rapidly growing, stolonlike local outgrowth, which constricts at the base and then is detached from the body wall when the outgrowth is four or five times longer than it is wide.[1] The third type of budding gives rise to a medusa exactly like those produced from the manubrium of a medusa, except that in the genera considered here only one medusa is produced at a time and always from the lower wall of a well-grown individual polyp.

So far as is known at present, the details of the histological process are es-

---

[1] The frustule's ability to establish a new individual or colony is discussed on pp. 188-190; in connection with similar phenomena of other hydroids.

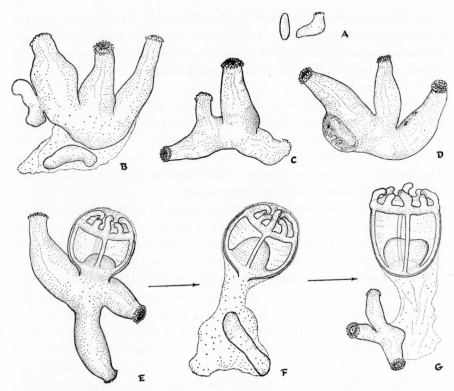

Figure 8-5. Developmental cycle in *Limnocnida*. **A.** Planula and first polyp. **B.** Colony of three individuals, showing frustule at base and another forming at side. **C.** Colony of four individuals, showing initiation of medusa bud rudiment. **D.** Colony of three individuals, showing medusa at more-advanced stage. **E.** Medusa completely formed. **F.** Medusa becoming detached and polyp greatly reduced and transformed into frustule. **G.** Medusa about to be liberated and polyps reduced. (After Bouillon.)

sentially the same as in the formation of a medusa from another medusa. The first sign is the congestion with cells of a local area which always appears where the polyp trunk is sufficiently wide to permit the segregation of a tangentially placed disk of a certain minimum size; i.e., the smaller diameter of smaller polyps may be the prime cause of inhibiting initiation, although other factors are almost certainly involved. As the medusa bud nears complete development, a marked resorption of the conjoined polyps occurs, as though other tissues had to yield to the medusa's paramount demand for nutrient. Eventually all that is left of the original polyp is a featureless residue little larger than a frustule, although capable of reconstituting a group of frustules or of forming smaller polyps.

In such anthomedusan hydroids as *Syncoryne* and *Bougainvillia,* medusa buds form from the end of slender outgrowths of the hydranth wall and of the subjacent stem, respectively, but in both genera the morphogenetic process is much the same: the initial evagination ceases to grow distally along the linear axis and begins to broaden at the tip. At the same time, the more-distal epidermal cells undergo a relatively less-distal growth and delaminate basally to form an inner plate of cells between the endodermis and the outer residual layer of epidermis. This plate of cells grows rapidly, most rapidly at its center. In *Bougainvillia* the plate is at first a solid mass growing inward from the tip, but eventually it forms a slitlike cavity at its core; in *Syncoryne* the growing, or expanding, platelike mass folds inward, and the lumen is in evidence from the beginning, much as in *Limnocnida.* Whatever the exact manner of the early growth of this inwardly growing or folding plate (the entocodon), in later growth its cavity becomes large, central, and square-sided. In both genera, 4 endodermal extensions appear along the 4 flat sides of the entocodon, an endodermal spadix evaginates from the center, and the tentacle cavity grows inward toward the entocodon from the distal epidermis. In a leptomedusan, such as *Obelia,* the process is essentially the same, though the scale throughout is

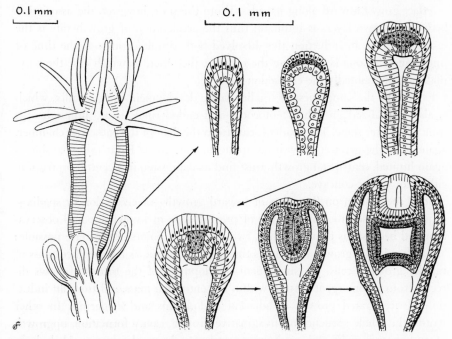

O.1 mm

O.1 mm

Figure 8-6.   Origin of gonophores at base of hydranth in *Bougainvillia superciliaris* and origin and development of ectocodon of medusa bud.

smaller and, consequently, the details are harder to follow. In all three orders—Limnomedusae, Anthomedusae, and Leptomedusae—whenever an outgrowth develops an entocodon, it subsequently becomes a medusa of a recognizable kind; whenever an entocodon is not formed, a medusa never appears, although some other organized structure may develop in the same site.

THE ENTOCODON

Several problems arise, the role of the entocodon being the most important since medusa formation of any kind seems to depend on the presence of an entocodon. The entocodon first appears as a local center of growth without initial direction or polarization. Its separation from the superficial epidermis, of which it was originally a part, seems to be primarily a segregation of a rapidly growing layer, or group, of cells from a relatively slow-growing layer. Once segregated, the rapidly growing cell layer is spatially restricted, with little or no room for lateral expansion, so that it can grow only inward. That the growth is inward may well be an extension or continuation of the earlier concentration of highest proliferation toward the center and away from the distal extremity.

As the entocodal invagination grows inward, the 4 endodermal extensions in effect grow outward along its sides. From the start, however, the association between the two layers is intimate; thus the acceleration of growth rate is the result either of an induction that involved both simultaneously at the time of initiation or of one layer setting the pace for the other, in which case the pacemaker is undoubtedly the entocodon tissue.

Essentially the same question applies to the development of the spadix, which is also a localized growth of entocodal and endodermal tissue, although of a somewhat later stage. The spadix forms from the floor of the entocodal chamber, again with the greatest growth occurring in the center. If lateral expansion is again limited, centripetal growth must find its expression in a central protrusion, either convex or concave.

Such speculation on the cause of inward growth—of entocodon or spadix—cannot be pressed too far, because of present gaps in knowledge and observation and because, as implied above, two alternative points of view are possible:

1. After initial growth establishes the entocodal tissue as a segregated mass of rapidly growing cells, the subsequent development of the medusa follows directly from an emergent pattern within the entocodal mass and from an induction of accelerated growth in adjacent endodermis and epidermis. In other words, the whole sequence of invaginative growth, cavity formation, upgrowth of endodermal radii and spadix core, and—not least—the square-sided shape

of the well-developed entocodon are all properties of consequences of the growth rate and inherent pattern of the entocodon.

2. Both endodermal and entocodal layers are activated together, so that the upgrowth of 4 endodermal radial canals is an expression of an initial growth impulse no more dependent on the entocodon than is entocodal growth on the endodermis, although a mutual adjustment of rates may well occur. Moreover, the specific number of 4 canals is no more part of a fixed pattern than are the first 4 tentacles that arise during the development of a hydra bud: rather, a circular territory divides into 4 presumptive local evaginations more readily than into fewer or more parts—a probability supported by the occasional formation of a 5-part medusa bud from *Limnocnida* medusae. In other words, the entocodon assumes its square-sided shape under stress from the endodermal upgrowths, and the general pattern of presumptive radial canals and spadix evagination is carried out primarily by the endodermis, although activated by the presence and growth of the entocodon.

The distinction between these two hypotheses may be less real than apparent. There is little doubt that the highest growth rate is exhibited at first by the recently segregated entocodal tissue, which may in turn induce comparable growth rates in adjoining tissues, so that it soon becomes difficult to distinguish an initial induction from a continuing, dependent one.

The matter of the actual focus of the initial growth acceleration, however, becomes accentuated, for growth seems to be limited at first not only to the distal epidermis but to the basal part of that epidermis. How, then, can an acceleration of growth affect a single-layered epithelium in such a way that a basal slice of high growth rate is segregated from a less-involved distal layer? Growth in the affected area is accomplished first by an elongation of the epidermal cells and then by cellular division, in a plane at right angles to the customary tangential plane, which merely extends the area of the epithelium. Consequently, the problem of the site of the organizational, or growth-controlling, agency once more arises.

The endodermis by itself hardly qualifies as that site, although it may be an influential, secondary contributor. Even more clearly the outer, peripheral surface layer of the epidermis is a poor candidate since this layer is so obviously a remnant tissue in the entocodal type of medusa development. The only remaining possibility is the basal half of the epidermal sheet or, more plausibly, the basal continuity itself either as the conjoined basal surface alone or together with the basement membrane, if such exists as a separate entity.

This basal surface, or layer, which becomes the outer surface of the entocodon, is in continuous contact with the new structures that develop from the endo-

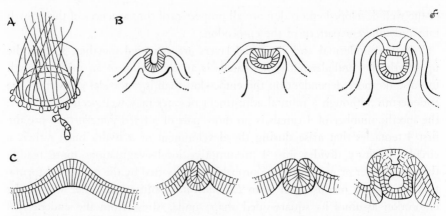

Figure 8-7. Formation of medusa buds and entocodon. **A.** *Limnocnida* with medusa buds forming at bell margin. (After Boulenger.) **B.** Formation of entocodon by invagination. (After Boulenger.) **C.** Invaginative origin of entocodon in *Gonionemus.* (After Payne.)

dermis. Thus the problem at least narrows a little, and the situation may be likened to mapping a field that contains a pool: the pool's outline is clarified, but its contents are not made known.

A few facts of a comparative nature, however, throw some more light on the problem. Although Bouillon (1955b) found that the entocodon appeared in his strain of *Limnocnida tanganyicae* virtually as a solid mass of tissue segregating from the distal epidermis, his drawings exhibit a slight indication of an invaginative activity. The segregating cells move toward the base from the central region of the epidermal tip. In older accounts—of *Limnocnida tanganyicae* by Boulenger (1911) and of *Craspedacusta sowerbii* by Payne (1924 and 1926)—the entocodon is pictured as an actual invagination of the distal epidermis, forming an open cavity, in the first two accounts, but without an actual lumen in the third. Bouillon's description supplies a further step in the same direction by his virtually complete obliteration of the infold, although the cavity reappears within the entocodon a little later in development. If there is some variability among the medusae of the *Limnocnida* species in Lake Tanganyika, either as different strains within a species or according to environmental conditions, and the different organisms are considered as a whole, the entocodon is essentially an invagination, not so much of the entire distal epidermis as of the basal surface of the epidermis. If the epithelium as a whole or its outer surface were the site of the activity, no such gradation from lumen to mere slit to no cavity at all could occur; but with the basal surface as the site of extension, gradation is entirely compatible.

The self-sufficiency of the formed entocodon is indicated in some incidental observations by Mayer (1910) on the origin of medusa buds from the manubrium of medusae of *Bougainvillia niobe* and by Chun (1896) of *Rathkea octopunctata*. In both species the endodermis plays no part at all, at least while the principal morphogenetic events are taking place. The entocodon, spadix, radial canals, and tentacles all arise from the manubrial epidermis: the entocodon first and the other structures from adjacent epidermal tissue; in other words, endodermal growth itself is induced by the entocodon.

Accordingly, a conclusion—tentative perhaps—is that (1) the basal surface or the basal cortex of the epidermis is the region most closely associated with the invaginative activity which gives rise to the entocodon; and (2) the pattern and shape of the medusa as a whole is inherent in the entocodon, so that the development of growth and structure in adjoining tissues is an induction. Despite the apparent importance of the epidermis, the endodermis is undoubtedly essential to the initiating conditions since all nutrient for the epidermis is drawn from or through the endodermis. Moreover, the endodermis of a presumptive bud increases in cytoplasmic density and basophile properties to much the same degree as does the epidermis. In fact, the basal surface of the epidermis can be regarded as the normally activated layer, and the subjacent endodermis as the activator, although in a restricted metabolic sense (see tissue interactions, p. 524). If this viewpoint is sound, variation in the extent to which

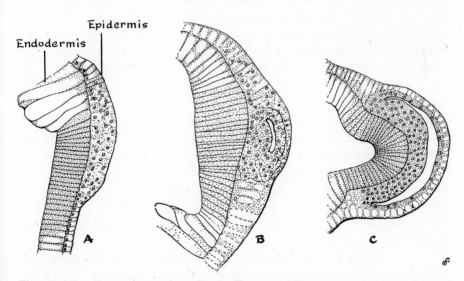

Figure 8-8. Entocodon and spadix in *Eugymnanthea ostrearum*. **A.** Origin of entocodon within thickening epidermis. **B.** Formation of entocodon. **C.** Subsequent induction of endodermal spadix. (After Mattox and Crowell.)

each layer may be involved in the morphogenetic process is to be expected, the activity ranging from an exclusively epidermal to approximately equal to exclusively endodermal.

This last type of activity is typical of many hydroids; for example, it is very clear in living *Tubularia*, whose mesogloea is a persistent demarcation between the epidermis and endodermis. The growth activity is mainly restricted to the distal end of outgrowths which are only subtly different from those of other species that produce epidermal entocodons; and a comparable segregation of an entocodal mass between epidermis and endodermis appears. The local growth and cell proliferation that give rise to this mass, however, are exclusively endodermal. The entocodon no longer develops, as it did before segregation, to form typical medusae; instead, the entocodon cells undergo precocious differentiation into male or female reproductive cells (Berrill, 1952).

The consequences of this endodermal activity are profound, quite apart from the primary question concerning the nature or significance of the shift of the basic growth activity from the epidermis to the endodermis. The consequences are so manifold that a separate discussion in a wider context is required (p. 206). The important aspect here is the restriction of growth to the endodermis, as though the essential locus of the growth-controlling agent were at the junction of epidermis and endodermis and involved either or both of the adjoining surfaces.

The location of medusa buds in hydromedusan hydroids and medusae is extremely variable. Among medusae such buds grow from the wall of the manubrium in *Limnocnida tanganyicae, Sarsia prolifera, S. gemmifera, Podocoryne fulgans, Bougainvillia grondosa, B. niobe, Rathkea octopunctata,* and *Cytaeis atlantica;* from the walls of the radial canals in *Proboscidactyla ornata* and *Eucheilota paradoxica;* from tentacle bulbs in *Hybocodon prolifer* and *Niobia dendrotentaculata;* and by fission of a medusa, commencing with longitudinal fission of the manubrium, in *Cladonema radiatum* (Berrill, 1950a). Among hydroids, medusa buds form from the upper surface of creeping stolons in species of, e.g., *Bougainvillia, Campanulina,* and *Leuckartiara;* from the lateral wall of stems in species of, e.g., *Bougainvillia, Podocoryne, Syncoryne,* and *Campulina;* from the walls of functional hydranths in species of, e.g., *Ectopleura, Tubularia, Corymorpha, Pennaria, Eudendrium, Coryne,* and *Syncoryne;* and from the walls of abortive hydranths called gonangia in, e.g., *Obelia.*

It is noteworthy that the radial canals of medusae of *Phialidium mccrady* produce typical gonangia, similar to those of the hydroid, which in turn produce medusa buds from their walls, thus combining the two phases of the life cycle within a free-swimming, individual medusa; the gonangial growths de-

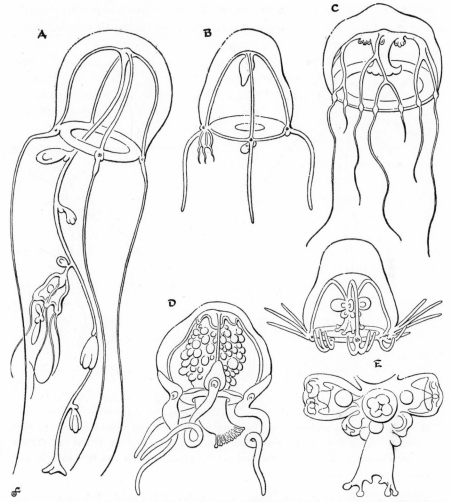

Figure 8-9. Medusa bud sites in medusae in Hydromedusae. **A.** From elongating manubrium in *Slabberia catenata*. (After Chun.) **B.** From base of tentacles in *Sarsia prolifera*. **C.** From radial canals in *Proboscidactyla ornata*. (After Mayer.) **D.** According to available space on large manubrium in *Cytaeis atlantica*. (After Haeckel.) **E.** In spiral sequence on manubrium in *Rathkea octopunctata*. (After Mayer.)

velop from the region of the radial canals that otherwise differentiates into male or female gonads. Moreover, in those species whose medusa buds develop directly from the radial canals of the medusa, the canals are the site of any gonads that form; whereas, in species whose buds develop from the manubrial wall, that wall is the site of gonadal formation, although buds and gonads do

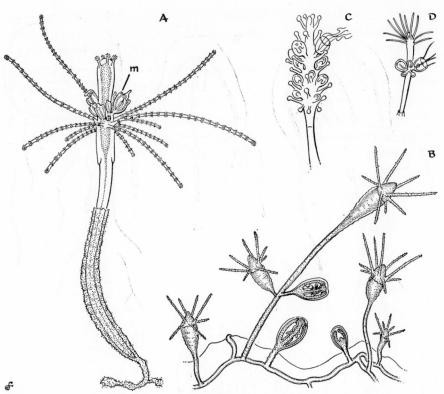

Figure 8-10. Medusa bud sites in hydroids in Hydromedusae. **A.** From zone immediately distal to tentacle ring in *Corymorpha annulicornis.* (After Rees.) **B.** Both hydranths and medusae forming from stems and stolons in *Leuckartiara octona.* (After Rees.) **C.** From between tentacle zones in *Syncoryne eximia.* (After Hincks.) **D.** From zone midway between tentacles and hydranth base in *Podocoryne cornea.* (After Hincks.)

not form at the same time. In both cases, under different conditions, primarily of temperature, the same tissue can give rise to gonads or to medusa buds.

In hydroid colonies, medusa buds arise almost indiscriminately on hydranths and stems and, to a lesser degree, on stolons. A bud arising from basal stolons invariably forms at a considerable distance from a stolon tip, usually where stolon tissue is not undergoing general extension; and invariably only one medusa forms at one locus—a feature which may be readily related to the small diameter of the stolon. On stems, also, one medusa bud forms at one locus, although in many species a succession of buds is produced from a single zone of growth. From the wall of a hydranth, however, several medusa buds may form, either directly from the wall or as clusters produced by a single ring of outgrowths. Similarly, in medusa buds formed in a medusa, a succession may grow from a narrow, elongate manubrium such as that in *Sarsia* species, or a large number may form almost simultaneously from the narrow zone of a wide manubrium as in *Limnocnida* or from the broad zone of the wide manubrium in *Cytaeis*. Clearly, there is no specific correlation of medusa-bud production with any particular site; in fact, in the same species, buds can arise from the hydranth base, the stem, or the creeping stolons.

Several related problems emerge from the foregoing discussion of medusa bud formation: What histological or environmental conditions, common to all sites, are correlated with medusa bud production? What determines the number or sequence of medusa buds? What is the histological and environmental basis of

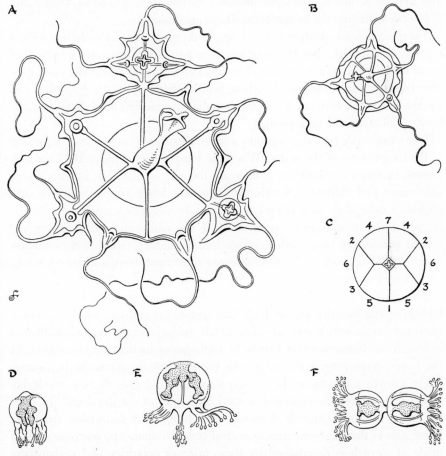

Figure 8-11. Medusae forming from medusae in *Niobia dendrotentaculata*. **A.** Tentacle bases transforming into medusae. **B.** Newly liberated medusa. **C.** Diagram, showing order of development and liberation. (After Mayer.) **D–F.** Three stages in division of manubrium and medusa bell. (After Pasteels.)

the change from medusa bud production to differentiation into gonads? What determines whether an outgrowth develops into a medusa, hydranth, creeping stolon, or simple terminal growth?

Polymorphism in Hydromedusae

The last question above leads to the general topic of polymorphism in Hydromedusae. Since almost every hydroid represents a variation of this theme, the phenomena are best presented through several examples: in *Bougainvillia* and *Syncoryne* among the gymnoblasts (Anthomedusae) and in *Obelia* and *Schizotricha* among the calyptoblasts (Leptomedusae).

The *Bougainvillia* hydranth has a single distal girdle of simple tentacles, a tapering trunk, and a basal zone of linear growth, which successively adds new tissue to the stem as in *Tubularia*. The basal growth zone can at the same time give rise to a lateral outgrowth, whose position becomes relatively farther and farther down the stem as new stem tissue is continually added distally to the junction of stem and outgrowth.

This outgrowth may develop into a stolon, hydranth, or medusa, according to the temperature of the water: When the temperature exceeds 20°C, the outgrowth becomes a stolon; i.e., it does not increase in diameter, but undergoes continuous and extensive growth. At somewhat lower temperatures the outgrowth is most likely to develop into a hydranth and grows much more slowly. At still lower temperatures, particularly if the temperature falls during the critical period, the outgrowth develops into a medusa, either from a solitary outgrowth on one side of the stem or as one of a cluster from a single encircling growth of 4 or 5 outgrowths.

Both growth rate and scale of organization are significant at this stage. At high temperatures the rate is high and growth is entirely linear. At lower temperatures growth is still intensive, but is appreciably greater in width than in length—a change associated with the initiation of hydranth organization. At the lowest temperatures the outgrowth is much smaller from the beginning, linear growth soon approaches a near-standstill, and growth as a whole becomes progressively concentrated toward the center of the distal end. In sum, the three modes of growth demonstrate a progressive conversion from the linear axis to the transverse axis to central concentration. The outcome of each mode of growth is determined by these primary polarities, and perhaps the prospect of locating their physical basis is greater in investigations of this type of material than in investigations of any other material.

Syncoryne has an additional variant. Stolons grow as in *Bougainvillia,* but

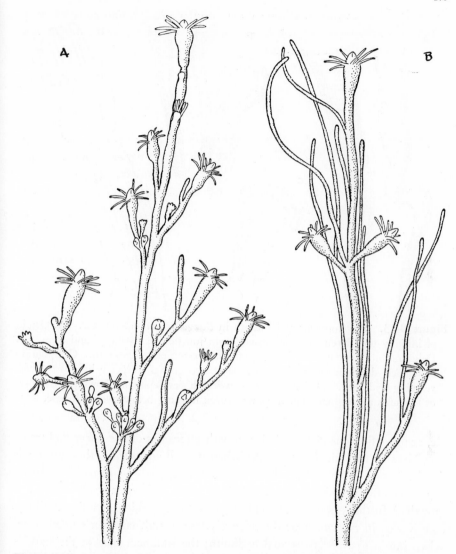

Figure 8-12. Part of main stem in *Bougainvillia superciliaris.* **A.** Typical appearance during low or falling temperatures, showing lateral branches bearing developing hydranths, medusa buds, and stolonic outgrowths. **B.** Typical appearance during falling temperatures, showing that all outgrowths except terminal hydranth are stolonic.

two kinds of hydranths can be produced: one is longer and narrower than the other and has 4 separate girdles of capitate tentacles; the other's tentacles are more numerous and crowded, and this hydranth is the one that produces medusa buds from its basal region. The main differences between the hydranths

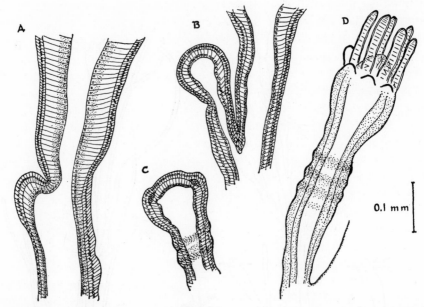

Figure 8-13. Development of hydranth in *Bougainvillia superciliaris.* **A.** Origin of initially undifferentiated outgrowth from junction of hydranth and stem. **B.** Delimitation of distal part of outgrowth into presumptive hydranth. **C.** Initiation of tentacle development and appearance of two growth surges below base of developing hydranth. **D.** Late stage in hydranth development, showing second set of tentacles arising and fourth growth surge forming below hydranth base.

are, again, reflections of the relative growth intensities of the linear and transverse axes of the initial outgrowth; whereas in *Bougainvillia,* the outgrowth that becomes a full-grown medusa—although apparently no alternative destiny is possible after a bud appears—is relatively small and typically tip-centered in growth. A further example of two kinds of hydranths exhibiting a similar difference is seen in *Eudendrium:* the relatively longer and narrower rudiment develops into a so-called nonsexual hydranth; the rudiment that, at the start, is relatively short and wide develops into a short and wide hydranth whose base bears numerous medusoid outgrowths which, in fact, may even abort growth of the hydranth structure. Accordingly, a close analysis of the early growth differences among the several possible outcomes is relevant.

Stolonic Growth

In species of *Obelia* and *Campanularia,* the tip of a stolon shows epidermis and endodermis adjoining without appreciable mesogloea, and basal interstitial cells are not discernible. Chitinous perisarc is visible only at a certain distance from the distal end, is present as a sticky film on the slopes of the distal end, and is absent at the tip itself. Cell proliferation seems to be confined to the naked tip of the epidermis and to the subjacent endodermis. At its basal end adjoining the epidermis, each endodermal cell of the proliferating tip tissue contains a large vacuole, which slowly disappears after cell growth and division have ceased. Consequently, the endodermis exhibits a gradient in cell vacuolation: the size of the vacuole diminishes with increasing distance from the stolon tip, until none is left. The extent of the endodermis exhibiting the vacuolation gradient is accordingly an index of the rate of production of new distal cells, if it is assumed that the rate of recovery from the vacuolate condition is uniform.

Not only may the vacuolate condition extend as many as 60 cells from the tip, of a very rapidly growing stolon, compared with about 20 in a slow-growing stolon, but the endodermal cells at the tip are small and dense, divide rapidly, and lack vacuoles. Immediately behind this meristematic center, however, the endodermal cells enlarge, acquire vacuoles, and undergo one or two further divisions comparable to those occurring at the very tip of a more slowly growing stolon. Thus there is an active terminal center of growth in the endodermis, with a peripheral decrement, so that newly produced cells inevitably become aligned with those already present in the stolon, which thereby grows in length but not in diameter. There is also obvious conformity between the rate of

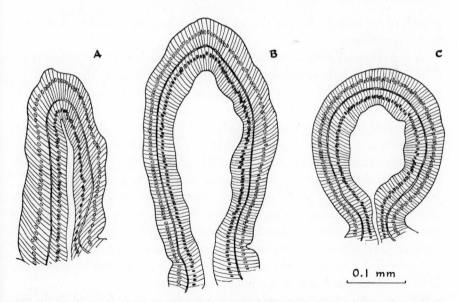

Figure 8-14. Initial stages of growth terminals in *Syncoryne eximia*. **A.** Stolonic growth. **B.** Nonsexual hydranth. **C.** Sexual hydranth.

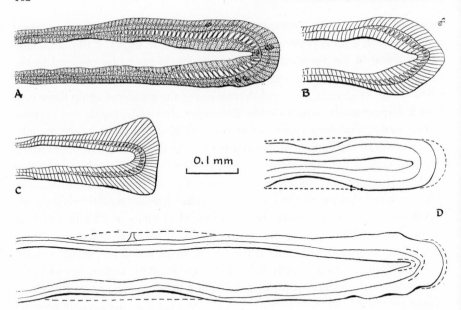

Figure 8-15. Stolon in *Obelia*. **A.** End of fast-growing free stolon. **B.** End of slow-growing free stolon. **C.** End of slow-growing attached stolon, showing laterodistal spread of epidermis. **D.** Free stolon tips, showing amplitude and extent of rhythmic contraction and expansion in recovery zone, and slight terminal pulsation of fast-growing stolon above and slow-growing stolon below.

growth of the endodermis and that of the epidermis; the endodermis seems to be the pacemaker.

At the proximal limit of endodermis, where vacuolation finally disappears, there is tissue which is histologically indistinguishable from the rest of the stolonic tissue, except for a peculiar rhythmic activity. Immediately behind the the vacuolated region and for a similar distance (i.e., long in rapidly growing stolons, short in slowly growing stolons), the stolon wall undergoes a regular contraction and expansion, causing a rhythmic hydroplasmic streaming within the stolon tissues. The maximum amplitude occurs midway in the contractile region. The rhythmic period, which varies with the temperature, may last 3-7 minutes: a contraction of 1-3 minutes, a quiescent period of 1 minute, and a relaxation of 1-3 minutes. Whether or not the streaming so induced serves a useful purpose is somewhat beside the point; the significance of this rhythmic contractibility is that it is clearly an aftermath of the process of proliferation and growth, for contractibility and growth are always present in association, in temporal sequence, and in quantitative relationship.

What actually effects the contraction is another question. It could be the epidermis, the endodermis, both in unison, or perhaps only one surface of one layer. Some indication is given by the behavior of the epidermis. In the contractile zone there is a perisarcal chitinous layer, to which the epidermis tends to adhere. During contraction the epidermis withdraws, leaving a space between itself and the perisarc. Frequently, however, some epidermal cells fail to detach from the perisarc, giving the unmistakable impression that the epidermis is being pulled away from its outer attachment by some force other than itself.

Accordingly, the active agent must be either the endodermis or the basal layer of epidermal protoplasm. Because the free epidermal surface, facing the lumen of the stolon, is irregular in outline and the constituent cells in the

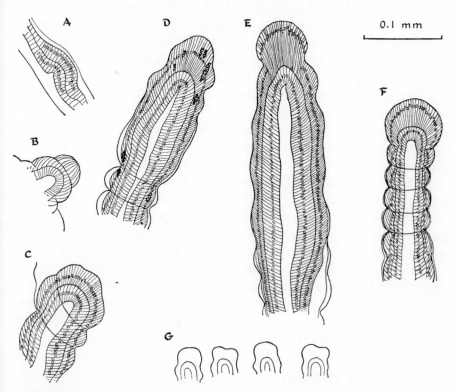

Figure 8-16. Pulsative or annulated growth in vertical stem growth in *Obelia*. **A.** Initiation of outgrowth: enlargement and division of cells of coenosarcal wall. **B.** Early stage, showing formation of second annulus; epidermis forms blob in advance of endodermal growth. **C.** Two-annulus stage, showing endodermis entering blob. **D.** Later stage, after first phase of annular growth and near end of stolonic phase, showing epidermal blob again forming (glandular cells also shown). **E.** Similar stage, showing fully formed epidermal blob and lagging endodermis. **F.** Final blob, at end of second sequence of annular growth, forming hydranth rudiment. **G.** Four terminal contours of epidermal blob, shown at one-minute intervals.

contractile region project in a lobular form, only the inner, or basal, layer of this tissue seems to qualify. The basal cortical layer of either the epidermis or endodermis or of both must be the initiator of the rhythmic contractile activity.

In rapid stolonic growth the endodermis and epidermis keep pace, with the tip of the endodermis driving against, so to speak, the tip of the epidermis. In other types of growth, mutual pace is not characteristic.

Unattached outgrowths which are at first essentially stolonic can originate either from the upper surface of an attached stolon or from the side of a vertical stem or of one of its branches. Initially the growth process is comparable to the initiation of a hydra bud. Soon, however, the growth of the two layers is somewhat out of step: the initiative seems to be in the epidermis, and the endodermis lags or is a little out of phase. Only after the epidermis has formed a hemispherical mass does the endodermis gather enough substance to follow. Epidermal protoplasm as a whole (i.e., ignoring its subdivision into discrete cells) seems to flow as it grows. Moreover, the growth and flow seem to be intermitttent, producing a series of approximately hemispherical masses; and since newly secreted chitin polymerizes on the curved external surfaces of these successively produced masses, a chitinous cast of the fluctuations of growth is preserved after other evidence of such fluctuations has disappeared. One of two explanations is possible: (1) Epidermal growth is an alternating series of surges and pauses or at least of rapid and slow rates; and the endodermis grows steadily at an even pace, pushing into an epidermal mass when epidermal growth is at an ebb and lagging behind again during the succeeding epidermal surge. (2) The epidermis grows at a steady rate, and the endodermis at an alternating fast and slow pace. At all times, however, the basal surfaces of the epidermis and endodermis are in contact. When the combined growth is comparatively slow, deep chitinous grooves, or annuli, mark the epidermal pauses; when growth is rapid, the epidermal masses are apparently stretched before chitinous polymerization is complete, so that annuli become indistinct.

Development of Hydranth

In all species of *Obelia* and *Campanularia,* and in many other related genera, a certain sequence of events in a short stolon invariably leads to hydranth initiation from the end of that stolon: namely, several initial annulations, a phase of simple (nonannulated) stolonic growth, and a second and distal series of annulations.

The first series of annulations results from the initial head start of the epidermis; however, as the endodermis catches up, the annuli become smaller and smaller, until finally, when growth rates and terminal tissues coincide,

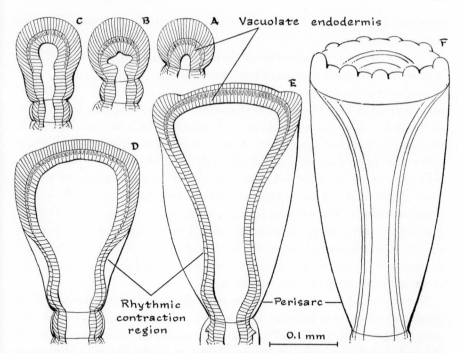

Figure 8-17. Development of hydranth in *Obelia* (continued from **Fig. 8-16F**). (Chitinous cuticle in heavy line.) Note extension of vacuolate area of endodermis. **A.** Initial semihemispherical rudiment, showing terminal epidermal blob fully entered by endodermis. **B–C.** Enlargement and elongation as cells become added from anterior disk to proximal cylinder. **D–E.** Progressive growth, limitation of vacuolated cells to anterior disk, proximal formation of rhythmic contraction region—causing detachment of coenosarc from perisarc. **F.** Slightly later stage (at same scale), showing complete separation of coenosarc and perisarc and conversion of marginal rim of disk into rudiments of primary and secondary tentacles.

typical stolonic growth gets under way. After a variable period of stolonic growth, the growth rate of the endodermis begins to lag behind that of the epidermis, and terminal epidermal masses and annuli successively reappear, although in a different form from the first series. The first series is a diminuendo, the epidermal masses becoming successively reduced in size at the time the endodermal tip pushes into them, until none is formed and the two layers grow in unison. In contrast, the second series is a crescendo, the epidermal masses successively increasing in size until at the formation of the third, fourth, or fifth mass a crisis is reached: an unusually large epidermal mass forms, as the result of either a greater surge of epidermal growth or a relative slowdown of endodermal growth, which permits the epidermis to accumulate in hemispherical form. The outcome of the second series is a hydranth rudiment of distinct configuration.[2]

Compared with a typical stolon tip, the terminal mass is relatively wide. There is no sign of the rapid division of small cells characteristic of the endodermis of a fast-growing stolon tip, but there is slow division of vacuolate cells in a comparatively wide transverse area. Vacuolation begins to diminish at the edge of this disk, where the tissue curves to form a shoulder; behind the shoulder the tissue narrows toward the base. As growth proceeds, the distal disk becomes wider and more distinctive, although marginal disk cells continuously take their place in the shoulder and, with diminishing vacuolation, become progressively displaced down the column toward the base, behaving exactly like the tissue proximal to the growing tip of a stolon and undergoing the rhythmic contractions characteristic of stolonic growth. In time, however, growth within the disk and at its margin becomes progressively slower and finally ceases, after formation of a long, wide column that is separated from its chitinous perisarc as a result of its disengaging contractility.

The primary difference between hydranth development and typical stolonic growth, in calyptoblast hydroids, is therefore in the activity at the extreme distal end: it is as though an expanding disk in the transverse plane were inserted in the center of a stolon tip, all else remaining essentially the same. In stolonic growth the center of growth is very small, and the new tissue at its periphery continuously displaces that already formed; i.e., the oldest cells are farthest from the growth center. In the developing hydranth the central region of the initial growing mass, corresponding to the growth center of the stolon, is more extensive in the transverse axis, with about 40, instead of 3 or 4, cells of endodermis or epidermis. The area then grows as a unit without significant age or time difference between center and margin. Only beyond the margin of this area—i.e., after most cells have ceased to divide—does recognizable decrement or aging reaction occur. As long as divisions take place, senescence cannot begin; thus the unitary character of the terminal disk is paramount.

The circular disk expands until it attains a certain maximum width. This diameter is variable, however, and correlated with the diameter of the original mass at the time hydranth development began: the stem, or free stolonic process, from which a hydranth develops may be slender or stout, depending on the diameter of the stem from which it arises. A narrow stem gives rise to a small primary mass, a wide one to a large mass; but, regardless of mass size, shape and tissue proportions are the same and the final diameter is about fourfold

[2] L. Belousov, of the Cathedra of Embryology, Moscow University, reports (1960, personal communication) that after the period of rhythmic growth is finished, the ectodermal cells of the annulated zone begin to migrate distally to form the ectoderm of the new hydranth, resulting in a rapid devastation of the annulated zone.

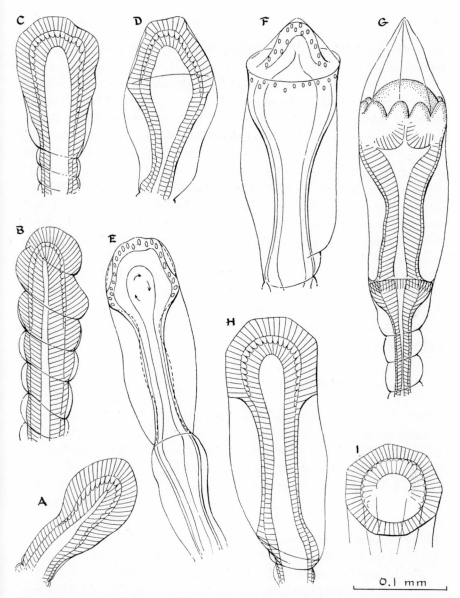

Figure 8-18. Development of hydranth and operculum in *Opercularella pumila*. (Chitinous perisarc in heavy line.) **A.** Free terminal, showing extent of vacuolation in distal endoderm in presumptive hydranth stage. **B.** Free stolon undergoing spiral annulation. **C.** Early hydranth stage, showing terminal growing plate, recovery zone, and pulsation zone. **D.** Later stage, showing manubrium and tentacle zone. **E.** Similar stage, showing distal and proximal pulsation zones (shown by broken lines). **F.** Full-grown hydranth, showing main distribution of chitin-secreting cells around tentacle ridge and manubrium. **G.** Later stage, showing manubrium partly withdrawn, operculum and tentacle rudiments present, and diaphragm forming below. **H.** Optical profile of stage preceding tentacle initiation, showing plate, slope, and side of distal region. **I.** End view of same stage, showing polygonal shape of circumference.

that of the initial disk. At this time expansion ceases and pattern emerges. The circumference of the disk now becomes polygonal, with the number of sides 8-12 depending on the size of the disk. The distance between two adjacent angles, or the length of a side, seems to be constant despite disk size, and the number of sides and angles is in direct ratio to the circumference; the number is therefore a function of the size of the disk at this critical stage when disk expansion is ending. Each angle, or corner, is the locus of a primary tentacle, which at the outset consists of a local hypertrophy of epidermal cells and a subjacent group of small cells derived by local proliferation from the endodermis. With the completion of disk expansion, a secondary, intermediate tentacle rudiment appears between each pair of primaries as available space and material permit, and the central region of the disk inevitably curves distally and differentiates as the oral cone, or manubrium.

Opercularella pumila, representative of the Campanulidae, undergoes a similar growth sequence, as do the Campanularian species of *Campanularia, Eucopella, Clytia,* and *Gonothyraea.* In all calyptoblastic species the hydranth ceases growth after it attains the functional state, in contrast to continuing hydranth growth in gymnoblastic species.

Development of Planuloids

When the water temperature rises above a certain threshold, which varies according to the species, the course of growth usually changes. In both *Bougainvillia* and *Tubularia* at temperatures higher than about 20°C, the hydranth usually resorbs, the rhythmic character of distal stem growth ostensibly disappears, and free stolon growth of impressive magnitude proceeds. In both genera long stems with terminal centers of stolonic growth replace short stems surmounted by organized hydranth structure. The temperature effect is thus twofold: the organized structure with its relatively specialized cell types is not maintained, and the growth rate of the distal zone is greatly increased.

A similar replacement of hydranths by long, rapidly growing free stolons occurs under comparable circumstances in *Syncoryne eximia* and *Obelia geniculata* (Berrill, 1949c and d; 1953b), but in the former species the stems and branches are relatively massive and such growth leads to a kind of asexual reproduction. The terminal region of each stolon continues to grow rapidly, but the stolon's basal connection becomes progressively more attenuated and finally breaks. Apparently, the terminal growth is so vigorous and its nutritive demands so great that excessive numbers of cells are resorbed from the stem. Whatever the cause, sausage-shaped pieces about ½ mm long pinch off from the end of the stem. Each piece continues to grow rapidly, as a closed system, at the distal

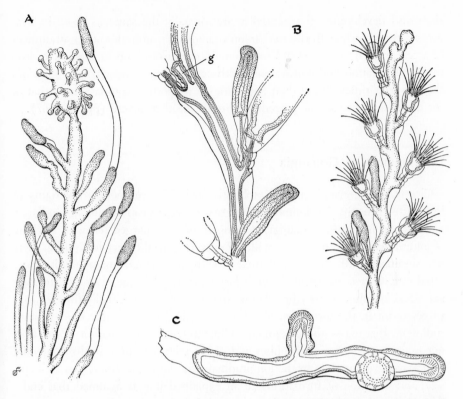

Figure 8-19. Formation of planuloids as response of branch terminals to high temperature. **A.** From primary branches (normally presumptive hydranths) in *Syncoryne eximia*. **B.** From axillary branches (normally presumptive gonangia **g**) in *Obelia geniculata*. **C.** Development of reattached planuloid in *O. geniculata,* showing terminal and lateral stolonic terminals and top view of differentiating hydranth.

end and to resorb cells at an equivalent rate at the proximal end, so that in effect the mass moves forward while it maintains a constant shape, leaving in its wake a perisarcal tube, which indicates the duration or extent of this particular type of growth. The masses produced in this manner are known as planuloids, whose size, shape, and tissue construction are similar to those of planular larvae, lacking only the cilia of the outer surface of the latter. Planuloids settle and give rise to a small stolon which supports a single, small hydranth.

In *Obelia geniculata,* although at relatively high temperatures the presumptive hydranth rudiments grow into long, slender stolons without producing planuloids, at the same temperatures the presumptive gonangial rudiments grow into more-massive stolonlike growths which produce planuloids whose dimensions and developmental potentiality are virtually the same as those in *Syn-coryne*. Apparently both the free stolon's rapid linear growth and its attainment of a certain girth are necessary to planuloid production and in themselves lead to the segregation and constriction of the critical mass, which is a unit from three to four times longer than it is wide—a proportion comparable to that of the dimensions of stolonic reproductive units in other genera (pp. 339, 367).

Development of Gonangia

Clearly, the gonangia of calyptoblastic hydroids are modified variants of hydranths; thus the development of the two structures can be closely compared. Similarly to hydranths, gonangia develop from the enlarged terminal swelling of an annulated outgrowth, or stalk. However, from the very beginning, there is a significant difference: Epidermal growth in gonangia maintains its more rapid rate relative to that of the endodermis, and each surge of growth forms a perisarcal annulus, thereby broadening the epidermal cap. Usually three or four surges result in the formation of a broad terminal mass; unlike the hydranth rudiment, this mass—the gonangial rudiment—is the culmination of the first, not second, series of growth pulses and grows relatively more in the tangential or transverse plane than in the longitudinal axis. Thus the differences in shape, size, and direction of formation are all explained if it is assumed that endodermal growth in the linear axis is slow relative to epidermal growth.

The gonangial rudiment, like the hydranth rudiment, continues to grow as a distal cap, or disk, which expands in area and whose proximal trunk extends in length. There the resemblance ends, for the gonangial disk fails to develop either tentacles or manubrium; instead, the lateral walls, after detaching from the perisarc in typical manner, produce medusa buds.

If hydranth and gonangium are regarded as homologous, the gonangium is clearly abortive in its distal development, and the hydranth may be considered defective in failing to produce medusa buds from its walls. Both the gonangium's abortive development and its medusa-bud formation call for explanation.

The failure of gonangia to develop tentacles and a manubrium seems to result from arrested disk development. In *Obelia* the diameter of the hydranth disk rudiment increases approximately twofold—whether estimated by cell count or by actual measurement—before tentacle core cells segregate from the endodermis and maximum width of disk is attained. By contrast, the gonangial disk expansion is less than one-third; thus it may be said that the gonangial disk fails to expand to the stage at which tentacle initiation might be expected.

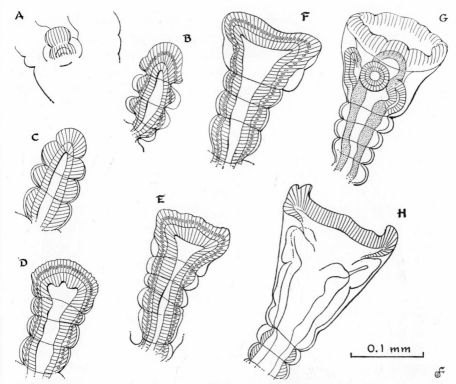

Figure 8-20. Development of gonangium in *Obelia* (cf. **Fig. 8-18**). **A.** Initial stage. **B–D.** Primary annulation growth phase, showing expanding diameter culminating in gonangial rudiment. **E–F.** Growth of rudiment, especially of disk. **G.** Initiation of medusa buds from wall of anterior cylinder (endodermis stippled). **H.** Consequent shrinkage of distal endodermis.

The second gonangial difference, from one standpoint at least, is related to the first. Later stages of gonangia show that arrest of disk development coincides with medusa-bud initiation in the column and that the distal endodermis of the disk shrinks as tissue immediately proximal to it becomes involved in medusa bud formation. Shrinkage is very marked in *Obelia,* but less so in *Campanularia, Gonothyraea,* and *Clytia;* and in all species the terminal disk becomes progressively smaller.

Closer analysis of the developing *Obelia* gonangium, however, reveals a further distinction. The column from immediately below the disk down to the perisarcal annulus shows evidence of having developed as the result of two or three separate growth surges, but surges too brief for annulus formation, and is clearly an enlarged stem rather than an integral part of the terminal developmental unit. At the time of enlargement, this section of column also undergoes

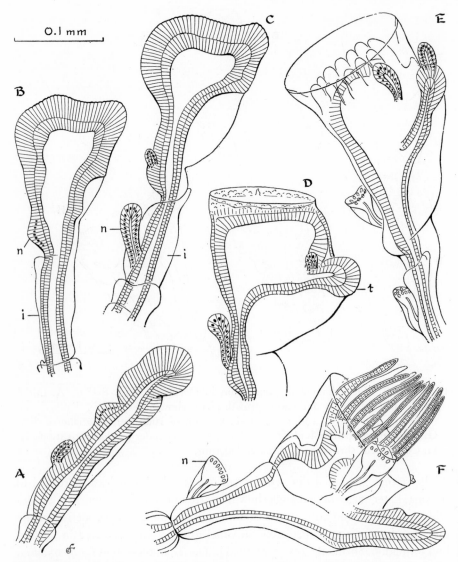

Figure 8-21. Development of hydranths, nematophores, and growing terminals in *Schizotricha tenella*. **A.** Early stage, showing terminal hydranth rudiment and internode bearing two young nematophores. **B.** Later stage, showing internode **i** without nematophores **n**, hydranth unit forming nematophore at base, and new growing point at right edge of distal plate. **C.** More-advanced stage of similar unit. **D–E.** Later stages, showing new growing terminals **t**, advanced stages in nematophore development, and segregation of tentacles. **F.** Fully developed hydranth.

much more active growth than does either more distal or more proximal tissue. It is this region of growing column that produces medusa buds.

Development of Nematophores and Tentacles

A contrasting kind of hydranth dimorphism is seen in the Plumularidae, e.g., *Schizotricha,* in the form of the so-called nematophore, which is attached to the wall and stem of a typical hydranth. A nematophore arises from the epidermis of either a stolon or the narrow basal stalk of a hydranth. Whatever the derivation, the tissue is in the same physiological state—i.e., rhythmic activity has ceased and a normal relaxed state, so to speak, is established—and the tissue is a certain critical distance proximal to the region of high endodermal vacuolation and approximately at the junction of contractile tissue and more-proximal noncontractile tissue. The region where nematophores originate can be compared with that area of a gonangium where medusa buds or gonophores arise or of the stem of a *Bougainvillia* hydranth where outgrowths which usually become medusae develop.

Of immediate significance, however, is the nematophore's origin as a small local thickening resulting from a progressive lengthening of the epidermal cells, with little or no indication of endodermal participation. This thickened area grows outward from the center as though it were a typical stolonic growth except for its minute size and almost exclusively epidermal constitution. Usually, but not invariably, a few endodermal cells are drawn out within the lumen of the outgrowth, and its essentially epidermal nature is clear. Very little growth occurs; yet it is sufficient for the outgrowth to acquire the form of a miniature hydranth, complete with stalk, cup, and peritheca, with the rim of the cup bearing a ring of large nematocysts. Histologically the epidermis bearing the nematocysts is comparable to the distal regions of the tentacles of a hydranth; the cup, to the body from which the tentacles arise; and the stalk, to its obvious counterpart. In other words, hydranth form and organization develop to the extent possible on such a small scale and in the absence of endodermal participation—a situation which suggests that the primary pattern somehow resides in the epidermis, that endodermal participation is responsive rather than determinative, and that absolute size (or, alternatively, cell number) is a limiting factor.

This conclusion is supported by a study of hydranth development in *Pennaria.* In this genus, tentacles are typically present as a basal girdle of long filiform tentacles, similar to those of *Tubularia,* and as 2 or 3 more-distal rings of shorter tentacles which are usually capitate, though occasionally filiform. The difference between filiform and capitate tentacles results from the relative extent of epidermal and endodermal participation in tentacle development. In the long, filiform tentacle the epidermal cells comprising the initial tentacle tip are comparatively few in number and differentiate later to form very small nematocysts; whereas the area of endodermis involved in the segregation of tentacle core cells is relatively large, and the aggregate divisions and subsequent growth of these cells produce a very long outgrowth with which the subterminal epidermis keeps pace. In the short, capitate tentacle, however, the opposite relationship holds: The initial epidermal cap is relatively large, and the subjacent endodermal involvement relatively small; in fact, the epidermal cap, which becomes the tentacle tip, is formed as an entity *before* the underlying endodermis has shown any response or activity at all. Furthermore, large nematocysts are already differentiated at this early stage, and endodermal core cells become visible only after the epidermal component has already attained a hemispherical shape; even then, endodermal response is minimal since the core cells segregate from the basal surface of the endodermis instead of invaginating.

The conclusion—tentative at least—is that the initiation of a tentacle is essentially an epidermal activity, but that the length of a tentacle depends on the degree of endodermal response. In *Pennaria* there is evidence that the epidermal activity at first is somewhat diffuse and subsequently becomes progressively localized; since before there is any sign of local epidermal hypertrophy, large nematocysts differentiate in the epidermis of the hydranth rudiment in a broad zone corresponding to the total territory of the tentacle rings.

A somewhat similar contrast is that of the polymorphic hydranths in *Hydractinia.* Gastrozooids develop from relatively large rudiments in which the endodermis is more than adequate, and the filiform tentacles that develop have a typical endodermal core. In the narrow, defensive type, on the other hand, the endodermis is clearly greatly reduced both absolutely and relative to the epidermis and seems to be incapable of responding to the several epidermal centers which are simultaneously established. The results are that the tentacles are in effect coreless and capitate and that epidermal cells differentiate into large nematocysts instead of being stretched over an extending core to form a sheath.

As individual cells, both endodermal core cells and epidermal cells seem to conform to, rather than determine, a situation. The shape of individual epidermal cells is that necessary for them to constitute a sheath of proportionate thickness for a particular level, and the shape of endodermal core cells is correspondingly adjusted to the width and shape of the tentacle as a whole, much as in myxomycete sporangia. Once again the process seems to be a local, limited expansion of a supracellular unit, possibly involving the basal surface of both the epidermis and endodermis.

Polymorphic growth within a colonial system is more highly evolved and differentiated in the siphonophores than in any other hydrozoans. Medusoid and

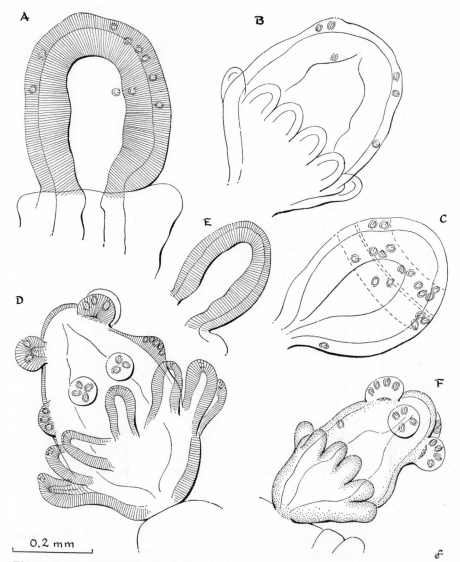

0.2 mm

Figure 8-22. Hydranth development, relative to initial size and shape, and tentacle initiation in *Pennaria*. **A–B.** Hydranth rudiments of main terminal, showing relative size, origin of nematocysts distally in advance of tentacle initiation, and formation of proximal filiform tentacles in advance of capitate tentacles. **C–D.** Comparable stages in development of hydranth from subterminal branch, showing smaller initial size and narrower basal region, precocious formation of nematocysts arranged in broad distal girdle, simultaneous appearance of capitate and filiform tentacles, and formation of epidermal capitate component in association with groups of nematocysts in advance of endodermal outgrowth. **E–F.** Two stages in development of smaller subterminal hydranth, showing size correlations.

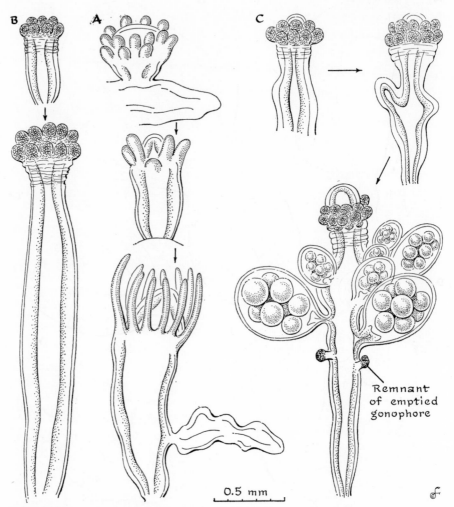

Figure 8-23.   Polymorphic development of polyps in *Hydractinia*.   **A.** Gastrozooid from relatively broad initial with filiform tentacles containing endodermal core. **B.** Dactylozooid from narrow initial with exclusively ectodermal, capitate tentacles. **C.** Gonozooid with tentacles similar to dactylozooid, but with endodermal growth zone in trunk, successively producing gonophores from upper region.

hydroid organismal units of various kinds develop as a single, functional, free-swimming colony, with every genus exhibiting phenomena and raising problems of pattern and development which are peculiarly its own. Unfortunately these genera, though abundant and diverse, are primarily pelagic, oceanic organisms, which are difficult to obtain for laboratory study, so that knowledge of their development is fragmentary and inconclusive. Garstang (1946) has made an extensive pictoral and expository survey of development in the Siphonophora, although more from a phylogenetic than from an ontogenetic point of view.

Determination of Developmental Type

The general question that emerges from the foregoing analyses of various types of development is: What determines the nature of the organization that develops from a local center of growth? This question is most sharply confronted in the case of medusa versus hydranth formation, particularly in species which can develop either structure from evaginations on both creeping stolons and hydranth stems. The temperature effect is suggestive: At relatively high temperatures epidermis and endodermis grow rapidly and conformably together and give rise to stolonlike growth; at lower temperatures epidermis and endodermis grow differentially and give rise to relatively wide and slow-growing terminal growths from which the tentacle rings and manubrium of the typical hydranth develop; and at still lower temperatures linear growth, especially of the endodermal component, is virtually suppressed and epidermal proliferation becomes concentrated at a distal growth center from which the entocodon develops. Thus the relative growth rate of the two tissues is all-important; and apparently depending on the temperature—though nutrient level or other factors may be equally influential—the rate of one tissue predominates over the other, permitting or inhibiting certain developmental outcomes.

Besides environmental determination, there is clearly a tendency for like to produce like. Planula medusa larvae growing within the nutritive medium of the parental gastrovascular cavity form buds of cellular masses whose dimensions and stage of development are the same as those of the planula; i.e., planulae give rise to planulae. Actinula medusa larvae growing attached to the parental manubrial lips, somewhat like transient parasites, subdivide to give rise to actinulae similar to themselves. With one or two exceptions, medusae bud to produce typical medusae; moreover, if the parent medusa is approaching sexual maturity, its medusa buds are at the same stage of development, although this similarity in stage may be the bud's response to the same external gonad-inducing agent to which the parent had responded. It is particularly noteworthy, however, that planulae do not form buds that become actinulae nor do actinulae produce planulae, although the definitive planula larva may be destined to pass through the actinula stage during its later development.

In all organisms whose medusa buds—whether destined to be free medusae or sessile gonophores—are formed directly from the manubrium of either a medusa or a hydranth, availability of space seems to be an important factor. In

Sycoryne, for example, gonophoral buds form on the lower part of the hydranth manubrium, first as a regular ring around the base and then everywhere, as additional space resulting from general growth is available between buds already formed or between tentacles. Similarly, the number of medusa buds formed on the medusa manubrium depends on availability of space: a very small manubrium has only 2 or 3 buds at any one time; a larger manubrium can have a close helix of about 12 buds; and the exceptionally large manubrium of *Cytaeis* has many more buds, each new one forming wherever there is space available between the stalks of older ones. It should be emphasized, however, that space alone is merely a permissive factor and not a determinant. Where space is inadequate or lacking, potential developments are inhibited; where space is sufficient, buds form if the tissue is in a state of growth and if that growth is in a plane no longer dominated by the original axial polarity.

Chapter 9

Pattern and Growth: Development
of Pattern (Continued)

Sexual Differentiation in Hydromedusae

Gonads in Hydromedusae are produced either in hydroids or in medusae, according to the species, but not in both. Medusae may produce mature gonads precociously or after a prolonged phase of growth as free-swimming organisms. Hydroids may produce gonads indirectly by means of gonophores which may represent all degrees of medusoid structure or, as hydras form gonads, directly within hydranth tissue. Moreover, gonads may form within either the epidermis or endodermis, again according to the species. The various types of gonad formation are described below.

Typical Reproduction

1. Medusae are liberated without a trace of gonads. During subsequent growth of free-swimming medusae, gonads appear and mature. Typical examples are *Bougainvillia superciliaris* and the various species of *Obelia*. The development of the medusa of *B. superciliaris* (pp. 169-171) exhibits no trace of oocytes in the invaginating entocodal epithelium.

2. *Bougainvillia carolinensis* oocytes are recognizable early in the invaginating entocodal layer; the largest oocytes in this layer are in the distal region where, presumably, the growth rate is most rapid. This region subsequently evaginates over the evaginating endodermis to form the stomach and manubrium, and 4 ridges of oocytes graded in size are visible in the epidermis of the manubrial wall. At every stage, from the initial entocodal stage onward,

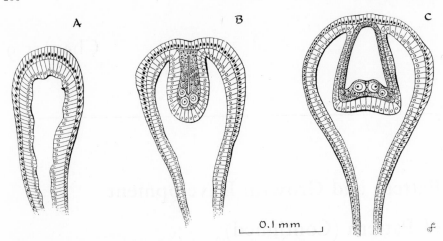

Figure 9-1. Early development of medusa bud in *Bougainvillia carolinensis.* **A.** Typical appearance at onset. **B–C.** Differentiation of oocytes within ectodermal entocodon, showing location on presumptive manubrium.

the size of individual oocytes in *B. carolinensis* is about one and a half times greater than in *B. superciliaris*—a difference which is reflected in cell number at each stage. The earlier emergence of oocytes in *B. carolinensis* is almost certainly related to this increase. It is also significant that the oocytes thus segregated cease to grow until after the medusa has been liberated; then they grow in proportion to the growth of the parent medusa. It is noteworthy, however, that the medusae of *B. carolinensis* begin to mature sexually when they are smaller than those of *B. superciliaris*.

Facultative Reproduction

1. In certain species, such as *Syncornye mirabilis* and *S. eximia,* the onset of medusa and gonad formation depends on temperature. In both species, at the low temperatures of late winter and early spring, the medusae produced are similar to those of *Bougainvillia superciliaris*—with no sign of gonads at the time of liberation—and grow to be fair-sized *Sarsia*-type medusae before sexual maturity is attained.

With the rising temperatures of late spring, the gonads form in the entocodal layer early in its development; and the ova or spermatozoa, depending on the individual, mature histologically at the same time as the medusa. The medusa produced in this manner usually fails to become free, is much larger than the first type of medusae at the time of the latter's liberation, and contains eggs of the same size as those produced by the latter. Since the medusa, which was rela-

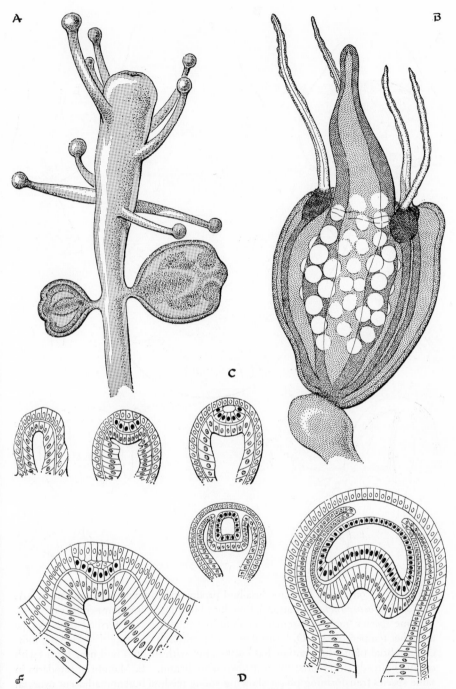

Figure 9-2. Facultative development in *Syncoryne*. **A.** Medusa buds near completion of development, attached to hydranth. **B.** Medusa buds (at same scale) at same stage, but at higher temperature, showing that hydranth has been resorbed and showing large scale and precocious formation of ova. **C–D.** Development of entocodon of **A** and **B**, respectively, showing relative scale of organization.

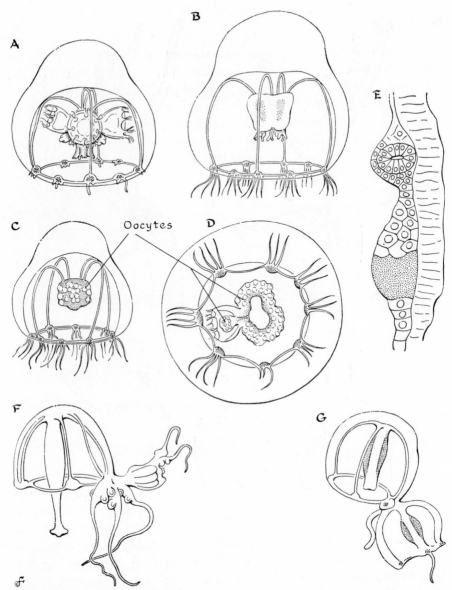

Oocytes

Figure 9-3. Conversion from medusa-bud production to sexual differentiation. **A.** In *Rathkea octopunctata,* manubrial medusa budding at low temperatures. **B–C.** In same species, male and female nonbudding medusae transformed from **A**-type at higher temperatures. **D.** Ventral view of medusa in same species, showing most of manubrial tissue differentiated as oocytes, but with medusa bud near term (with oocytes) already differentiating in its own manubrium. **E.** Manubrial ectoderm in same species transforming phase, showing young medusa bud and adjacent ovocytal differentiation. **F.** In *Hybocodon prolifer,* nonsexual medusa buds forming in profusion at one tentacle base of immature medusa. **G.** In same species, single, sexually differentiated medusa bud forming from tentacle base of sexually maturing medusa.

tively large at liberation time, ceases to grow at that time, its precociously mature eggs are far fewer in number than those ultimately produced by the *Bougainvillia* type of medusa during its long period of postliberation growth.

2. Somewhat similar is the reproductive process of *Hybocodon prolifer*, in which medusa buds are produced successively from a single, growing, subdividing tentacle bulb. Medusae budded from sexually immature medusae do not possess sexual tissue; those from medusae with differentiating gonadal tissue in the manubrial epidermis form gonadal tissue of much the same degree of differentiation, so that sexual maturity is attained at, or soon after, liberation.

3. During late winter and early spring, manubrial epidermis produces a succession of medusae buds in both *Rathkea octopunctatus* and *Limnocnida tanganyicae*. As the temperature rises above a critical degree, medusa budding ceases and the manubrial epidermis proceeds to differentiate into gonadal tissue. Though the transition is not known in histological detail, it is like the change in the processes described above, except for the complete suppression of medusa-bud formation rather than a restriction of medusa growth.

Effect of Increase in Mass of Epidermal Entocodon

The entocodons of medusa buds forming in hydroids vary greatly in mass. Since the range in mass greatly exceeds the range in initial extent or area, much of the variation must be due to differences in rate of growth. In *Pennaria*, for example, the massive initial epidermal growth becomes a multilayered mass, which sinks inward as a multilayered entocodon and, in conjunction with the adjoining endodermis, gives rise to typical medusoid structure except for the precocious differentiation into gonadal tissue of the entocodal tissue around the manubrium. There is a slight difference between the spring and summer forms of *P. tiarella*: the former, developing at the lower temperature, are liberated as free-swimming, mature medusae, soon shed their eggs or sperms, and then die; the latter, developing at the higher temperature, remain attached and shed their germ products in place—a procedure suggestive of somewhat less well-accomplished medusoid development.

In *Aselomaris*, a genus closely related to *Bougainvillia*, the initial scale of gonophore development is relatively small, the growth rate of the entocodal layer is high, and the whole entocodon differentiates into either a group of about 8 ova or an equivalent mass of spermatozoa, surrounding a typical endodermal spadix. The total expansion of material from the primary entocodal layer to the final mass of histologically mature gonadal tissue is obviously much greater in *Aselomaris* than the expansion from initial to liberated medusa in *Bougain-*

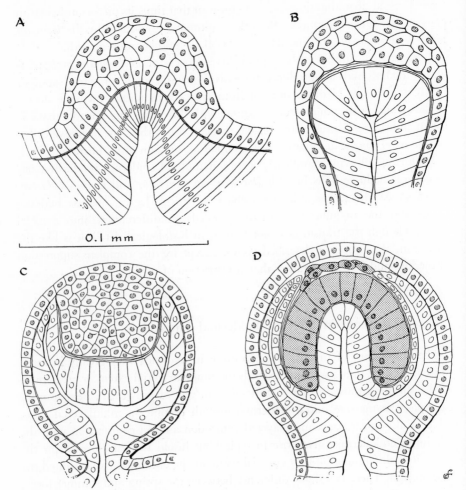

Figure 9-4. Early development of gonophore in *Pennaria*. **A.** Initiation, showing multilayered epidermal cone and single-layered endodermal protrusion. **B.** Slightly later stage, showing gonophore established but still consisting of single layer of endodermis and multilayered cap of epidermal cells. **C.** Growth by proximal extension of epidermal cap to form entocodon, showing distal region of endodermis pushed inward except for 4 marginal regions (2 shown), which are destined to become radial canals. **D.** Differentiated gonophore, showing entocodon completely segregated and differentiating primarily into gonad, gonophore epidermis completely reconstructed, and endodermis extending distally as central manubrium and lateral radial canals.

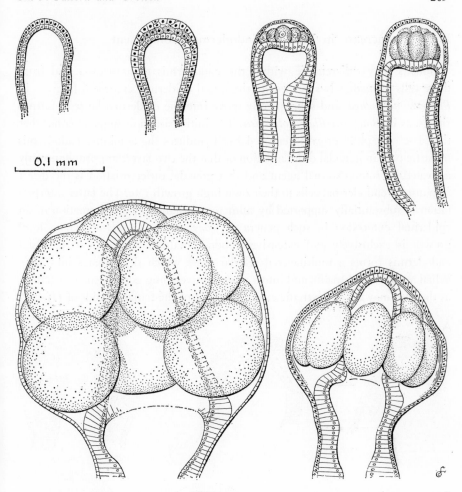

0.1 mm

Figure 9-5. Direct conversion of ectodermal entocodon into oocytes in *Aselomaris*.

villia. The medusoid structure in *Aselomaris* is thus reduced to the outer epidermal envelope and the inner spadix.

Similarly, many Campanulariidae genera, such as *Campanularia* and *Gonothyraea,* develop massive epidermal entocodons compared with those of *Obelia,* which differentiate mainly into gonad. Thus in *Campanularia* a gonophore consisting of epidermal envelope and endodermal spadix replaces the medusa, and mature germ cells are liberated after development of the gonophore is complete. In *Gonothyraea* a cluster of abortive tentacles that form from the distal epidermis emphasizes the medusoid nature of the gonophore.

Effect of Increase in Mass of Endodermal Component

Gonads may differentiate within the endodermis of a medusa bud from those cells immediately subjacent to the initial epidermal entocodal layer, as in *Podocoryne carnea,* and subsequently move into the epidermal layer to form 4 ridges, as in *Bougainvillia carolinensis.* The initial situation suggests either that the plate of rapidly growing entocodal cells induces the adjoining endodermis to participate in gonadal differentiation or that the two layers are simultaneously activated by some external agent and that gonadal differentiation is the direct response of endodermal cells to their own high growth rate. The latter interpretation is substantially supported by numerous instances of nonformation of an epidermal entocodon in such genera as *Coryne* and *Tubularia,* whose local growth is exclusively and extensively endodermal. In both genera the distal endodermis forms a multilayered cellular mass which delaminates into a residual, endodermal spadix and into a mass of enveloping gonad—male or female as the case may be. An extreme example is that of the female zooids of *Hydrac-*

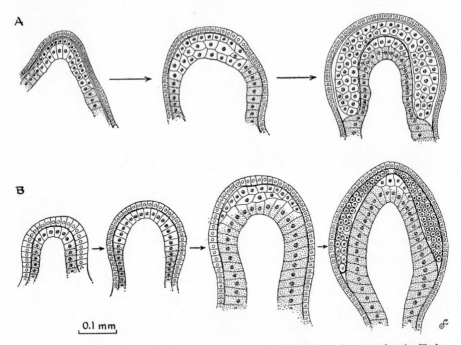

0.1 mm

Figure 9-6. Development of gonads directly from endodermal entocodon in *Tubularia.* **A.** Female. **B.** Male. Note relative stages of entocodal initiation in male and female; earlier start has greater growth potential and is correlated with differentiation of oocytes.

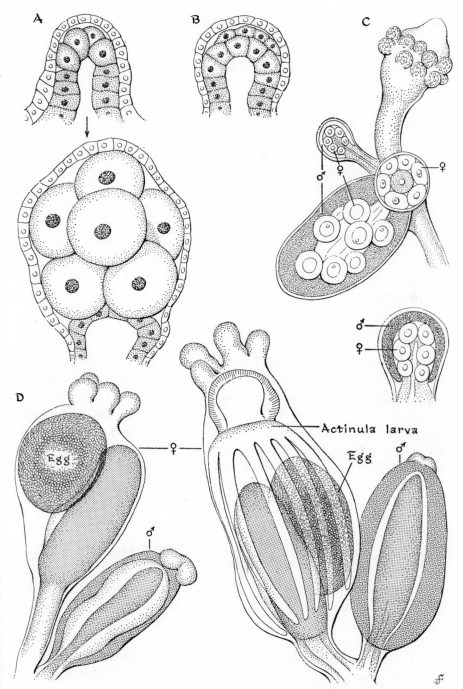

Figure 9-7. Initiation of gonads. **A.** Direct conversion of apical endodermal cells into oocytes in *Hydractinia echinata*. **B.** Mixed type which will give rise to hermaphrodite gonad with peripheral testis. **C.** Hermaphrodite gonad with peripheral testis. **D.** Two female gonophores of old female *Tubularia* polyp, each with new male gonophore grown from gonophore stem.

tinia, in which about 8 distal, endodermal cells start to grow but, instead of proliferating to form a mass of delaminating tissue, differentiate directly into oocytes and proceed with their individual growth and maturation.

Determination of Sex

Whether endodermal cells differentiate into male or female germ cells seems to depend primarily on growth rate. In *Tubularia, Hydractinia,* and *Aselomaris,* for example, initiation of female differentiation undoubtedly occurs at an earlier stage in the developmental process as a whole than does male differentiation. Moreover, the ultimate size of the female gonophore is considerably larger than the male. In other words, if female and male growth curves are assumed to be similiar in shape, if not in dimension, ovocytes are produced lower on the curve than are spermatogonia. The difference in final gonophore size suggests that the female grows faster from the outset than does the male and, accordingly, the endodermal cells respond to the higher rate by becoming female rather than male germ cells (for further discussion of this subtle, important problem, see p. 387).

It is of interest that in female *Tubularia* gonophores the residual gonad remaining in the endodermis after the primary mass has segregated and matured usually differentiates into male tissue—an occurrence which is in keeping with the retarded growth rate of the older tissue. Somewhat similarly, *Hydractinia* gonophores in certain regions of a colony are neither purely male nor female but are mixed because the originating tissue underwent some endodermal proliferation which gave rise to a small mass of segregated cells that differentiate into spermatogonia and to some endodermal cells that differentiate in place into ova, according to their location.

Terminal Growths and Morphogenetic Fields

Local outgrowths in hydroid colonies form either creeping stolons or free growths; the latter can arise from a stolon or from the stem of a hydranth. A free growth from a hydranth stem can give rise to a stolon, a hydranth, a medusa, or, presumptively, a combined hydranth and growing terminal. Thus, apart from the alternative developmental courses, the determination and properties of an outgrowth are of interest. Factors of dominance and inhibition are clearly involved in determination.

Whatever the ultimate form of an outgrowth, its initial location seems to be relatively fixed. Thus in *Obelia* or *Campanularia* a center of local growth appears at a certain distance behind the tip of the growing creeping stolon. Similarly, a lateral outgrowth from a free hydranth-bearing stem appears only at a certain distance from the distal end of the developing hydranth. In both the creeping stolon and the hydranth, the outgrowth forms from tissue which has completed the terminal growth cycle of growth, cell division, vacuolation, and recovery, including the rhythmical contractions.

Growth of a free terminal can be either rhythmic or spiral. Rhythmic growth, most clearly observed in campanularian hydroids, can consist of a series of growth surges which either are regularly spaced in time and in hydranth initiation or are interrupted by a phase of continuous growth, as in *Obelia*.

The question is whether a true growth pulse really occurs or whether surface-tension forces alone account for the apparent rhythmic activity. If epidermal tissue at the tip grows at a somewhat faster rate than does the endodermis and flows ahead of the endodermis, it might well form a succession of approximately spherical masses, as does water barely dripping from a collecting surface, except that, in the epidermis, tissue cohesion and polymerization of the chitin cuticle prevent actual separation of the successively produced units. This interpretation is supported by the fact that a terminal growing in contact with a smooth substratum does not show morphological evidence of growth surges, for the epidermis flattens out at the advancing tip. On the other hand, rhythmic growth in a growth zone below the neck of a fully formed hydranth of gymnoblastic hydroids generally, e.g., *Bougainvillia* and *Pennaria*, is clearly the same; i.e., the pulses occur whether or not a hydranth surmounts the growing tip. Thus when a hydranth is present, the surface-tension interpretation must give way to the conclusion that a true rhythm of growth takes place.[1]

Perisarcal Formation As a Record of Growth

The chitinous perisarc of a hydroid is as effective a record of past growth as is the shell of a mollusk. The chitin is, apparently, secreted by specialized cells in the surface of the advancing epidermis, largely at the growing tip, but to a lesser extent throughout the length of the stem or stolon, and also by the disk of a developing hydranth or gonangium. The material as a whole assumes the curvature of the surface of the epidermis. When epidermal growth is continuous or very rapid, a continuous sheath of polymerized material is produced. When the epidermis grows by intermittent surges, the material coating a bulge polymerizes before the next successive bulge is formed, and a series of perisarcal annuli result. The degree of pointedness of an annulus is an indication of the

---

[1] The relative roles of epidermis and endodermis in growth was discussed on p. 183; the more-general topic of growth rhythms is considered on p. 321, in relation to strobilation and segmentation.

length of the interval between two growth surges. After the perisarcal rings
have thus been formed, the tissue within usually stretches, narrows, and
withdraws from the cuticle, so that without the perisarc there would be no
record of past events.

Similarly, during the development of hydranths, gonangia, and other structures, the various hydrathecae and gonothecae, apart from their functional value
as protective devices, are records of the developmental process. As in epidermal
perisarc formation, chitin is secreted mainly by specialized cells concentrated
toward the distal end of the rudimentary structure, and the material polymerizes
as it is produced, progressively assuming the shape of the expanding distal end
of the rudiment. As the rhythmic contractions of the recovery growth phase

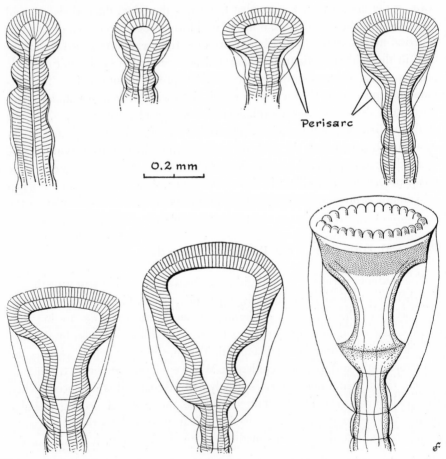

Perisarc

0.2 mm

Figure 9-8. Development of hydranth in *Eucopella caliculata,* showing formation
of double sheath of polymerized chitin.

commence, the underlying tissue withdraws from the polymerized perisarc, leaving it as a protective and supporting external wall. In one genus, *Eucopella*, the secretion occurs at two stages of hydranth growth, so that two separate walls are produced. However, if growth is rapid before polymerization is complete, the cuticle may be greatly stretched before it hardens; by contrast, a local growth may induce a reversal or softening of cuticle already hardened, as at the base of a gonangium.

Forms of Terminal Growth

Although the series of complete annuli seen in *Obelia, Campanularia,* and various gymnoblasts demonstrates the pulsative character of activity in either a terminal or subterminal growth zone, the same perisarcal formation in other organisms, e.g., the Campanulidae *Calycella* and *Opercularella,* indicates the spiral character of their growth. In the latter, there is no reason to suspect that the rhythmic type of growth does not take place in the epidermis, but the spiral continuity of the perisarcal groove definitely indicates an eccentric course of the terminal growth. The terminal growth region is off-center and revolves around the linear axis as growth proceeds. In *Sertularia* the spiral is steeper because the revolution of the growth center around the axis is relatively slower. Whether growth is rhythmic or spiral, it is significant that the hydranth rudiment culminating from such growth seems to be symmetrical. Once morphogenetic territories commence to form, however, growth eccentricity may appear as a lateral oscillation.

Terminal differentiation is more complex in the families Sertularidae (e.g., *Sertularia* and *Sertularella*) and Plumularidae (e.g., *Schizotricha*), and even in the genus *Coryne* as compared with the closely related *Syncoryne*, in which terminal growth is simple (p. 200). *Coryne corrigata* terminal growth is much faster in one transverse axis than in the other and results in an elliptical area which subdivides into two almost circular areas: one may give rise to a hydranth, and the other continue to grow in the same transverse axis and to subdivide; or both areas may give rise to a hydranth.

A similar growth phenomenon occurs in the Sertularidae in a more striking form. In *Sertularia* the terminal growth becomes much wider in one direction than in the other, and the disklike distal end—which in *Obelia* (Campanularidae) would form the tentacle-bearing shoulder of a single hydranth—subdivides into two lateral formative areas and a smaller central region. The three areas continue to develop independently of one another, the lateral ones to form hydranths and the central as a continuing terminal growth. Matched pairs of hydranths are thus formed successively as the stem continues to grow. The

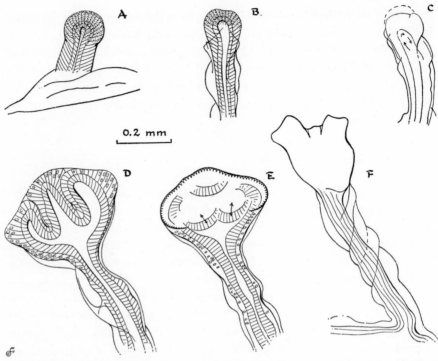

Figure 9-9. Spiral growth of free stolon in *Sertularia pumila* and its terminal
transformation into bilateral hydranth initiation zone. **A.** First stage, showing
erect stolon. **B–C.** Young erect stolons in spiral-growth phase (in **C**, broken line
indicates extent of growth in 4 hours; arrows denote circulation in hydrocoele).
D–E. Formation of first pair of hydranths with residual growing terminal between
them (in **E**, arrows indicate pulsative amplitude of hydranth wall). **F.** Slightly
younger stage, showing record of spiral growth in stem perisarc and terminal conversion to bilateral growth.

relative rate of growth in the two transverse axes is fixed, and the formation
of hydranth pairs is always in the same plane; the central terminal region grows
to a critical width each time a hydranth pair is formed, thereby creating space
for territorial segregations.

In the related genus *Sertularella,* hydranths form alternately on one side and
then on the other, although, as in *Sertularia,* always in the same plane. In both
genera the residual growing area, where the growth rate is presumably highest, seems to be approximately central. The pattern of growth seems to be due,
in *Sertularia,* to a stationary, central growth process; in *Sertularella,* to a growth
center revolving round the geometrical center (i.e., potentially forming a spiral
course); and in both, to the faster growth in one transverse axis than in the
other. The growth rate in both genera is highest at the center and lower on

the shoulders, where the primordia appear (cf. formation of leaf primordia, p. 420).

The determination of the relative axes is a separate matter, which seems to relate to a basic polarity. In *Campanularia* for certain, and probably in other

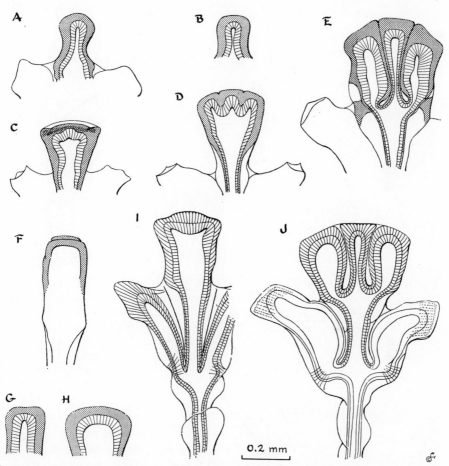

Figure 9-10. Development of hydranth and central terminal growth zone in *Sertularia pumila*. **A.** Young terminal, in same plane as paired hydranths, with terminal expansion. **B.** Same from side, showing growth uniformity. **C–D.** Later stages, showing subdivision of oblong or elliptical terminal apex into three distinctive areas: lateral hydranths and central terminal zone. **E–F.** Front and side views, respectively, of later stage. **G–H.** Narrow and broad sides, respectively, of single growing terminal, showing growth differential. **I.** Laterally developing hydranths, showing central terminal close to maximum phase. **J.** Central terminal at minimum phase. **I** and **J** represent renewal of growth from old stem tip. Note resumption of spiraling of stem as girth is reacquired.

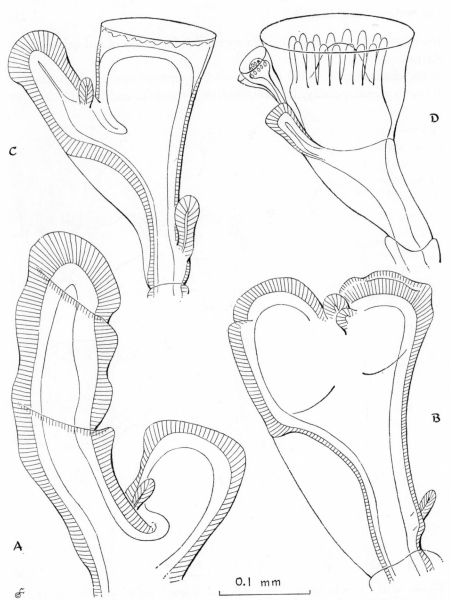

Figure 9-11. Relative growth of terminals and hydranths in *Schizotricha tenella*. **A–B.** Two stages in growth of main stem terminal with hydranth and associated nematophores forming on right in each stage; area of terminal growth exceeds that of hydranth primordium. **C.** Branch terminal, showing relatively advanced hydranth stage and relative reduction of growth terminal. **D.** Secondary branch terminal, showing hydranth and nematophore fully developed and growth terminal reduced to minimal size.

genera, new terminal growths, from which hydranths form, appear as either branches or vertical stems only on the side facing the growing end of an adjacent, attached, basal stolon.

The terminal growths of hydroids generally, and especially of the calyptoblastic families, offer ample material for the study of morphogenetic fields. The terminal hydranth-forming growth of *Schizotricha,* for instance, in the Plumularidae family, carrying out the type of pattern in the Sertularidae, grows much faster in one transverse axis than in the other. However, in contrast to *Sertularella,* the terminal growth area remains on one side, and the presumptive hydranth area on the other side successively segregates first an internodal nematophoral area, then a more-distal nematophoral area, and finally the hydranth rudiment. The residual growth terminal—of the one that remained intact—is relatively small; and shortly after it has become discernible as a distinct entity, it segregates at its base a nematophoral area which consequently develops into a rudiment attached to the side of the hydranth.

In all such hydroids, a shifting ratio between amount of space occupied by the presumptive hydranth area and that occupied by the residual terminal area is evident as stem growth proceeds. At early stages in the formation of a free stem with its lateral hydranths, the residual terminal area is relatively large, but as successive hydranths are formed, this area becomes smaller and finally ceases to grow. This seems to be a consequence of diminishing size of the terminal area, starting at the time morphogenetic areas are segregated and progressing as growth of the stem as a whole slows down. However, the areas segregated or delimited as presumptive hydranths attain much the same size as before, and it is the residual area that becomes progressively curtailed.

A rather extreme example of this diminution phenomenon is found in *Obelia.* When growth of a frond begins early and is vigorous, a massive growth terminal alternately segregates a presumptive hydranth first on one side and then on the other, and the growth rate of the terminal is such that extensive growth of the main distal stem has occurred before the development of a hydranth rudiment is even near completion. A terminal at a subapical or lateral location grows less rapidly, and segregation of a hydranth rudiment may leave a residual area the same size as the rudiment, so that the hydranth may be well on its way to completion before the residual terminal approaches it in length or mass. Finally, usually at the extreme apex of a main stem or on tertiary branches, virtually the entire terminal area is involved as a presumptive hydranth area, in which a hydranth develops completely and there is no sign of a residual terminal. With subsequent growth and widening of the stem, however, a new terminal may eventually appear at a point below the annular region, which is

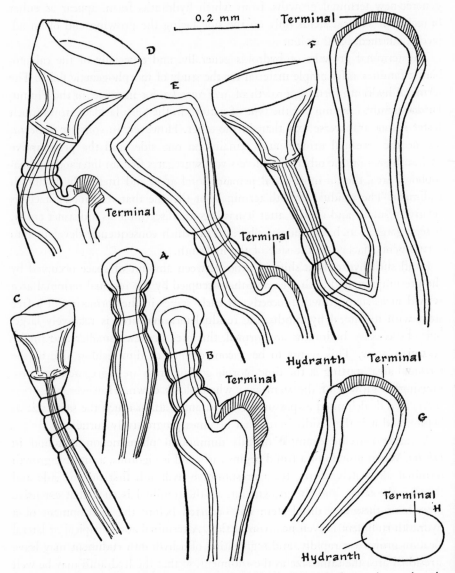

Figure 9-12. Differential growth of hydranths and terminal growing points in *Obelia geniculata.* (Chitinous perisarc in heavy line.) **A** and **C.** Hydranths developing from slender stems, with no residual terminal. **B, D,** and **E.** Hydranths at various developmental stages with terminal growing points in subterminal locations. **F.** Hydranth at stage comparable to **D,** with growing terminal massive and prolonged—typical of early and vigorous growth. **G–H.** Side and end views, respectively, of terminal growing point similar to **F,** showing segregation of small lateral presumptive hydranth (terminal growing region shaded).

immediately below the hydranth, where it would have been earlier if tissue or area had been available.

Thus it seems evident that terminal growth areas vary in shape from almost perfect circles to extended ellipses, that those morphogenetic areas which are presumptive hydranth areas must be essentially circular and have certain minimum dimensions if they are to form at all, and that their segregation may leave residual growth areas of varying size and shape. It also seems evident that in a terminal growth region the highest growth rate occurs at one place, which may be geometrically centered and stationary, may revolve around the geometric center in a spiral course, or may oscillate from one side of the region to the other.

Crowell (1953 and 1957), studying the influence of nutrient level on the growth zones in colonial *Campanularia,* found that once a terminal zone becomes active, its proliferation is just as rapid and vigorous in poorly nourished colonies as in the best-fed ones (nutrient level was experimentally controlled by varying the supply of newly hatched brine-shrimp larvae). The ability of the zone to become active, however, was directly related to the level of nutrition: at the lowest level the zone never becomes active; at intermediate levels it becomes active after a variable period of time—the less food, the longer the period. Yet the zone of growth, even when inhibited, remains potentially present, since it can be reactivated by raising the nutrient-level, and retains its essential position within the system (cf. *Botryllus,* p. 392).

As mentioned above, whenever a growing or developing unit has been well launched, it continues to grow or develop. Hydranth rudiments once established develop to completion even in starving colonies, as a rule at the expense of older stem or stolon tissue or of hydranths already fully developed. The same phenomenon often occurs during the development of medusa buds from hydranth walls, both in the hydroids *Limnocnida* and *Syncoryne* (p. 168) and, as Huxley (1921) found, in the colonial ascidian *Perophora,* in which young buds on a stolon continue to develop while inducing the degeneration and resorption of older buds. Actively growing tissue is clearly in a different state of vitality from histologically mature tissue: buds of both hydroids and ascidians develop to histological and functional maturity under conditions of, e.g., temperature and nutrition which cause the dissolution of fully developed individuals, only to dissolve upon attainment of histological maturity and cessation of growth.

Polarity, Orientation, and Pattern

Although local centers of growth, once the developmental course is established, develop into complete organismal units that seem to be virtually self-contained, there is usually evidence that the parental tissues which gave rise to the growth center impose upon it their own essential polarities or symmetries. In the hydroid *Campanularia* the hydranth rudiments, which develop from a vertical branch growing from the upper surface of an attached stolon, form on the side of the stem facing the stolon tip or on the upper side of secondary branches that in turn face the stem tip; and in general, Crowell (1953) found, when secondary branches form on two sides of a stem, the plane of symmetry of all the branches is in line with the basal stolon from which the stem has grown. In other words, although the pattern of growth is presumably inherent in the zone of growth, the initial orientation of the pattern is derived from the tissue polarity of the parental stolon—a comparable relationship, perhaps, to that between leaf primordium and apical meristem in vascular plants (p. 422). Similarly, the budding stolon of the tunicate *Salpa* (p. 344) produces numerous sexually mature and eventually free individuals toward the distal end; the dorsal, ventral, right, and left sides of the stolon and its progeny are directly imposed by, and in conformity with, the parent's organization. The same relationship is also evident in the buds of the ascidian *Botryllus:* each bud rudiment originates as a disk in the atrial epithelium of the parent, and although it grows and develops into an individual separate from its parent, its symmetries derive from, and conform to, those of the parent.

There is another phenomenon which is puzzling: Although the axes and symmetries of developing zooids in a colony are clearly derived from the parent organism whenever zooid and parent are connected, in certain ascidians tissue continuity with the parent is lost either temporarily or permanently; and at the time continuity is lost the developing buds orient in groups to form distinctive systems. In *Botryllus* and *Botrylloides,* for instance, the zooids of a colony are arranged in star-shaped and ladderlike systems, respectively, connecting with common cloacal cavities. Similarly, in polyclinid ascidians such as *Circinalium* and *Distaplia,* free, developing zooids group around common cloacal apertures to form separate systems. In all such separate groupings, the zooids are fully embedded—except for the siphon opening—in a common test (of tunicin and cellular material), which is either highly viscous or actually gelatinous.

The questions which arise are: What induces the mutual orientation? How is the orientation effected? As a rule, buds which group were close together and at about the same stage of development before they grouped; conversely, great distance and marked differences in stage of development militate against cooperation. In *Symplegma,* for instance, despite its close relationship to *Botryllus,* buds are produced in a series whose intervals are relatively long and whose buds are widely separated in space from one another, so that systems are not formed.

The developing zooids from a preceding zooid generation in *Botryllus* are at least somewhat oriented. In polyclinids and *Distaplia* the buds lie at all angles at first and assume mutually oriented positions in the common matrix only as development proceeds. According to Brien (1937), they turn not only so that their distal ends all point one way but also around their own axis as much as 90° so that their dorsal, or atrial, surfaces are all directed toward a common center. There is, undoubtedly, a reciprocal gradient in carbon dioxide, oxygen, and other metabolites between the center of a group and its periphery, but it is unknown whether this gradient is an inductive or directive agent effecting the ultimate orientation. Moreover, whatever the cause, the means by which a seemingly inert developing bud can turn toward a particular direction within such a viscous matrix remains a mystery. Mutual attraction and orientation are most marked during development, although Oka and Usui (1944) report that, in the polyclinid *Polycitor,* zooids in a colony continually change their position relative to each other. The questions remain: How can such a turning process be effected? What induces it?

Chapter 10

Pattern and Growth: Growth
and Differentiation

Hydra, the standard introduction to the study of metazoans in general, has long been the object of intensive analysis of growth and form, beginning with Trembley's pioneering work in the eighteenth century and culminating with the classical studies of Brien and Reniers-Decoen in the mid-twentieth century.

At the differentiated and fully organized multicellular level, *Hydra* exhibits about the simplest structure known. Unlike the difficulty of perception of more-complex organisms, *Hydra* form can be perceived readily in its three dimensions; and its tissues can be mentally pictured as proliferating sheets or centers of growth. Essentially *Hydra* form consists of an oral cone, or hypostome, surmounting a ring of tentacles below which a hollow, cylindrical column tapers to form a pedicel which terminates in a basal disk; and at all levels the body wall consists of two epithelial layers of tissue joined back to back and separated by a barely discernible layer of collagenous mesogloea. Despite the simplicity of structure, the organism raises most of the familiar problems concerning the maintenance and development of form—as Trembley's account (1744) of his experiments made abundantly clear.

Fundamental Experiments

Using the simplest of lenses and operating with exquisite delicacy on hydras contained in a drop of water in the palm of his hand, Trembley discovered that perfect polyps grow from small pieces of the body, although isolated tentacles do not regenerate; that multiheaded individuals are produced after splitting of the distal ends; that hydras turned inside out, by inserting a horse-hair in the mouth and then withdrawing it, can live for several months; and that pieces cut from the same species could be grafted.

Peebles (1897) found that the smallest piece of a hydra capable of regeneration, a sphere of tissue about ⅛ mm in diameter, forms an oral cone and one tentacle; and that although isolated tentacles do not regenerate, a tentacle whose base includes a small fragment of oral cone tissue can eventually regenerate a whole polyp. In addition, this investigator found that pieces of column cut successively closer to the base regenerate at a progressively slower rate—an observation which was confirmed and extended by Rulon and Child (1937). Undoubtedly, there is an axial gradient in regenerative capacity.

Brien and Reniers-Decoen (1949, 1950, and 1955) analyzed the process of growth of hydras in great detail—keeping individual hydras isolated from one another, feeding them daily with *Daphnia* or fragments of *Tubifex,* and recording the history of each individual both before and after operative manipulation or other experimentation. Much of the detail about individual hydras results from this procedure. On the other hand, changes occurring in dense populations have been studied by Loomis (1957), who used mass methods of culture.

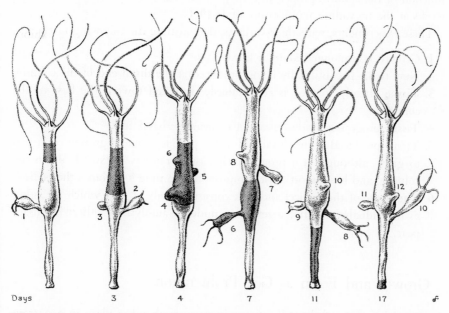

Figure 10-1. Basipetal growth of *Hydra* column, shown by shift and extension of stained segment grafted between head and trunk. Tissue successively becomes mid-column, budding zone, pedicel, and disk. Also shown is series of 10 buds produced during 17-day experiment. (After Brien and Reniers-Decoen.)

Brien and Reniers-Decoen's initial attempts to study the process of growth in nonsexual, nonbudding individuals by using silver markers and local stains were virtual failures. The technique finally employed consisted of staining some individuals with neutral red or Nile blue and leaving other specimens unstained, cutting a stained and an unstained individual in half, threading the upper half of a stained and the lower half of an unstained specimen on a thread of fine tungsten wire so that they were in contact and then fused, and finally withdrawing the wire. Despite the fact that the stain does not cross cell boundaries, it shifts slowly down toward the base until, about 10 days later, all the unstained tissue had disappeared and stained tissue had replaced it, the conjoined hydra maintaining the typical *Hydra* shape throughout the process.

A similar procedure was employed to graft a narrow segment of stained distal column into its equivalent position between the distal and basal regions of an unstained individual. The colored band traveled slowly down the column, the more-distal unstained tissue underwent extension, and finally the stained material successively became the budding region, the pedicel, and the basal disk. Moreover, as the stained tissue moved downward, it increased to between two and three times its initial length by the time its lower edge had become the junction of the column proper and the pedicel; the whole process took about 3 weeks at the prevailing temperature of 18-19°C.

These experiments, with their variants, demonstrate the following facts:

1. Growth is prolonged along a linear axis.
2. New tissue is continually being produced at the distal end of the column.
3. Tissue already formed is continuously displaced toward the base of the column.
4. Tissue progressively disappears at or near the basal end.
5. Tissue produced at the distal end in the primary zone of growth continues to grow, although at a progressively slower rate, as it extends down the column, and has almost ceased to grow by the time it becomes the pedicel.
6. The tissue of the pedicel and disk consists mainly of cells which are aging as individuals and after a few days undergo autolysis or otherwise disappear.

Growth and Form as Cell Proliferation

Analyzed in terms of the cells of a column, growth takes place in a narrow zone just below the ring of tentacles and is accomplished by cells relatively rapidly dividing in such a way that tissue growth is linear, with no significant increase in diameter. To accomplish this almost exclusively linear growth, the division spindles evidently lie parallel to the long axis of the column. Also, in the zone of growth, the distal daughters of dividing cells apparently maintain the maximum division rate while the basal, or posterior, offspring divide at a slower pace; otherwise, there would be no gradient, or decrement, down the column.

Growth of the column, according to Kanajev (1930), McConnell (1933b), and Brien and Reniers-Decoen (1950), is the result of mitotic activity of the large epitheliomuscular cells of the epidermis and of the large secretory endodermal cells and is not in any way the direct consequence of proliferation of the interstitial cells at the base of the epidermis. However, Tardent (1954) found that the number of interstitial cells is greatest at the distal end of the column and progressively decreases toward the basal end.

The form of a hydra column would be maintained even if the distal growth zone consisted solely of a narrow ring of dividing, polarized cells proximally producing successive generations of nondividing cells, which after a certain period die. However, the demonstrated progressive expansion of narrow bands of stained tissue intercalated below the maximal growth zone and the evidence of continuing mitosis down the column indicate that the cells which are progressively shunted along undergo, on an average, two additional divisions before protoplasmic growth ceases.

In sum, there persists a growth gradient which is expressed in the rate of tissue formation, the frequency of cell division, and the rate of general protoplasmic growth—all of which are maximal at the distal, or anterior, end of the column and negative at the opposing end. Thus the column of a hydra is essentially a continuing expression of growth; and the shape and length of the column can be regarded as the products of orientation, proliferation, and aging of the constituent cells. Any factor affecting only the rate of cell production at the distal end or the rate of cell senescence lengthens the column; and any modification of the polar orientation of division spindles widens the column at the expense of its length.

The role of the interstitial cells in regenerative growth has long been in dispute. Tardent and others assign them a dominant part, but irradiation experiments of Brien and Reniers-Decoen (1955) indicate that these cells are of secondary importance in regeneration. In nonirradiated individuals an amputated head is regenerated at first by large epithelial cells of both ectoderm and endoderm at the cut edge; these epithelial cells are the primary elements of both cicatrization and regeneration. About 6 hours after the regeneration, primordial interstitial cells accumulate at the margin of the wound; migrating between the epithelial ectodermal cells, they proliferate and give rise to multiple histogenic lines, i.e., to all types of ectodermal cells (especially cnidoblasts) and to the glandular cells of the endoderm. The question naturally arises: Are these interstitial cells essential?

Similarly, in X-irradiated individuals, the epitheliomuscular cells of the ecto-derm, together with the endodermal cells, accomplish first cicatrization of the wound and then regeneration of a normal, though somewhat small, specific hydra head, without subsequent congregation of interstitial cells. Yet the re-generated individual is not viable: nematocysts are not replaced, and there is a general cellular disaggregation after the tenth day. In other words, reconstitu-tion of form takes place without interstitial cells and without cell division gen-erally, but neither further growth nor tissue replacement occurs without cell proliferation.

However, irradiated individuals can be revitalized by grafts of normal tissue. A narrow band of nonirradiated tissue grafted, by the tungsten-wire method, between the upper and middle region of the column of an irradiated individual saves the combined individual from otherwise inevitable extinction. The re-suscitation process proceeds slowly, hour by hour, with some amoeboid cells transforming directly into stenotelar and holotrichal nematocysts or reconstitut-ing pockets of desmoneme cnidoblasts and others mixing with, and reinforcing, existing epithelial cells. Discernible in the living tissue are a penetration gradient and two encircling grooves: one is the original line of suture between irradiated and nonirradiated tissue, and the other is the line between tissue recolonized and that not yet colonized by interstitial cells. About 24 hours after the graft, inter-stitial cells start to migrate, gliding along the mesogloea at the base of the ectoderm and endoderm and passing beyond the suture line of the graft. The interstitial cells in the nonirradiated graft tissue apparently do not undergo usual differentiation, but remain or become large, basophile, amoeboid, pro-liferating cells—as if preparing for migration, even though it does not take place. The actual migration is from the graft in both distal and proximal regions, including even the tentacles; but only and always in intraspecific homo-grafts, never in interspecific heterografts (i.e., between tissues of different species).

Other questions arise: To what extent can the phenomena of growth be accounted for in terms of cells alone? Certainly the phenomena can be *de-scribed* adequately in terms of the constituting cells, in the same way that the construction of a building can be described in terms of the placement and quantity of bricks. The basic problem in both cell growth and building con-struction lies in the planning. Are the cues to which the cells respond in such an orderly way "built in," within the cells, or are the activating cues supracellular? If the controlling agency is supracellular—as seems likely—what is its material basis?

Growth as a Whole

Describing the growth within a hydra column in terms of cell division, growth, and aging merely sharpens the issue; for the cells are apparently merely the medium of the event and are subjected to controls to which cells as individuals may contribute, yet remain subordinate. Indeed, the problem becomes clearer only if the cell as a unit is ignored and the event regarded as a whole, for in noncellular terms the process of column growth is simply a linear extension (or expansion) of material at a rate progressively decreasing from its highest value at the distal region to zero at the base. The process recalls the growth of liquid crystals in one direction from a center of production, or a progressive unfolding of protein chains along one axis, with the rate of unfolding most

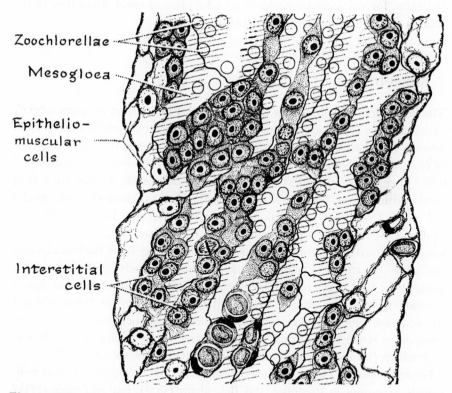

Figure 10-2. Longitudinal tangential section through base of ectoderm of *Hydra* column. Large, sinuous-lined areas are limits of epitheliomuscular cells based on mesogloea. Between these cells, trains of interstitial cells form helical bands. (After Brien.)

rapid in the zone of primary growth and becoming increasingly slow as the limits of extension are reached. Eventually it may be possible to describe such growth events in terms of supramolecular architecture, but the present state of knowledge suggests that such an attempt is somewhat premature. Nevertheless, at a purely biological level of inquiry, a further analysis of *Hydra* is at least suggestive.

There is no evidence that the growth zone at the distal end of a hydra column contributes to the maintenance or growth of the oral cone surmounting the column. Any cell replacement that occurs in the cone undoubtedly derives from its own tissue, either by division of surviving cells or by substitution of interstitial basal cells. As long as the total oral cone persists, so does the zone of growth below the ring of tentacles, and therefore the organism. Anything which destroys the integrity of the cone usually results in the progressive destruction of the adjacent growth zone and therefore eventually of the entire organism, for cell senescence and death take place as usual at the base of the column without compensatory cell birth at the top.

Reconstruction

The key to individuality seems to be localized. Finding that attempts to graft pieces of green *Hydra viridis* into brown *H. fusca* were invariably unsuccessful, Browne (1909) modified the discovery, by Whitney (1907), that the green color of *H. viridis* can be removed by keeping the hydras for 3 weeks in a mixture of 0.5% glycerin in water. Pieces of these artificially white hydras were then combined in various ways with the naturally green hydras of the same species (*H. viridis*). A reciprocal relationship between graft and host tissue became evident. Given the correct graft stimulus, a new head consisting of oral cone and tentacles formed anywhere in the body, except in the region of the tentacles and oral cone; but the peristomial stimulus was supplied only by the peristomial tissue at the base of the tentacles. Peristomial tissue, which adjoins and almost certainly includes the anterior limit of the growth zone, but apparently not the oral cone proper, evokes new head structure anywhere except in its own territory; whereas tissues from all lower levels of the column are ineffective everywhere (for comparable graft behavior in flatworms, see pp. 271-275).

Isolated masses of either ectoderm or endoderm do not reconstitute new individuals, according to Papenfuss and Bokenham (1939), but an earlier (1934) report by Papenfuss attests that even very small fragments consisting of both layers—25-60 fragments cut from the body of a single hydra—unite to form typical hydra structure. Furthermore, isolation experiments show that endoderm readily unites with endoderm and with mesogloea, but not with ectoderm; whereas ectoderm does not fuse with mesogloea or, apparently, with another layer of ectoderm. Beadle and Booth (1938) found that in the brackish-water hydroids *Cordylophora lacustris* and *Obelia gelatinosa,* neither layer in isolation reconstitutes structure, but ectoderm unites with ectoderm and endoderm with endoderm, although the endodermal fusions do not hold together for long and the ectodermal masses become rounded, forming hollow vesicles.

In *Hydra* when two fragments, each consisting of the two layers, join together, the initiative for joining comes from the endodermal cells, which send out protoplasmic strands that interweave with one another. Gaps in the ectodermal layer then seem to be covered by regenerative growth, whereby ectoderm spreads over the united endodermal surface. Thus first the fragments unite to form a plate with endoderm on one side and ectoderm on the other; then the edges roll together, forming a cylinder with the endoderm inward; and finally a head and basal disk appear.

Tissues and cells that are displaced from their normal positions relative to one another usually resume their correct positions if the distances to be traversed are of the same order as the cell dimensions. Roudabush (1933) found that in everted hydras that are prevented from undergoing neuromuscular reinversion, the ectodermal and endodermal cells seem to slip past one another to occupy their proper respective places.

When several fragments from various parts of the body are brought in proximity, an equally striking migration occurs, as vital-staining techniques by Papenfuss have shown. In fact, reconstitution of new individuals from a number of fragments takes place by a combination of regulation and migration. Coalescence occurs only between like parts—multiple heads, feet, or tentacles—thus eliminating supernumerary parts. Two like parts at a considerable distance from each other migrate as though mutually attracted and finally coalesce. When the number of tentacles is doubled by the coalescence of two heads, that number is reduced by further coalescence of tentacles, so that the final number is proportionate to the new peristomial circumference.

Brien and Reniers-Decoen (1955) effected longitudinal fusion of a pair of hydras of the same species: one individual was stained with neutral red, and the other was either stained with Nile blue or left unstained; then each was slit lengthwise through one side of the body wall and opened; one was spread ectoderm down in water on the bottom of a paraffin-covered dish, and the other endoderm down on top of the first; the two were held together for 2-3 hours by the weight of paraffin-covered platinum wire. The two individuals fused and became as one, behaving and growing with perfect unity. When two individuals of different species were employed, they remained together for a

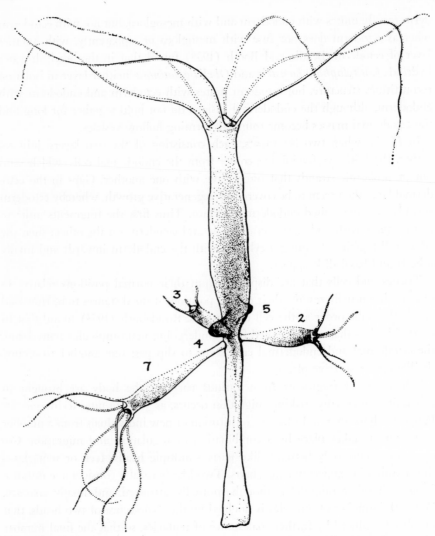

Figure 10-3. Succession and disposition of 5 buds in helical spiral in budding zone of *Hydra* column. (After Brien.)

while and then separated by a simple process of pinching along the two lines of suture, beginning at either end.

As in experiments with two like parts, in homografts between two hydras, the number of tentacles is at first the sum of the two original sets and the supernumerary tentacles are gradually eliminated. The final number of tentacles, proportionate to the peristomial circumference, is brought about by a merging of adjacent tentacles first at their base and progressively toward their tip. Not only does the total pattern and number of tentacles represent a dynamic state of structural equilibrium but individual tentacles seem to be in a state of growth, with new cells being added at the base and older cells dying at the distal end.

Burnett and Garofalo (1960) confirmed this dynamic state of growth in experiments in which the possible effects of graft wounds and any uncertainty concerning diffusion of vital dyes were avoided. An artificially white individual of the green *Hydra viridis* (obtained by Whitney's glycerin method) was cut in half through the center of its budding zone. The piece containing the head and tentacles was then grafted to a piece containing the base, stalk, and lower half of a normal green hydra. Buds produced from such combined forms were occasionally half-white and half-green, but were usually white with irregular blotches of green. When these buds reproduced, many of the several hundred new individuals had only a single patch consisting of one or two alga-laden cells, whose subsequent growth and ultimate fate were then studied. In all new individuals examined there was both a distal and a proximal migration of colored cells from the growth region. In many of these individuals the colored material moved down only one side of the body column—a situation showing that diffusion did not take place—and took 2-3 weeks to reach the basal disk, as Brien and Reniers-Decoen (1949) had found; on the other hand, colored cells took somewhat less than a week to move from the growth region to the tips of the tentacles.

These phenomena of reconstitution emphasize the importance of surfaces. Contractile tension at the basal surface of ectodermal sheets may account for the formation of hollow vesicles. Extension, as expressed by endodermal cells sending out cortical strands to unite endodermal fragments to one another seems to be equally characteristic of the endoderm. In other words, the basal surfaces both of individual ectodermal and of individual endodermal cells comprise a separate, continuous surface which has distinctive tensile properties, the former contractile and the latter extensive. Not only do the two surfaces differ significantly from each other but the migration and coalescence of similar parts within single reunion masses suggest that there are important differences between the uniting surfaces of head tissue on the one hand and foot tissue on the other. Whether the responsibility for this distinction belongs to either the endoderm or the ectoderm or to the two layers jointly is not clear.

The Seat of Organization

In *Hydra* and its close relatives the entire body wall consists of the comparatively simple tissue arrangement of a layer of ectoderm lying base to base with a layer of endoderm, with a very thin layer of noncellular mesogloea between the two layers as though it were cement holding them together.

If the organizational continuum demands significant material continuity, several possibilities arise. For instance, there is obvious continuity in the mesogloea, which seems to be a product of the endoderm rather than of the ectoderm; then, too, there is probably continuity of some kind in the free and basal surfaces of both endoderm and ectoderm in the form of united surface coats. In addition, in *Hydra* at least, a nerve net consisting of anastomosing protoplasmic strands lies in the basal region of the ectoderm. In other words, continuous surfaces formed by the conjoined surfaces or strands of adjacent individual cells may in some degree supersede the individuality of the contributing cells. The question is whether any of these possible extensive surfaces is the seat of the organizational controls.

The mesogloea seems to be the least likely candidate. Neither free outer surface—of the ectoderm or the endoderm—is probable: growth and repair in both layers proceed from the basal layer outward, and the free surfaces are relatively unstable because they constantly undergo deterioration and replacement. The nerve net cannot be the basis of organization, since it is not always present. Thus, unless the basal surfaces, possibly including the contiguous cellular interfaces, of the two epithelia—one or the other or both—are accepted as the material seat of organization, some metabolic, electrical, or even less-tangible agency would be the sole recourse.

Accordingly, as a provisional hypothesis, it may be postulated that there is a material continuity of cortical or ectoplasmic substance consisting of a continuous layer at the base of both the ectoderm and endoderm, and of less easily observed but nonetheless existent cortical continuities at intermediate levels, a continuity which does not negate cell individuality except that it ignores cell boundaries at the surface. Supporting evidence and argument for this concept of a supracellular layer as the basis of pattern and tissue behavior pervade this book.

At this juncture an analysis of the process of asexual and sexual reproduction in *Hydra* affords further insight into the nature of the organism.

Asexual Reproduction

Initiation and Development of Buds

Buds form as protrusions of the body wall of the lower middle region of the column. The very existence of a bud indicates a localized acceleration of growth. Such a growth center, which arises in succession at a certain distance from the anterior end of the column, represents a local return to a condition of maximal growth. Thus the larger region surrounding this local center is itself no longer growing at a maximum rate (otherwise, no differential in growth rate would be detected).

Most of the tissue at the lower end of a hydra column becomes involved in the production of buds; only that tissue which does not become thus involved becomes the pedicel. Consequently, the demarcation between column and pedicel is sharper in budding individuals than in juveniles. Questions that arise are: What determines the distance from the anterior growth zone to the local center in which a bud arises? What initiates the actual outgrowth of the bud?

Initiation

Both ectoderm and endoderm participate in the incipient stages of bud initiation in *Hydra viridis,* according to Brien and Reniers-Decoen (1950). Growth of the endoderm first appears as an increase in density of cytoplasm and number of chlorellae. Ectodermal growth is seen mainly as intense proliferation of

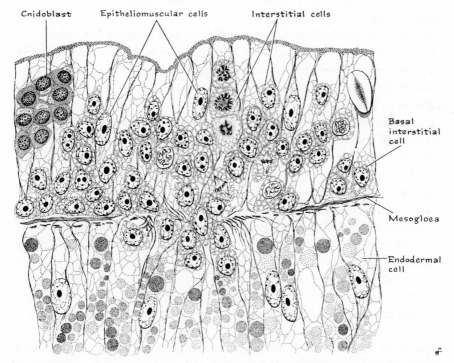

Figure 10-4. Onset of blastogenesis, showing multiplication of basal ectodermal interstitial cells and invasion of epidermis. (After Brien.)

the basal interstitial cells, which give rise to cnidoblasts at the margin of the presumptive bud area, then to large epitheliomuscular cells in the central region of the area, and after migration across the mesogloea, to glandular cells in the endoderm. Mitotic figures are almost exclusively in the horizontal plane relative to the polar axis of the organism, so that a local bulging of the body wall results rather than a linear extension. Subsequent growth, essentially a radial extension from the new center, involves a spreading outward of the interstitial cells and a coextensive growth of ectoderm and endoderm as a whole. The growth activity has a unitary quality: the epitheliomuscular tissue, interstitial tissue, and endoderm all respond as a unit in that they proliferate simultaneously although independently—each according to the nature of its cells. The interstitial cells seem to take the lead in growth of the local bud region, perhaps because they have less inertia owing to their small size; they play a very subordinate role in the growth of the column as a whole.

Studies of the histophysiology of the endoderm—e.g., that by Semal-Van Gansen (1954)—and of bud initiation demonstrate that the richness of protein inclusions directly relates to endodermal structure and indicates the fundamental importance of the endoderm as the functional tissue more sensitive to external factors.

One of three relationships between endoderm and ectoderm is possible:

1. Ectodermal variations induce endodermal variations. This possible relationship seems to have little weight, since the most obvious variable in external factors is the content of digestive inclusions and it is known that the endoderm is the more sensitive to external factors.

2. A chain reaction resulting from a number of internal factors (e.g., age) and external factors (e.g., food) determines the value of the protein reserves. Richness of protein in the reserve spherules of the endoderm induces multiplication of the reserve cells. One objection to this possibility comes from the discovery by Brien and Reniers (1955) that irradiated hydras totally deprived of interstitial cells still form buds.

3. A given content of reserve spherules, rich in RNA, corresponds to a particular state of the endoderm. The ectoderm as a whole then reacts to the stimulus received from the nourishing layer. One reaction is the multiplication of interstitial cells, and another probable reaction is the modification and multiplication of the epithelial ectodermal cells themselves. A hydra is thus forced to grow outward, and this outgrowth eventually detaches itself as a new polyp.

The third hypothesis at least fits the observations and is the most probable as the relationship between endoderm and ectoderm.

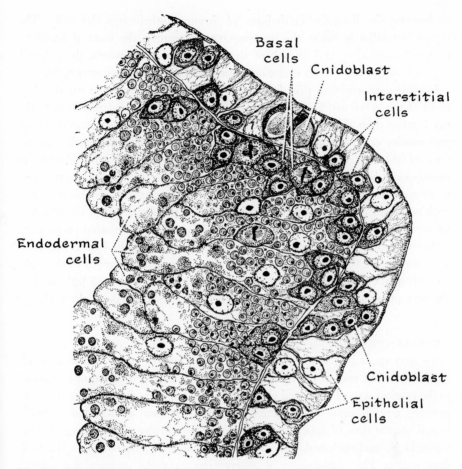

Figure 10-5. Section through bud evagination at beginning of bud formation, showing activity of interstitial basal cells (now on both sides of mesogloea) and intense proliferation of zoochlorellae—indicating growth metabolism in endodermis. (After Brien.)

Development

When the small, polystratified mass of proliferating interstitial cells is first evident, the underlying endoderm exhibits a slight fissure, which is the future gastric cavity of the bud. The fissure enlarges in proportion to the swelling of the ectoderm until the evagination looks like a cone whose base is much enlarged and merges tangentially with the wall of the column. The summit of the cone, which is the future oral cone of the bud, corresponds to the site where the interstitial cell mass first appeared. The polystratified plate of interstititial cells disperses and becomes greatly reduced in cell number, and the ectoderm thins to become the irregular epithelium of young epitheliomuscular cells. The region immediately below the presumptive oral cone is the zone of proliferation, differentiation, and growth of the column. In other words, the primary organization consists of an oral cone region and a subjacent growth zone.

In close conformity to this description of ectodermal growth is the development of the nerve net. Developing from the initial plate of interstitial cells, the net is formed by the growing together of protoplasmic strands from ganglion and sensory cells; but it remains undifferentiated until just before the appearance of the first tentacle rudiments. The anastomosing strands advance as pseudopodia between the epitheliomuscular cells. As the net grows, it seems to sweep downward from the head toward the foot of the emergent hydra, with new cells arising and anastomosing close to the distal end of the bud and progressively open and extended as its containing tissue is shunted proximally.

As the bud elongates, the first pair of tentacles arises at the base of the cone at an angle of 120° from each other and in symmetrical placement to the right and left of the polar axis of the parent. A third tentacle appears a little later in the very plane of the polar axis and on the most spacious side of the bud.

CONCLUSIONS

To sum up, a bud is essentially a secondary growth center which expands like a spreading pool at the base of the ectoderm and in a direction tangential to the original axis of polarized growth. The bud appears at a distance from the primary growth zone, where linear expansion or axial growth has virtually ceased. At the time of appearance, tissue nutrition in the primary growth zone seems to be undiminished, and cytoplasmic congestion characterizes the budding region. The combination of a persisting metabolic potential for growth and of the attainment of full linear extension in the organism as a whole may thus inevitably result in such a shift in the direction and expression of growth as is seen in the initiation of a bud.

Sexual Reproduction

The initiation and growth of the gonads in *Hydra* are in marked contrast to growth of the individual and to the process of budding. The following account of gonadal development is that by Brien and Reniers-Decoen (1949 and 1950) of *Hydra oligactis* and *H. viridis,* a dioecious and hermaphrodite species, respectively.

Budding and the Sexual Phase

During the sexual phase of *H. oligactis,* linear growth of the column seems to be arrested. With the onset of ovogenesis the ectoderm thickens in that region of the column which corresponds to the budding area in budding individuals. The thickening process gradually proceeds toward the head until the entire column from pedicel to peristome is tumescent. Eggs appear successively in various low regions of the column distally, and budding is inhibited. In *H. viridis,* however, the ovocytal growth processes occur in only a comparatively restricted transverse zone midway between the tentacles and the base of the column so that a considerable amount of column tissue is not involved in the processes. Such tissue shifts down the column to a site where it may take part in the process of budding. Thus budding and the sexual phase are compatible in this species.

Development and maturation of testes in *H. viridis* are complete long before ovogenesis commences. The same general process is discernible. Testes form in isolated areas in the distal region of the column in rapid succession, so that as many as 10 testes may be evident at once, although usually only the lowest 2 are functional. As the column grows, the 2 functional testes become displaced more and more toward the base while new testes form in the distal region, so that eventually testes actually regress in the pedicel region.

Brien and Reniers-Decoen (1950) conclude that the sexual phase is essentially an epiphenomenon arising from the inevitable competition between the general growth process and blastogenetic activity of the column. Accordingly, how the sexual phase and blastogenesis compare histologically becomes important.

Histology of Sexual Differentiation

In *H. viridis* each testis consists of a number of cysts known as testicular follicles, each of which exhibits a gradient in spermatogenesis, with spermatogonia at the base and spermatozoa at the apex of each follicle. The whole follicular and spermatogenetic differentiation is confined to the ectoderm and primarily to the interstitial cells.

Interstitial cells at the base of the epitheliomuscular cells become arranged as coarse strands in an irregular network stretched out on the mesogloea; both the cell strands and the mesogloeal meshes have a helical pattern.

The cells proliferate and at the same time the tissue condenses to form a thin layer, one cell thick, as if the meshes were being filled in. As a whole the cell plate acquires an elliptical shape, whose longer axis is parallel to the longitudinal axis of the column. The existence of the basal network of interstitial cells and the mode of formation of the spermatogenic cell plate thus seem to account for the location of the testes in helicoidal rows. The cells of each initial plate become spermatogonia, which give rise successively to more-peripheral layers of proliferating spermatocytes and differentiating spermatids.

During ovogenesis the ectoderm becomes profoundly thickened by an excessive interstitial-cell proliferation, which coincides with a total and permanent disappearance of cnidoblasts from the area of proliferation. An egg forms from an ovocytal plasmodium which has absorbed all the cells of a certain area of the thickened ectoderm; at the same time the ectoderm becomes reduced to only those epitheliomuscular cells which have not participated in the ovogenetic process.

Brien and Reniers-Decoen (1950) divide the differentiation process into four stages:

Stage 1. A preliminary phase typical of the normal column, during which

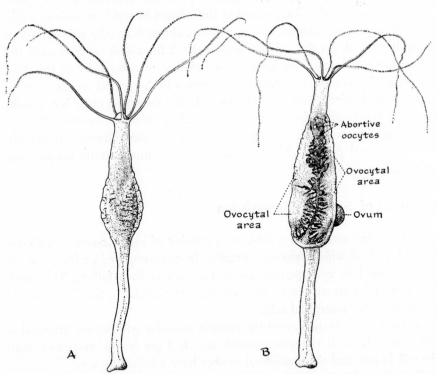

Figure 10-6. Ovogenesis in *Hydra.* **A.** Onset, showing ectodermal thickening. **B.** Extensive thickening of ectoderm and delimitation of ovocytal areas. (After Brien and Reniers-Decoen.)

basal interstitial cells proliferate and produce strands of small, differentiating cnidoblasts which migrate toward the free ectodermal surface.

Stage 2. The onset of ovogenetic thickening, during which strands (or columns) of interstitial cells extend almost to the ectodermal surface without differentiating into cnidoblasts.

Stage 3. The interval of differentiation of interstitial cells into primary

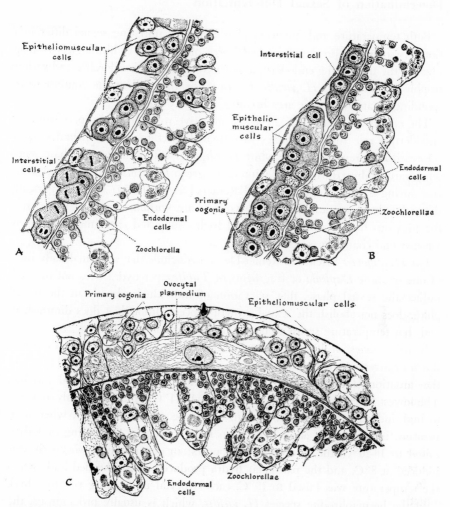

Figure 10-7. Histology of ovogenesis in *Hydra viridis*. **A.** Onset, showing interstitial cells proliferating. **B.** Interstitial cells giving rise to primary oogonia resting against mesogloea at base of epitheliomuscular cells, and symplasmic union of oogonia. **C.** Transformation of symplasmic mass into uninuclear ovocytal plasmodium, other nuclei having been resorbed. (After Brien.)

oogonia; during this stage the size of individual cells progressively decreases in the direction of the ectodermal surface.

Stage 4. The stage during which oogonia in contact with the mesogloea enlarge to form secondary oogonia; the peripheral layers remain unchanged and still covered by an attenuated residual layer of epitheliomuscular cells.

Determination of Sexual Differentiation

Both temperature and nutrition play a part in initiating sexual differentiation. As soon as a presumptive sexual area of ectoderm is histologically recognizable, the underlying endoderm appears congested with chlorellae and various trophic inclusions. In *H. viridis,* for instance, the wall of the column corresponding to an ovogenetic area becomes intensely green.

The endoderm clearly has an active nutritive role, although apart from the endoderm's role the two species investigated react to environmental circumstances in different ways: the white *H. oligactis* enters the sexual phase when temperatures drop during early winter; the green *H. viridis,* when temperatures rise during early spring. Brien and Reniers-Decoen (1950) state categorically that related observations by earlier workers are without statistical value since the previous history of any individual hydra subjected experimentally to environmental change was not recorded.

For *H. oligactis* Brien and Renier-Decoen conclude that a normal daily meal of one or more *Daphnia* or fragments of *Tubifex* per hydra does not by itself induce the sexual phase, that starvation of individuals already in the sexual phase does not abolish the sexual growth process although it does diminish it, and that temperature seems to be the decisive factor.

In this species, at 10-11°C, sexuality develops and coincides with the slowdown or arrest of the budding process, although the inhibition of budding and the initiation of ovogenesis at this temperature is both gradual and variable. The lowest lethal limit is 6-7°C, but even at 10-11°C some individuals continue to bud and seem to have difficulty in becoming sexual. However, when temperature is lowered to about 8-12°C to induce ovocytal thickening and then raised to 18-20°C, the thickening disappears; in one instance ovogenesis was induced at 8°C, and the particular hydra produced double-headed buds when the temperature was raised to 18°C, each ovocytal unit giving rise to a bud.

In the hermaphrodite species *H. viridis,* which is usually protandrous, the situation is not so clear. According to Whitney (1907), when individuals are exposed to 11°C for several weeks and then to 26°C without nourishment, sexual tissue appears—testes after 6-15 days at the high temperature and ova after 10-12 days. Abundance of good food immediately following the temperature rise impeded the formation of gonads and immediately caused large individuals to give rise to hermaphrodites and small ones to males only.

Loomis (1957) found that the sexual state is induced in hydra cultures when the population exceeds a certain density and that the critical factor in sexual initiation was a volatile gas, identified as carbon dioxide. However, carbon-dioxide tension may not be the only effective agent, particularly under circumstances other than those of mass culture. Nevertheless, the question remains whether the carbon-dioxide tension in mass cultures operates directly or, as seems more likely, indirectly on the interstitial cells to promote their differentiation as sex cells. In the light of the analysis by Brien and Reniers-Decoen (1949) of the histological growth processes involved in column growth and sexual differentiation, carbon dioxide probably reduces the growth rate in the zone of growth at the head of the column; i.e., the gas affects growth mainly where the growth rate is highest. In other words, sexual differentiation seems to be made possible by retardation or cessation of normal column growth; such a slowdown results from increased carbon-dioxide tension, which in turn causes a reduction of oxidative growth metabolism. If so, other factors, singly or in combination, are probably indirect determinants of the sexual condition.

Interpretation and Summary of Reproduction

1. The helical pattern of the mesogloea and of the bands of interstitial cells indicates that growth from the primary zone of growth takes a spiral course, which together with availability of space may well account for the spiral arrangement of both buds and testes.

2. An adequate nutrition level is necessary for budding and particularly for sexual differentiation.

3. Budding occurs only above a certain critical temperature level and depends on the active participation of both the endoderm and ectoderm as well as of the interstitial tissue.

Low temperature seems to (a) retard or inhibit the primary basipetal growth of the column; (b) inhibit endodermal cell division; (c) affect neither growth and proliferation of the interstitial cells at the base of the ectoderm nor division of the ectodermal epitheliomuscular cells; (d) consequently reduce or inhibit budding (which depends on endodermal participation), permit a massive response of interstitial-cell proliferation, and encourage epitheliomuscular cells to divide successively to form interstitial cells.

4. Cell specializations seem to be selections among absolute alternatives. If interstitial cells in a particular organism undergo differentiation at all; they can

differentiate into cnidoblasts, spermatocytes, oocytes, epitheliomuscular cells, or ganglion and sensory cells. The alternatives both between differentiation and nondifferentiation and among the differentiated forms seem to be clear-cut: there are no intermediate possibilities.

5. The relative growth rate, which may well be the decisive factor, depends on temperature, carbon-dioxide tension, and nutrition level. Any particular tissue probably has its own maximum rate, as indicated by tissue-culture studies generally. The rate as measured by an objective scale is not very significant; it is the rate relative to the tissue's maximum under specified conditions that is all-important. Interstitial cells growing and dividing at a relatively low rate seem to differentiate into cnidoblasts; at a higher rate, into spermatocytes; and at the highest rate, into oocytes. Moreover, epitheliomuscular cells growing and dividing relatively slowly give rise to similar epitheliomuscular cells; at a higher rate, to typical interstitial cells. Thus it seems that both interstitial cells and epitheliomuscular cells are generalized rather than specialized types.

6. Initial growth activities are indeterminate. For instance, when the temperature is high enough to allow endodermal proliferation, bud fields result; otherwise, ovarian or testicular fields result instead.

Low temperature seems to inhibit the polarized proliferation characteristic of the peristomial growth zone and of the basipetal growth of the column posterior to it and may be solely responsible for both sexual development and cessation of budding. Normal basipetal growth of the column precludes sexual development simply because progressive displacement of tissue is generally too rapid to allow for its sexual differentiation: at 18-20°C column tissue is entirely replaced by new tissue within 2 weeks. At a lower temperature, only after budding slowly ceases does ovogenesis become noticeable. Moreover, budding and sexual differentiation may be mutually exclusive at least in a region common to both. In both asexual and sexual reproduction, growth is primarily in the transverse plane of the column and interstitial cells proliferate at the base of the ectoderm; but the area involved in budding is very small compared with that typical of the early stages of sexual differentiation. A greater difference between the two types of reproduction lies in tissues other than the interstitial: in budding, proliferation continues in both the large ectodermal cells and the large endodermal cells, which are approximately orientated toward a common center; in sexual differentiation, however, the epitheliomuscular cells of the ectoderm and the gastroepithelial endodermal cells do not divide, and growth is confined to the original basal layer of interstitial cells at the base of the ectoderm. This sheet of interstitial tissue grows without extending its basal area, so that proliferation gives rise to a polystratified mass of cells; differentiation takes place in the peripherally displaced strata, the type of differentiation depending on the residual growth rate.

The question arises: can these various growth phenomena be interpreted in terms of a continuously growing sheet of protoplasmic material lying at the base of the ectoderm and endoderm and determining directly the fate of the adjacent cells and indirectly that of their descendants? Further investigations, particularly with the aid of electron microscopy, are necessary for proof or disproof of this hypothesis.

REFERENCES FOR
Pattern and Growth

HYDROIDS AND MEDUSAE

Agassiz, L. 1862. *Contributions to the Natural History of the United States of America.* Boston, Mass.: Little, Brown and Co. **4**:1-180.

Berrill, N. J. 1948. "A New Method of Reproduction in Obelia." *Biol. Bull.* **95**:94-99.

———. 1949a. "Developmental Analysis of Scyphomedusae." *Biol. Rev.* **24**:393-410.

———. 1949b. "Growth and Form in Calyptoblastic Hydroids. I. Comparison of a Campanulid, Campanularian, Sertularian and Plumularian. *J. Morphol.* **85**:297-336.

———. 1949c. "Growth and Form in Gymnoblastic Hydroids. I. Polymorphic Development in Bougainvillia and Aselomaris." *Ibid.* **84**:1-30.

———. 1949d. "Polymorphic Transformations in Obelia." *Quart. J. Micro. Sci.* **90**:235-64.

———. 1950a. "Development and Medusa-Bud Formation in the Hydromedusae." *Quart. Rev. Biol.* **25**:292-316.

———. 1950b. "Growth and Form in Calyptoblastic Hydroids. II. Polymorphism within the Campanularidae." *J. Morphol.* **87**:1-26.

———. 1952. "Growth and Form in Gymnoblastic Hydroids. II. Sexual and Asexual Reproduction in Rathkea. III. Hydranth and Gonophore Development in Pennaria and Aeaulis. IV. Relative Growth in Eudendrium." *Ibid.* **90**:1-32.

———.1953a. "Growth and Form in Gymnoblastic Hydroids. VI. Polymorphism within the Hydractiniidae." *Ibid.* **92**:241-72.

———. 1953b. "Growth and Form in Gymnoblastic Hydroids. VII. Growth and Reproduction in Syncoryne and Coryne." *Ibid.* pp. 273-302.

Bouillon, J. 1955a. "Le bourgeonnement manubrial de la méduse Limnocnida tanganyicae." *Acad. roy. sci. coloniales* (Brussels). **1**:1152-80.

———. 1955b. "Le cycle biologique de Limnocnida tanganyicae." *Ibid.* Pp. 229-46.

Boulenger, C. C. 1911. "On Some Points in the Anatomy and Bud Formation of Limnocnida tanganyicae." *Quart. J. Micro. Sci.* **57**:83-106.

Child, C. M. 1933. "Reconstitution in Haliclystus auricula Clark." *Sci. Reports Tohoku Univ.* Series 4. **8**:75-106.

Chun, C. 1896. "Atlantis I. Die Knospungsgeschichte der prolifierenden Medusen." *Bibliotheca zool.* **19**:1-51.

Crowell, S. 1953. "The Regression-Replacement Cycle of Hydranths of Obelia and Campanularia." *Physiol. Zool.* **26**:319-27.

——. 1957. "Differential Responses of Growth Zones to Nutritive Level, Age, and Temperature in the Colonial Hydroid Campanularia." *J. Exp. Zool.* **134**:63-90.

Dalyell, J. G. 1834. "On the Propagation of Scottish Zoophytes." *Edinburgh New Philos. J.* **17**:411.

Garstang, W. 1946. "The Morphology and Relations of the Siphonophora." *Quart. J. Micro. Sci.* **87**:103-93.

Gilchrist, F. G. 1937. "Budding and Locomotion in the Scyphistomas of Aurelia." *Biol. Bull.* **72**:99-124.

Hauenschild, C. 1954. "Genetische und entwicklungsphysiologische Untersuchungen über Intersexualität und Gewebeverträglichkeit bei Hydractinia echinata Flemm." *Arch. entwick. Org.* **147**:1-41.

Kuhn, A. 1914. "Entwicklungsgeschichte und Verwandtschaftsbeziehungen der Hydrozoen. Teil I. Die Hydroiden." *Ergeb. Fortschr. Zool.* **4**:1-284.

Mayer, A. G. 1910. "Medusae of the World." *Carnegie Inst. Washington.* Pub. 109. Pp. 1-148.

Payne, F. 1924. "A Study of Freshwater Medusae Craspedacusta ryderi." *J. Morphol.* **38**:387-430.

——. 1926. "Further Studies on the Life History of Craspedacusta ryderi." *Biol. Bull.* **50**:433-43.

Rees, W. J. 1939. "Observations on British and Norwegian Hydroids and Their Medusae." *Marine Biol. Assoc. U.K. J.* **23**:1-42.

Sars, M. 1841. "Ueber die Entwicklung der Medusa aurita und der Cyanea capillata." *Arch. naturges.* **7**:9-42.

Uchida, T. 1926. "Anatomy and Development of Mastigias papua." *Tokyo Univ. Fac. Sci. J.* Series 4. **1**:45-57.

ASCIDIANS

Brien, P. 1937. "Formation des coenobies chez les Polyclinidés." *Ann. Soc. Roy. Zool. Belgium.* **67**:63-73.

Huxley, J. S. 1921. "Studies in Dedifferentiation. II. Dedifferentiation and Resorption in Perophora." *Quart. J. Micro. Sci.* **65**:643-97.

Oka, H., and M. Usui. 1944. "On the Growth and Propagation of the Colonies in Polycitor mutabilis." *Sci. Reports Tokyo Bunrika Daigaku Tohoku Univ.* Series 4. **7**:23-53.

HYDRA

Beadle, L. C., and F. A. Booth. 1938. "Reorganization of Tissue Masses in Cordylophora." *J. Exp. Zool.* **15**:303-26.

Brien, P., and M. Reniers-Decoen. 1949. "La croissance, la blastogenèse, l'ovogenèse chez Hydra fusca Pallas." *Bull. Biol. France-Belgique.* **84**:293-386.

——. 1950. "Étude d'Hydra viridis (Linnaeus)." *Ann. Soc. Roy. Zool. Belgium.* **81**:33-108.

——. 1955. "La signification des cellules interstitielles des hydres d'eau douce et le problème de la réserve embryonnaire." *Bull. Biol. France-Belgique.* **89**:258-325.

Browne, E. 1909. "The Production of New Hydranths in Hydra by the Insertion of Small Grafts." *J. Exp. Zool.* **7**:1-24.

Burnett, A. L., and M. Garofalo. 1960. "Growth Pattern in Green Hydra, Chlorohydra viridissima." *Sci.* **131**:160.

Chalkley, H. W. 1945. "Quantitative Relation between the Number of Organized Centers and Tissue Volume in Regenerating Masses of Minced Body Sections of Hydra." *Nat. Cancer Inst. J.* **6**:191-5.

Ewer, R. F. 1948. "A Review of the Hydridae." *Zool. Soc. London Proc.* **118**:226-44.

Hyman, L. H. 1928. "Miscellaneous Observations on Hydra with Special Reference to Reproduction." *Biol. Bull.* **54**:65-108.

Kanajev, J. 1930. "Zur Frage der Bedeutung der interstitiellen Zellen bei Hydra." *Arch. entwick. Org.* **13**:135-77.

Loomis, W. F. 1954. "Environmental Factors Controlling growth in Hydra." *J. Exp. Zool.* **126**:223-34.

———. 1957. "Sexual Differentiation in Hydra." *Sci.* **126**:735-9.

McConnell, C. H. 1931. "A Detailed Study of the Endoderm of Hydra." *J. Morphol.* **52**:249-75.

———. 1933a. "The Development of Ectodermal Nerve in the Buds of Hydra." *Quart. J. Micro. Sci.* **75**:495-509.

———. 1933b. "Mitosis in Hydra." *Biol. Bull.* **64**:86-102.

Papenfuss, E. J. 1934. "Reunion of Pieces of Hydra with Special Reference to the Role of the Three Layers and the Fate of the Differentiated Parts." *Biol. Bull.* **67**:223-43.

———, and N. A. M. Bokenham. 1939. "The Fate of Ectoderm and Endoderm of Hydra When Cultured Independently." *Ibid.* **76**:1-6.

Peebles, F. M. 1897. "Experimental Studies on Hydra." *Arch. entwick. Org.* **5**:794-819.

Roudabush, R. L. 1933. "Phenomena of Regeneration in Everted Hydra." *Biol. Bull.* **64**:253-8.

Rulon, O., and C. M. Child. 1937. "Observations and Experiments on Developmental Pattern in Pelmatohydra oligactis." *Physiol. Zool.* **10**:1-13.

Semal-Van Gansen, P. 1954. "L'histophysiologie de l'endoderme de l'hydre d'eau douce." *Ann. Soc. Roy. Zool. Belgium.* **85**:217-78.

Tardent, P. 1954. "Axiale Verteilungs-Gradienten der interstitiellen Zallen en hydra und Tubularia und ihre Bedeutung für die Regeneration." *Arch. entwick. Org.* **146**:593-649.

Trembley, A. 1744. *Mémoires pour servir à l'histoire naturelle d'un genre de polypes de douce à bras en forme de cornes.* Leiden.

Turner, C. L. 1950. "The Reproductive Potential of a Single Clone of Pelmatohydra oligactis." *Biol. Bull.* **90**:285-99.

Whitney, D. 1907. "The Influence of External Factors in Causing the Development of Sexual Organs in Hydra viridis." *Arch. entwick. Org.* **24**:524-37.

Chapter 11

Reconstitution: Reconstruction and Axial Gradients

Reconstruction from Different Levels

Although in some organisms, e.g., the hydroid *Tubularia* and the polychaete worm *Sabella,* pieces isolated from any level along the apicobasal or antero-posterior axis can reconstruct the complete organism, in most organisms the capacity varies according to the position along the axis, but a different agency is responsible for the graded differences in position.

Fundamental questions that arise are: What is the basis of the gradient in regenerative capacity along the polar axis? What is the physical or chemical basis for the phenomenon of dominance exhibited in regeneration following oblique cuts? Child (1933) suggests that the relations between areas in development are primarily transmissive in character. In his discussion of regeneration from different levels in the scyphomedusan *Haliclystus,* instead of invoking metabolic gradients as in his earlier work, he considers regional differences as essentially electrochemical in character, with more rapidly developing regions transmitting electrical effects almost continuously to less rapidly developing regions within a certain distance. With this theory, Child anticipated the electromotive theories, concerning the nature of organization, of Burr and of Moment (p. 314), although undoubtedly he was influenced by the work of Lund (1923) on the electrical basis of polarity in hydroids. Whatever the merits of this concept, the question remains: What is the material basis of the controlling system? Is there, for instance, underlying the visible pattern, a symplasmic or exoplasmic fibrillar system that is responsible for regenerative phenomena and capable of producing and transmitting electrical differentials?

Regeneration distally but not proximally is a general phenomenon particularly of appendages or appendagelike structures. For example, both starfish and brittle stars regenerate distally any ray or part thereof which has been amputated, but an amputated distal part does not regenerate proximally unless part of the central disk is present. Similarly, the tentacles of anemones are readily regenerated, but isolated tentacles do not regenerate proximally to form the whole unless some basal tissue is present. It is probably significant that *Hydra* tentacles grow outward from the base even when regeneration is not involved, as in the process of reduction of supernumerary tentacles when adjacent tentacle bases fuse and a pair of tentacles become confluent progressively from the base to the tip—a process indicating that tentacles are normally maintained by a flow of organizational substance from the base toward the tip. In other words, if peripheral pattern is dynamically maintained acropetally, reconstruction of missing structure in the opposite direction may be of necessity precluded.

Regeneration in Hydroids and Flatworms

Hydroids

Regeneration and reconstitution have been intensively studied in the larger hydroids, particularly with regard to level of amputation, size of piece, and orientation of recombined fragments. The findings of such studies should be appraised in the context of hydroid sensitivity to physical environmental factors, which may play unsuspected roles in otherwise apparently simple experiments. Temperature (p. 179) is known to influence polymorphic development during normal growth. Light may be equally important, as Nakamura (1941) and earlier workers have demonstrated. Thus in *Syncoryne nipponica,* as in *Pennaria* species, stems kept in the dark do not regenerate amputated hydranths, but exposure to direct sunlight for 1 hour on each of 6 successive days is sufficient to induce regeneration, primarily effected by the blue region of white light. In view of the influence of light on the growth of *Phycomyces* (p. 27) and in producing striking reorganization in the polychaete *Sabella* (p. 302), the action and significance of visible light in relation to growth and morphogenesis is a major problem.

Clava

Clava, as does *Corymorpha* and *Tubularia,* forms clusters of comparatively large individuals which are usually united basally by stolonic growth. The process of wound closing and subsequent regeneration has been studied in *Clava* probably more closely than in any other hydroid, principally by Brien (1942 and 1943).

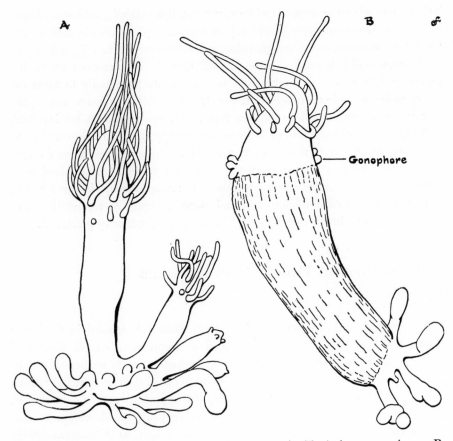

Figure 11-1. Regeneration in *Clava squamata.* **A.** Typical young colony. **B.** Regeneration from two cut ends of piece of column, including reappearance of gonophore at base of new hypostome. (After Brien.)

The general form of *Clava* is similar to that of a hydra. At the upper end of the long and tubular column is the mouth, surrounded by an oral cone; at the base of the cone is a peristomial zone, which bears several whorls of tentacles. Immediately below the tentacles is a gonophoral zone, and at the basal end of the column a variable number of blind strands grow out either to form anchoring stolons or to bend upward from the substratum and eventually become new individuals.

The exact extent to which basipetal growth of the column occurs is unknown, but it is known that growth does not approach the senility and dissolution characteristic of the *Hydra* pedicel, for a piece cut out of a *Clava* column regenerates typical basal strands from the lower cut end. The upper, or distal, end—whether cut high or low on the column—regenerates a conical mass that differentiates successively into oral cone, tentacle region, and gonophoral zone.

The histological processes involved in the attainment of this significant regenerative pattern is simple and comparable to those in hydras. The epidermis, essentially a single layer of cells, varies only slightly throughout the length of the individual *Clava*. The epidermis is thickest in the distal region adjacent to the gonophores and is progressively thinner toward the column base; the variation in thickness results mainly from the presence of a relatively large number of cnidoblasts in the gonophoral region. Interstitial cells in abundance at the base of the oral cone apparently constitute a fine, cambial layer from which other cell types arise, notably large epidermal cells, so that epidermal growth seems to be most concentrated in this region.

The endodermis also consists of a single layer of cells, larger than those of the epidermis, but variable in size depending on the state of vacuolation of an individual cell. The cell size increases progressively toward the base of the column as the cells become more specialized and distended. Cell division is in evidence only in the peristomial and gonophoral region, where endodermal cells are relatively small and proliferate rapidly. Accordingly, if the base of the oral cone, or hypostome, is the primary zone of growth for both layers—as seems probable—such would account for the youngest tentacles always being

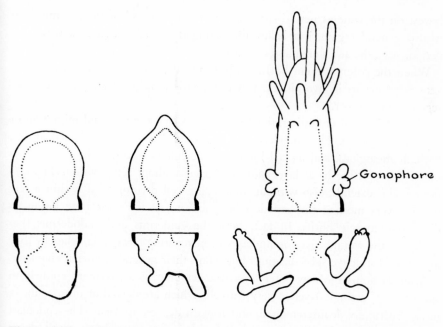

Figure 11-2. Three stages in regeneration of aboral hypostome and aboral stolons bearing new individuals. (After Brien.)

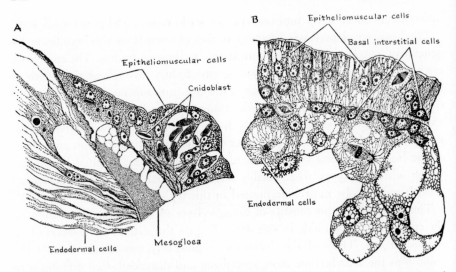

Figure 11-3. Sagittal sections in *Clava*. **A.** Through closing wound of transected individual, showing migration of epidermis over extended endodermis and mesogloea left behind. **B.** Lateral part of sagittal section of hypostome during regeneration, showing epitheliomuscular cells in division and basal interstitial cells passing into endodermis to become aligned against mesogloea. (After Brien.)

lowest on the cone as new tissue forms at the upper end of the column. It is in this general region, just below the youngest tentacles, that at a particular period, sex cells and gonophores originate.

When the column is cut horizontally below the gonophoral zone, distal regeneration occurs in two phases: first cicatrization and then reorganization, or regeneration proper.

About 1 hour after cutting, the wound closes by extension and migration of endodermal tissue. The endodermis at the wound is consequently naked, i.e., without mesogloea, and its cell limits are difficult to discern. The extension is mainly that of the basal layer and is effected without cell division. The epidermis also extends, also without undergoing cell division, and within a few hours covers most of the endodermal syncitium, at the same time losing its distinctive muscle fibrils. In fact, both the epidermis and endodermis that constitute the wound covering undergo a rejuvenating metaplasia, whereby they lose their distinctive specializations but retain their general epithelial character.

After 24 hours the new outer layer, the ectoblast, is a pavement epithelium consisting of a single layer of very flat cells which are devoid of muscle fibrils, have structurally dedifferentiated, and contain dense cytoplasm. The endoblast underlying the ectoblast is also epithelial and consists of relatively small, dense endodermal cells which possess neither vacuoles nor muscle fibrils. Cell multiplication occurs in both layers during the second day. Ectodermal cells which regain their columnar shape reacquire muscle cells; those that remain flat persist as a cambial layer of interstitial cells.

Cambial interstitial cells lying at the base of the new epidermis migrate in large numbers into the endodermal layer of the newly constituted oral cone,

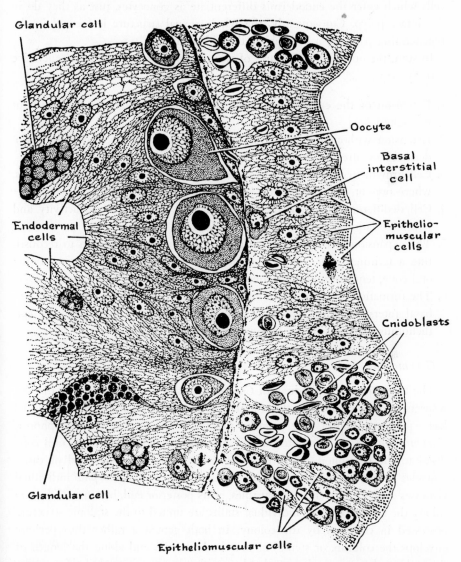

Figure 11-4. Portion of transverse section through base of regenerated hypostome in region of germinal reconstruction, showing endodermal crypt with young geno-cytes. Note no cnidoblasts in germinal area. (After Brien.)

where they differentiate into mucus-gland cells. Similar, but less-intense migration continues on the flank of the cone, where the interstitial cells slide between the large gastric cells to become granular gastric cells. A similar, but even slower replacement of gastric cells also occurs in the original column.

Below the regenerating oral cone and tentacle region, the migrating basal cells which enter the endodermis differentiate as gonocytes, just as they do in the intact polyp. Consequently, the whole sexual structure is formed in the regenerating part just a little above the old tissue.

In sum, the outstanding features of wound closure and onset of regenerative activity are:

1. Extension of the endodermal layer without cell proliferation and without mesogloea.
2. Extension without cell proliferation of the epidermal layer over the endodermal base already formed.
3. Cell growth in both layers, as expressed by loss of vacuoles and inclusions, where present, and by increase in cytoplasmic density.
4. Cell dedifferentiation, not merely modulation, during growth activity and redifferentiation according to new positions.
5. The expansion of tissue by extension, growth, and proliferation to reconstitute a terminal cone with a specific pattern of successive differentiations: oral cone, tentacle zone, and gonophoral zone.
6. The formation of gonocytes of either sex in the regenerating epidermis from basal interstitial cells, which in turn result from dedifferentiation of epitheliomuscular cells in the original epidermis.

Corymorpha

The large, solitary individuals of *Corymorpha* species are very similar in most respects to those comprising the colonial clusters of *Tubularia*. Each individual has a proximal ring of long tentacles just below the level of the gonophores: *Corymorpha* has an indefinite number of whorls of tentacles on the oral cone distal to the gonophores; whereas *Tubularia* forms a single whorl of distal tentacles at the anterior extremity. Furthermore, a *Corymorpha* individual possesses anchoring strands, or frustules, at its posterior end, but in a *Tubularia* colony the bottom ends of individual stalks are united to the stolonic structure possessed in common by the colony. In both genera a rather thin perisarc envelops the column, or stem, but not the hydranth, and along the length of the column there are endodermal ridges with grooves, or canal-like spaces, between adjacent pairs of ridges. Moreover, both genera possess a red pigment and a yellow pigment, which are apparently related to the nature of pattern.

Thus, because the two genera are so similar, the phenomena of regeneration, or reorganization, in each are discussed below in the light of the other; the differences between them are striking, but seem to be variations on a single theme.

The first analysis of regeneration in *Corymorpha,* by Torrey (1910) on *C. palma,* summarized below, is a necessary foundation for understanding the experimental results of later workers.

When a *Corymorpha* column is transected in the naked region distal to the perisarc, the large parenchymal cells, comprising the axial core and filling in the central region between the peripheral endodermal canals, protrude in the general shape of a dome. Marginal tissue at the cut edge extends beyond the old mesogloea and rapidly covers the dome. The old axial tissue then shrinks, but persists for a time in the form of endodermal ridges between the canals. Peripheral canal epithelium alternating with ridge epithelium extends as bands over the newly constituting oral cone. Canals and ridges then fuse distally and disappear.

Formation of the first set of proximal tentacles usually begins after, though occasionally before, the wound has closed, one tentacle between each pair of adjacent ridges. A second set then arises between, and at the same level as, the first.

Distal tentacles develop in successive groups, usually after the first proximal set forms. When canal bands are discernible in the oral cone region, the first distal tentacles arise, one on each band, and form a single whorl, as did the proximal tentacles. Succeeding tentacles form somewhat irregularly below the first distal set after the canals have fused. Accordingly, the initial regularity and subsequent irregularity of the distal tentacles may be related to the initial presence and progressive disruption of distal canal pattern.

Viewed from above, tentacles are first visible as an increased opacity, resulting from increased endodermal granulation at the sites where tentacles will appear. Endodermal cells at these sites multiply rapidly and push outward to form the tentacle core.

Gonophores develop as 8 rounded structures protruding immediately distal to the proximal tentacles when the latter are little more than conical elevations, and are produced as endodermal evaginations from the canal tissue between the ridges.

When a column is cut horizontally through the perisarcal region, new perisarc is formed within the old at a certain distance from the old edge. The epidermis left naked by this shift rapidly acquires the characters that readily distinguish naked ectoderm from that normally covered by perisarc. Normal adult proportions are then attained as the length of the stem, chiefly the naked part,

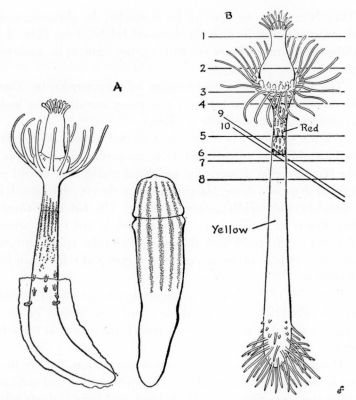

Figure 11-5. Hydranth regeneration in *Corymorpha*. **A.** Two stages in development of hydranth in *C. palma* after cut made just distal to edge of perisarc. (After Torrey.) **B.** Red and yellow regions and lines of cuts in *C. tomoensis*. (After Okada.)

increases and the diameter decreases. The shape of the adult after regeneration is always changed, the degree of change from normal varying with the initial departure from the normal. At the same time, new tentacles and gonophores form in typical fashion.

When the cut is immediately in front of the old perisarc, so that the distal tentacles and the greater part of the oral cone arise from naked stem, the proximal tentacles and base of the oral cone arise from covered stem. The edge of the perisarc then moves back while the new hydranth and neck extend forward.

Child's observations (1926 and 1927) on multipolar organisms developing from short pieces of *C. palma,* of which the commonest type consists of a single basal ring of proximal tentacles and gonophores surmounted by a multiple oral cone, each ring with its own set of distal tentacles, are in accordance with the basal-to-distal sequence of reconstruction. In other words, the basal part of the regenerating hydranth is determined in accordance with the patterns residual in the margin of the old tissue, but before the more-distal oral cone has been determined, if not fully formed, when the latter is still labile and subject to environmental disturbance but the basal region has reached an irreversible state.

Torrey's work has been continued directly—with repetition—by Okada (1927) on *C. tomoensis,* a species which in some ways is significantly different from *C. palma.* In this species the stalk, or column, consists of an upper red region and a lower yellow region, the former extending a little more than one-quarter of the way down the column. The red pigment is in the form of longitudinal bands, or stripes, which extend distally to the end of the oral cone and proximally to the junction of naked and perisarc-covered column. The bands seem to be in the same line as the tentacles since both proximal and distal tentacles are pigmented, and the unpigmented, transparent bands pass between the tentacle bases. Accordingly, the bands can be considered as the tissue of, or corresponding to, the peripheral endodermal canals, as in *C. palma.* Yet, the unpigmented bands extend into the yellow region of the column, so that pig-

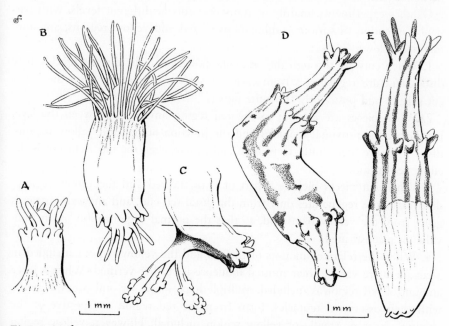

Figure 11-6. Regeneration in *Corymorpha* (see **Fig. 11-5B** for levels). **A.** From transverse cut at level 1. **B.** From transverse cut at level 2. **C.** From oblique cut at level 2–3. **D.** Of red piece cut from levels 4–5. **E.** Of red-yellow piece cut from levels 5–8. (After Okada.)

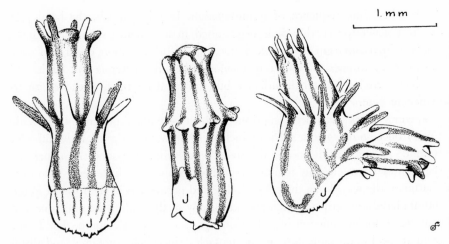

Figure 11-7. Three examples of *Corymorpha* regeneration in short pieces of red column with yellow tissue at their basal extremity. (After Okada.)

mentation apparently has a significance different from that of *C. palma* endodermal canals.

Okada's experiments, mainly of transverse cuts at different levels, with particular attention to various combinations of red and yellow regions, are summarized below:

1. When cuts are through the oral cone posterior to the distal tentacles, new distal structure regenerates from each cut surface—anteriorly from the lower cut surface and posteriorly from the upper.

2. When pieces are cut out of the red region anywhere between the level just below the proximal tentacles and the junction of red and yellow regions, distal structure regenerates from both ends, even when the piece cut out is the entire red region.

3. When cut pieces include a part of both the red and the yellow regions, distal structure regenerates only from the distal red end, and anchoring strands grow from the yellow basal end, so that the normal organization of an individual is reconstituted.

4. When the relative amounts of the red and the yellow regions included in a cut piece are varied, their mutual influence is further verified. When a large amount of red region is included, multiple hypostomes, or oral cones, together with their associated tentacles, form from the red region irrespective of the amount—large or small—of yellow region included. However, yellow region in the new individual is supplied only by the yellow region of the piece. When the amount of red is small and of yellow even smaller, a single oral cone usually forms from the red region, and the yellow forms the typical basal end.

5. When cuts are below the junction of red and yellow regions, i.e., only yellow, regeneration does not directly occur, as in *C. palma;* instead, the yellow region lengthens enormously about 12 hours after section and begins to twist, until about another 12 hours later it transforms into a long, tangled knot. The torsion becomes further accentuated; and finally the column constricts and divides into numerous small pieces, which vary in size and shape: some are spherical, some ovoid, and some claviform. All but the smallest (less than 0.4 mm long) develop into small, but perfect hydroid individuals and have been called planuloids since they are comparable in size and development, though they lack an external ciliary coat, to planula larvae.

On an average, pieces of yellow region originally about 1 cm long divide into 60-70 pieces; this fragmentation takes place apparently simultaneously throughout the region, and not, as in *Hypolitus peregrinus,* by commencing in the lowest part and progressing upward. Elongation of the yellow region, which may result from oxygen deficiency in the water and not only from section, involves proliferation of both epidermal and endodermal cells: those of the endodermis form an irregular, proliferating mass; and those of the epidermis merely proliferate and shorten, without forming a mass.

6. When a piece of yellow column long enough to exhibit fragmentation has a trace of red column attached to its upper end, no elongation or fragmentation takes place, a small area of oral cone forms from the red tissue, and the bulk

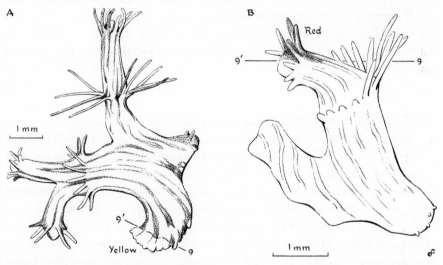

Figure 11-8.   Regeneration of colored columns in *Corymorpha.* **A.** Formation of multiple hydranth in piece of red column whose base is incompletely covered by yellow tissue. **B.** Regeneration of part of yellow column which has red tissue at its apical extremity. (After Okada.)

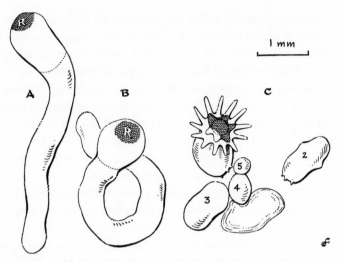

Figure 11-9. Regeneration of piece of yellow column with minute fragment of red tissue at apical end from transverse cut at level 10. Column elongates and subdivides, but apical unit regenerates hydranth at once. **A.** Elongation phase. **B.** Beginning of torsion. **C.** Fragmentation. (After Okada.)

of the hydranth forms from the yellow material. When the red component is proportionally further reduced, however, it does not inhibit the usual reaction of the yellow part of the column; thus the yellow region elongates and fragments in a typical manner, except that the anterior fragment, which contains the red material, is somewhat larger than the rest and reconstitutes a new hydranth with 4 out of 6 distal tentacles red and the other 2, together with the proximal tentacles, yellow.

Tubularia

Of all hydroids the several species of *Tubularia* have received the most attention, from perhaps the earliest observations, those of Dalyell (1847) on *Tubularia divisa,* that the head drops off and is replaced by a new one and that the stem behind the hydranth progressively elongates until the new hydranth dies, when the process is repeated. Whether this process of replacement and elongation results simply from age or time is uncertain, but there is no doubt that the head is relatively susceptible to temperature above a critical level. Both Morse (1909) and Moore (1939) attest that *T. crocea,* the species most studied in the west Atlantic, sloughs off its head when water temperature exceeds approximately 20°C, and regenerates a new head only when the temperature falls below 20°C. When the temperature does not fall, the distal end of the stem continues to grow, forming a stolonlike process (Berrill, 1948).

The most-detailed and best-illustrated account of *Tubularia* structure is still that of Agassiz (1864). *Tubularia* resmbles *Corymorpha* superficially, particularly in the general nature of the hydranth, which consists of a ring of tentacles at its distal end and a proximal circle of tentacles at its widest part, with 8 branching groups of gonophores growing out from the hydranth wall immediately in front of the proximal tentacles. Below the tentacles the hydranth body tapers to a narrow base, or neck, which joins the upper end of the stem. The stem is long and covered by perisarc, which is thin at its commencement just below the junction with the hydranth neck and gradually thickens down the stem. Throughout the length of the *T. crocea* stem, there are two opposing ridges, which were formed by thickening of the endodermis and which correspond to the more numerous endodermal ridges of some other species of *Tubularia* and *Corymorpha*.

As the organism grows, the hydranth grows larger without a significant change in shape; but both distal and proximal tentacles increase in number as the circumference at the two levels increases. At the same time, the initially simple gonophoral outgrowths lengthen and branch to produce complex clusters of progressively ripening gonophores.

Stem growth occurs within a relatively narrow zone between the hydranth neck and the stem's upper region, where perisarc is least discernible because it is thinnest; this zone of stem growth corresponds to the primary zone of maximum column growth in *Hydra*. Elongation of a tubularian stem is therefore mainly a process of the addition of new tissue at its distal end. The exact extent of the relatively retarded growth throughout the stem is not known, but the increase in stem diameter accompanying the increase in length indicates at least some growth in the transverse plane and possibly some growth along the linear axis as well.

The uppermost stem tissue seems to be uncovered by perisarc; below that tissue, the perisarc is very thin, but as it progresses down the stem the perisarc imperceptibly thickens and increases in number of constituent laminations. There is little doubt that perisarcal material is continuously, though slowly secreted by stem tissue throughout the length of the stem, so that the lower region of a stem consists of old tissue enveloped by perisarc that is relatively thick compared with more-distal regions. This double gradient in tissue age and envelopment is surely significant in relation to the phenomena of axial gradients in regenerative behavior. The decreasing gradient in oxygen consumption down the stem, as reported by Hyman (1926) and Barth (1940), may well be the consequence of both age of tissue and thickness of confining perisarc.

Red pigment is as conspicuous in *Tubularia* as it is in *Corymorpha*, and in the intact individual is diffuse throughout the body of the hydranth (including

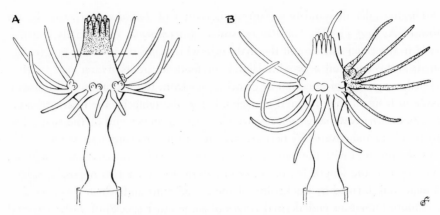

Figure 11-10. Regeneration in *Tubularia*. **A.** Distal regeneration from cut through manubrium. **B.** Lateral regeneration from cut through proximal tentacle region. (After Davidson and Berrill.)

the neck) and gonophores but inconspicuous in the tentacles. Unlike *Corymorpha, Tubularia* pigment is not arranged in bands alternating with unpigmented bands; however, it is concentrated—at least in certain varieties of *Tubularia*—in the tissues of the two longitudinal ridges of the stem. Morgan (1901) observed that when the stem is split longitudinally, the two endodermal ridges appear as two long red stripes. In *Corymorpha,* however, the pigment lies only in the peripheral cells of the longitudinal canals, and not in the intervening ridges.

Apart from the possible importance of both the red pigment and the perisarc in regenerative phenomena, the pigment is unquestionably a valuable index of metabolic events, as grafting experiments have demonstrated.

The functional hydranth of *Tubularia* regenerates missing distal regions 1-2 days after removal. When the oral cone including the distal tentacles is cut through at any level, the residual tissue rapidly regenerates a new cone. Removal of lateral parts of a hydranth, consisting of a number of the proximal tentacles, gonophores, and a fragment of the lateral body wall at its widest location, is also followed by rapid replacement of the amputated tissue (Davidson and Berrill, 1948). (To date, a histological analysis of hydranth regeneration has not been made for either *Tubularia* or *Corymorpha,* and cytological information about either genus is virtually nonexistent.)

Regeneration from hydranth fragments is in sharp contrast to the reconstitution of a complete new hydranth after transection of the stem or loss of the original head as a result of temperature fluctuations. In fact, the process of stem reconstitution seems to be the only method by which an entire hydranth ever develops (other than from an egg, of course). Stemlike branch terminals develop into hydranths by exactly the same procedure as do transected stems; neither terminals nor stems require a special stimulus, but terminal stolonic growth of the branch must have virtually ceased (Berrill, 1952). It is of particular interest, however, that as soon as the terminal growth of a branch or primary upgrowth has ceased and a hydranth has formed from the subterminal tissue, a secondary growth is established at the junction between the newly formed hydranth and the more-proximal region of the branch.

The process of reconstitution, as studied in transected stems (to date, analyses of normal growth of young colonies have not been made), consists of the following stages (Davidson and Berrill, 1948; and Rafferty, 1955).

Stage 1. A very brief stage, from time of cutting to healing of open end of stem.

Stage 2. Prepigmentation stage, from the completion of healing to the first appearance of diffuse red pigment.

Stage 3. Diffuse-pigmentation stage, from the appearance in the endodermis of diffuse red pigment as a broad zone commencing a short distance below the clear, unpigmented healed tip and extending down to a variable distance usually 2-3 times the diameter of the stem), with the pigmentation less diffuse proximally.

Stage 4. Defined-pigmentation stage, during which the pigment zone is sharply defined at its proximal limits. At this stage the whole primordium is delimited.

Stage 5. Tentacle-band stage, during which the pigmented zone as a whole is

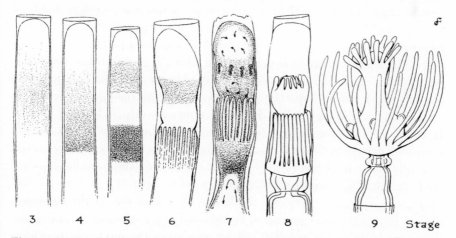

Figure 11-11. Normal stages in reconstitution of hydranth in piece of *Tubularia* stem.

differentiated into a distal and a proximal band, separated from each other by a clear, or unpigmented, region.

Stage 6. Proximal-striation, or proximal-tentacle-ridge, stage, during which tentacle ridges, indicated by red striations, appear in the proximal pigmentation band.

Stage 7. Distal-striation, or distal-tentacle-ridge, stage, during which distal tentacle ridges similar to those in the proximal band appear; at the same time, the proximal ridges progressively separate from the stem until they are united only by their proximal end.

Stage 8. Constriction stage, during which distal ridges separate in the same manner as did the proximal ridges; at same time, a constriction appears just below the proximal tentacles at the junction between pigmented and unpigmented tissues.

Stage 9. Final stage, at which time the completed hydranth protrudes beyond the apical limit of the stem perisarc.

This general course of reconstitution raises many questions: What is the significance of the red pigment? To what extent is cutting of the stem or removal of perisarc a stimulus to reconstitution? What determines the scale of organization with regard to the length of the pigmented zone or to the number of tentacles produced? What is the histological basis of the progressive differentiation? After sixty years of intensive experimental though not histological, work concerning such problems, only partial answers have been forthcoming.

The immediately striking phenomenon in tubularian reconstitution is the appearance of red pigmentation delimiting the formative area and progressively concentrating into regional bands—concentrations which foreshadow structural developments. As long ago as 1891, Jacques Loeb suggested that the red pigment was a formative substance that played an important part in the regeneration of tubularian stems—a hypothesis that Driesch strongly advocated a few years later, in 1897. The concept of formative pigment and other specific formative substances was an important one in the nineteenth century and persists in more subtle phraseology today, despite its critics as early as Morgan in 1901. In denouncing the formative role of the red pigment in *Tubularia,* Morgan contended that the red color is that of red granules observed in endodermal cells, especially of the basal part of the hydranth and of the stem immediately below the hydranth. The granules vary in size upward to about 50 μ in diameter. Red globules seen circulating within the lumen of a stem are for the most part derived from local disintegration of endodermal ridges. Morgan concluded that the polyp produced the pigment, and not vice versa. He also noted that pieces cut from very white stems reconstituted extremely pale, but otherwise normal hydranths, with virtually no red pigment discernible during the reconstitutive process.

In recent years the problem has been re-examined with new techniques. Cohen (1952) fused pieces of red stems and yellow stems by inserting a narrow yellow piece into a wider red piece, keeping the gastric ridges of the two fragments aligned. Whenever a hydranth formed in the yellow region of such a combination, it had more red than did the control hydranth formed in two fused yellow pieces. The gastric ridge of a yellow component joined to red was lined with red particles; but when the ciliary circulation of the coelenteron was blocked with oil drops, the yellow tissue remained as yellow as in controls. Evidently, granules in some form are picked up from the circulation by endodermal cells; and from this observation Cohen concluded that pigmented granules thus extracted came originally from the breakdown of endodermal tissue, specifically that of the gastric ridges, which are known to disappear in the reorganization region after section. The pigment was identified by Cohen— confirmed by Goldman (1953)—as the carotenoid astancene.

Cohen also concluded that the concentration of pigment as stripes, corresponding to the tentacle ridges, resulted from differential phagocytic activity of special endodermal cells. Spectrophotometric assays of pigment concentration in short pieces made at successive intervals in the reconstitutive process show that during regeneration the amount of pigment definitely increases, in the form of anisotropic crystals in the endodermi. Goldman (1953) draws the conclusion that the crystals are synthesized *in situ* and that the reddening of the hydranth primordium results from increased crystal concentration. The evidence, then, is that pigment is synthesized by the endodermal cells under certain poorly known circumstances, may be lost into the lumen when these cells disintegrate, and may be picked up again phagocytically by other endodermal cells; the evidence to date does not indicate that pigment granules or crystals play any direct formative role. Nevertheless, the pigment seems to be an index of the metabolic state of endodermal tissue and primarily associated with growth. The endodermis of the tip of a growing stolon—whether that stolon is long (i.e., full grown) or an emergent outgrowth from the side of a stem—is intensely red, as are the growing wall of the hydranth and the endodermal fingers which form the nutritive cores of the gonophores. In a stolon or branch tip the intensity of redness is correlated with the actual extent of the region containing growing and dividing cells—a region which extends down from the tip proper for a distance roughly twice the diameter of the stem. When cell growth and division cease, the pigmentation fades; in other words, the red pigment (carotenoid crystals), produced within the endodermis, is indicative of growth, either of the endodermal tissue itself or of its products when they are shifted outward to support growth of adjacent epidermis or gonocytes. Thus knowledge of the particular relationship of these crystals to growth metabolism or protoplasmic synthesis is most desirable. Moreover, if pigment connotes growth generally, as it seems to in growth cycles of *Tubularia* colonies as a whole, it probably holds for the reconstitution phenomena as well.

The major problem in *Tubularia* restitution concerns the progression of differentiation, which is somewhat illuminated by the following isolation procedures summarized from Davidson and Berrill (1948): Up to and including stage 3, when an individual is cut through the middle of the pigmented zone— whether the pigment is visible or presumptive—the small, isolated distal fragment forms distal tentacles at both ends (this is the usual procedure of any such short pieces of stem); the other fragment, which is proximal region of the pigmented zone, differentiates into both proximal and distal bands as though it were whole. When a similar separation is made at stage 4, the proximal fragment continuous with the stem reconstitutes a complete hydranth; but the isolated distal piece forms distal tentacles only at the distal end and forms typical gonophores at the other (this is normal procedure: these structures would have formed from this region had it remained in the original whole). When an equivalent cut is made at stage 5 and 6, the distal piece differentiates as it did in stage 4, but the proximal region proceeds as though it were still a region in the original whole: it develops proximal tentacles only, with a ring of gonophores adjacent to them on the distal side.

Thus differentiation is progressive and also seems to proceed according to a pattern since the determinate condition is reached in the distal region before it is confirmed in the proximal region. On the other hand, the gonophoral region, which lies between the distal and proximal pigment zones and is associated with the two groups of tentacles, is determined only in a general way since splitting the presumptive zone results in duplicate sets of gonophores.

The fate of the partial structures thus produced is especially interesting. Those formed from distal pieces isolated at stage 3 and developing bipolar rings of distal tentacles, or at stage 4 and developing a ring of distal tentacles and a ring of proximal gonophores, never proceed from either stage of development: no subsequent growth, regeneration, or reorganization occurs. On the other hand, after section at stage 5 or 6, the partial organization typical of the proximal region continues to differentiate as a partial structure until the normal time for extrusion of a hydranth, when the tissues have become histologically functional; only then are the missing distal structures regenerated. Such regeneration is true regeneration in that new tissue is grown distally from proximal, differentiated structure and acquires typical distal differentiation in the process. Accordingly, the phenomenon is typical of distal, or anterior, re-

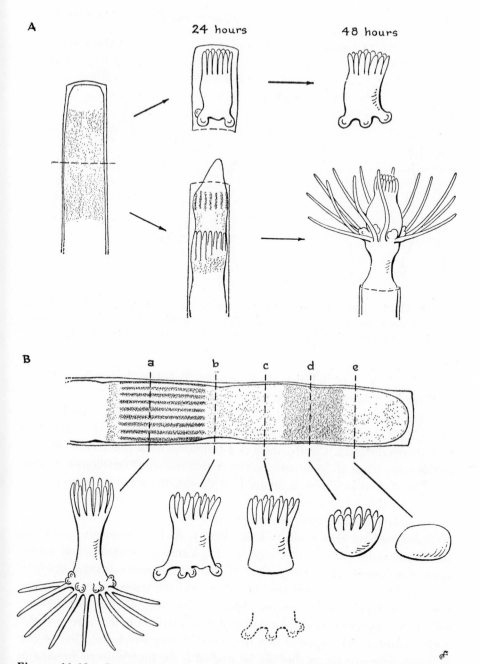

Figure 11-12. Regeneration of pieces of reconstituting stem in *Tubularia*. **A.** Cut through middle of pigmentation stage (stage 3), showing complete development of proximal piece and formation of gonophores from cut surface of distal piece. **B.** Fate of pieces isolated to right of lines of cut: piece **cd** forms gonophores belatedly. (After Davidson and Berrill.)

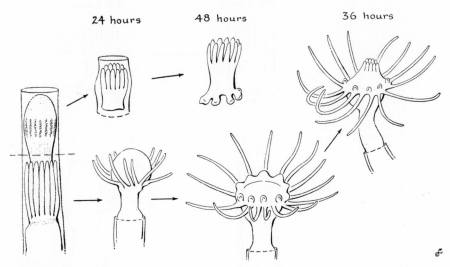

Figure 11-13. Cut between two tentacle-ridge zones (stage 7) and subsequent fate of distal and proximal pieces, showing regeneration of gonophore ring from both cut surfaces and of distal tentacles later. (After Davidson and Berrill.)

generation in general, whether of the mature *Tubularia* hydranth, of *Clava,* or of *Hydra.* The two significant features are that whereas proximal tissue can eventually regenerate distally, distal tissue never regenerates proximally; and that a differentiating partial structure does not regenerate until it completes its own development. Clearly, there is a transition from one state to another at the time when differentiation ceases and histological function—e.g., contractility of muscle fibrils—is attained, as though the structure were passing from a closed to an open system.

Experiments combining distal and proximal parts of regenerating primordia of *Tubularia* have been made by Rose (1957b). As in experiments concerning the origin and significance of pigment, fragments of red stems and yellow stems were combined by using stems of slightly different diameters and telescoping the perisarc of one over the other until the tissues met. Subsequent fusion was rapid and permanent. The question was whether organization spreads by suppression of the distal type of differentiation in the more-proximal regions. Rose (1955) reported that the period during which the effect of one differentiating part on another can be studied is the time between appearance of a delimited primordium (stage 4) and appearance of both proximal and distal tentacle ridges (stage 7). The main results of Rose's experiments were as follows:

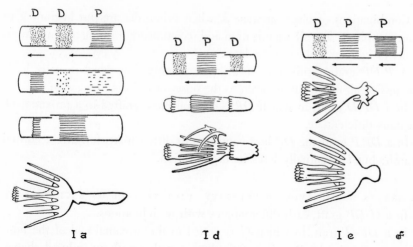

Figure 11-14. Grafts of red and yellow stem pieces, with wider piece fitted over narrower and without change of polarity. (For details, see Ia, d, and e in text.) (After Rose.)

I. GRAFTS WITHOUT CHANGE OF POLARITY

a. A few hours after union of the distal primordium at stage 6 (*D*) to the distal end of a whole primordium at stage 4, 5, or 6 (*DP*), the tentacle ridges of the host begin to fade, and new ones appear, in a new position at the former *D* level of the host, to give rise to a complete, typical hydranth. Thus the *D* in *D-DP* grafts inhibits the distal component of *DP*.

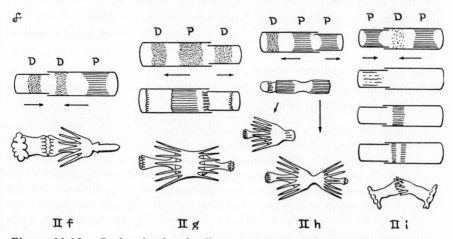

Figure 11-15. Grafts of red and yellow stem pieces, with reversed polarity—face to face or back to back. (For details, see IIf, g, h, and i in text.) (After Rose.)

b. Combinations of pieces at stage 3, when primordia are just beginning to appear, completely lose all organization and recommence forming a primordium at either or both ends about a day later.

c. A *P-DP* graft forms a most unstable combination which usually leads to the loss of all organization, even to the onset of cell disaggregation.

d. In a *DP-D* graft, in which *DP* is a distal piece grafted to a proximal *D*, all regions differentiate.

e. In a *DP-P* graft the *DP* host continues to differentiate, forming a normal hydranth; but *P* eventually loses its organization.

II. GRAFTS WITH REVERSED POLARITY, FACE TO FACE OR BACK TO BACK

f. In a *ᗡ-DP* graft, each differentiates with no inhibitions.

g. In a *DP-ᗡ* graft, in which *ᗡ* is joined to the proximal end of the host, doubling occurs in the *P* region of the host, together with an induced change of polarity.

h. A *DP-ꟻ* graft forms two complete hydranths back to back.

i. Most *ꟻ-DP* unions lead to a complete loss of organization. In a few, graft regression occurs and is followed generally by reorganization or reconstitution at both ends; the organization spreads in a distoproximal direction, so that anything missing is always proximal.

Since an isolated fragment from an adult stem, regenerate, or adult hydranth regulates itself to produce any missing distal regions, and since in experimental combinations distal structures inhibit distal structures and proximal inhibit proximal in the distoproximal direction, the rule, according to Rose, governing differentiation is: Differentiation depends on specific inhibition moving in a distoproximal direction in an originally equipotential system, since any level of a stem or hydranth can transform to any level not already present or forming distal to it. (Unusual combinations, however, can institute different polarities.) Organization is attained without inductors, but by each region suppressing a like expression proximal to it.

If a host, DP, receives a D graft so that the combination is either DP-D or DP-P there will be two independently differentiating systems. Originally the DP host would have formed in disto-proximal sequence: hypostome with distal tentacles, gonophores, proximal tentacles and base of hydranth. But now that region of the host which would have been the base of hydranth produces gonophores. Without the graft the region in question would have received inhibitory information from the left in the diagram and would have been reduced to something lower on the disto-proximal scale than gonophore production. Now with an additional disto-proximal relationship instituted by grafting, the region in question is part of the system with new disto-proximal polarity. Instead of its level of differentiation being controlled by information from the left it is controlled from the right. In

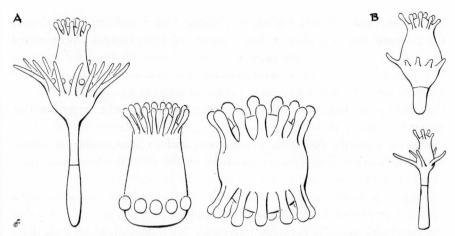

Figure 11-16. Regeneration in *Corymorpha* relative to length and mass. **A.** Forms resulting from differences in scale of organization in pieces of same length. **B.** Differences in scale of reorganization of aggregates of dissociated cells of different mass. (After Child.)

the right-hand system there are no gonophores distal to it and it follows the rule that any region may produce those structures not already differentiating distal to it. According to this view the region in question is not induced to form gonophores. Instead it is being allowed to do so.[1]

Rose's interpretation is supported to some extent by the nature and sequence of phenomena associated with reconstitution after tissue disaggregation in *Tubularia, Corymorpha,* and other hydroids. Child (1928) had found that in *Corymorpha* both in short pieces and in reaggregation masses the organizational process—as far as it went—evidently proceeded from the apical end basipetally. Distal tentacles always developed; but proximal tentacles and the perisarc-covered column developed only if space permitted. Often, even in large masses, the initial scale of organization proceeding from the apical region was far greater than could be accommodated by the available tissue. Lack of space seems at least the partial explanation for the formation of apical, or head, structure at each end of a reorganizing short piece without the reconstruction of the rest of the body—not only in hydroids, such as *Tubularia* and *Corymorpha,* but also in flatworms and ascidians, such as *Clavelina.* With apical organization proceeding simultaneously from both ends, neither end involves the whole piece; so that depending on the scale of organization, a varying amount of the proximal region does not differentiate.

The nature of the inhibitory agent in *Tubularia* has been investigated further

[1] Rose, *J. Morphol.* **100**:203 (1957).

by Tardent and Eymann (1959), who found that a hydranth grafted onto the proximal end of a piece of stem inhibits the reconstitution process at the distal end, although a fragment of stem grafted onto the proximal end of a hydranth does not inhibit reconstitution of either end. Grafting of all possible combinations (49) between pieces at 7 different stages of reconstitution indicates that there is no interaction between regenerates of the same stage, but that young regenerates are inhibited by older stages. The inhibiting capacity of the regenerate increases during its development and its susceptibility decreases; therefore, Tardent and Eymann conclude, in the normal primordium these two factors must be equilibrated quantitatively in order to prevent any self-inhibition. According to these workers, the inhibiting factor can be extracted by water or alcohol from *Tubularia* hydranths and is thermostable. Whether the extractable agent is normally responsible for inhibitory effects and, if so, whether it exerts its influence by diffusion are undecided questions; however, experiments by Barth (1944), in which the inhibitory effect of distal reconstitution on reconstitution at the proximal end was blocked by a drop of oil injected into the intervening stem hydrocoele, suggest that a diffusing agent may be involved (for a discussion of these questions in relation to diffusion theories generally, see p. 283).

The ultimate significance of regeneration-inhibiting substances in inhibitor water-and-hydranth extracts has been questionable since the discovery by Fulton (1959) that no specific effect is observed when antibiotics are used to maintain bacteriostasis and that the inhibitors otherwise present in inhibitor water are by-products of bacterial growth, for which the hydranth material is nutrient.

THE PERISARC AND TISSUE MOVEMENT

A tubularian stem always reconstitutes after a transverse cut. Pieces of stem usually reconstitute at either end unless one end or the other is subjected to inhibitory conditions. Since all of the stem is covered by perisarc (of greater or lesser thickness), it has been natural to suspect the perisarc of being an inhibitor. Moreover, it is significant that the sloughing off, or the disintegration, of the hydranth as a result of temperature rise or simply of age is followed by reconstitution within the stem without an open end in the perisarc having been formed and that the virtual cessation of terminal growth is also followed by reconstitution in growing branches, again without perisarcal disturbance.

On the other hand, there is some evidence that the immediate effects of cutting the stem are the escape of inhibiting substances—in particular, carbon dioxide and other metabolites—from the lumen and the corollary increase in penetration of oxygen. Local removal of perisarc, even without injury to the underlying tissue, is followed by reconstitution within the exposed area, particularly when the ends of the stem have been ligatured to eliminate dominating influence by those regions. According to Zwilling (1939), whether a single, partly double, or double reconstitution occurs within the exposed area depends on the size and shape of the exposed area. Local removal of perisarc does not, however, result in local increase in rate of ciliary activity of the endodermis; thus if such removal permits significant access to oxygen, the metabolic effect must be primarily on the epidermis. Although the necessity for oxygen in reconstitution has been demonstrated by many workers—especially by Barth (1940), who has shown the existence of a decreasing distoproximal gradient in oxygen consumption down the stem—it is equally certain that carbon dioxide is a powerful direct or indirect inhibitor and that, according to Miller (1942), the perisarc a barrier to the diffusion of both these gases. Since the perisarc is progressively thicker toward the base of a stem, its inhibiting effect is increasingly pronounced in that direction and at least in part affects the scale of organization of a reconstituting end. Barth emphasizes that both primordium size and regeneration rate can be decreased by exposure to low-oxygen tensions during the process of reconstitution—that both size and rate are determined by the rate of energy production. The size, or mass, of the tissue does not affect either the rate or the duration of the reconstitutive processes; i.e., temperature, oxygen tension, various respiratory inhibitions have the same effect on large and small reconstituting units—a phenomenon investigated both experimentally and mathematically by Spiegelman and Moog (1944). Furthermore,

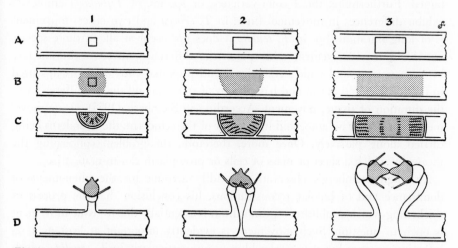

Figure 11-17. Effect of increasing size of exposed area on side of *Tubularia* stem. 1, 2, and 3. Three size groups. **A.** Amount of perisarc removed. **B.** Beginning of pigment deposition. **C.** Constriction of regenerate from rest of stem and appearance of tentacle ridges. **D.** Condition after emergence. (After Zwilling.)

low temperature induces increased primordium length; and high temperatures, reduced. The conclusion is that, whatever the particular details of the reconstitutive situation, the opposing responses of mass and time may be the means by which an open system approaches a steady state.

According to Steinberg (1955), variation in primordium size—in actuality, primordium *length* since width is predetermined by stem width—depends on the extent of tissue movement. As a general rule, reconstitution at the distal end of a stem is completed before that at the proximal end. However, once a thickened zone of tissue has been established at either end, the process of reconstitution proceeds at much the same rate at both ends. Thus the proximal end requires more time only in the establishment of a primordium, for subsequent development does not differ in rate.

Steinberg (1954) observed that after the breakdown of the endodermal ridges, tissue shifts toward the distal end of the perisarc to provide additional cells in that region—a movement which compresses the distal end of the coenosarc and stretches the proximal end—and leads directly to the establishment of the primordium. The extent of such tissue movement determines the primordium size. Experiments comparing the length of time between the cutting of the stem and the appearance of the thickening (i.e., the duration of tissue movement) and that between the appearance of the thickening and the appearance of the constriction which marks the base of the new hydranth show that only the duration of movement varies, even when the same conditions are maintained. Furthermore, the 4 color-varieties, or species, of *Tubularia* employed exhibit differences in movement time: in *T. crocea* and two others movement is of long duration; but in *T. tenella* it does not occur at all. In those species in which tissue movement takes place, the differences in reconstitution rate between narrow and wide stems, distal and proximal ends of stems ligatured in the middle, and apical and basal levels of stems are all due to differences in the duration of tissue movement. According to Steinberg (1955), tissue movement is primarily an amoeboid movement of the ectoderm, the endoderm being carried along passively. Once more, therefore, the problem concerning the means by which a sheet or mass of cells or protoplasm can migrate arises.

Although Steinberg's observations readily account for the phenomena of dominance and of bipolar reconstitutions, his conclusion that the process of thickening which establishes a primordium is essentially due to tissue movement is open to question. Such movement is evidently important at least as a reinforcing process; but it is reasonably certain, particularly in *T. tenella*—which is reported to exhibit no such movement—that cell growth and multiplication at the primordium site are also important, although tissue movement elsewhere may possibly interfere.

Flatworms

Flatworms, particularly the turbellarians, whose growth and form have been most extensively investigated, are noteworthy for their polarized structure and bilateral symmetry. The distinctive body regions are the head, or cephalic region, which contains simple eyes and a ganglionic brain; a prepharyngeal region, which lies between the head and pharynx; a pharyngeal region, which contains the muscular pharynx; and a postpharyngeal region, which forms a tail of variable length. The branching digestive tract ramifies in both directions from the base of the pharynx. Relative growth, i.e., growth which results in an elongation of the body relative to its width, occurs mainly in the postpharyngeal region anterior to the hindmost part of the tail; and Child (1910) found that in those species which reproduce by fission, fission zones appear in this anterior postpharyngeal region.

Flatworms are well known for their capacity to reconstitute new individuals from body fragments and for the fact that fragments and partial individuals can be grafted to one another. As early as 1814, Dalyell wrote that planarians are "almost to be called immortal under the edge of the knife"; and since that time more than five hundred papers have been published concerning planarian responses to operational procedures. Present understanding begins with Morgan's classical account (1900) of regeneration in *Planaria*. He found, for instance, that, although a decapitated planarian rapidly regenerates a head, two decapitated individuals joined at their anterior cut surfaces did not regenerate heads and that two individuals joined at their posterior cut surfaces failed to regenerate tail tissue.

The application of quantitative methods to the study of regeneration, in flatworms in particular, was made primarily by Child. As early as 1906 he observed that heads regenerated at progressively posterior levels of the body are correspondingly smaller and surmised that this gradient in head size was due to

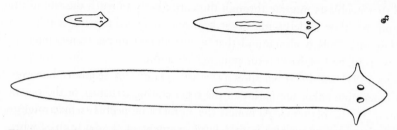

Figure 11-18. Three sizes of flatworm *Planaria (Dugesia) dorotocephala,* showing relative position and proportions of pre- and postpharyngeal regions for different absolute body size. (After Morgan.)

physiological conditions graded along the main axis. Later, finding that depressive substances such as alcohol, ether, and chloretone lowered the head frequency, he concluded that decreased oxygen consumption in the tissues was the underlying cause, and so launched the metabolic gradient theory (for a detailed review of this theory, see Brønstedt, 1955). Whatever the merits of the metabolic theory, there is an axial gradient with regard to regenerative phenomena: Regenerated heads are progressively abortive at successive levels along the antero-posterior axis, until at a certain posterior level, only a healing process takes place; however, tails regenerate posteriorly at any level. Besides the gradient in head size, there is a gradient in the rate of regeneration of head tissue along the same axis. Moreover, Child (1911 and 1929), Abeloos (1930), Sivickis (1930-31), Watanabe (1935), and Brønstedt (1942) confirm a similar, though less-striking gradient from the median line of symmetry to the margin on both sides. In contrast to Morgan's theory that the polarity and the gradients are morphological expressions of diversified physical or chemical structures, Child regards them as gradients in metabolic quantities—an interpretation supported by Watanabe and Yamagishi (1955).

Blastemata

The regeneration of a head or tail occurs after the establishment of a blastema. The origin of blastemata has been a controversial question, as attested by an extensive literature. According to Stevens (1907), wandering parenchymal cells, or neoblasts, migrate to a wound surface, where they play a predominant role in cicatrization. Curtis and Schulze (1934) stated that these formative cells are not only a persistent, unspecialized embryonic stock but also proportional in number to the regenerative power of the species; but Brønstedt (1956) has reported that counts of neoblasts in species whose head-frequency curve is sharp do not differ significantly in neoblastic number along the axis. According to Dubois (1949), the entire blastema, including the epidermis, consists of neoblasts. By irradiating various areas of a planarian body while screening other areas, he observed that neoblasts migrate through the entire body to reach the site of the wound and that they undergo mitosis at least once before differentiation begins in the blastema. Dubois also found that nonirradiated pieces transplanted in completely irradiated planaria resuscitate regeneration.

Brønstedt (1954 and 1955) has shown that the quantity of neoblasts, however, does not determine either the nature of the regenerating structure or the rate of regeneration of the blastema. Although the number of neoblasts that migrate is greater when the cut surface is in a large segment of the body than when the cut is in a small segment, the rate of regeneration is the same so long as the cuts are at equivalent locations in the body. When a wound surface is very

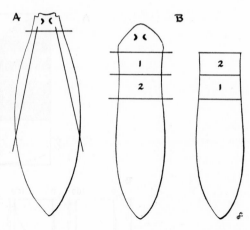

Figure 11-19. Rate of head regeneration. **A.** Very large wound on two sides of *Bdellocephala* at time of head amputation does not affect rate of head regeneration. **B.** *Euplanaria lugubris* pieces 1 and 2 are interchanged by transplantation; piece 2, however, continues to regenerate head at its own intrinsic rate. (After Brønstedt.)

extensive—e.g., when there are both transverse and longitudinal lateral cuts—the neoblasts become distributed among all areas of the injured surface, so that the number reaching the wound area is reduced; yet the head regenerates at the same rate as it does in planaria with only transverse cuts. Regenerated eyes are somewhat smaller in short segments than in long segments—a phenomenon indicating that the scale of organization is at least partly dependent on the quantity of neoblasts reaching the wound. Clearly, neoblasts are subordinate to the regenerative controls so far as size of regenerated structure and rate of regeneration are concerned, although a limited quantity of neoblasts may exert a restrictive influence on scale of organization.

Grafting Experiments

Differences along the primary axial gradient have been the focus of most of the earlier and some of the more recent grafting experiments on *Planaria*. The results of such experiments are summarized below according to the nature of the graft and the site of implantation in the host or according to the position of union when the two components are equivalent in size and axial derivation.

GRAFTS BETWEEN TRANSVERSE LEVELS

Whether pieces from various regions induce or fail to induce the growth of new tissue in the host depends on the site of implantation.

1. A piece grafted into a homologous position in the host does not induce new tissue growth between implant and host, regardless of relative sizes of the two components and of any independent reorganization of host structure. Miller (1937) and Okada and Sugino (1938) implanted cephalic pieces in the head region, prepharyngeal pieces in the prepharyngeal region, and postpharyngeal

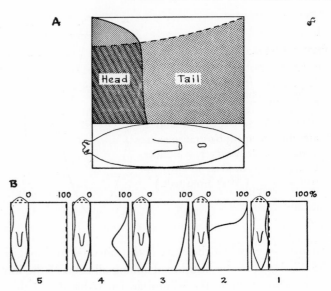

Figure 11-20. Frequency curves. **A.** Head and tail frequency curve of *Bdello-cephala.* **B.** Head frequency curves of *Bdelloura* groups: 1. *Dendrocoelum.* 2. *Phagocata.* 3. *Planaria dorotocephala.* 4. *Planaria velata.* Broken line in 1 indicates absent power of head regeneration; in 5, absolute power. (After Sivickis and Brønstedt.)

pieces in the postpharyngeal region. All three types of grafts unite with host tissue without inducing new tissue to grow at the site of junction.

2. A piece grafted into a heterologous position induces growth of new host tissue at the site of junction. Santos (1931), Miller (1937), and Okada and Sugino (1938) implanted cephalic pieces in the prepharyngeal and the post-pharyngeal region, postpharyngeal pieces in the prepharyngeal region, and tail pieces in the prepharyngeal region. All four types of grafts induce growth of new host tissue, usually as a tubular outgrowth toward the posterior side of the implant. The greater the discrepancy between level of implantation and original level of donor tissue along the body axis, the greater the growth induced at the graft junction.

3. According to Brønstedt (1955), when a segment of the body consisting of most of the prepharyngeal region is cut out and the head grafted to the re-mainder of the body, i.e., to the pharyngeal and postpharyngeal region, approxi-mately as much new tissue grows at the site of union as was originally cut out.

4. When a large, rectangular piece wide enough to include the ventral nerve cords is cut out and reinstated in the same site but in reversed orientation, com-plete reunion takes place. Sugino (1938) observed that the original pharynx re-mains within the reversed piece, but faces forward instead of backward. Within

a few days two rudimentary pharynges appear: one in new tissue anterior to the reversed tissue, and the other in new tissue posterior to the reversed tissue; both are in normal orientation to the body as a whole. In control experiments in which similar rectangles were cut out and reinserted in their original orientation, no new tissue appeared and the normal worms which resulted had ventral nerve cords united from end to end.

In the experiments described above, Santos and Miller used the species *Planaria dorotocephala* (also known as *Euplanaria* and *Dugesia*), Okada and Sugino used *Planaria gonocephala,* and Brønstedt used *Bdellocephala punctata.* Irrespective of the particular species and of any differentiation of gross structure, the results as a whole affirm that position is a factor affecting growth and that there is both a lack of fit in unions between pieces from different levels and a stimulus to regenerate new tissue—a stimulus sufficient to establish a proper fit. Each level seems to have its own intrinsic specificity, whose basis— though obscure—is clearly fundamental to an understanding of the nature of organization.

FATE OF GRAFTED TISSUES[2]

1. A cephalic fragment develops into a head and a postcephalic outgrowth forms behind it regardless of the region or orientation of the grafting—except as Miller (1937) reported, when grafted into the cephalic region. In the latter

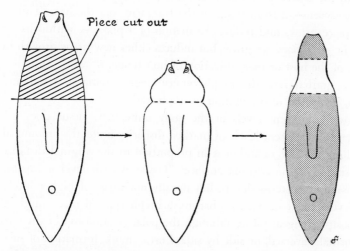

Figure 11-21. Prepharyngeal piece in *Bdellocephala* is cut out, and head is grafted to rest of body. New regenerated tissue is intercalated between head and body, distinguishable by unpigmented appearance. (After Brønstedt.)

[2] Unless noted otherwise, all material in this section is from Okada and Sugino (1938).

type of graft, the piece becomes fully incorporated in, and the graft epithelia and eyes become replaced by, the host tissue. When a cephalic fragment is transplanted into the postpharyngeal region, the cephalic fragment does not grow, but new tissue appears along the line of union while two new pharynges appear—either in the new tissue or in the original host tissue—one anterior and the other posterior to the graft, so that the polarity of the host tissue anterior to the graft is thus reversed.

2. When a small piece is cut from the prepharyngeal region and transplanted into the postpharyngeal region, new tissue appears along the line of union. When a subsequent cut is made anywhere along the line of union to expose some of the original fragment, a head regenerates from the exposed cut surface and two new pharynges form, one anterior and the other posterior to the graft. When a fragment is intimately united on all sides with host tissue, a head does not regenerate but an anterior and a posterior pharynx form. Thus pharyngeal induction and anterior reversal of polarity takes place without head or brain tissue forming.

3. When a small fragment is cut from immediately in front of the pharynx, without including any pharyngeal tissue, and is grafted into the prepharyngeal region, a new pharynx develops in the host tissue. Similar fragments transplanted into postpharyngeal regions do not induce pharynx formation.

4. When a postpharyngeal piece is transplanted into the prepharyngeal region, a tail grows from the posterior end, a head from the anterior end of the graft, and a pharynx between—all from tissue of the fragment, not the host. Thus a postpharyngeal piece grows and induces formation of a pharynx within itself; whereas a cephalic graft does not grow, but induces either new tissue growth in adjacent regions of the host or reorganization of host tissue.

5. A tail piece grafted into the prepharyngeal region remains a tail and grows in length, regardless of orientation of implantation.

Thus when two equivalent levels of the body unite, new tissue does not appear; but when two different levels of marked disparity unite, the amount of tissue formed along the line of union is in proportion to the original distance between the positions of the two cut surfaces. Tissue at each level is capable of organizing tissue of posterior levels, but usually not tissue of higher levels. Thus for the development of a new pharynx a prepharyngeal piece must be united with a postpharyngeal piece, although the polar relationship of the two pieces can be normal, reversed, or side by side. Furthermore, formation of new tissue is not a necessary preliminary for pharynx development since, after excision of the pharynx from a normal worm, a new pharynx is formed along the line of healing.

GROWTH AND REORGANIZATION IN LATERAL UNIONS

1. When two lateral halves are combined at corresponding levels, no new tissue forms between them. Okada and Sugino (1938) observed that when the original pharynx has been excised, all the territories and organs of the complex combine to form a single formation except that two pharynges develop, one from each median cut surface.

2. According to Sugino (1938), when two lateral components each consisting of three-quarters of an individual are conjoined at equivalent levels, no new tissue forms, and the complex becomes a single worm with double internal structure.

3. Sugino also found that when two lateral components each consisting of the marginal quarter of an individual are conjoined at corresponding levels, no new tissue forms along the line of junction; a single new, but small pharynx develops; and considerable growth in length occurs.

4. When two lateral halves are conjoined in reverse relationship, new tissue forms along the whole line of union, and each component behaves as though it were driving away the other component in order to replace the missing half by itself. Consequently, according to Okada and Sugino (1938), the complex becomes a Siamese twin.

Time-graded Regeneration Field

The concept of head frequency, developed by Child (1911) for *Planaria dorotocephala,* was applied to a number of triclad species by Sivickis (1930-31). He cut planarians of a number of species into a series of transverse segments, in order to obtain an average for the head-forming capacity at the different levels, the dark pigment of the eyespots being the criterion of subsequent full regeneration. Then for each species he drew a curve representing the capacities for head and tail regeneration along the main axis. In certain species (e.g., those of the genus *Dendrocoelum*) the capacity for head regeneration does not exist at any level; in others (e.g., *Planaria velata*) the capacity extends throughout the length of the body; and in most species (e.g., *Phagocyta* and *Belloura* species, *Planaria dorotocephala,* and *P. gonocephala*) the capacity is strong anteriorly and diminishes toward the posterior end; the rate of diminution in capacity strength varies among species. The time required for head formation is correlated with the head-frequency value.

This rather simple concept has been elaborated somewhat by Brønstedt to mean "time-graded regeneration field" since the capacity for head formation in most species decreases not only posteriorly along the main axis but also laterally

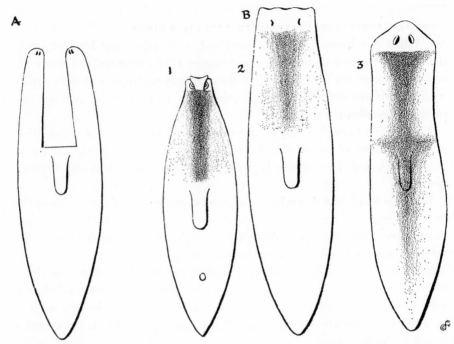

Figure 11-22. Head regeneration. **A.** Median part of *Bdellocephala* body is cut away, and two arms regenerate heads. **B.** Time-graded regeneration fields. 1. *Bdellocephala.* 2. *Dendrocoelum.* 3. *Euplanaria lugubris.* (After Brønstedt.)

from the median line toward both margins and because a head can regenerate anywhere within this field of head-forming potency. Thus when the middle of the anterior region is excised, each of the two remaining lateral regions regenerates a head, as does the anterior margin of windows cut in the prepharyngeal region—whether or not the original head has been excised. The intrinsic regeneration rate, moreover, is maintained in pieces transplanted to other positions in the same body, so long as a new head has not yet appeared. When a segment from immediately behind the head is cut out and rejoined to the transected postpharyngeal body, it regenerates according to its original status; and a similar segment transposed with one immediately posterior to it, without change in orientation, regenerates after healing and subsequent decapitation, at the rate typical of its original position. Similarly, in *Euplanaria lugubris* after decapitation and transposition of a median and a lateral square-shaped piece of tissue of the same prepharyngeal level, two heads regenerate at the rate and to the extent of development typical of the original location of each, the median piece performing faster and more completely—as it would in the original location.

So long as the primary axial orientation is not disturbed, the degree of co-operation between the right and left components of a lateral union depends on the degree of disparity. When a piece from the right side of the prepharyngeal region is grafted to a piece from the left side of the postpharyngeal region, perfect union takes place, but head regeneration occurs only from the right side of the united anterior margin. Similarly, when a planarian is split longitudinally, one half is rejoined posteriorly about one-quarter of the body length distant, and the individual is then decapitated, the right and left sides each regenerate a head; but then when a second transection is made at the level of the pharynx of either the left or the right side, the anterior cut surface regenerates a single head which embraces the entire wound area. Brønstedt (1955) sums up cooperation between lateral components as follows: When the levels (of the time-graded field) are equivalent, one symmetrical head regenerates; when the levels are somewhat disparate, one half starts its regeneration before the other; when the levels are greatly disparate, a head regenerates only from the half whose transverse cut is in a region where head-forming capacity is strong—although the other half goes so far as to form a new blastema, which merges with its original one.

The head of a planarian has often been called an "organizer," either in comparison with the so-called organizer of vertebrate embryos or in its own context. According to Child (1914), the formation of a new head is not merely reconstitution but is the first step of its dominant organizational role in developing a new individual. However, Brønstedt's new grafting technique has cast doubt on

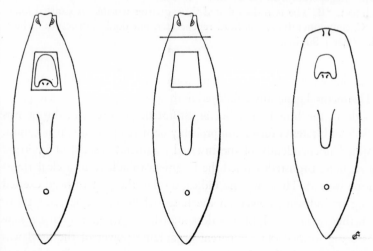

Figure 11-23. Window is cut out of forepart of *Bdellocephala* body, and head regenerates within it whether or not original head is amputated. (After Brønstedt.)

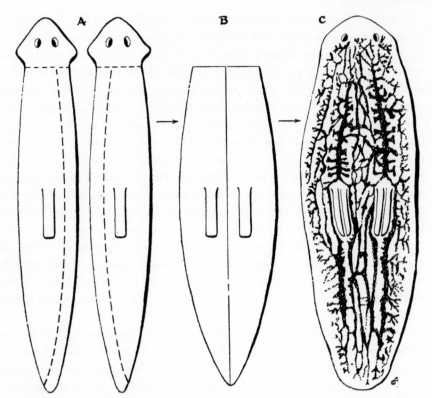

Figure 11-24. Regeneration in flatworms. **A.** Two flatworms are cut (as shown by broken lines). **B.** The two decapitated three-quarter worms are grafted longitudinally. **C.** Subsequently, single head regenerates, but double structure of body is retained. (After Sugino.)

this interpretation, which had major significance for the metabolic gradient hypothesis. The method generally employed in the past was to insert small pieces into holes made in the host tissue; but as Brønstedt pointed out, the fit was rarely exact, so that blastema formation probably occurred in many, if not most, such grafts quite independently of the inserted fragment. Instead of inserting the piece into a hole, Brønstedt caused the fragment to adhere to a clear transverse or oblique wound. There was no indication that the grafted head exerted any organizing or inducing power on the host—whatever the level of transplantation whether or not the host was decapitated, and whether or not a new blastema was being formed. The apparent organizing power of the head was probably surmised from the fact that first eyes and ganglia appear in a blastema arising at an anterior cut surface—that thereafter other organs form sequentially from head to tail. Despite the fact that in a blastema forming at a posterior wound surface, regeneration proceeds sequentially from tail to head, no one has ever named the tail "organizer."

Formation of Head

The regeneration of the planarian head is largely the regeneration of the brain and eyes. The time between appearance of the pigmented eyes and that of their final development, together with the dark pigment of the eyespots, are indications of the rate and degree of differentiation of head development as a whole. As described earlier, rates and degrees of development vary with the level of a cut along the anteroposterior axis and with various chemical depressants, such as Child and his co-workers used to demonstrate susceptibility curves in support of the theory of metabolic gradients.

Rulon (1940) has classified heads for *Planaria dorotocephala* according to appearance as follows:

1. Normal: triangular; auricles at lateral margin; and two symmetrically placed eyes.
2. Teratophthalmic: triangular; auricles at lateral margins; but eyes ranging in degree of proximity from near-normal spacing to complete cyclopia.

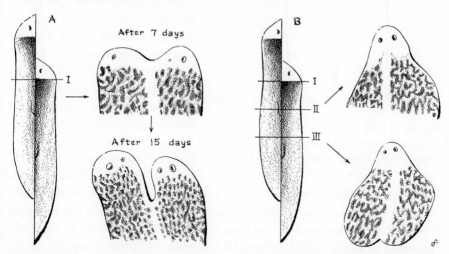

Figure 11-25. Regeneration in *Euplanaria lugubris*. **A.** Two halves of split individual are joined together with shift of about one-quarter of body length, and then both heads are amputated. Double head regenerates with partial reduction on conjoined side. **B.** Similar experiment, but with transverse cuts made at 9 posterior levels. Single head regenerates with greater development from originally more-anterior tissue. (After Brønstedt.)

3. Teratomorphic: approximately rounded; auricles in close proximity; and one small eye in the center of the head.
4. Anothalmic: rudimentary; neither auricles nor eyes.
5. Acephala: no head.

This classification clearly reflects a sequence of diminishing organizational scale, which corresponds to level along an axis or to progressively increasing depression. Since the differentiating structures are influenced by the initial scale, it is equally clear that scale restriction is expressed first in the central area of the blastema and extends progressively toward the margin.

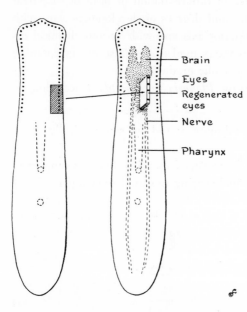

In *Polycelis nigra* the eyes extend as a row along the anterior margin on each side of the worm. When the eyes and the brain are excised simultaneously, the eyes normally regenerate in 7 days; the brain, in 3 or 4. When the new brain is excised before the appearance of the eyes, eyes do not form —a phenomenon which suggests that the brain is necessary for eye formation. This hypothesis has been confirmed by transplantation of a piece of the eye rim into the dorsal region near the brain of a host and by subsequent excision of the eyes: new eyes form at the normal time and develop at the normal rate. However, when a similar transplantation is made in the caudal region or when an eyeless piece from immediately behind the eye rim is grafted near the host's brain, eyes do not form until a brain has formed in the graft. According to Lender (1952), when, after a brain transplanted into the caudal region of the host has healed, an eye rim is grafted close to, but not touching, the brain, and the eyes are then excised, new eyes regenerate.

Figure 11-26. Regeneration in *Polycelis nigra* of excised eyes on slab of border tissue transplanted near brain. (After Lender.)

Postulating a diffusible substance, Lender initiated the use of both crude broths of crushed heads or tails and, alternatively, the supernatant fluid of centrifuged mixtures. In the presence of a head broth, eyes regenerate but brain differentiation is rather strongly inhibited. In the presence of tail broth, brain differentiation is almost normal. Lender (1956a) concludes that an inhibitory, diffusible substance is distributed throughout the body in a gradient which decreases from head to tail. Tucker (1959), employing regional tissue broths (brei) in regeneration studies on nemertean worms, reports results that seem to support the concept of differential inhibition.

Specific Inhibition during Differentiation

Elaborated as a formal theory independently by Turing (1954) and Rose (1952, 1955, and 1957a), the concept that the establishment of pattern may result in part from specific inhibitory effects which proceed from more-developed to less-developed regions undoubtedly has some basis. At face value, i.e., without modification in view of Fulton's discovery that bacterial metabolites have a differential effect, the various experimental results indicate that specific tissues contain specific substances which are more capable of inhibiting the growth of the homologous tissue in which they appear than other kinds. Saetren (1956) has reported organ-specific inhibitors of mitosis in homogenates for mammalian liver and kidney; Clarke and McCallion (1959) have reported similar inhibitors for developing frog and chick brain and heart.

The dual question remains, however: Are organ- or tissue-specific substances the internal regulators of growth? Even if they are, do such substances exert any inhibitory effect only on tissues of a lower degree of differentiation, and not on a higher? It may be significant that chemical extracts of malignant tumors typically induce the qualities of tissue specificity and general malignancy simultaneously, in host cells or tissues that are histogenetically related to the particular tumors. Chemical controls generally, however well substantiated they may be in the future, are an unsatisfying explanation of the basis of pattern. Diffusion gradients, no matter how complex, are not the complete answer; and although it is tacitly agreed that all protoplasm exhibits fine structure of some kind, it is unsafe to assume that any such structure is secondary to familiar chemical interactions. Hence the view in this book is that though knowledge is at present lacking for an understanding of some instrument which organizes matter in novel macromolecular ways, it is nevertheless wise to assume that such a structural and dynamic entity is fundamental—that diffusible agents may be capable of modifying its expression, but not its predominant role.

REFERENCES FOR
Reconstitution

HYDROIDS AND MEDUSAE

Agassiz, L. 1862. *Contributions to the Natural History of the United States of America.* Boston, Mass.: Little, Brown and Co. **4**:240-71.

Barth, L. G. 1940. "The Relation between Oxygen Consumption and Rate of Regeneration." *Biol. Bull.* **78**:366-74.

———. 1944. "The Determination of the Regenerating Hydranth in Tubularia." *Physiol. Zool.* **17**:355-66.

Berrill, N. J. 1948. "Temperature and Size in the Reorganization of Tubularia." *J. Exp. Zool.* **107**:455-64.

———. 1952. "Growth and Form in Gymnoblastic Hydroids. V. The Growth Cycle in Tubularia." *J. Morphol.* **90**:583-602.

Brien, P. 1942. "Études sur les deux hydroides gymnoblastiques: Cladonema radiatum, Clava squamata." *Acad. roy. Belgique* (Classe sci.) *Mem.* **20**:1-114.

———. 1943. "Études de la régénération et de la rénovation de l'appareil sexuel chez les hydroides (Clava squamata O. F. Müller)." *Arch. biol.* **54**:410-75.

Child, C. M. 1926. "Studies in the Axial Gradients of Corymorpha palma. III." *Biologia Generalis.* **2**:771-98.

———. 1927. "Modification of Polarity and Symmetry in Corymorpha palma. I." *J. Exp. Zool.* **47**:343-83.

———. 1928. "Axial Development of Aggregates of Dissociated Cells from Corymorpha palma." *Physiol. Zool.* **1**:419-61.

———. 1933. "Reconstitution in Haliclystus auricula Clark." *Sci. Reports Tohoku Univ.* Series 4. **8**:75-106.

———. 1935. "Dominance of Hydranths Induced by Grafts in Corymorpha." *J. Exp. Zool.* **71**:375-87.

Cohen, A. I. 1952. "Studies on the Pigmentation Changes during Reconstitution in Tubularia." *Biol. Bull.* **102**:91-9.

Davidson, M. E., and N. J. Berrill. 1948. "Regeneration of Primordia and Developing Hydranths of Tubularia." *J. Exp. Zool.* **107**:465-78.

Driesch, H. 1897. "Studien über das Regulations vermögen der Organismen. I. Von den regulativen Wachstum und Differentzierungs fähigkeiten der Tubularia." *Arch. entwick. Org.* **5**:389-418.

Fulton, C. 1959. "Re-examination of an Inhibitor of Regeneration in Tubularia." *Biol. Bull.* **116**:232-8.

Goldman, A. S. 1953. "Synthesis of Pigment during the Reconstitution of Tubularia." *Biol. Bull.* **105**:450-65.

Hyman, L. H. 1926. "The Axial Gradients on Hydroza. VIII. Respiratory Differences along the Axis in Tubularia with Some Remarks on Regeneration Rate." *Biol. Bull.* **50**:406-26.

Lund, E. J. 1923. "Normal and Experimental Delay in the Initiation of Polyp Formation in Obelia Internodes." *J. Exp. Zool.* **37**:69-83.

Miller, J. A. 1942. "Some Effects of Covering the Perisarc upon Tubularian Regeneration." *Biol. Bull.* **83**:416-27.

Moore, J. A. 1939. "The Role of Temperature in Hydranth Formation in Tubularia." *Biol. Bull.* **76**:104-7.

Morgan, T. H. 1901. "Regeneration in Tubularia." *Arch. entwick. Org.* **11**:346-81.

Morse, M. 1909. "The Autotomy of the Hydranth of Tubularia." *Biol. Bull.* **16**:172-82.

Nakamura, N. 1941. "Effect of Light on the Regeneration in Syncoryne nipponica." *Japan. J. Zool.* **11**:185-93.

Okada, Y. K. 1927: "Études sur la régénération chez les Coelentérés." *Arch. zool. exp. gén.* **66**:497-551.

Pyefinch, K. A., and S. F. Downing. 1949. "Notes on the General Biology of Tubularia Larynx." *Marine Biol. Assoc. U.K. J.* **28**:21-43.

Rafferty, K. A. 1955. "The Fates of Segments from Tubularia Primordia." *Biol. Bull.* **108**:196-205.

Rose, S. M. 1955. "Specific Inhibition during Differentiation." *N. Y. Acad. Sci. Ann.* **60**:1136-59.

———. 1957a. "Cellular Interaction during Development." *Biol. Rev.* **32**:351-82.

———. 1957b. "Polarized Inhibitory Effects during Regeneration in Tubularia." *J. Morphol.* **100**:187-206.

Spiegelman, S., and F. Moog. 1944. "On the Interpretation of Rates of Regeneration in Tubularia, and the Significance of the Independence of Mass and Time." *Biol. Bull.* **87**:227-41.

Steinberg, M. 1954. "Studies on the Mechanism of Physiological Dominance in Tubularia." *J. Exp. Zool.* **127**:1-26.

———. 1955. "Cell Movement, Rate of Regeneration, and the Axial Gradient in Tubularia." *Biol. Bull.* **108**:219-34.

Tardent, P., and H. Eymann. 1959. "Experimentelle untersuchungen über den Regenerationshemmende Faktor von Tubularia." *Arch. entwick. Org.* **151**:1-37.

Torrey, H. B. 1910. "Biological Studies on Corymorpha. III. Regeneration of Hydranth and Holdfast." *Univ. Calif. Pub. Zool.* **6**:205-21.

Tweedell, K. S. 1958. "Inhibitors of Regeneration in Tubularia." *Biol. Bull.* **114**:255-69.

Zwilling, E. 1939. "The Effect of the Removal of Perisarc on Regeneration in Tubularia crocea." *Biol. Bull.* **76**:90-103.

FLATWORMS

Abeloos, M. 1930. "Recherches expérimentales sur la croissance et la régénération chez les Planaires." *Bull. Biol. France-Belgique.* **64**:1-140.

Brønstedt, H. V. 1939. "Regeneration in Planarians Investigated with a New Plantation Technique." *Kong. danske vidensk. selsk. Hist.-filos. medd.* **15**:1-39.

———. 1942. "Further Experiments on Regeneration Problems in Planarians." *Ibid.* **17**:1-28.

———. 1946. "The Existence of a Static Potential and Graded Regeneration Field in Planarians." *Ibid.* **20**:1-31.

———. 1955. "Planarian Regeneration." *Biol. Rev.* **30**:65-126.

———. 1956. "Experiments on the Time-graded Regeneration Field in Planarians." *Kong. danske vidensk. selsk. Hist.-filos. medd.* **23**:1-39.

Child, C. M. 1902a. "Studies on Regulation. I." *Arch. entwick. Org.* **15**:187-238.

———. 1902b. "Studies on Regulation. II." *Ibid.* Pp. 603-37.

———. 1910. "Physiological Isolation of Parts and Fission in Planaria." *Ibid.* **30**:159-205.

———. 1911. "Studies on the Dynamics of Morphogenesis, etc. II. The Axial Gradient in Planaria dorotocephala as a Limiting Factor in Regulation." *J. Exp. Zool.* **10**:265-320.

———. 1914. "Studies on the Dynamics of Morphogenesis, etc. VII. The Stimulation of Pieces by Section in Planaria dorotocephala." *Ibid.* **16**:413-42.

———. 1929. "Physiological Dominance and Physiological Isolation in Development and Reconstitution." *Arch. entwick. Org.* **117**:21-66.

———, and Y. Watanabe. 1935. "The Head Frequency Gradient in Euplanaria dorotocephala." *Physiol. Zool.* **8**:1-40.

Coe, W. R. 1934. "Analysis of the Regenerative Processes in Nemerteans." *Biol. Bull.* **66**:304-15.

Curtis, W. C., and L. M. Schulze. 1934. "Studies on Regeneration. I. Contrasting Powers of Regeneration in Planaria and Procotyla." *J. Morphol.* **55**:477-512.

Dalyell, J. G. 1814. *Observations on Some Interesting Phenomena in Animal Physiology Exhibited by Several Species of Planaria.* Edinburgh.

Dubois, F. 1949. "Contribution à l'étude de la migration des cellules de régénération chez Planaires dulcicoles." *Bull. Biol. France-Belgique.* **83**:213-83.

———, and T. Lender. 1956. "Correlations humorales dans la régénération des planaires paludicales." *Ann. sci. nat. Série zool.* **18**:223-30.

Kepner, W. A., and J. R. Cash. 1915. "Ciliated Pits in Stenostonum." *J. Morphol.* **26**:234-46.

Lender, T. 1952. "Le rôle inducteur du cerveau dans la régénération des yeux d'une Planaire d'eau douce." *Bull. Biol. France-Belgique.* **86**:140-215.

———. 1956a. "Analyse des phénomènes d'induction et d'inhibition dans la régénération des Planaires." *Ann. biol.* **32**:457-71.

———. 1956b. "Recherches expérimentales sur la nature et les propriétés de l'inducteur de la régénération des yeux de la planaire Polycelis nigra." *J. Embryol. Exp. Morphol.* **4**:196-216.

Loeb, J. 1891. *Untersuchungen zur physiologische Morphologie der Tiere. I.* Würzburg.

Miller, J. A. 1938. "Studies on Heteroplastic Transplantation in Triclads. I. Chephalic Grafts between Euplanaria dorotocephala and E. tigrina." *Physiol. Zool.* **11**:214-47.

Morgan, T. H. 1900. "Regeneration in Planarians." *Arch. entwick. Org.* **10**:58-119.

Okada, Y. K., and H. Sugino. 1937. Transplantation Experiments in Planaria gonocephala Duges." *Japan J. Zool.* **7**:373-439.

Ott, H. N. 1892. "A Study of Stenostonum." *J. Morphol.* **7**:263-304.

Rulon, O. 1940. "The Environmental Control of Regeneration in Euplanaria." *Am. Nat.* **74**:501-12.

Santos, F. V. 1931. "Studies on Transplantation in Planarians." *Physiol. Zool.* **4**:111-64.

Sivickis, P. B. 1930-31. "A Quantitative Study of Regeneration along the Main Axis of the Triclad Body." *Arch. zool. ital.* **16**:430-49.

Stevens, N. M. 1907. "A Histological Study of Regeneration in Planaria simplicissima, P. maculata and P. morgani." *Arch. entwick. Org.* **24**:350-73.

Sugino, H. 1938. "Miscellany on Planaria Transplantation." *Annot. zool. japoneses.* **17**:185-93.

Tucker, M. 1959. "Inhibitory Control of Regeneration in Nemertean Worms." *J. Morphol.* **105**:569-600.

Turing, A. M. 1954. "The Chemical Basis of Morphogenesis." *Roy. Soc. London Philos. Trans.* Series B. **237**:37-72.

Watanabe, Y., and S. Yamagishi. 1955. "Differential Susceptibility and Redox Indicator Patterns in Polycelis sappora." *Physiol. Zool.* **28**:1-18.

MISCELLANEOUS

Clarke, R. B., and D. J. McCallion. 1959. "Specific Inhibition in the Frog Embryo by Cell-free Homogenates of Adult Tissue." *Can. J. Zool.* **37**:129-31.

Saetren, H. 1956. "A Principle of Auto-regulation of Growth. Production of Organ-specific Mitose-Inhibitors in Kidney and Liver." *Exp. Cell Research.* **11**:229-32.

Chapter 12

Rhythmic Growth: Regeneration, Reorganization, and Segmentation

THE PHENOMENA OF regeneration and reorganization in annelids are very similar to those in flatworms. The segmented form of the annelid body on the one hand makes possible a more precise measurement of events and on the other presents an additional phenomenon—segmentation—calling for analysis.

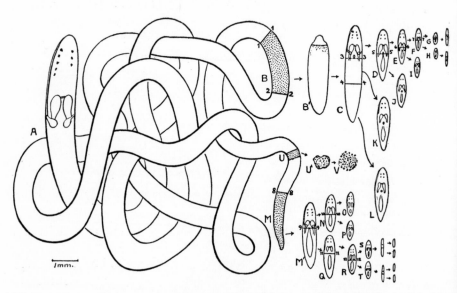

Figure 12-1. Diagram of entire nemertean *Lineus socialis* individual, showing regenerated individuals resulting from two isolated fragments that have undergone succession of cuts and subsequent regenerations. (After Coe.)

Fragmentation and Regeneration

The most spectacular expression of regenerative power in worms is the reconstitution of spontaneously or experimentally fragmented individuals. As Coe (1929) has demonstrated, certain nemerteans, e.g., *Lineus socialis* and *L. vegetatus,* reconstitute into a perfect individual from a minute fragment so long as the length of the fragment exceeds the width. The reconstituted worm can be subdivided into 3 or 4 pieces, which when reconstituted can also be subdivided, and so on. By repeating this procedure six or seven times, Coe succeeded in obtaining reconstituted worms approximately 0.15 mm long and 0.04 mm wide; theoretically, 200,000 such pieces could be obtained from one adult worm.

Among a number of annelids, polychaetes in particular, an isolated single segment can reconstitute the complete worm. Single segments of *Chaetopterus variopedatus* (Berrill, 1928) and *Myxicola aesthetica* (Okada, 1934b) have been isolated experimentally; but single segments of *Ctenodrilus monostylos* (Korschelt, 1919), *Dodecaceria caulleryi* (Dehorne, 1933), and *D. fimbriatus* (Martin, 1933) are isolated by spontaneous fragmentation. *Ctenodrilus* and *Dodecaceria fimbriatus* reconsti-

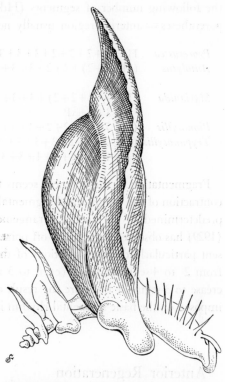

Figure 12-2. Anterior and posterior regeneration from single fan segment in *Chaetopterus.*

tute a small, but perfect worm; and the substance of the original segment becomes so reduced by transfer to the regenerating tissue that it is difficult to distinguish from the new. However, in *D. caulleryi,* each of the comparatively massive original segments gives rise to relatively slender anterior and posterior structures, which break off and become two separate individuals: one to grow a head, and the other a tail. The remaining original segment, with its two stumps of regenerated tissue, repeats the process, as a rule two or three times.

Spontaneous fragmentation in syllid polychaetes has been described by Allen (1921 and 1927) and Okada (1929). *Procerastea halleziana* first undergoes rapid multiplication by a process of fragmentation; then each fragment regenerates an anterior and a posterior end. In this species, and in others which may not fragment spontaneously, orderly fragmentation can be induced by subjection of the worms to sea water diluted to 60 per cent with distilled water. In each syllid investigated, the worm breaks up into a specific and constant series of pieces with the following number of segments (Hd 7 = head piece, xP = pygidial piece; parentheses = anterior region usually not fragmented):

Procerastea   Hd   7+2+2+2+3+3+3+4+4+4+4(+4)+3+3+3+3+3+xP
Autolytus   (Hd 7+2)+2+2+3+3+3+4+4+4+4+4+4+3+3+4+4+4+4
                 +xP
Myrianida   (Hd 7+2+2+2)+3+3+3+4+4+4+3+3+4+4+4+4+4+4
                 +3+3+xP
Pionosyllis   (Hd 7+2)+2+2+3+3+3+4+4+4+4+4+4+xP
Trypanosyllis (Hd 7+2+2+2+3+3+3+4)+4+4+4+3+3+4+4+4+3+3
                 +4+4+4+4+4+3+3+3+xP

Fragmentation in these genera seems to be provoked by an unusually strong contraction of the longitudinal segmental muscle, but the breaking positions are predetermined by the special arrangement or sequence of megasepta. Okada (1929) has observed these break-off points as white transverse lines which represent particularly deep constrictions of the alimentary canal. Evidently, the rise from 2- to 4-segment units (even to 5 according to Allen, 1927) and the decrease to 3-segment units are the result of a secondary rhythm which is superimposed on the basic segmental rhythm in the prepygidial zone of growth.

Anterior Regeneration

Because annelid regenerative processes at the anterior end are usually very different from those at the posterior end, they are discussed separately below.

Size of Blastemata

As manifested in the polychaete *Autolytus* and in other syllids, there is little doubt that the size of the initial blastema, i.e., its maximum size before the onset of differentiation, varies and that the final size of the differentiated structure and the extent of differentiation vary accordingly. In syllids there is usually a quantitative anteroposterior gradient in degree of anterior regeneration—

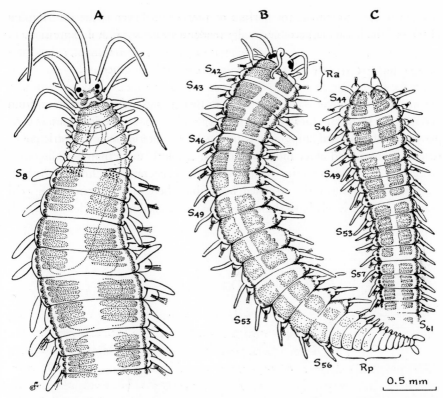

Figure 12-3. Regeneration in *Autolytus pictus*. (White transverse bands in original part indicate fragmentation levels.) **A.** Anterior regeneration **Ra** from cut in front of setigerous segment 8. **B.** Anterior regeneration from cut in front of setigerous segment 42 and posterior regeneration **Rp** from segment 56. **C.** Anterior regeneration from segment 43. (After Okada.)

a gradient which, according to Okada (1929), is perhaps most clearly found in *Autolytus*. At the cut surface of segment 8, a head with stomodaeum and 8 setigerous segments is regenerated; from segments 8-13, a head with stomodaeum and 3 or 4 setigerous segments regenerates; from segments 13-42, only a head (no stomodaeal invagination or setigerous segments) regenerates; and posteriorly from segment 42, only a small, rounded cap of tissue forms. The initial blastema progressively decreases, as does the quantity of tissue derived from it.

Even more striking is the fact that with reduction in initial scale of organization, there is successive inhibition of certain developmental features. The disappearance of setigerous segments from the regenerate is correlated with the suppression of a zone of growth at the peripheral margin of the blastema, and the absence of a stomodaeum is also correlated with reduction in scale. Furthermore, posteriorly to segment 42, the surface or mass—whichever is the determining factor—of the blastema becomes totally inadequate for structural expression.

Postcephalic Growth

Annelids differ greatly among species in their capacity for regeneration from anterior cut surfaces. *Autolytus pictus* is typical of many species: complete replacement of head and amputated segments in the anterior region; restriction of regeneration to head alone more posteriorly; and little more than a healing-over process in the most posterior region. Sabellid regeneration is limited to the 3 segments which form the so-called head; but there is no regenerative axial gradient, and all levels have the same capacity. According to Sayles (1932), all

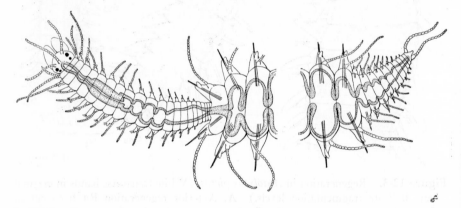

Figure 12-4. Anterior and posterior regeneration from original 6-segment piece in *Syllis gracilis.* (After Okada.)

segments in the maldanid *Clymenella torquata* are exactly replaced from all levels, with regard to both number and characteristic features. Certain syllids—in particular *Syllis gracilis,* according to Allen (1921), and *Procerastea halleziana,* according to Langhammer (1928)—regenerate all missing segments either exactly or almost exactly; for example, segments 17-19 of *Procerastea* regenerate 16 segments anteriorly and 20 posteriorly. In other syllids—e.g., *Syllis spongicola, S. prolifera, Trypanosyllis zebra,* and *Autolytus edwardsi,* according to Okada (1929)—anterior regeneration is limited to 2 head segments only or to the 2 head segments plus 2 setigerous segments. On the other hand, in *Chaetopterus variopedatus* (Berrill, 1928) all missing segments are replaced from the segment 14 forward, but no anterior regeneration occurs from segment 15 backward. An isolated segment 14 becomes segment 14 on the new worm, and each new segment possesses the structure typical of its relative position.

Oligochaete regeneration is somewhat similar to that of polychaetes, although the former's regenerative capacity is on the whole more restricted. Moment (1950) observed that in *Eisenia foetida* 3 segments are usually regenerated when the anterior 3-5 segments are removed; 4, when 6-10 are removed; and the number regenerated falls again when more than 10 are removed, until none is replaced when 20 or more are removed. Haase (1898), Abel (1902), and Stone (1933) found that in *Tubifex* 3 segments only are regenerated when no more than 12 are amputated, and that no anterior regeneration occurs posterior to segment 15. *Limnodrilus* is similar, accord to Krecker (1910). In *Lumbricillus variegatus von Wagner* (1900) observed that 5-9 segments are regenerated from

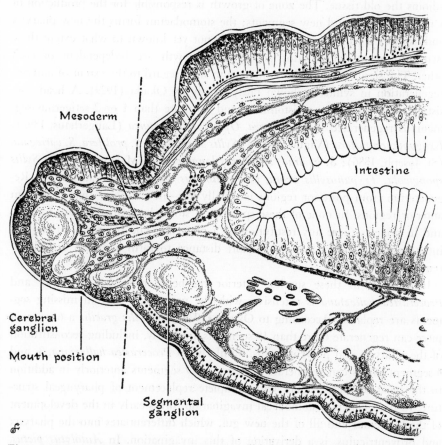

Figure 12-5. Anterior regeneration from posterior level in *Syllis prolifera,* showing regenerated head and position of one setigerous segment (in front of broken line). Anterior end of intestine remains undeveloped, and ectoderm does not form invagination corresponding to position of mouth, as it does from more-anterior levels. (After Okada.)

virtually all levels of the body; according to von Haffner (1928), all are replaced when no more than 8 segments are removed, and no more than 8 are regenerated when more than 8 are removed.

The progressive reduction of the blastema and its development in *Autolytus pictus* have already been discussed. When the blastema is fully developed, as invariably occurs in the sabellids, it gives rise to a prostomium, peristomium, and 1 setigerous segment, together with the stomodaeal invagination. The formation of further segments between this 3-segment head and the parental tissue is associated with both extension of the stomodaeal invagination and the presence of a zone of growth at the blastema's posterior margin, which adjoins the old tissue. The zone of growth is responsible for the production of a new body wall and new segments; the stomodaeum forms the new pharynx and associated parts of the intestine. It is not yet known to what extent these two components of anterior postcephalic growth are independent of each other. The significance of the stomodaeum with regard to the extent of anterior regeneration in syllids is shown fairly clearly by Okada (1929). A head consisting of prostomium, peristomium, and no more than 1 or 2 setigerous segments is regenerated in *Syllis rosea, Opisthosyllis brunea* (Langerhans, 1897); *Exogone gemmifera* (Viguier, 1902); *Syllis alternoseta, S. prolifera, S. variegata,* (St. Joseph, 1886); *S. hyalina, Autolytus longeferiens* (Malaquin, 1893); *Syllis spongicola, Trypanosyllis zebra,* and *Autolytus edwardsi* (Okada, 1929). Regeneration from posterior regions is usually less—never more—than that from anterior regions. In none of these species is a stomodaeal invagination formed, and in none is there any restitution of a pharynx. The front end of the old intestine either grows forward a short distance to form a mouth directly or remains undeveloped.

In contrast to these species, anterior regeneration in *Syllis gracilis* and *Procerastea halleziana* is extensive, with the probability that all missing segments are replaced. According to Okada (1929), in *Syllis gracilis* a 6-segment piece can regenerate more than 18 segments anteriorly, including reconstitution of the pharynx and proventriculus. Similarly, in *Procerastea halleziana*, 3- and 4-segment pieces can regenerate as many as 19 segments anteriorly in addition to the head segments, again with complete replacement of pharyngeal structure. In both genera a stomodaeal invagination forms early in the development of the blastema; and all of the new gut, which differentiates into the pharynx and proventriculus, is a derivation of this invagination. In *Autolytus pictus,* the stomodaeal invagination and its derived pharynx appear only in regenerations anterior to the junction between segments 13 and 14; posteriorly to this level neither stomodaeum nor pharynx develops.

Okada further correlates lack of posterior regeneration by anterior fragments with pharyngeal structure. A chitin-lined pharynx extends through a peripharyngeal chamber, which is without septa, to 5 segments in species of *Procerastea, Virchowia, Pterosyllis,* and *Exogone;* 7 in *Syllis longocirrata;* 9 in *Pronosyllis* and *Autolytus;* 13 in *Eusyllis;* 16 in *Syllis spongicola* and *S. prolifera;* and 22 in *Trypanosyllis zebra.* Only posterior to these levels does the intestine become monoliform and pass through septa. *Procerastea halleziana* requires a head plus at least 7 segments for posterior regeneration; *Autolytus pictus,* a head plus 11 segments; *Syllis spongicola* and *S. prolifera,* a head plus 18 segments; and *Trypanosyllis zebra,* a head plus 24 segments.

Thus it seems that: (1) only a stomodaeal invagination can reconstitute pharyngeal structure; (2) a stomodaeum cannot regenerate intestine posteriorly; and (3) intestine can regenerate intestine posteriorly but not pharyngeal structure anteriorly and can regenerate intestine anteriorly for only a very short distance, so that intestine seems to be polarized for posterior growth.

In oligochaetes the old intestine seems to be relatively more plastic. Hescheler (1898) states that in the Lumbricidae, the gut of the first 3 segments is derived from a stomodaeal invagination, but the pharynx is derived from the old intestine; Kroeber (1900) demonstrated both derivations experimentally in *Eisenia foetida.* Similarly, the buccal cavity is said to arise from a small stomodaeum, and the pharyngeal cavity from the old intestine, in *Lumbriculus variegatus*—according to von Wagner (1900) and Ivanow (1903)—and in *Tubifex rivulorum*—according to Hasse (1898) and Abel (1902).

There are several possible interpretations of the absence or presence and extent of postcephalic growth. With regard to presence or absence, it is reasonably certain that a full-scale blastema, before the onset of specific differentiation, is large enough for the delimitation of head pattern, including a presumptive stomodaeum and a posterior zone of growth; whereas a substandard blastema, produced on a smaller scale or at a slower rate, may have territorial scope for presumptive head pattern but lacks sufficient cellular material for a presumptive stomodaeum and growth zone. Furthermore, when the blastema fails to attain a certain critical size within a given time, head differentiation is inhibited. As to the extent of secondary or later postcephalic growth, the pacemaker may be the posteriorly growing stomodaeum or the posterior growth zone of the blastema as a whole; or there may be a unitary growth activity involving both regions.

That the stomodaeum is the pacemaker is unlikely, since there are many examples of stomodaeal invagination unaccompanied by anything more than head formation. It seems more likely that a stomodaeum grows posteriorly for the same reason that the epidermis and mesoderm do.

The posterior growth zone of the blastema as a whole seems more probable as pacemaker. First, there is great variability in the extent of growth from this zone—from 3 or 4 segments to 20 or more. The segments differentiate anteriorly; i.e., they are successively cut off from the anterior part of the zone of growth. After demarcation a segment continues to elongate to a certain critical length, as shown very clearly in *Procerastea halleziana* by Allen (1921), who recorded also the duration of each successive growth stage. Second, new segments are added from behind more and more slowly, and accordingly, new segments become successively longer since the time interval progressively increases between successive demarcations.

In other words, the rate of growth in this zone falls progressively: the posterior residual growing zone itself grows progressively more slowly, segregating new segments anteriorly more and more slowly, until it ceases to grow and differentiates into the last segment to be formed. Thus the number of segments regenerated in any particular species is the product of the initial growth rate in the growth zone and the value of the decrement. Since the decrement is not a variable but seems to be a general property of developmental growth, the extent of linear growth and therefore the number of segments formed are likely to be a function of the initial rate of growth. What determines the initial growth rate is another question (discussed in another context on page 386).

Postcephalic Reorganization

Internal Regulation

In all annelids in which anterior regeneration is limited to a very small number of segments and in which the gut is normally regionally differentiated, either regulation of the gut occurs, giving rise to normal development, or it does not occur, in which case the reconstituted worm does not develop normally.

In *Stylaria,* the pharynx is followed by esophagus and stomachal dilatation, the stomach occupying segment 8. According to Harper (1904), when the worm is sectioned behind segment 8, and since no more than 5 segments regenerate anteriorly, the new esophagus and stomach formed behind the site of section must arise from transformation of the old intestine. A similar change occurs in *Stylaria* after normal fission and a comparable process of intestinal reorganization occurs during reorganization of abdominal pieces of the ascidian *Eudistoma* (p. 369).

According to von Haffner (1928), in *Lumbriculus variegatus* no more than 8 segments, each of which has a character of its own, are regenerated. Behind the 8 regenerated segments, no matter how many have been amputated, the next 10 segments of the old worm are modified to bring them into line with segments 9-18 of a normal worm: they undergo such changes as formation of dorsoventral vascular commissures, regression of nephridia in segments 9 and 10 (i.e., 9 and 10 of the worm as newly constructed), and reduction in size of the branched vascular caeca in segments 9, 10, and 11.

External Regulation

Regulation of external structure is most readily observed in the sabellid polychaetes, and has been studied intensively in *Sabella pavonina* (Berrill, 1931; Gross and Huxley, 1935; and Berrill and Mees, 1936a and b). Anterior regeneration, which can take place from any level of the worm, is precisely limited to a head consisting of a tentacle-bearing segment, a collar segment, and 1 setigerous segment of thoracic character. The postcephalic body is divided into (1) a short thorax of 5-11 segments, each with dorsal setae and ventral uncinigerous hooks; and (2) an extensive abdominal region, of as many as 300 segments, each with ventral setae and dorsal uncini. Moreover, in the body wall of the normal worm, a deep groove, which is midventral throughout the length of the abdominal region, passes up the right side at the junction with the thorax and continues forward to the head along the dorsal side. The topographical relationship between thorax and abdomen appears as if the anterior part of the worm had been twisted counterclockwise 180°. When the worm is sectioned through the abdominal region, a typical 3-segment head is regenerated. All new thoracic-type segments posterior to the regenerated one are acquired by direct transformation of original abdominal segments.

Posteriorly regenerated segments are abdominal in character, unless posterior regeneration is from origi-

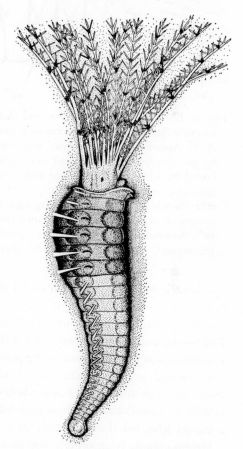

Figure 12-6. Four original abdominal segments in *Sabella pavonina* regenerate 3-segment head and thoracic-type collar anteriorly and 18 abdominal segments posteriorly and transform from abdominal- to thoracic-type segments.

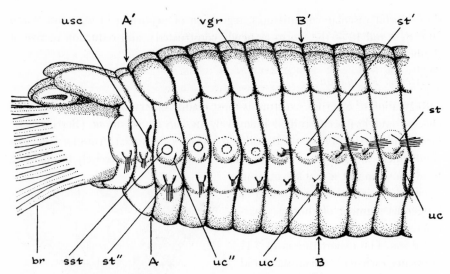

Figure 12-7. Regeneration from, and metamorphosis of, anterior abdominal segments in *Sabella,* showing gradation in time of transformation of affected segments. **A-A′.** Anterior limits of original segments. **B-B′.** Junction between last segment to undergo transformation and first segment to remain unchanged. **br.** Regenerated branchia. **st.** Unchanged abdominal setigerous appendage. **st′.** Setigerous appendage commencing regression. **sst.** Setigerous appendage almost resorbed. **st″.** Newly formed thoracic setigerous appendage. **uc.** Unchanged abdominal uncinigerous groove. **uc′.** Abdominal uncinigerous groove undergoing regression. **uc″.** Newly forming thoracic uncinigerous groove. **usc.** Uncinigerous groove of regenerated segment. **vgr.** Ventral groove.

nal thoracic segments, in which case 1 or 2 thoracic segments are regenerated directly before abdominal segments are formed, or from abdominal levels, in which case outgrowth of the intestine is delayed and usually 1 or 2—occasionally 7 or 8—original abdominal segments at the posterior end of the piece transform into thoracic segments, commencing posteriorly and progressing forward.

In normal anterior regeneration from abdominal segments, the blastema which develops differentiates as a whole—first as a bilobed projection and then rapidly as a small unit in which each lobe exhibits tentacle rudiments—with a common base, but distinguishable as 2 segments.

Metamorphosis is an orderly step-by-step process which proceeds anteroposteriorly. The first metamorphic change, which occurs in the original abdominal segments at about the time the blastema is recognizable as a bilobed entity, is a degenerative change, consisting of loss of uncinigerous hooks. Positive development, of new hooks and setae, in reversed positions, occurs 2 or 3 days later. For example, old abdominal segments 1 and 2 lose uncini on the fifth day after cutting; 3-6, on the sixth; 7-8, on the tenth; and 9-11, on the thirteenth. New setae appeared 3 days later in each temporal group. Thus there is a gradient in the time at which the metamorphic changes commence in the series of segments, but the rate of change seems to be constant for successively affected segments. There is, in fact, a progressive extension in a posterior direction of the metamorphic effect, which commences with the first sign of anterior regeneration and is complete in the last affected segment at about the time anterior regeneration is complete. One case of extremely oblique section has been reported in which anterior metamorphosis took place without recognizable accompanying regeneration.

Therefore there is a correlation between limitation of anterior regeneration to 3 segments and the occurrence of metamorphosis of abdominal structure—a correlation with two possible interpretations:

1. Limitation of anterior regeneration to a 3-segment head in some way disturbs equilibrium, which is restored by a posterior extension of the whole morphogenetic field. In other words, there is something which expresses the whole pattern, and since anterior regeneration accounts for only a part of the

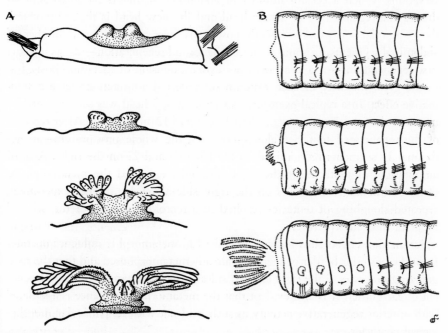

Figure 12-8. Regeneration and transformation in *Sabella pavonina*. **A.** Four stages in regeneration of new anterior structure. **B.** Three stages in transformation of abdominal- to thoracic-type segments, correlated with stage of anterior regeneration. (After Berrill and Mees.)

missing component, the thoracic "morphogenetic field"—for want of a better
term—slips posteriorly to induce metamorphosis instead of building new thoracic
tissue anteriorly.

2. The alternative interpretation is more in keeping with the observations.
Anterior metamorphosis commences with the onset of anterior regeneration and
posterior metamorphosis occurs in its absence. Therefore when metamorphosis
of the most anterior abdominal segment is initiated either by section or by
the initial unspecific healing and growth stimulus, conversion of this segment
to the thoracic character, quickly followed by conversion progressively of more-
posterior segments, may itself be responsible for the limitation of anterior regeneration; for it is known that injury to any abdominal parapodium usually
results in restoration of a thoracic character. An objection to this interpretation
is that although it applies to sabellids, it does not adequately explain internal
regulation in *Lumbriculus.*

Extended Metamorphosis

Further light is thrown on this subject by a number of experiments and observations. When a certain number of abdominal segments have reorganized
during the regeneration of a new head and the new head is then amputated,
the reorganization process extends posteriorly over an additional number of
abdominal segments. In some instances the second regeneration—just as in some
first regenerations—of a head does not evoke extension of metamorphosis, but
in the great majority of such experiments repeated amputations have a sum-
mative effect. In a typical example, regeneration of a head was accompanied by
metamorphosis of 7 segments on the left side and 12 on the right. After removal
of the regenerated tentacles alone—not even the whole prostomium—an additional 19 segments reorganized on the left side and 22 on the right. A third
amputation, of tentacles of the right side only, extended the metamorphosis
of an additional 16 segments on the right side alone. In another experiment,
repeated sloughing of tentacles resulted in a tenfold extension of the original
range of thoracic structure.

Summation effects thus indicate that the full metamorphic influence operates
from the posterior border of segments already metamorphosed and that the reorganizing processes of the right and left sides of the worm are virtually independent of each other. It is also evident that the metamorphic influence is associated
with anterior regenerative activity as such, and not with its particular morphological character.

It has already been noted that in the metamorphosis of the usual 5-10 segments, transformation takes place usually in three successive steps, of 2-3 segments each, progressing posteriorly. When the extent of metamorphosis is

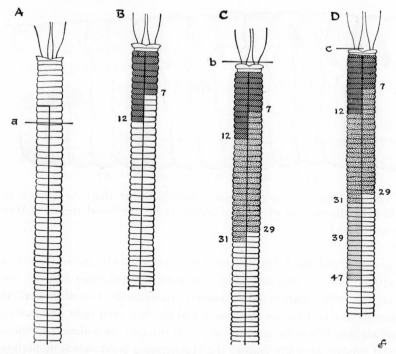

Figure 12-9. Summation effect and bilateral independence in *Sabella*. (Numbers refer to segments posterior to original amputation level **a.**) **A.** Intact worm from ventral side, showing turning of ventral groove and level of amputation. **B.** Extent of reorganization, different on two sides, at end of regeneration period. **C.** Extension of reorganization of abdominal to thoracic segments after amputation of branchiae **b.** **D.** Extension on one side only after amputation of branchia **c** of that side. (After Berrill and Mees.)

much greater, as in 30-40-segment pieces which transform completely, the morphological gradient is much more gradual—a situation implying much shorter time intervals between commencements of change in adjacent segments, for the conversion of 40 segments takes no longer than that of 5. When in such extended metamorphoses the process does not reach the posterior limits of the old abdominal segments, a curious condition generally appears at the thoraco-abdominal junction. Instead of a sharp transition, from the last thoracic segment to the first abdominal segment, there are two anomalies: parapodia of mixed type; and irregular segment skipping—i.e., metamorphosis occurs in some but not all the segments in the transitional zone, so that both thoracic and abdominal segments are isolated from their own kind.

Metamorphosis of parapodia is accompanied and apparently preceded by changes in the epidermal ciliary field. In the intact *Sabella* this field is essen-

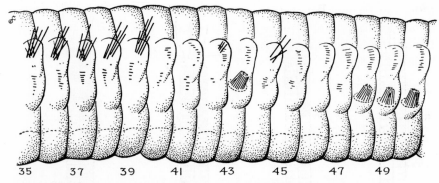

35   37   39   41   43   45   47   49

Figure 12-10. Thoracoabdominal region established after third cut (shown in Fig. 12-9D), showing mixed types of parapodia in transitional region. (After Berrill and Mees.)

tially a generally polarized field on the dorsal surface, with cilia of the entire dorsal surface beating anteriorly, and a seemingly unpolarized field on the ventral surface, with ciliary activity confined mainly to the ventral groove. At the thoracoabdominal junction the groove and the beat turn right and pass to the dorsal surface. When the worm is sectioned through the abdominal region, the ventral groove naturally reaches the regenerating head, where it persists as a groove, though usually losing its ciliary polarization. Usually about 24 hours before the first visible morphological change—the loss of the first uncini— in the parapodia of the first metamorphosing segment, the anteriorly directed ciliary beat ceases, and a more strongly directed beat to the right commences at the junction of segments 1 and 2. Step by step, corresponding to, but pre- ceding the metamorphic steps, the anterior ventral surface becomes depolarized and the right-hand shift becomes displaced progressively posteriorly until the depolarization process ends and the beat to the right remains fixed. Even in thoracic segments which are isolated among abdominal segments, thoracic ciliary pattern is present.

The process of extended metamorphosis is not affected by temperature changes, but is apparently influenced by visible light. In the absence of light, at 11-25°C, the metamorphic range in over 250 worms averaged less than 4 seg- ments, with a maximum of 6. The rate of regeneration and reorganization was in no way related to the range of reorganization. Visible light alone (with ultra- violet and heat rays screened), apparently the effective agent in causing exten- sion of the usual metamorphic process, had a twofold simultaneous effect: (1) it induced a retardation in the rate of anterior regeneration, the retarded rate being accompanied by a diffusion of the pigment pattern; and (2) visible light acting only on anterior regenerating tissue resulted in the metamorphosis, after a single amputation, of as many as 53 segments in one worm and 79 in another. The effective light intensity was about 15,000 meter-candles (Berrill and Mees, 1936b).

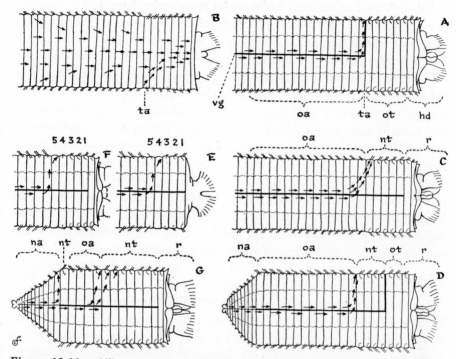

Figure 12-11. Ciliary currents associated with thoracic and abdominal regions in *Sabella pavonina*. **A.** Anterior end of intact worm from ventral side, showing original head **hd** and ventral groove **vg** dividing ventral-gland shields in abdominal region **oa** (original abdominal segments), but not in thoracic region **ot** (original thoracic segments), and turning of groove toward right side at thoracoabdominal junction **ta**. The only ciliary current present is associated with groove. **B.** View of same from dorsal side, showing diffuse posteroanterior currents and continuation of local current from right extremity of ventral groove. **C.** Worm with regenerated head segments **r** and new thorax **nt** reorganized from abdominal segments **oa**, showing ventral groove extending to anterior end, but ciliary current passing to dorsal side at thoracoabdominal junction. **D.** Piece with regenerated head **r**, 2 original **ot** and 4 reorganized thoracic **nt** segments, and posterior regenerated abdominal segments **na**, showing displacement of current from groove to new junction. **E–F.** Progressive shift of lateral current, from segments 4 to 5, with extending reorganization. **G.** Piece with right-half last posterior original segment transformed to thoracic type, as well as 7 anterior segments, showing characteristic lateral current at both thoracoabdominal junctions.

Posterior Regeneration

Formation of Blastemata

Although most of the studies of posterior regeneration have been of annelids, Coe (1934) has studied posterior reconstitution in nemerteans, both in pieces containing part of the original intestine and in pieces cut anterior to the mouth. In fragments containing intestine, posterior regeneration consists essentially of posterior proliferative elongation of the original organ systems and of posterior incorporation of mesenchymal cells. Coe's figures suggest that the elongation results primarily from a conical, terminal posterior zone of growth, so that new tissue is continuously added posteriorly until typical body proportions are regained. In pieces cut anterior to the mouth there is no trace of the original intestine to give rise to the new intestine; here, posterior reconstitution is necessarily more profound. Epidermal epithelium heals the wound, and cephalic mesenchyme accumulates beneath it to form a blastema. Mesenchymal cells become arranged in a vesicular layer to form the mid-gut, and a median invagination of epithelium from the border of the old epidermis meets the vesicular layer to form the mouth and the buccal cavity.

Origin of Tissues

Annelid posterior reconstitution encompasses the formation of new epidermis, intestine, nerve cord, and segmented mesoderm with its various derivatives. The details of the production of these tissues vary considerably among species.

Possibly the simplest—though not necessarily primitive—process occurs in the sabellid *Euratella chamberlini*. According to irradiation analysis by Stone (1933), the epithelial layer (of epidermal origin) of the terminal region gives rise to all new structures; old mesoderm and apparently the old intestine play no part. Epidermal cells push into the body cavity from ventrolateral areas of active proliferation at the epidermal tip to form solid bands of dividing cells, the bands extending forward as the mesoderm, and other epidermal cells move from a mid-ventral site somewhat more anterior to give rise to new nerve cord.

In other annelids—*Spirographis* (Ivanow, 1908), *Autolytus* (Okada, 1929), *Tubifex* (Stone, 1932), *Lumbriculus* (Turner, 1934), and *Lumbricillus* (Herlant-Meewis, 1947)—the epidermis seems to play a much more restricted role, although there is general agreement that the new nerve cord arises, as in *Euratella*, by ingrowth of cells from the mid-ventral surface of the new epidermis a short distance in front of the newly formed pygidium. In all annelids, however (with the doubtful exception of *Euratella*), new intestine is produced by growth posteriorly from the cut end of the old intestine. Haase (1898) states that in

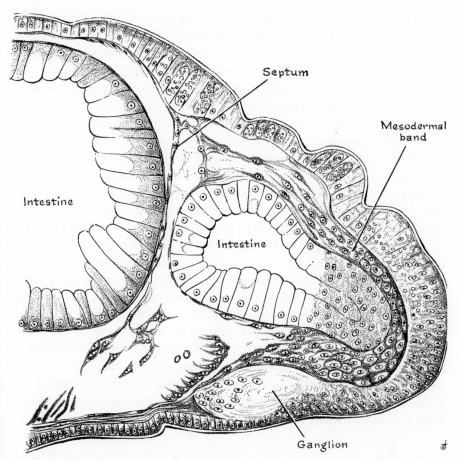

Figure 12-12. Posterior end of regeneration piece in *Autolytus edwardsi,* 3 days after cutting, shown in longitudinal section. Epidermis is proliferating, posterior end of intestine is growing out, and mesodermal bands are forming. (After Okada.)

Tubifex the terminal part of the new gut is ectodermal: the old gut grows out toward the closed body wall and unites with it; the newly formed gut perforates; and then the ectodermal lining slowly grows inward as a proctodaeum.

There is also general agreement that new epidermis and new intestine arise from migration and multiplication of the basal cells of the epidermis and of the intestine, respectively, and not from the larger differentiated cells. According to Herlant-Meewis (1947) in *Lumbricillus* the wound is closed by the migration of epidermal cells from the basal layer to cover the wound area; the cut intestine closes to form a cul-de-sac; and then an ectodermal proctodaeum invaginates to meet the cul-de-sac. The new epidermis proliferates; the basal cells of the intestine, for a distance of 2 or 3 segments from the cut end, multiply, leave their original position, and migrate to the inner surface of the posterior epidermis to form new intestine. In this genus the parietopleural cells at the posterior surface of the closed wound proliferate to form the mesoderm. In *Autolytus* and *Owenia,* according to Okada (1929) and Probst (1932), respectively, new mesoderm arises from dedifferentiated muscle tissue.

In the remaining annelids which have been sufficiently studied, neoblastic cells migrate posteriorly for a considerable distance to constitute a posterior mesodermal anlage. According to Faulkner (1932), in *Chaetopterus,* neoblasts lie between the pair of nerve cords and, when activated, migrate posteriorly. In *Aricia,* Probst (1931) describes the neoblasts as syncitial nests of small mesenchymal cells lying at the angle of septum and nerve cord; when activated, these cells enlarge and migrate posteriorly parallel to the nerve cord. In *Nais* and *Lumbriculus,* according to Herlant-Meewis (1947), large numbers of neoblasts lie ventrally along the whole length of the parietopleura, and, as in the others, migrate posteriorly when activated. Stone (1932) observed *Tubifex* neoblasts on the posterior face of the septa; these neoblasts migrate along the ventral body wall, parallel to the nerve cord, to form new mesoderm posteriorly; new circular muscle, however, originates from the basal layer of the epidermis.

The activities at the cut posterior end of an annelid can be considered as specific responses to a general stimulus—a particular response varying according to the histological state of the particular tissue within the range of the stimulus. The general response of all affected tissues through several segments from the cut end seems to be a combination of posterior migration, growth,

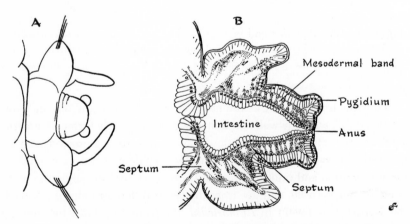

Figure 12-13. New tail in *Autolytus edwardsi.* **A.** External view. **B.** Horizontal section. Segmentation proceeding in new mesodermal bands. (After Okada.)

and proliferation, with some variability in the sequence of events. Neoblasts are essentially multipotent cells which, because they are not otherwise employed, are free to make a rapid response, so that in most cases they reach the blastema before mesoderm can be segregated from epidermal sources. Why neoblasts migrate posteriorly in response to a posterior cut but not anteriorly to an anterior cut is an unanswered question.

Influence of the Intestine[1]

Krecker (1910) found that when the posterior end of the old intestine is removed from segments adjacent to the posterior cut surface in either *Tubifex* or *Limnodrilus,* there is no posterior regeneration until the remaining intestine has grown posteriorly to meet the epidermal cap. New gut epithelium is formed entirely from old epithelium by proliferation at the closed posterior end of the intestine. Okada (1938) demonstrated in *Autolytus* a similar dependence of posterior regeneration as a whole on the posterior growth of the intestine.

The great majority of annelids can regenerate posteriorly from any level of the body, except from the extreme anterior end, though there is commonly some difference between the posterior reconstitutive process at anterior levels and that at posterior levels. In posterior regeneration from anterior regions in *Lumbriculus,* Turner (1934) observed that the transected intestine is closed by overgrowth of ectoderm and an ectodermal proctodaeum invaginates to meet the intestine. At more-posterior regions the gut opening is not obliterated, and the ectoderm fuses with the margin of the intestine. Turner also found that the growth rate as a whole is faster at more-anterior regions than at posterior regions. When both epidermis and intestine grow posteriorly at the faster rate, the epidermal growth, or migration, is relatively faster and completes the closure before the intestine can elongate sufficiently to prevent closure. Abel (1902) detected a similar process in both *Tubifex* and *Stylaria:* closure, then proctodaeal invagination at anterior levels, but persistence of the original gut at posterior levels.

Whether posterior growth of the intestine results from extension of the old intestine or from formation of new tissue posteriorly, posterior regeneration as a whole is basically dependent on intestinal outgrowth; whereas anterior regeneration at least in its earlier stages is independent of both the intestine and the mesoderm. A dual question arises: Are there two competing, or opposing, fields of influence emanating from the two morphogenetically active ends of the worm? If so, what is the material basis of such conflicting and perhaps independent morphogenetic forces? Some light is thrown on this question by

[1] For a discussion of the influence of nerves on regeneration, see Chapter 20, p. 313.

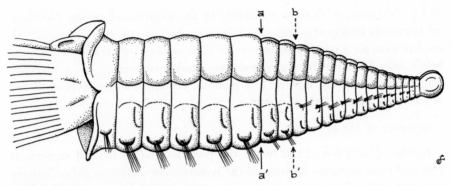

Figure 12-14. Posterior regeneration from original head and thoracic segments in *Sabella.* **a–a′.** Junction between original and regenerated thoracic segments. **b–b′.** Junction between regenerated thoracic and regenerated abdominal segments.

additional observations on *Sabella,* in which thoracic (anterior) segments are formed during reconstitution primarily from the transformation of old abdominal segments. However, when the reconstituting piece consists of the original head and some thoracic segments, so that only posterior regeneration is required, the first 1 or 2 of the new posterior segments are likely to be thoracic and later ones abdominal, as though the thoracic field already present was sufficiently assertive during the early stages of posterior regeneration but was suppressed soon afterward. In contrast, abdominal segments are formed only in the pygidial growth zone, which is maintained at the posterior region and dependent on the intestine.

Perhaps the anterior (thoracic) field is primarily an ectodermal phenomenon; and the posterior field, an endodermal or an endoderm-mesodermal phenomenon—not as germ layers, but as the outer and the inner components of the system as a whole. This interpretation seems to be supported by several facts: (1) When any fully differentiated abdominal segment is injured on one side, it reconstitutes as thoracic structure, i.e., with reversed dorsoventral polarity. (2) When for any reason the intestine of an abdominal piece fails to grow or is delayed in growing posteriorly, no posterior regeneration occurs, but the adjacent posterior abdominal segments transform to thoracic segments; i.e., any activation of the ectoderm, when there is no competition, establishes or re-establishes thoracic polarities. (3) When there is a greatly extended thoracic influence from the anterior end, the thoracoabdominal junction is drawn out and of a mixed character, as though external segmental structure was deorganized rather than reorganized.

Segmentation

Commonly found among annelids and other groups are two types of rhythmic growth: one leads to the subdivision of mesodermal and associated tissues without interfering with the integrity of the individual organism; the other, resulting from epidermal activity, culminates in the subdivision of the individual into a number of units each of which has the properties of an individual organism. The term segmentation denotes the former, and strobilation or fission is employed in the following chapter to designate the latter.

Annelids

Posterior growth as a whole leads to a definite sequence of events, or formations. After the primary layers of the initial posterior cap of tissue, whether regenerated or original, are established, the cap divides into a terminal and subterminal part. The terminal part is the presumptive pygidium and grows in a limited manner very different from the subterminal (more-anterior) part, which remains as the zone of growth.

In polychaetes the pygidium is a well-formed structure from the start and clearly undergoes very little subsequent growth. In oligochaetes the demarcation between terminal and subterminal parts is less well defined, but there is also a terminal, relatively nongrowing region. In both orders the zone of growth proper lies immediately anterior to the pygidial region. Growth in the zone is for the most part linear along the anteroposterior axis and involves epidermis, endodermis, and mesoderm. Therefore the dividing cells in all three tissues must be polarized so that cleavage planes are perpendicular to the main axis; accordingly, spindles must be orientated in line with the axis.

Posterior growth as a whole, as indicated by the extirpation experiments of Krecker (1910) and of Okada (1938), depends on the posterior growth of the intestine. Whether intestinal growth is no more than a necessary condition for the growth of the other tissues or whether it is the actual pacemaker throughout posterior regeneration cannot be determined from the evidence.

Formation of Segments

Segments are produced successively forward from the zone of growth primarily as a mesodermal activity. The mesoderm, whatever its immediate origin, lies in the elongating blastema as a ventrolateral band on each side of the intestine. As the mesoderm grows, each band develops a transverse constriction which separates an anterior part (a presumptive segment) from a posterior part (the persistent zone of growth). The posterior part repeats the process, so that as long as axial growth persists, presumptive segments are successively segregated anteriorly from the growing region proper. Linear, or axial, growth in each segmental part thus delimited slows down considerably, so that growth of each newly formed segment is as much transverse as axial. In this region, only the mesoderm seems to be involved in segment production; both the epidermis and the endodermis maintain their smooth contour with no indication of territorial division. Segment initiation, by successive appearances of transverse grooves in the mesoderm, is associated with a decrement from the maximum growth rate in the zone of growth.

According to Okada (1929), during caudal regeneration in syllid polychaetes, all mature mesodermal tissues can be mesenchymatized, and from these the mesodermal bands arise anew on each side of the intestine, as in embryonic development. The bands spread forward from the posterior end and then divide backward from the anterior; and at no time during caudal regeneration does the ectoderm participate.

In segment formation in earthworms, Gates (1948) distinguishes three methods (which seem to be minor modifications of one process): (1) In embryogeny the mesoderm forms as a pair of separate ventrolateral bands, and the anus is terminal. (2) In regeneration the anus is dorsal, and the two ventral bands divide almost simultaneously into a number of segments. (3) Also in regeneration the anus is terminal, and the growth zone is either large and forms several segments almost simultaneously or small and forms one segment at a time, presumably very slowly; for the most part this variability of size and segment formation can be correlated with varying rates of growth of the mesodermal bands.

Differentiation of Segments

In subsequent segment differentiation there is, therefore, a necessary growth of each mesodermal unit on each side laterally and dorsally, to join with its fellow in the mid-dorsal region. This pattern of growth leads to a consideration of the relative independence of the two sides.

Many polychaetes and oligochaetes have been found to possess so-called compound metameres, i.e., a metamere which consists of 1 segment on one side and 2 on the other (Morgan, 1895; Berrill, 1931). As long as the basic segmentation of the two mesodermal bands remains independent, discrepancy in the matching of the right and left sides can readily occur. Commonly, however, a metamere with 2 segments on one side is balanced by a similar, though distant metamere on the other. The latter situation suggests that the basic rhythm of

Figure 12-15. Anterior regeneration from, and metamorphosis of, abdominal segments, showing compound metameres and possible "balancing" of segmentation. **A–A′.** Limit between abdominal and newly formed thorax. **a.** Segment single on left, double on right. **b.** Segment single on left and right, double in middle. **c.** Segment single on right, double on left. In **a, b,** and **c,** the division extends to ventral aspect of segments; view shown is dorsal.

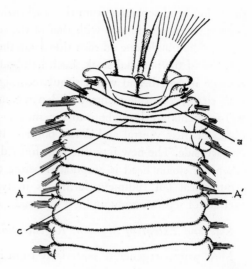

segment demarcation is undisturbed, but that coupling of balanced metameres is inaccurate.

The essential independence of the right and left sides is shown in *Syllis prolifera* by Okada (1929), *Trypanosyllis krokni* by Marion and Bobretsky (1875), and *T. zebra, Syllis vittata,* and *S. cirropunctate* by Michel (1909). The posterior part of these syllids transforms into a stolon with a stolon head. Reconstitution of a new tail by the posterior end of the stock takes place before

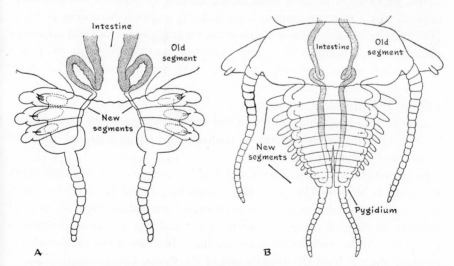

Figure 12-16. Regeneration of tail after detachment of stolon in *Typosyllis prolifera.* **A.** Independent regeneration of right and left halves. **B.** Subsequent progressive fusion of two halves. (After Okada.)

the stolon head separates from the stock, while the intestine is still in line and approximately continuous with that of the stolon. The new tail develops from a pair of half-buds, one on each side, from the posterior lateral border of the last segment of the parental stock. Each half-bud is at first club-shaped and as they increase in length and width simultaneously, each consists of a half-pygidium with cirrus, 3 half-segments, and 3 parapodia. When the stolon separates, the half-segments are brought into the median position, and by the time fusion occurs at their ventral and dorsal borders, 10 half-segments have been formed on each side. The new intestine thus formed is of ectodermal origin. However, when the stolon head is amputated before tail reconstitution begins, a single, symmetrical tail rudiment forms, as in typical posterior regeneration, with no trace of separate right and left half-buds.

Rate and Extent of Segment Regeneration

Most worms regenerate posteriorly from any level posterior to the pharynx, but rate and quantity of posterior regeneration vary with the length of the regenerating piece and with the level from which regeneration takes place. For example, the rate varies slightly with the length of the fragment and greatly with the level in *Lumbriculus,* which exhibits a marked axial gradient in anterior regenerative capacity; but slightly or not at all in *Sabella,* which has no axial gradient.

In most polychaetes, posterior growth continues throughout the life of the worm. In some, as in most oligochaetes, increase in segment number ceases early, so that a constant number is precociously attained. According to Sayles (1932), the polychaete *Clymenella torquata,* for example, precociously acquires a constant segment number of 22, and postlarval growth is primarily elongation of individual segments already formed. In both anterior and posterior regeneration the number of lost segments is exactly replaced, with the exception of anterior levels posterior to segment 10 or 11; i.e., the majority of segments regenerate readily in either direction and eventually assume the same relative position as in the original worm, as confirmed for Clymenella and extend to *Axiothella* by Moment (1951a).

In *Eisenia foetida,* segment formation is either complete at a very early stage, according to Moment (1946), or almost complete, according to Gates (1950). Moment (1950) finds that posterior parts cut off are replaced until the original segment number of the whole worm is restored: the number of segments regenerated in the posterior direction is a linear function of the number of segments of the cut from the anterior end of the worm. Every time the level of amputation is moved 10 segments toward the hind end of the worm, the average number of segments regenerated decreases by 10. The questions which

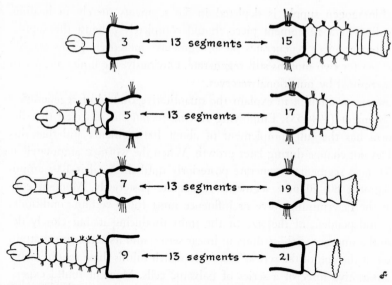

Figure 12-17. Anterior and posterior regeneration from 13 *Clymenella* segments taken from different positions along axis, showing exact replacement. (After Moment.)

arise are: In complete replacement, what determines the limits of regeneration? In incomplete replacement, what is responsible for the deficiency?

Liebmann (1942 and 1943) attempts to correlate the varying capacity for regeneration in *E. foetida* with a single factor—the chloragogen, or lymphotrophic, system—the free elements of which are thought to be nutritive and activating substances which are carried and discharged to all regions of the body. He describes two kinds of such eleocytic bodies: one, in the head region, is involved in cephalic restitution; the other, formed only during regeneration, participates in the reconstitution of the rest of the body and the tail. Typical reserve cells and substances undoubtedly play a part in limiting total capacities for growth and regeneration, but Liebmann's concept is essentially that there are special formative substances for specific types of structure.

The fact is that wherever regeneration is possible at all, it can be repeated by the same piece of original tissue a number of times and at each time regeneration attains a predetermined degree of completion. Thus according to Müller (1908), in *Lumbriculus* the head can regenerate as many as 21 times, the tail 42 times, and both in the same individual 20 times. In *Nais elingius,* Consoli (1923) found that the head could regenerate 12 times and the tail more than 12 times. Gates, commenting on Liebmann's statement that chloragogen depletion in 10-15 segments is necessary for posterior regeneration, said that the chloragogen supply is depleted in 5-6 segments merely in healing, so that healing in a 10-segment piece should completely exhaust the supply. Yet a 10-segment piece regenerates 30-40 segments; moreover, pieces starved of chloragogen for two months still regenerate. Obviously, the limits of growth are not determined by nutritional reserves.

The most recent attempt to explain the quantitative control of regeneration is that of Moment (1946-53). Employing *Eisenia foetida,* he observed that the young worm has the full complement of about 100 segments and that this number does not change during later growth. When the worm is amputated at segment 50, new segments regenerate posteriorly until the total is 100; amputated at segment 80, 20 new segments regenerate posteriorly. Moment concludes that the determining force or influence must fulfill certain conditions: it must be independent of the size of the units producing it, but closely dependent on the number of those units in linear series; and its action must attain its full force at the end of the series. The force which fulfills these requirements is the electromotive force of a series of galvanic cells. Moment found experimentally that the posterior growing tip is electropositive to the remainder, the mean value being about 15.1 mv. This positive value becomes negative at the time of cutting; as posterior regeneration progresses, it becomes positive again and gradually rises until, at the time segment addition ceases, the value is again 15 mv, at which it remains. Moment correlates the voltage either with the number of segments or, on the hypothesis that individual segment growth is the result of cell enlargement and not of further cell multiplication, with the number of cells in linear series.

The evidence thus shown undoubtedly indicates a correlation between replacement of lost segments and acquisition of a certain maximum electrical potential between the anterior and posterior ends of the worm, and it is clear that this electromotive force is not related to the absolute size or length of the organism. Moment concludes further that the electrical potential controls or determines the total segment number in the intact or in the regenerating worm, and his formal theory is that a critical inhibitory voltage causes the cessation of linear growth by segment addition or cell multiplication. Undoubtedly there is a correlation between voltage and linear growth, but whether it is a cause-and-effect relationship is at present debatable. The distinctive, individual character of the various segments of the worm *Chaetopterus,* for instance, is so great that structural make-up as well as segment number and position must participate in the determination process; and segments that seem alike in other worms may on closer examination be found to possess individual qualities of equal significance.

Arthropods

Segmentation as a dynamic process of rhythmic growth, as distinct from segmental homologies and specializations, has been studied in arthropods almost exclusively in the brine shrimp *Artemia*. In studies by Weisz (1946 and 1947), the postembryonic development is divided into 19 stages, each stage number being defined by the number of body segments present during that stage. Larvae of the same stage have the same total length, irrespective of the time or number of molts required to attain it. A body segment is considered to be established as soon as an externally visible transverse constriction separates it from the posterior, nonsegmented zone, which is homologous with the prepygidial zone of the annelid.

All segment rudiments arise in the same way: A patch of cells arranged as a single layer is segregated from the ventrolateral ectoderm on each side into

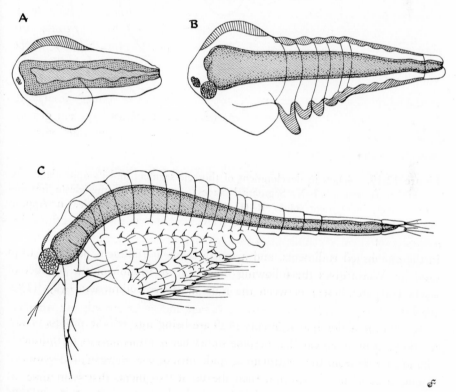

Figure 12-18. Three stages in larval development in *Artemia*. **A.** Just hatched. **B.** Stage 5. **C.** Stage 16. (After Weisz.)

the interior of the body to form a mesodermal layer. A second mesodermal
layer is then similarly formed from the ventrolateral ectoderm and displaces
the first layer toward the gut. While the second layer of mesoderm is being
formed in a rudiment, another new rudiment appears posteriorly, leaving a
small, interrudimental space free of mesoderm. Then a third layer of mesoderm
forms in the first (anterior) segment rudiment, displacing the first two layers
still further toward the gut. At the same time, a second layer arises in the
second (posterior) rudiment, a third rudiment begins to form still more caudally,
and a single layer of mesoderm is formed to occupy the space between the
first two rudiments. After these events an additional sheet of cells appears

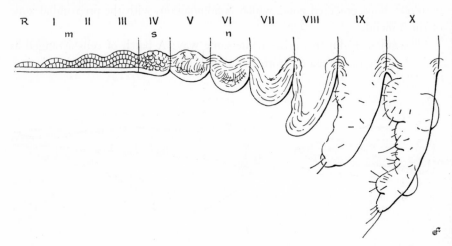

Figure 12-19. Stages in development of thoracic segment and its appendage. **R.**
Segment rudiment. **I–X.** Segmental stages. **m.** Layers of mesoderm forming
successively in segment stages, **R–III. s.** Segment externally visible. **n.** Appearance of ganglionic rudiment. (After Weisz.)

in the established rudiments, and a fourth segment rudiment begins to form
caudally. Weisz gives the following numerical description of the number of
layers (in parentheses) between the four successive rudiments: 4-(2)-3(1)-2-
(0)-1-0.

By the time abdominal segments 14-19 are being formed, the process of laying down trunk segments has become somewhat modified. Segment rudiments
14 and 15 arise from the ventrolateral epidermis, or ectoderm, by the formation
of one layer of mesoderm and then the usual two more mesodermal layers;
but presumptive segments 16 and 17 form only the first two layers by the time
the segments have been formed, and presumptive segments 18 and 19 only
one. In other words, relative to pygidial or terminal axial growth, the process

of tangential inward growth slows down. In fact, there is a decreasing quantitative gradient from the anterior trunk to the posterior abdomen in the relative amount of mesoderm produced within each rudiment by the time of segment formation. Furthermore, Weisz suggests that the ratio of mesoderm to ectoderm in each rudiment determines to a great extent the size and nature of the segmental appendage and also the absence of abdominal appendages.

The process of segment formation in arthropods, as indicated by this account and particularly by Weisz' numerical formula, is significant for several reasons. First, the segregation of mesoderm from the ventrolateral epidermis is essentially the same as the process of mesoderm formation in the annelid *Euratella* (p. 304). Second, it is evident that there is a rhythm of growth in arthropods. The posterior zone of ectodermal growth increases linearly to a certain extent and then gives rise ventrolaterally on each side to a mesodermal mass of a critical area. A mid-ventral ectodermal ingrowth gives rise to a ganglion rudiment during the formation of all but the final few segments. When the terminal region has grown to an equivalent extent, the process is repeated.

The question arises: How is this repetitive process of relative growth to be interpreted in terms of absolute growth or extension? There seem to be two possible answers:

1. The outer ectodermal layer of the posterior growing region grows continuously at a steady or a declining rate; at the same time, the inner side of this layer undergoes a rhythmic pulse of fast and slow growth, giving rise to the segmental masses of mesodermal and neural tissues.

2. The alternative is that there is no pulse in the rate of growth as measured by the mitotic index, but there is a pulse in the rate or extension of the tissue which is growing; i.e., tissue extension can be slow or rapid irrespective of the growth or proliferation rate of constituent cells. When extension ceases or slows, the tissue continues to grow, but in thickness only; when extension takes place, any growth or cell multiplication is accommodated, so that thickening does not occur.

The difference between the two interpretations is that the first postulates a rhythm of protoplasmic growth which is embodied independently in two ventrolateral areas and one mid-ventral area of tissue, so that the only tissue extension that takes place results from the protoplasmic growth; by contrast, the second interpretation postulates unitary growth, but with tissue extension exhibiting a rhythmic and regional quality. Until an experimental test—perhaps by means of markers—determines which interpretation is valid, the latter seems the more plausible. Extension, or stretching can be regarded as a kind of growth, but, as a phenomenon different from cell proliferation and mass increase, extension should perhaps be distinguished from these typical kinds of growth by a different term.

If the terminal extensive growth occurs as a gradual waxing and waning rather than as an abrupt start-and-stop phenomenon, peaks of maximum and minimum thickness would naturally appear in the inner layer. Since the terminal growth pursues a declining course, the differential between maximum and minimum inevitably decreases through the body, and the thickness of the inner layer progressively decreases as succeeding segmental units form, until finally there is insufficient inner tissue available for either ganglion formation or appendage rudiments. Once formed, each segmental rudiment, such as that of an appendage, grows to an extent dependent on its initial mesodermal mass and growth rate; thus the growth potential progressively decreases posteriorly as successive segmental units form.

Growth as a whole may therefore be regarded as two phases: (1) a progressive posterior extension which decreases in rate until a certain relative state is attained, possibly comparable to the detorsion phenomenon in ciliates (p. 68); and (2) an expansion in all dimensions once the final state has been reached. Since full extensive linear growth is attained progressively along the antero-posterior axis, a relative form of growth typifies the early phase, when expansive growth is under way anteriorly at the same time that extensive growth is still in process posteriorly.

This two-phase interpretation is not meant to account for the basic pattern of, for example, a limb: such a pattern is innate in the segmental rudiment that gives rise to the structure. It does, however, seem to account for the various pattern modifications which occur in successive segments, principally as reductions and partial or complete suppressions resulting from inadequate material or rates of growth, i.e., as forms of inhibition (see also pp. 282, 291).

Vertebrates

The developmental process of segmentation in vertebrates properly involves consideration of the morphogenetic movements in both gastrulation and neurulation and so goes beyond the immediate interest in segmentation *per se.* It is, nevertheless, pertinent to a limited discussion that the process of segmentation in vertebrates seems to result from extension, and not from regional growth. According to Pasteels (1943), the rate of cell proliferation as shown by the mitotic index is no greater in tail buds of embryos of all classes of vertebrates than it is in adjacent tissues, and tail buds cannot be regarded as centers of local growth from which segments are successively produced.

This view conforms to the conclusions of Bijtel and Woerdeman (1928),

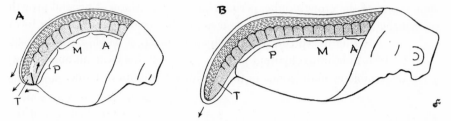

Figure 12-20. Process of tail formation in urodele *Triturus*. **A.** Early tail-bud stage. **B.** Elongation of tail. **A, M,** and **P.** Somites of anterior, middle, and posterior regions of trunk. **T.** Tail. (After Nakamura.)

Vogt (1929), Nakamura (1938), and Ford (1950) that the formative tissue, though growing in proportion to growth of the whole, develops primarily by stretching, not proliferation. The presumptive dorsolateral mesoderm, whether invaginated early through the blastopore or invaginated belatedly as part of the medullary plate, undergoes linear extension and in so doing successively delimits one segmental unit after another, commencing with the anterior region. If this is so—i.e., if segments are not formed as the result of local spurts of growth from a posterior center—the phenomenon as a whole becomes comparable not only to the process of mesodermal segmentation in *Artemia* but to strobilation processes encountered in other organisms, although strobilation is primarily an epidermal and not mesodermal event.

In an experimental approach to segmentation in urodele embryos, Waddington and Deuchar (1953) altered the dorsoventral thickness of the mesoderm sheet by adding or removing material on the ventral side at the beginning of gastrulation and found that when the height of somites was changed, their anteroposterior length was altered in the same sense. These investigators suggest that a mechanism dependent on forces acting in the surface of the mesodermal sheet may be responsible—by analogy with the breaking up of a jet of drops under the action of surface tension.

Summary

Whatever the specific cause of segmentation, it is apparent that the tissue involved is growing with a regular periodicity, as a kind of pulsative growth either comparable to that of free hydroid stolonic terminals or perhaps more intimately associated with the periodic growth of cortical-tissue microstructure of a paracrystalline nature that overrides cell boundaries. The problem of rhythmic growth as evidenced in segmentation leads to, or includes, the phenomenon of cephalization as seen in vertebrates, insects, crustaceans, and other organisms. If the more simple concept of pulsative mesodermal growth or initiation is accepted, it is possible to conceive of a partial, but not complete suppression of the growth pulse, perhaps by the acceleration of general growth, so that there is fusion to a greater or lesser degree between successively forming units, resulting in cephalization. Such an acceleration would be comparable to that which in terminal hydroid growths suppresses the clear-cut annulations, yet is discernible as undulations in the continuous column of tissue.

Chapter 13

Rhythmic Growth: Fission and Strobilation

FISSION AND strobilation are virtually the same phenomenon, strobilation being fundamentally a multiple or serial form of simple fission. By custom, however, each term has been employed for certain groups of animals: "fission" for flatworms, annelids, and polychaetes; "strobilation" for coelenterates and tunicates. Although this customary usage is adhered to in this chapter, no significant differences are implied.

In contrast to segmentation, which seems to be basically a rhythmic, extensory mesodermal activity which does not disrupt the essential unity of the organism, fission eventually divides the organism in two or more parts, is primarily ectodermal or epidermal, and is accompanied or followed by a process of reconstruction of each part into a new whole. In its serial form, i.e., as strobilation, it is seen in flatworms, annelids, tunicates, and coelenterates.

Fission

Flatworms

The phenomenon of fission in flatworms seems to be basically the same as that in annelids, although the lack of metameric septa in flatworms makes analysis somewhat more difficult.

In each species of *Stenostomum*, which has been most studied, individuals exceeding a certain critical size, especially in length, develop a second head near the middle of the body; the posterior part together with the head finally separates from the body to become a free individual. According to Kepner and Cash (1915), the first sign of the second head is a paired ectodermal thickening or bud, which gives rise directly to a pair of ciliated pits and to an associated pair of pit ganglia. Ott (1892) observed that the bud finally separates from the stock by means of a circular constriction of the ectoderm alone, without the formation of a parenchymal thickening.

Sonneborn (1930), analyzing fission in *Stenostomum,* maintains that the two products of division are different: lines of successive anterior products show many of the characteristics of a single individual that develops, grows old, and then dies; lines of successive posterior products of division show the characteristics of newly produced juvenile individuals. The successive anterior products of division are therefore considered as a single individual which successively gives off young from its posterior end. As it ages, it becomes progressively abnormal in form and structure, and in varying degree these abnormalities are transmitted to the successive posteriorly budded individuals—an inevitable transmission, since the posterior half of an aging individual becomes the body of a new individual. Such abnormalities are useful markers for analyzing the process of growth. Examination of Sonneborn's copious illustrations of continuous budding lines of various types of abnormal individuals reveals that both the anterior region (including the head) and the posterior region (including the tail) persist in fairly unchanged forms and that the middle third of the body grows, elongates, and acquires the fission zone. This regional analysis is confirmed by the position and sequence of an individual's successive fission planes, which give rise to chains of 2-7 individuals in various stages of formation.

Stenostomum chains have been studied by Child (1902) and Van Cleave (1929) with particular reference to the fate of zooids when individuals are cut anterior to a fission zone or between adjacent fission zones. In general, anterior partial zooids and younger zooids are resorbed by an older, posterior head region. When no fission plane is present at the time of section, a new head is reconstituted at the anterior cut surface.

The first fission plane appears in a single *Stenostomum* individual about two-thirds posterior to the anterior end. As the new head differentiates, both the anterior and the posterior individual continue to elongate. A second plane appears, but relatively more-posterior in the anterior individual than was the first fission plane in the original individual. The third plane appears in the posterior individual, but relatively anteriorly, i.e., anterior to the mid-body level. A fourth plane appears closely anterior to the second; and, about the same time, a fifth appears anterior to the third, and a sixth appears between the first and second. The presence of the original tail in the posterior zooid alone partly accounts for the apparent relative shifting of the fission planes, but there is no doubt that new fission planes develop closer to new or developing heads than to old heads, as though the organic system undergoes condensation at the region of fastest growth.

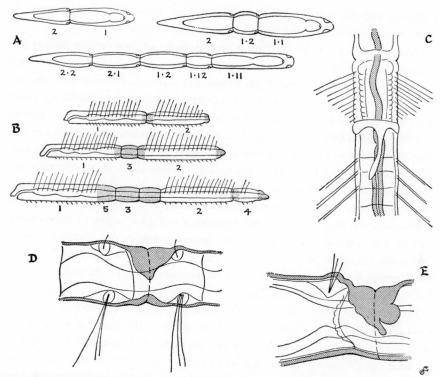

Figure 13-1. Fission zones. **A.** Chain formation and sequence of buds in *Stenostomum leucops*. (After Child.) **B.** Succession of fission zones in *Nais*. (After Stolte.) **C.** Differentiation of divided fission zone of *Stylaria lacustris* into head of posterior individual and segmented posterior regenerate of anterior individual. (After Dehorne.) **D–E.** Fission zone of *Pristina longiseta* and *Chaetogaster crystallina*, respectively, showing epidermal thickening and subdivision in anterior half of segment. (After Dehorne.)

Annelids

Oligochaetes

Fission and related phenomena have undoubtedly originated independently in oligochaetes and polychaetes, and probably have a multiple origin within the polychaetes alone. Since both groups are characterized by segmentation of the body into metameres separated by septa, analysis of growth processes can be much more precise than that in flatworms.

Among oligochaetes the simplest kind of growth is that of *Lumbriculus variegatus,* in which an individual divides without previously producing new segments and the two individuals are completed by anterior and posterior regeneration from the posterior and anterior pieces, respectively.

Dehorne (1916) divides the fission process, in which one definite fission zone appears before separation, into two types: slow and rapid. In the slow type, e.g., *Dero, Aulophorus,* and *Ophidonais,* the individual shows no sign of a second fission zone at the time of separation; a period of growth follows separation; and a second fission zone appears only after both individuals have attained a certain length. In the rapid type, i.e., all other Naididae, a second fission zone appears anterior to the still-embryonic segments produced in the previous zone. This rapid type further subdivides into a naidian type, in which each new zone appears at the anterior limit of the previously produced zone and at exactly the same segment count from the anterior end, and a stylarian type in which the third zone forms behind segment $n - 1$ ($n =$ segment number position of the original zone). Successive fission zones in the stylarian type of fission are therefore at $n - 2$, $n - 3$, etc., and each new individual acquires one segment from the parent. In *Nais elinguis, n* can be 12-20; in *N. variabilis,* 13-29, according to Stolte (1922).

There is, of course, a limit to the extent this process can continue, for the stock becomes progressively shortened. In *Stylaria* the limit is about $n - 7$, after which the anterior stock elongates as in normal posterior growth until the total number of segments exceeds 40 and a new fission zone is intercalated in the middle region. According to Hempelmann (1923), in *Pristina* the limit is $n - 12$.

The distinction between the naidian and stylarian types of individuals is not hard and fast, for a worm usually forming one type may on occasion form the other. In the naidian type the fission process may result in chains of individuals comparable to those of *Stenostomum.*

Stolte (1922) reports that the order in which successive fission zones, especially the fourth and fifth, appear differs among the different genera. Once the first fission zone has appeared near the middle of the worm, a second zone appears within the limits of the first; and a third may form anterior to the second, still within the active growth region as a whole. At the same time, an independent fission zone—which, like the first, is followed by successive zones—usually appears in the posterior half of the original worm some distance anterior to the pygidial segment.

A proliferation zone first appears in every case—not in the plane of a septum nor midway between septa, as once thought—but a short distance behind a septum, according to Dehorne (1916). In the early stages proliferation is essentially an epidermal activity. The epidermis of an anterior half-segment thickens as an inwardly proliferating ring; then a slight constriction in the middle of this narrow band distinguishes an anterior and a posterior part, from which an anterior and a posterior individual will develop.

The fission plane is the region of greatest proliferation, with the quantity of new tissue diminishing both anteriorly and posteriorly. Almost immediately after the onset of proliferation in *Chaetogaster* and *Pristina,* the new epidermal tissue posterior to the line of future separation forms masses recognizable as brain and nerve collar, buccopharyngeal structure, and rudiments of the setigerous sacs of the buccal segment. (The intestine gives rise to its own pharyngeal component considerably later.) The new epidermal tissue anterior to the fission plane gives rise to the regenerating tail, including nerve cord and ganglia, of the anterior individual.

Galloway (1899) observed that in *Dero* when a large number of posterior segments are removed from a worm whose budding, or fission, zone has just started to develop, the budding process continues; but there is no posterior regeneration from the cut end—possibly because the intestine fails to grow posteriorly. Furthermore, bud differentiation is abnormal, and what should become the head of the posterior zooid remains amorphous. However, the anterior half of the budding segment reconstitutes a normal posterior piece for the anterior zooid. On the other hand, when only a small number of posterior segments immediately anterior to the preanal zone of growth are removed, posterior regeneration occurs at the level of the cut; but the epidermal thickening at the fission zone disappears. In this type of cut, when tail regeneration is complete, the thickening reappears in the original segment (segment 18), and separation eventually takes place. In some way the individual regains complete integrity and posterior regeneration occurs, perhaps because the intestine grows posteriorly. When the posterior region is re-established, instability in the form of thickening reappears at the original site.

There is an individual variability and a seasonal variability in the position of a fission zone; and the factors influencing position have been studied in *Nais* by Stolte, *Pristina* by Hempelmann (1923) and Van Cleave (1937), and *Stylaria* by Eckert (1927), Chu (1945 and 1946), and Tsui, Chu, and Pai (1946).

In *Nais* Stolte found that from January to March n decreases (i.e., shifts anteriorly), rises rapidly (i.e., shifts posteriorly) and is highest—on an average by 1.4, according to Piquet (1906)—during the sexual period in midsummer or absent in fully sexual individuals, and then decreases later in the year. Stolte concludes that the position is determined primarily by the state of nutrition.

Van Cleave, investigating *Pristina,* found that any factors—such as high nutritive level or high temperature (20-25°C)—conducive to rapid growth of new segments, produce intense fission, during which n shifts anteriorly. With poor nutrition at high temperature there is a posterior shift; with good nutrition at low temperature there is an anterior shift. He concludes that environmental factors alter the range of dominance by the head end of the zooid, and *n* shifts accordingly.

On the basis of numerous experiments on *Stylaria,* Chu and his co-workers conclude that there are two regeneration gradients, possibly with distinct material bases, and that natural fission proceeds in three successive stages. In the first stage the fission zone is unstable, as shown by the fact that cutting is sufficient to induce its disappearance, and it is assumed that fission takes place in the segment possessing equal anterior and posterior regenerative potencies.

In contrast to *Stylaria* and *Pristina,* in which multiple fission, i.e., strobilation, alternates with normal growth and simple fission, strobilation occurs in *Aeolosoma* only at one time in the most anterior bud of the young chain which forms by successive budding from the pygidial zone, according to Herlant-Meewis (1951). The strobilation is always toward the middle of the worm at a certain distance from the head or from the initial pygidium. There are 2 or 3 such strobilation units, or buds ("*strobiles*") in *Aeolosoma,* and 4 or 5 in *Ctenodrilus.*

The anterior bud of the chain subdivides secondarily while it is still in contact with the pygidial zone from which it formed, but ceases to subdivide as soon as a new bud is formed from the pygidium. Thus subdivision of the strobile, or bud, occurs if the rate of pygidial budding is rapid, but does not occur if budding is slow. Herlant-Meewis does not explain the variability of rate in pygidial budding, but concludes that the number *n* of segments of the anterior zooid of *Aeolosoma viride* is fixed by the position of the strobiles and under normal conditions varies little. Strobilation, however, is localized in a zone extending over several segments and under poor nutritive condition is displaced posteriorly, so that the anterior zooid becomes longer and *n* becomes higher.

These phenomena raise questions rather than supply answers. In general, a fission zone is essentially a region of relatively extensive linear growth. The question arises: What initiates such a zone? to which there is at present no answer. It seems evident that the high rate of growth in that region in some way brings about a discontinuity in the basis of organization. If, for instance, there is a degree of symplasmic continuity in the epidermis, particularly at the more basal level, with a continuous, polarized microstructure which is the core of an individual's individuality, and when in an area of maximal growth this microstructure loses its fundamental structural character simply as the result of that growth, such a region could be expected to be at least potentially an independent entity. When the disruption occurs along the length of a polarized individual such as a worm, both the anterior and the posterior sections react as though the other were missing. This situation corresponds to what Child in the early years of this century called "physiological isolation." It is noteworthy, however, that although the actual growth disruption, so far as is visible, is primarily epidermal, there is evidence that intestinal growth, on which posterior regeneration in great part depends (p. 307), may play a very important role, and that inducing or disturbing growth in one place effects comparable activity in another.

Polychaetes

In their bewildering variety of budding processes, the syllids stand alone. Possibly the simplest budding process is that in which the stolon head appears before separation from the stock, as in *Procerastea halleziana,* according to Allen (1921) and *Autolytus* (syn. *Proceroea*) *pictus* and *A. cornutus,* according to Okada (1929 and 1933). In each of these species the stolon head develops on the anterior half of segment 14—counted from the anterior end. It may also be significant that in each the junction between segments 13 and 14 is the site of change of the fragmentation megasepta from 2-segment to 3-segment series and that in *A. pictus* at least, in anterior regeneration, stomodaeal invagination and associated renewal of the pharynx, etc., take place at every level down to the end of segment 13; but posteriorly from segment 14 no such invagination occurs.

Thus the presumptive fission plane is recognizable both in the morphology of the intact worm and in the regenerative capacity at different levels. At the junction between segments 13 and 14, there is a critical transition, which is correlated with the later development of an intercalated head at this level. According to Okada (1934a), the activity giving rise to the new head occurs in the outer layer of the body, for both intestine and longitudinal muscles remain uninterrupted and inactive and even the outer circular muscles seem to be undisturbed. Moreover, it is doubtful that connective mesenchymal tissue is involved, so that the seat of change seems to be basically epidermal.

The formation of the stolon head on segment 14 in these genera is constant: sectioning a worm in the anterior region fails to displace the stolon head posteriorly. Thus in *Autolytus pictus* when the worm is sectioned through the anterior half of segment 13 anterior to the parapodia, the head is regenerated from this level; but at the same time, a stolon head develops on the anterodorsal face of segment 14, so that two heads are formed, one in front of the other, separated by no more than half a setigerous segment.

In syllids, as in other annelids, a metamere commonly consists of 2 segments on one side and 1 segment on the other. When this occurs in the anterior

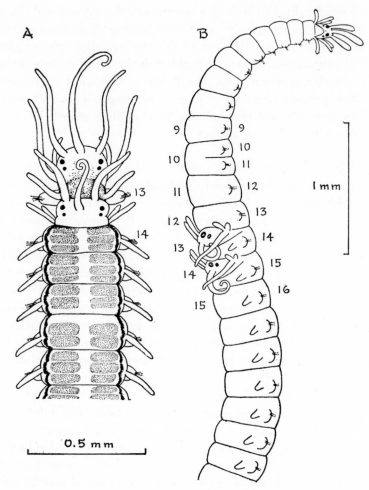

Figure 13-2. Head regeneration in *Autolytus pictus*. **A.** Regeneration of head from surface of anterior half of segment 13, and simultaneous induction of second new head from anterodorsal surface of segment 14. **B.** Individual with compound metamere (segment 10 on left side, segments 10 and 11 on right), with two stolon heads in relation to segment 14 of left and right side, respectively. (After Okada.)

region, segment 14 on one side may be matched with segment 13 or 15 instead of 14 on the other. Okada describes such an *Autolytus* with two stolon heads: one from the anterodorsal surface of segment 14 as counted from the right side, and one on segment 14 counted from the left, which are not at the same level.

The first sign of stolonization in normal individuals is the appearance of a white line, which is presumptive new tissue, between segments 13 and 14.

When worms—e.g., *Dero,* according to Galloway (1899)—in this stage are sectioned behind segment 14, the white line disappears, and a tail regenerates posteriorly. On the other hand, according to Okada (1934a), when a worm is sectioned through the anterior part of segment 14 anterior to the parapodia, a head forms posteriorly instead of a tail.

Evidently, whatever the basis of organization, the right and the left side of a bilaterally symmetrical organism, such as an annelid, are virtually independent, although the effects of distance and the phenomenon of two heads suggest a condition far more subtle and complex than can be explained by any simple concept of dominance by metabolic gradients or another factor.

In the syllid subfamily Syllinae the location of the stolon head is not rigidly predetermined as in *Procerastea* and *Autolytus.* In *Typosyllis prolifera,* Okada (1934a) observed as the first sign of stolonization a slight growth of new, unpigmented tissue on the anterodorsal border of one of the segments in the middle of the body and, a little later, a small spot of red pigment (presumptive eyes) on each of the two sides. Immediately afterward, a small protuberance, representing a half-bud of a tail, appears on each side of the ventroposterior face of the adjacent anterior segment. Unlike *Procerastea* and *Autolytus,* posterior segments are added to the *Typosyllis* stock at the same time that the head develops on the stolon. That the head is not predetermined becomes evident when a *Typosyllis* worm is cut obliquely through the presumptive stolon head: a half-bud of a tail develops on one side (from tissue that in the intact worm would become a head) and carries with it the ocellus of that side. Sections at successively later stages of development show that the head is progressively determined.

Within a certain rather wide range any segment of *Typosyllis* can produce a head. Michel (1909) cut a 28-segment fragment from the middle of a *Typosyllis amica;* the fragment regenerated a head and several postcephalic segments anteriorly and a segmented tail posteriorly. A second head appeared later on segment 7, and successive heads appeared on adjacent segments until 12 heads had been formed, on segments 2-13, respectively. Okada confirms this regenerative capacity in a *Typosyllis prolifera* stolon in which heads formed successively in the four segments 39-42. In both species a pair of tail half-buds appeared at the ventrolateral border of the segment anterior to a head.

In other syllids more-complex budding of several kinds occurs. In *Myrianida pinnigera* and *Autolytus edwardsi,* according to Malaquin (1893) and Okada (1935 and 1937), chains of buds are produced. A new stolon head appears in a segment between segments 16 and 22 from the posterior end and does not separate from the stock until 2, 3, or more stolons have appeared anterior to it. The result, a chain of stolons, is likely to occur during the warmer months.

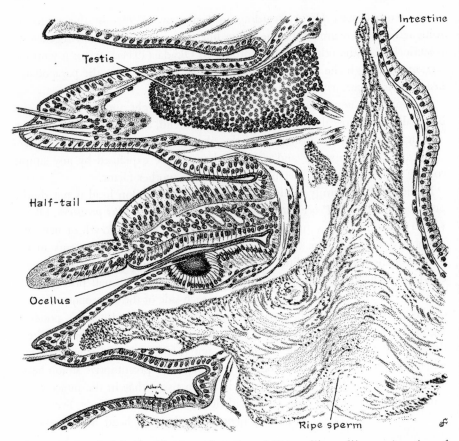

Figure 13-3. Left half of horizontal section of *Typosyllis prolifera* at junction of two individuals (stock and stolon). The half-tail, with segmenting mesoderm enclosed, and the left-side ocellus of head of posterior individual are developing from epidermis at junction between two segments. (After Okada.)

In some ways chain formation in *Autolytus* and *Myrianida* is comparable to that in naids. A new segment—if one can call it that—develops between segments 30 and 31, for example, of the original *Autolytus* or *Myrianida* worm. This segment rapidly constricts into three parts: the most posterior progressively differentiates as a pygidium; the middle undergoes progressive segmentation, commencing with a single division into the first setigerous segment and a zone of growth; and the most anterior persists in the form of the original segment and repeats the process. Gradually a chain of zooids is formed; the oldest is the most posterior and the residual formative segment remains contiguous with the stock. It is noteworthy that in each zooid the first segment to be differentiated is the pygidium and that the first segment produced by the prepygidial zone of growth is a typical setigerous segment, which slowly transforms into a stolon head only after segment formation has continued posterior to it for some time.

In *Myrianida,* when an individual in process of budding is cut so that the larger posterior part, including the chain of stolons, is removed, the missing

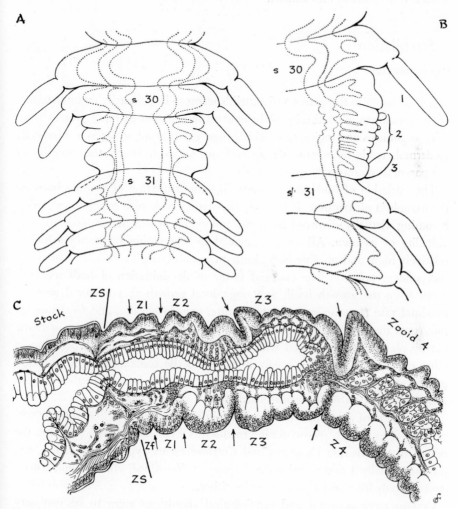

Figure 13-4. Subdivision of intercalated segment (between segments 30 and 31) at commencement of budding in *Autolytus edwardsi.* **A.** First triple division of new segment. **B.** More-advanced state. **C.** Longitudinal section through chain of buds. **ZS.** Posterior end of stock individual. **Zf.** Fission zone. **Z1–4.** Four successive zooids, or stolon individuals. Note mesodermal segmentation in **Z2.** (After Okada.)

segments regenerate immediately; after a while they develop directly into a new chain of stolons, with the most anterior of the new segments persisting as the source of further growth; i.e., a new chain is formed by direct metamorphosis. On the other hand, when the cut is outside the range of stolonization, i.e., anterior to setigerous segment 30, posterior segments regenerate, but there is no transformation into stolons.

Strobilation

Coelenterates

Just as fission in flatworms and annelids seems to be primarily an interruption of the fundamental continuity of the epidermis by a zone of rapid growth, so in other organisms any region of relatively high growth rate, so far as it is basically epidermal, tends to become set apart from, or to interrupt, the integrity of the parental organism.

The initiation of buds, for instance, like the formation of medusa buds on the manubria of medusae, is basically the setting aside of a certain area which, becoming virtually divorced from the individual organism, acquires an individuality of its own. All such segregated or segregating territories exhibit a relatively high, and possibly biologically maximal, growth rate and a somewhat independent polarity. The essential factor in the initiation of buds seems to result from the growth itself, as though local growth at a maximal or near-maximal rate (for the current temperature, oxygen tension, and food supply) interrupts the organizational continuity of the whole. That this is so is borne out by phenomena associated with the processes of fission and strobilation in other organisms. Although briefly discussed in the context of the strobilation process of scyphomedusan polyps (p. 163), strobilation deserves further analysis as an introduction to the growth activities of coelenterates.

Strobilation of a scyphistoma can result in the formation of 2-30 ephyrae successively liberated as free-swimming medusae. Of the more-common genera the polyps of the medusa *Cassiopea* and *Cotylorhiza* are typically monodisk, i.e., produce usually 1 and occasionally 2 ephyrae; *Nausithoë, Aurelia,* and *Chrysaora* are polydisk; and *Cyanea* may be either.

Certain environmental and physiological conditions seem to be necessary for ephyra production. All observers agree that a relatively long nutritive preparation is essential and that strobilation usually takes place during mid- or late-winter, when the temperature is lowest. It may be significant that the two monodisk genera cited, *Cotylorhiza* and *Cassiopea,* are warm-water organisms. Among the other genera there is considerable variation: *Chrysaora* produces

4-16 ephyrae per polyp; one *Aurelia* brood produced 20-30 per polyp, another produced 1-4, and another consistently produced 3; *Cyanea* produces from 1 or 2 to 10 or 11 according to environmental circumstances.

The transformation of the distal end of either a monodisk or a polydisk polyp into an ephyra presents problems distinct from the process of strobilation. As a general rule, rapid growth of the polyp occurs toward the end of the summer, after which there is a resting period of two or three months. According to Lambert (1935), after the rest-period, the polyp lengthens to several times its length attained during the summer growth period and commences strobilation.

The first sign of polydisk strobilation in *Aurelia, Chrysaora,* and *Nausithoë,* according to Percival (1923), Chuin (1930), and Komai (1935), respectively, is a series of external grooves encircling the column of the polyp. The grooves are produced serially in time and place by the epidermis and gastral endodermis, which undergo a local heightening and proliferation, resulting in folds that cut through the internal septa as far as the longitudinal muscles. After the muscle bands degenerate, there occurs a combination of new growth and destruction (involving histogenesis, histolysis, and phagocytosis) of old tissues—processes typical of metamorphic change in general.

Delap (1905) made a cerful record of the times of first appearance of the successive grooves and the times of liberation of the respective ephyrae of strobilating polyps of *Cyanea capillata* var. *lamarckii.* With daily temperature of about 7-11°C, grooves appeared in a very orderly manner: 11 formed successively from apical to basal end of the polyp at intervals of a little more than 24 hours, so that the whole process required 13 days. Ephyrae were also liberated in regular succession in time, 1 ephyra every 10 days after the appearance of each groove. In other words, the process of constriction and the conversion of the transverse disk into an ephyra constitute an orderly developmental process of a definite duration at a given temperature.

The type of strobilation depends on the size and shape of the head and column of the scyphistoma. When the head is shallow and the stalk long and slender, strobilation is usually monodisk and results in 1 or 2 ephyrae; when 2 are formed, there is a long interval between formation of the first and that of the second. When the head is approximately columnar, strobilation is typically polydisk. In both monodisk and polydisk types, the successive constrictions appear after the polyp has attained a critical column width. Thus the two significant features—the critical width of the column and the process of lengthening, which pursues an apicobasal direction and time sequence—seem to be correlated with growth and indirectly with food reserves.

Several processes are involved in the over-all strobilation event:

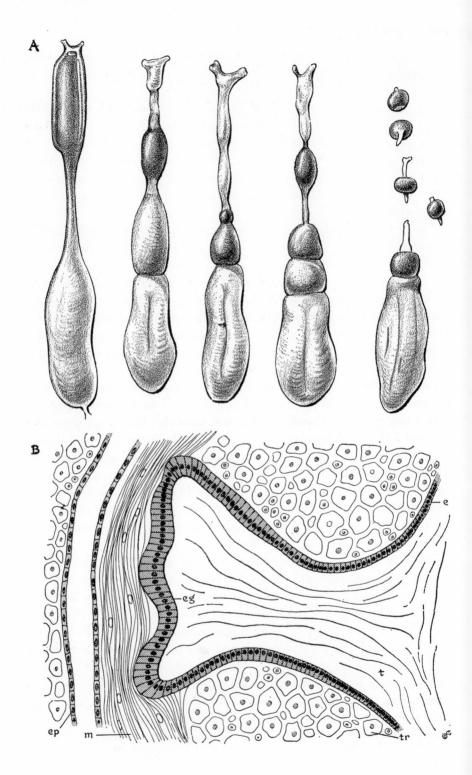

1. Conversion or transformation of an apical polyp to an ephyra. This process is essentially the same as the conversion of an actinula larva to a medusa in the development of intermediate-size eggs of *Aurelia* and some other genera whose attainment of a critical width or circumference seems necessary for expression of medusan structure.

2. Lengthening of the polyp immediately before the onset of strobilation.

3. Constriction.

4. Development of medusan structure in the constricting segment. This process is comparable to the direct development of medusae from the largest eggs of *Aurelia* or the eggs of *Pelagia* (p. 160).

5. Isolation. For this final process—as distinct from the morphogenetic events set in motion by such isolation—lengthening and constriction are of primary importance.

This is seen in striking form in the strobilation of the hydroid *Corymorpha* (p. 256), which is as follows: When the column of an individual is cut in the yellow region below the level of the red striped tissue, the yellow region lengthens to many times its original length and becomes correspondingly narrow. At the same time, its more-differentiated endodermal tissue undergoes dissolution, and a long, slender tube of narrow epidermis and endodermis is produced. This tube then constricts in serial manner in rapid succession posteriorly to form units about twice as long as they are wide. Subsequently each unit reconstitutes a small hydroid. Details of the constricting process are not known, but it is known that the constricting layer is the conjoined epidermis and endodermis which as a unit have undergone the elongation and histo-rejuvenescence process.

Tunicates

Ascidians

Strobilation is effected in much the same manner in some ascidians and in certain other tunicates. The constrictive agent is seen more clearly in tunicates than in coelenterates; for although constriction is accomplished by the outer body layer of the tunicate as it is in the coelenterate, the layer in the former

---

Figure 13-5. Strobilation in *Diazona violacea*. **A.** Five zooids, showing stages in resorption, posterior congestion by trophocytes, and strobilation. **B.** Longitudinal section through abdominal constriction showing active epidermal growth at constriction site, cutting through passive internal tissues. **e.** Normal epidermis. **eg.** Growing epidermis forming constriction. **ep.** Epicardium. **m.** Longitudinal muscle. **t.** Tunicin secreted in constriction groove. **tr.** Trophocyte.

consists of epidermis alone. Thus in *Diazona* and *Eudistoma* the zooid is divided anteroposteriorly into several segments by local annular growth of the epidermis to form folds, or constrictions, which cut through the various internal tissues such as muscle, gonad, epicardium, and intestine. This local annular growth is equivalent to local lengthenings, whether or not the zooid as a whole lengthens significantly.

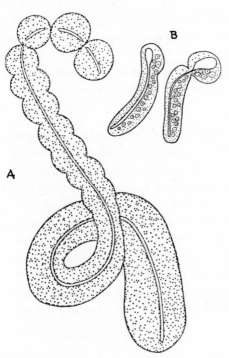

Figure 13-6. Extended ascidian postabdomen, containing epicardial tube and nutritive cells and exhibiting epidermal strobilation. **A.** *Colella pedonculata.* **B.** *C. cerebriformis.* (After Caullery.)

In related genera such as *Colella, Circinalium, Polyclinum, Amaroucium,* the postthoracic body wall with its contents lengthens considerably at a certain early phase of the life cycle and then subdivides by epidermal constriction into a series of short segments. The subdivision process begins anteriorly as in *Diazona* and *Eudistoma* and proceeds in regular fashion toward the posterior end—a situation suggesting that this direction may also be the course of the lengthening process. It is noteworthy that in polyclinids such as *Circinalium* and *Amaroucium,* whose lengthened column is relatively congested with trophocytes, the segments are from two to three times longer than they are wide; whereas in the polycitorid *Colella,* which is more sparsely supplied with trophocytes, the segments are not only smaller but nearly spherical—i.e., segment length and width are almost equal.

In all these genera, but particularly in *Circinalium* and other polyclinid zooids, the region which lengthens is postabdominal, i.e., between the lower end of the U-shaped intestine and the heart. This region contains the epicardium and gonadal tissue, both of which lengthen in proportion to the whole. However, since only the lengthened epidermis plays an active part in the process of subdivision or constriction, it is reasonable to suppose that the epidermis may be the pacemaker for the rest of the body. If so, then the lengthening process in the polyclinid *Aplidium zostericola,* as described by Brien (1925), is significant.

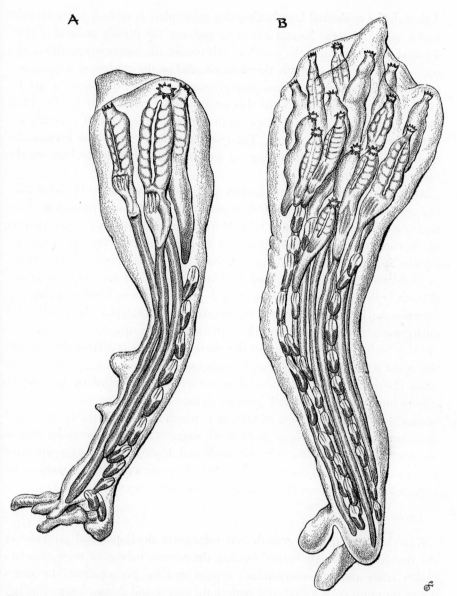

Figure 13-7. Young colonies in *Circinalium concrescens*. **A.** Lengthening of postabdomen. **B.** Strobilative subdivision into files of buds, all with same orientation. (After Brien.)

Although the epidermal lengthening that takes place is typical, it involves the region containing the looped intestine, and not the greatly reduced postabdominal region—because the zooids of this species are comparatively short. The region most affected contains the stomach, and as the epidermal lengthening proceeds, the stomach lengthens proportionately and its ridges become greatly extended in consequence. Since this organ cannot conceivably be the pacemaker for lengthening in other species of even the same genus, it is highly unlikely that it is in this species. Therefore it probably lengthens because the epidermis—adjacent or throughout—is lengthening, although nowhere are the two tissues in contact.

Accordingly, two phenomena assume particular importance: (1) the stretching and regular constricting of the epidermis; and (2) the induction of an equivalent stretching or growth (should stretching and growth be synonymous) in the contained tissues, even though at an appreciable distance from the epidermis. Stretching or lengthening of a tissue may or may not be a consequence of proliferative or any other kind of cellular growth. In *Diazona* and *Archidistoma,* epidermal cell enlargment and probably multiplication accompany constriction; but in *Colella* and the related genus *Distaplia,* there is no cell enlargement and no indication that cells are proliferating faster in the constricting regions than elsewhere, so that there is no doubt at all that the constrictive agent is the thin layer of simple epidermal epithelium. The question again arises: Is the extension of material—extension being implied by the fact of constriction—the result of cell growth, change in cell shape, or extension of surface coats or of a basement membrane to which the associated cells are forced to accommodate? The process in *Distaplia* suggests that cell growth by itself is not the answer, for although the epidermal tissue undergoes proliferative growth, such growth is not regional; rather, the constrictive process seems to be comparable to the waves of contraction in ascidian ampullae (p. 113).

Thaliaceans

Knowledge of stolonic growth and subsequent developmental processes in the thaliaceans—though limited because the oceanic habitat of these tunicates makes study difficult—encompasses certain striking phenomena. The genus *Salpa,* for example, is a challenge both in the nature and development of its egg and in the nature and strobilation of its budding stolon.

The *Salpa* stolon grows as a slender, cylindrical structure from the ventroposterior side of the individual. At its base the stolon consists of an epidermal or ectodermal epithelial tube enclosing an endodermal tube which derives as extension from the posterior end of the endostyle—and a mass of mesodermal tissue between the two epithelial layers. The three components (epidermal

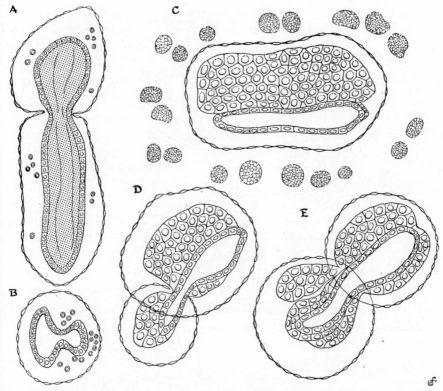

Figure 13-8. Subdivision of probuds in *Distaplia*, showing active constriction by epidermis. **A.** Probud in *D. bermudensis* tadpole, showing primary constriction. **B.** End view of same, showing doubling of epicardial component. **C.** Probud in mature *D. garstangi* blastozooid, showing doubled and differentiated epicardial component. **D–E.** Similar probuds in *D. garstangi* in process of unequal constriction.

tube, endodermal tube, and mesoderm) grow as one, despite the considerable space which eventually intervenes between the ectodermal and endodermal tubes in the established stolon. At the start, however—or at the earliest stage described to date, when the ectodermal cap of approximately columnar epithelium already protrudes slightly, the three layers are contiguous: the cap is in contact with a proliferating mass of mesodermal tissue, which in turn is in contact with evaginating endodermal epithelium of a columnar shape. As the blastema continues to grow, the combined tissue progressively protrudes as a hollow tube within a tube, with the mesodermal tissue growing as a solid strand.

As growth of the whole stolon proceeds, two virtually independent, though

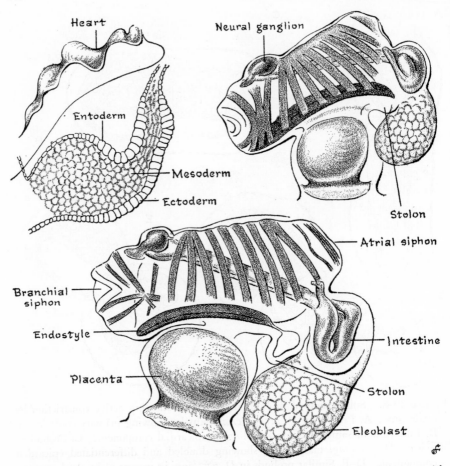

Figure 13-9. Embryos of *Salpa* (*Cyclosalpa pinnata*) attached to placenta, with prominent muscle bands, showing outgrowth from posterior end of endostyle, initiation of budding stolon, and early stage of young stolon.

related processes occur: subdivision, or strobilation, of the ectodermal tube to form a series of presumptive individuals which eventually break free successively from the end of the chain; and morphogenesis.

The greatest rate of stolonic growth takes place at the base, where the stolon is joined to the parental body. As new tissue is produced, older tissue is shunted outward. In other words, there is a basal zone of growth, and the farther along the stolon from the base, the older the tissue is—as in the growth of a hydra column from a distal growth zone. Also as in the hydra column, the tissue progressively shunted along the stolon continues to grow although at a progressively decreasing rate.

STROBILATION OF THE STOLON

The first sign of strobilation of the *Salpa* stolon—a striking example of ectodermal segmentation—occurs close to the stolon base where the ectodermal
epithelium is most columnar in shape. More basal than this the ectodermal cells
are flat and show no pattern of surface orientation. Where the epithelium
thickens, so that its constituent cells are relatively very numerous per unit area,
the cells become aligned around the stolon in transverse rows whose cellular
density progressively increases in a distal direction. To account for this appearance of increasing density, it may be assumed that when the growth rate of
the ectodermal tissue is close to its maximum and the tissue consequently advances to a critical position, the cells in a particular transverse segment divide
simultaneously two or three times in succession, so that parallel rows of closely
packed columnar cells are produced and encircle the stolon.

Since each single row of ectodermal cells can be regarded as an ectodermal

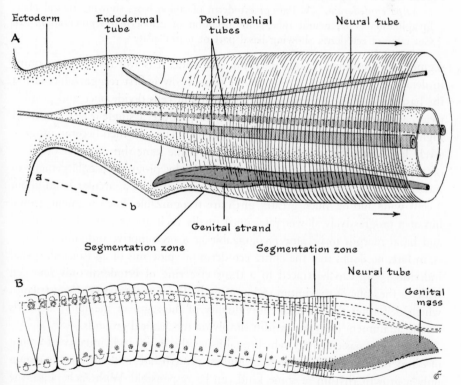

Figure 13-10. Base of *Salpa* stolons. **A.** *S. fusiformis,* showing zone of general
growth **a–b** and ectodermal segmentation zone. **B.** Relatively smaller stolon of
S. democratica.

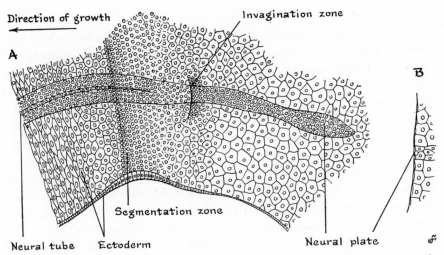

Figure 13-11. Segmentation of stolon ectoderm and development of neural tube in *Salpa confederata*. **A.** Part of ectoderm of stolon base, showing neural plate, invagination zone, neural tube, and inception of ectodermal segmentation. **B.** Cut edge of ectoderm, showing basal part of neural plate.

segment or presumptive individual one cell wide, it is worthwhile to examine in detail the fate of such a row. At the time a row is first recognizable, i.e., at the basal side of the growth zone, its constituent cells divide once and possibly twice, without disrupting the alignment, to produce a single-row segment of high, closely packed columnar cells. The constituent cells then become progressively less columnar and almost squamous. At the same time, further divisions occur, but more slowly, and are accompanied by progressive enlargement of individual cells, until ectodermal segments consisting of a circular band about 4 cells wide and of much greater girth are recognizable. Each segment grows, but at a progressively slower rate, until finally each grows inward at its distal and basal margin to become recognizable as a presumptive individual. There is, in fact, no doubt that the entire ectoderm or epidermis of an isolated, sexual individual salp can be traced to a transverse ring of ectoderm only one cell wide in the zone of maximum stolon growth. The actual process of constriction which delineates the essential individuality of the segment inevitably depends on the attainment of greater width simply on mechanical grounds.

Clearly, the basic activity which establishes the initial individuality of a ring of ectoderm lies within the zone of highest growth rate, and within this region only a growth rhythm of some kind can be responsible. With each pulse, the cells in a row respond by growing and dividing simultaneously to become permanently segregated in stage of development from those that received the growth pulse earlier or those that will receive it after them.

As long as the *Salpa* stolon continues to grow at its base, the growth zone continues to exhibit its rhythmic quality. Stolonic growth as a whole, however, has a larger rhythm, for growth in length from the basal zone, apart from the process of strobilation into presumptive individuals, is clearly intermittent. At any one time a stolon usually consists of 3 sections of greatly different diameters, which represent 2 or 3 growth surges separated by periods of rest. Since the terminal section breaks off when it reaches a certain size and stage of development, it is not possible to say how many such stolon blocks are produced during the life of the individual because *Salpa* has not yet been grown in captivity nor the entire life of an individual followed in nature, though Johnson (1910) has reported that as many as 8 are present simultaneously in the Pacific Coast variety of *Salpa fusiformis*. The probability is that stolonic growth consists of surges of growth of roughly equal extent, each surge terminating in a relatively sudden retardation of the growth rate to zero, and continues as long as the parent itself is still growing, much as in the sequence of buds produced by the buds of *Symplegma* (p. 395).

Salpa species differ greatly in size of the individual as a whole, and the dif-

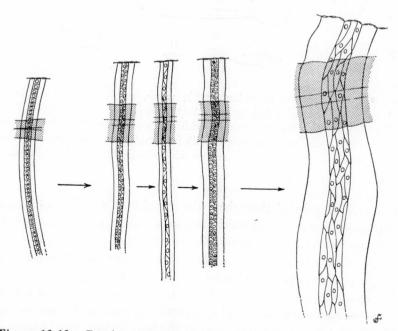

Figure 13-12. Ectoderm and neural tube of individual segments taken at successive intervals distally along stolon of *Salpa confederata*, showing multiplication and expansion of single row of ectodermal cells in relation to expansion of third dimension of prospective individuals.

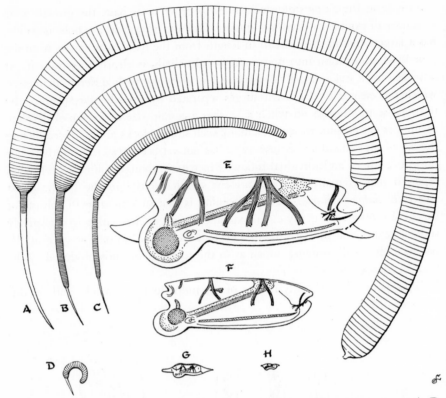

Figure 13-13. Comparison of stolon length, width, and segmentation. **A–D.** Stolons of two- or three-block lengths in *Salpa maxima, S. confederata, S. fusiformis,* and *S. democratica,* respectively; with end blocks consisting of about 30, 60, 100, and 200 double segments, respectively (see text), all at same magnification. **E–H.** Maximum-size nurse of same four species, respectively, all at same magnification.

ference in scale is reflected throughout the organism. Thus the large *Salpa maxima* produces a large embryo which has a large initial stolon whose diameter is several times that of *S. democratica;* the diameter of the *S. confederata* and *S. fusiformis* stolons is intermediate. The stolon rudiment first appears in a particular stage of the embryo, whatever the size of the embryo at that stage. A wide (initially large) stolon block grows to a greater length than does a narrow one, either because its period of formative growth lasts longer or because its growth rate is faster. The latter reason may be the more likely if it is assumed that the larger embryo which has produced the larger stolon has grown faster, thereby setting a larger initial scale of organization, so that such a stolon would naturally attain a greater length in a certain time than would one growing from a more slowly growing parent. The relatively slow growth of the small stolon of *S. democratica* is indicated by its relatively steep morphological axial gradient.

A block of stolon tissue in a large stolon is both wider and longer than in a smaller one. But whatever the diameter, each presumptive individual is first defined as an ectodermal segment one cell wide; and there are no observable differences in the size of the cells of the various species, including the ectodermal cells at the stolon base. Consequently, if the block length in one species is twice that in another at the same morphogenetic stage, the block in the former species consists of twice as many one-cell segments; thus the number per block is 40-60 in *S. democratica*, 100-140 in *S. fusiformis*, about 200 in *S. confederata*, and 400 in *S. maxima*. The rhythm within the growth zone which initiates the one-cell segments, accordingly, continues as long as the growth zone is active, and that rhythm is independent of the rhythm exhibited by the growth zone as a whole.

SUBDIVISION OF THE STOLON

Once established, each one-cell ring, or girdle, retains its essential unity. Passing outward along the stolon, certain critical stages of later morphogenesis of the individual salp, whose changes in size and form are relatively rapid and striking, are more clearly seen. The fact that there are small, but very definite differences in size and particularly in degree of development between adjacent segments of an extensive series in certain regions of the stolon can have but one of two explanations: Either there is a regular, constant interval between the initiation of one ectodermal segment and that of the next throughout the series; or there is a different rate of development for each segment. The ectodermal segments are all the same size at the start; and at the chain's far end, where development of individuals is virtually complete, differences in size and structure between adjacent salps are practically indiscernible. If it is assumed that the growth curve is the same for all units whose beginnings and ends are alike and whose environmental circumstances are similar, the steplike differences in degree of development of intermediate stages must indicate that such stages start at different times in different units. Although the actual times are not known, the greater the morphological change between one unit and the next, the longer the time interval—hours rather than minutes for the greatest changes. Whatever it may be, the same measure would be the measure of the initial rhythm in the growth zone.

What of the larger rhythm which produces the stolon blocks? First, the growth zone where the one-cell segments are initiated is not exactly at the base of a stolon, but a short distance from the parent-stolon junction. However, the segment-forming zone is approximately at the site where stolon growth is mainly increase in length rather than in width; i.e., ectodermal segmentation is associated with a process first of condensation and then of linear stretch. Second, the more-proximal short section of the stolon as a whole tapers relatively sharply and therefore represents growth of a somewhat different type. It seems likely that the initiating growth of the stolon as a whole, as distinct from its subdivision or from the continuing growth of tissue already formed, is a product of the parent-stolon junction, for the progressive widening of the stolon outward from this level is an index of growth. Therefore, a tentative hypothesis is that the protracted surges which give rise to the stolon blocks are successive surges in the growth of the stolon base and that ectodermal segmentation is a distinct process which operates a certain distance from the base.

MORPHOGENESIS OF THE STOLON

The morphogenetic process as a whole in the *Salpa* stolon is so peculiar and striking that a more detailed account and a comparison with the *Pyrosoma* and *Doliolum* stolons would be instructive.

The *Salpa* stolon, whose primary zone of growth is at or near its base, grows from the mid-ventral region of the embryo and retains the basic polarities of the parental organism with regard to the right and the left side and the dorsal and the ventral surface. Early in the stolon's growth, its tissues differentiate both morphogenetically and histogenetically, so that a transverse section shows the typical organization of a tunicate even before ectodermal strobilation begins —well before the third dimension is defined—with an ectodermal layer, a mesodermal layer, or presumptive genital mass, a central pharynx, a pair of perithoracic or peribranchial chambers, and a neural tube. This organization is exhibited in tissue that is in a state of dynamic extension in the third dimension and during neurulation, for instance, is a continuous process of morphogenesis. Thus stolon ectoderm becomes slightly thickened as a band along one side to form neural plate, 1 somewhat flat cell thick and 2 or 3 wide. This self-perpetuating band grows continually in the baso-distal direction. As this band of tissue reaches the critical level at the very beginning of the segmentation growth zone, it invaginates as a neural groove which becomes a tube and is cut off from the ectoderm by the time the ectoderm has grown forward to the level of definitive ectodermal segmentation. Similarly, the endodermal tube thickens on each side at the transverse level where the neural plate begins, and the thick walls progressively segregate into two perithoracic tubes as the tissue is shunted along.

Farther along the stolon—past the zone of ectodermal segmentation, but far short of the region where the strobilation constrictions are evident—each com-

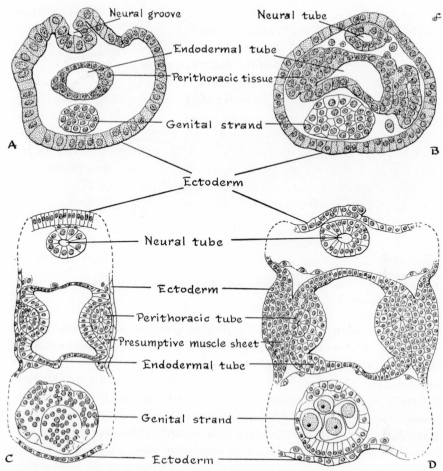

Figure 13-14. Sections through four levels of stolon in *Salpa fusiformis,* showing progressive differentiation along linear axis of stolon independently of delimitation of third dimension of prospective individuals. **A.** At level of neural invagination. **B.** Segregation of perithoracic (peribranchial) tube from endodermal tube. **C–D.** Progressive differentiation of genital tissue and lateral muscle sheet.

ponent continues to grow linearly and simultaneously proceeds with histogenesis and morphogenesis. The mesodermal, or genital, strand differentiates into central oocytes, spermatogonia, and follicular tissue. The neural tube thickens and induces a correlated thickening in the overlying adjacent ectoderm. The lateral walls of the endodermal tube, which gave rise at the base of the stolon to the perithoracic tubes, continue to thicken and eventually segregate a further mass laterally to form presumptive muscle tissue. In fact, the stolon as a whole continues to develop organization and tissue characteristic of a salpa at the same

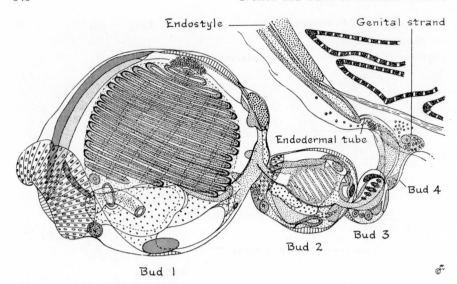

Figure 13-15. Chain of 4 buds forming from stolon of blastozooid in *Pyrosoma.*
(After Godeaux.)

time that it continues growing in length from new tissue continuously forming
at its base. A clear understanding of stolon growth as a whole would undoubtedly enlighten the whole question concerning the basis and nature of
organization.

Brooks, in a classical monograph (1893) on *Salpa,* stated that the internal
tissues are merely cut through by the infolding ectoderm of each presumptive
individual, but his theory has since been revised: The ectoderm does infold
between successive segments at a very early stage when a segment is no more
than 4 ectodermal cells wide; but before such infolding is in any way mechanically effective, the lateral walls of the endodermal tube together with the perithoracic tubes become pressed against the inner surface of the ectoderm (the
neural tube and genital strand have been close to the mid-dorsal and mid-ventral
ectodermal surfaces from the beginning). As an ectodermal segment becomes
distinguishable by the slight convexity of its outer surface, the neural tube in each
segment acquires a corresponding curvature, as does the presumptive muscle
tissue external to the perithoracic tubes. Examination of a series of such segments along the length of a stolon block reveals that in each segment all contiguous tissues grow as a unit, as though the individual growth rate of each
ectodermal segment is imposed on any tissue adjacent to it internally. The
growth activity of the inner tissues is also evident by the growth of a single
oocyte to a relatively large size within each stolonic segment *before* the genital
strand is cut or constricted by infolding ectoderm.

The conversion of a constricting stolon segment into a typical salp is a complex and fascinating phenomenon involving processes known as assembly and deployment (Brooks, 1893; Berrill, 1950). Essentially conversion consists of a complicated reorientation of the stolon segment during the process of constriction, but more importantly a relative growth of the linear, or third, dimension; i.e., the tissue that was initially but 1 cell in diameter grows until it becomes the linear axis of the fully developed individual. However, the total scale is set by the diameter of the stolon, i.e., by the scale and growth potential of a cross section; and the third, or linear, dimension progressively catches up by extensive growth, until standard proportions are attained. The phenomenon can be legitimately compared with the process of anterior and posterior reconstitution of a complete annelid from a single isolated segment.

In another thaliacean type, *Pyrosoma,* according to the species, the stolon gives rise to 4 or 7 individuals. The stolon, though clearly homologous to that of *Salpa,* appears in the blastozooid at a relatively later stage and develops apparently at a slower rate. Each surge of growth yields a single blastozooid individual, and it would seem that such a unit is comparable to *Salpa*'s major growth surges, which give rise to successive blocks of individuals, and not to definitive individuals.

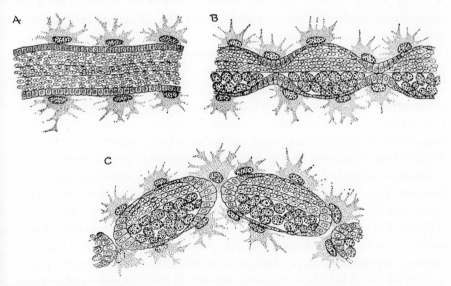

Figure 13-16. Strobilation of stolon in *Doliolum.* **A.** Typical appearance of stolon. **B.** Regular epidermal constrictions result mainly from change in cell shape; internal tissue strands are passive. External amoeboid cells are phorocytes, which play a secondary role in bud migration. **C.** Separation. (After Neumann.)

The *Doliolum* stolon is clearly homologous to that of *Salpa* and *Pyrosoma,* but it grows as a long, slender tube more like that of *Salpa* than of *Pyrosoma,* except that no morphogenesis occurs before units are constricted at its distal end; whereas in the other two genera units are fully differentiated functional individuals at the time of constriction. In *Doliolum,* as in *Salpa,* the stolon widens somewhat as it grows from the base, but thereafter maintains a uniform diameter. No subdivisions appear until, close to the distal end, units approximately one and a half times as long as the stolon is wide are successively constricted and separated. These units are not definitive buds, but are probuds similar to those of *Distaplia;* after a peculiar migration (see Neumann, 1906) up the side of the parent to become lodged on a dorsal spur, these probuds continue to elongate and then give rise to several definitive buds, as does the *Distaplia* probud.

Although there are important differences among *Salpa, Doliolum,* and *Pyrosoma* in the constitution of the internal tissues of the stolon and although in each an inner epicardial tube of endodermal origin is an active agent, the actual process of subdivision in all three genera is unmistakably an ectodermal or epidermal phenomenon. Two questions arise: Is subdivision attributable to the constituent cells, whose mitotic orientation relative to electrical or metabolic polarities results in linear, but not transverse, expansion of tissue; or is the agent of the subdivision process supracellular? Second, when the stolon constricts, what constituent of the epithelium is constricting?

In all the tunicates discussed in this chapter—i.e., the three thaliaceans, *Distaplia,* and the somewhat related ascidians, with or without postabdominal extensions of the body—the inner, tubular endodermal (epicardial) component of the strobilating region may well be responsible for the lengthening process. If it is, then, the epidermal tube enclosing it lengthens in response; i.e., extensive growth of one induces a matching extension in the other. Alternatively, if the epidermal component is responsible for lengthening, then extensive growth of the outer component (epidermis or ectoderm) induces a matching growth of the inner. Whichever proves to be the pacemaker, the basic problem of organization arises: What is the basis of tissue and organismal continuity and activity? If directed or determinative growth is regarded as resulting from orientation patterns of spindle axes in proliferating cells—an especially dominant attitude among students of the plant meristem (pp. 415-425)—there remains the question: What controls such orientation patterns? In the organisms under discussion there seems to be an exceptional opportunity to examine morphogenetic activities which apparently override cell individuality, yet seem nonattributable to diffusion or other metabolic gradients.

In all organisms considered thus far in this book, extensive linear growth of a simple epithelium in the form of a hollow cylinder is the primary process. Furthermore, as long as such growth is at the maximal or near-maximal rate, subdivision does not occur; but as the rate of extension slows down, there appear undulations or constrictions, which deepen until units are completely separated from the parent. On the topic of changes in form, Thompson has reported the

> interesting case . . . of a viscous drop immersed in another viscous fluid, and drawn out into a thread by the shearing motion of the latter. The thread seems stable at first, but when left to rest it breaks up into drops of a very definite and uniform size, the size of the drops, or the wave-length of the unduloid of which they are made, depending on the relative viscosities of the two threads.[1]

Obviously, strobilation is more than the phenomenon of viscosity in an extending thread; but on the assumption that there is in the epithelial cylinder some sort of cortical symplasmic continuity, which involves the inner or basal surface, and that high-polymer microstructure pervades it, strobilation can certainly be regarded as superseding individual cells.

Thus at a high, possibly near-maximal growth rate, cortical material in the form of a stable paracrystalline fluid flows at a decreasing rate from a proximal zone toward the distal end in the stolons of *Salpa* and especially *Doliolum* and from an apical zone toward the base in the column of *Hydra*. The flow is actually an active, progressive extension of material, as though of a packed fibrous matting, which may have certain liquid qualities. Whether or not this is a valid suggestion, the epithelial cylinder subdivides in a manner strikingly similar to that of an unduloid, particularly in the stolon of *Doliolum*, of the postthoracic body wall of merosomal ascidians, and of the isolated yellow column of *Corymorpha*. On the other hand, the ratio of length to diameter of the segregating units both in *Salpa* and in strobilating scyphistomae is so small that the unduloid properties are not comparable to any known state of matter outside the living protoplasmic system, so that evidently the properties at this emergent level are merely analogous to those of a lower material hierarchy.

The histological details of the subdividing epithelial cylinders, particularly in *Doliolum* and *Distaplia*, seem to indicate that subdivision is independent of cell shape and number—that the active agent has some other material basis. Neumann's precise, graphic description (1906) of the strobilating *Doliolum* stolon shows that the basal layer of the epidermis (as symplasmic ectoplasm, exoplasmic basement membrane, or both) extends and contracts in regular form and that the shape of the cells abutting this layer conforms to the amount of available territory. In other words, cells are columnar in the constricting zones, but cuboidal or intermediate in the intervening zones, although the cells throughout have approximately the same volume and apparently the same rate

---

[1] Thompson, *On Growth and Form*, p. 398.

of proliferation. The regular changes in cell shape might be considered responsible for the constrictions, were it not that in *Distaplia*, where the elongating probud subdivides into two or more units, an almost squamous epidermis constricts without discernible change in shape or proximity of cells. Yet, since the constricting process is clearly the same in both genera, the only possible conclusion is that changes in cell shape are merely conformant and not causative and that the significant dynamic process is associated with a continuous, approximately basal cortical layer of the tissue as a whole. In view of this conclusion, it is probably also significant that in *Salpa,* in which transverse units the width of only 1 cell are successively produced, the epidermal cells within the rhythmic zone are decidedly columnar, but those anterior and posterior to the zone are not; i.e., the essential material is relatively contracted or condensed in the rhythmic zone, but is progressively extended in other zones.

Summary

Strobilation is an epidermal or ectodermal process which occurs during or immediately after a slowdown period of active growth whereby tissue extends markedly along one axis more than along another. In organisms whose linear extension of tissue is approximately simultaneous throughout the column, the column subdivides almost simultaneously, as in the yellow column of *Corymorpha* and in the stolon of the *Pyrosoma* embryo but unlike the stolon of the *Pyrosoma* blastozooid. By contrast, in organisms whose extensive growth proceeds from one end of a column and decreases in rate toward the other, constrictions appear successively at a distance from one another—a situation which suggests that constriction is associated with a critical rate of extension, although this situation may signify only that the constricting activity is inhibited at a certain high rate and appears as soon as the rate slows to immediately below that value, not that it cannot occur at even slower rates.

REFERENCES FOR
Rhythmic Growth

WORMS

Abel, M. 1902. "Beitrage zur Kenntnis der Regenerationsvorgänge bei den limnicolen Oligochäten." *Zeitschr. wiss. Zool.* **73**:1-74.
Allen, E. J. 1921. "Regeneration and Reproduction of the Syllid *Procerastea.*" *Roy. Soc. London Philos. Trans.* Series B. **211**:131-77.

————. 1927. "Fragmentation in the Genus *Autolytus* and in Other Syllids." *Marine Biol. Assoc. U.K. J.* **14**:869-76.

Berrill, N. J. 1928. "Regeneration in the Polychaete *Chaetopterus variopedatus.*" *Marine Biol. Assoc. U.K. J.* **15**:151-8.

————. 1931. "Regeneration in Sabella pavonina (Sav.) and Other Sabellid Worms." *J. Exp. Zool.* **58**:495-523.

————, and D. Mees. 1936a. "Reorganization and Regeneration in Sabella. I. Nature and Gradient, Summation and Posterior Reorganization." *Ibid.* **73**:67-83.

————. 1936b. "Reorganization and Regeneration in Sabella. II. The Influence of Temperature. III. The Influence of Light." *Ibid.* **74**:61-89.

Child, C. M. 1902a. "Studies on Regulation. I." *Arch. entwick. Org.* **15**:187-238.

————. 1902b. "Studies on Regulation. II." *Ibid.* Pp. 603-37.

Chu, J. 1945. "A Genealogical Analysis of Fission in *Stylaria fissularis.*" *Zool. Soc. London Proc.* **115**:194-206.

————. 1946. "Further Studies on the Relation between Natural Fission and Regeneration in *Stylaria fissularis.*" *Ibid.* **116**:229-39.

Coe, W. R. 1929. "Regeneration in Nemerteans." *J. Exp. Zool.* **54**:411-59.

————. 1934. "Analysis of the Regenerative Processes in Nemerteans." *Biol. Bull.* **66**:304-15.

Consoli, L. 1923. "La rigenerazione in rapporto con la strobilazione negli Oligocheti limnicoli. *Boll. inst. zool. univ. Palermo.* **1**:23-47.

Dehorne, L. 1916. "Les Naidimorphes et leur reproduction asexuée." *Arch. zool. exp. gén.* **56**:25-157.

————. 1933. "La Schizometamerie et les segments tetragemmes de *Dodecaceria caulleryi.*" *Bull. Biol. France-Belgique.* **67**:298-314.

Eckert, F. 1927. "Experimentelle Untersuchungen über die Lage der Teilungszone von *Stylaria lacustris* L." *Zeitschr. wiss. Zool.* **129**:589-642.

Faulkner, G. H. 1932. "The History of Posterior Regeneration in the Polychaete Chaetopterus variopedatus." *J. Morphol.* **53**:23-58.

Galloway, T. W. 1899. "Observations on Non-sexual Reproduction in Dero vaga." *Mus. Comp. Zool. Harvard Bull.* **35**:115-40.

Gates, G. E. 1948. "On Segment Formation in Normal and Regenerative Growth of Earthworms." *Growth.* **12**:165-80.

Gross, F., and J. S. Huxley. 1935. "Regeneration and Reorganization in *Sabella.*" *Arch. entwick. Org.* **133**:582-620.

Haase, H. 1898. "Über Regenerationsvorgänge bei Tubifex rivulorum mit besonderer Berücksichtigung des Darmkanals und Nervensystems." *Zeitschr. wiss. Zool.* **65**:211-56.

Haffner, K. von. 1928. "Über die Regeneration der vordersten Segmente von *Lumbriculus* und ihre Fähigkeit, ein Hinterende zu regenerieren." *Zeitschr. wiss. Zool.* **132**:37-72.

Harper, E. H. 1904. "Notes on Regulation in Stylaria lacustris." *Biol. Bull.* **6**:173-90.

Hempelmann, F. 1923. "Kausal-analytische Untersuchungen über das Auftreten vergrösserter Borstein und die Lage der Teilungszone bei *Pristina.*" *Zeitschr. zellforsch. mikro. Anat.* **98**:379-445.

Herlant-Meewis, H. 1947. "Contribution à l'étude de la régénération chez les Oligochêtes." *Soc. roy. zool. Belgique Ann.* **77**:5-47.

————. 1951. "Les lois de la scissiparité chez les Aeolosomatidae: Aeolosoma viride." *Ibid.* **82**:231-84.

Hescheler, K. 1898. "Über Regenerationsvorgänge bei Lumbriciden." *Naturwiss.* **31**:521-604.

Hyman, L. H. 1916. "An Analysis of the Process of Regeneration in Certain Microdrilous Oligochaetes." *J. Exp. Zool.* **20**:99-163.

Ivanow, P. 1903. "Die Regeneration von Rumpf- und Knopfsegmenten bei *Lumbriculus variegatus.*" *Zeitschr. wiss. Zool.* **75**:327-90.

——. 1908. "Die Regeneration des vorderen und des hinteren Köpferendes bei *Spirographis Spallanzanii* (Viviani)." *Ibid.* **91**:511-58.

Kepner, W. A., and J. R. Cash. 1915. "Ciliated Pits in Stenostomum." *J. Morphol.* **26**:234-46.

Korschelt, E. 1919. "Über die natürliche und künstliche Teilung des Ctenodrilus monostylos Zeppelin." *Arch. entwick. Org.* **45**:602-85.

Krecker, F. H. 1910. "Some Phenomena of Regeneration in *Limnodrilus* and Related Forms." *Zeitschr. wiss. Zool.* **95**:383-450.

Kroeber, J. 1900. "An Experimental Demonstration of the Regeneration of the Pharynx of Allolobophora from the Endoderm." *Biol. Bull.* **2**:105-10.

Langerhans, P. 1879. "Wurmfauna von Madeira. *Zeitschr. wiss. Zool.* **32**:513-92.

Langhammer, H. 1928. "Teilungs- und Regenerations-vorgänge bei *Procerastea halleziana* und ihre Beziehung zu der Stolonisation von *Autolytus prolifer.*" Inaug.-Diss. Erlangung Doctorwürde (Thesis, Univ. Marburg).

Liebmann, E. 1942. "The Correlation between Sexual Reproduction and Regeneration in a Series of Oligochaeta." *J. Exp. Zool.* **91**:373-89.

——. 1943. "New Light on Regeneration of Eisenia foetida (Sav.)." *J. Morphol.* **73**:583-610.

Malaquin, A. 1893. "Recherches sur les Syllidiens." *Mém. Soc. Sci. Arts. Lille.*

Marion, A. F., and N. Bobretsky. 1875. "Études sur les annélides du Golfe der Marseille." *Ann. sci. nat.* Série zool. **2**:1-106.

Martin, E. A. 1933. "Polymorphism and Methods of Asexual Reproduction in the Annelid, Dodeccaria, of Vineyard Sound." *Biol. Bull.* **65**:99-105.

Michel, A. 1909. "Sur la formation du corps par la réunion de deux moitiés indépendantes, d'après l'origine de la queue de la souche chez les Syllides." *Acad. sci. Comptes rendus.* P. 1421.

Miller, J. A. 1938. "Studies on Heteroplastic Transplantation in Triclads. I. Cephalic Grafts between Euplanaria dorotocephala and E. tigrina." *Physiol. Zool.* **11**:214-47.

Moment, G. B. 1946. "A Study of Growth Limitation in Earthworms." *J. Exp. Zool.* **103**:487-506.

——. 1950. "A Contribution to the Anatomy of Growth in Earthworms." *J. Morphol.* **86**:59-72.

——. 1951a. "Simultaneous Anterior and Posterior Regeneration and Other Growth Phenomena in Maldanid Polychaetes." *J. Exp. Zool.* **117**:1-14.

——. 1951b. "A Theory of Growth Limitation." *Am. Nat.* **87**:139-53.

——. 1953. "The Relation of Body Level, Temperature, and Nutrition to Regenerative Growth. *Physiol. Zool.* **26**:108-17.

Morgan, T. H. 1895. "A Study of Metamerism." *Quart. J. Micro. Sci.* **37**:395-476.

Müller, C. 1908. "Regenerationsversuche an *Lumbriculus variegatus* und *Tubifex rivulorum.*" *Arch. entwick. Org.* **26**:209-77.

Okada, Y. K. 1929. "Regeneration and Fragmentation in the Syllidian Polychaetes." *Arch. entwick. Org.* **115**:542-600.

———. 1933. "Two Interesting Syllids, with Remarks on Their Asexual Reproduction." *Kyoto Univ. Coll. Sci. Mem.* **8**:325-38.

———. 1934a. "Formation de têtes dans la stolonisation des polychêtes syllidiens." *Soc. zool. France Bull.* **59**:388-404.

———. 1934b. "Régénération de la tête de *Myxicola aesthetica* (Clap.)." *Bull. Biol. France-Belgique.* **68**:340-81.

———. 1935. "Stolonization in Myrianida." *Marine Biol. Assoc. U. K. J.* **20**:93-8.

———. 1937. "La Stolonisation et les caractères sexuels du stolon chez les Syllidiens polychêtes (Études sur les Syllidiens, III)." *Japan J. Zool.* **7**:441-90.

———. 1938. "An Internal Factor Controlling Posterior Regeneration in Syllid Polychaetes." *Marine Biol. Assoc. U.K. J.* **23**:75-8.

———, and T. Kawakami. 1943. "Transplantation Experiments in the Earthworm, *Eisenia foetida* (Savigny), with Special Remarks on the Inductive Effect of the Nerve Cord on the Differentiation of the Body Wall." *Tokyo Univ. Fac. Sci. J.* Series 4. **6**:25-96.

Okada, Y. K., and H. Tozawa. 1943. "Supplementary Experiments of Transplantation in the Earthworm. The Induction of a Tail by the Transplanted Nerve Cord." *Ibid.* Pp. 635-47.

Ott, H. N. 1892. "A Study of Stenostoma." *J. Morphol.* **7**:263-304.

Pflugfelder, O. 1929. "Histogenetische und organogenetische Prozesse bei der Regeneration polychäter Anneliden. I. Regeneration des Vorderendes von *Diopatra ambionensis.*" *Zeitschr. wiss. Zool.* **133**:121-210.

Piguet, E. 1906. "Observations sur les Naididées et revision systématique de quelques espèces de cette famille." *Rev. suisse zool.* **14**:185-315.

Probst, G. 1931. "Beitrage zur Regeneration der Anneliden. I." *Arch. entwick. Org.* **124**:369-403.

———. 1932. "Beitrage zur Regeneration der Anneliden. II." *Ibid.* **127**:105-50.

Sayles, L. P. 1932. "External Features of Regeneration in Clymenella torquata." *J. Exp. Zool.* **62**:237-58.

Saint Joseph, M. 1886-1887. "Annélides polychêtes des côtes de Dinard." *Ann. sci. nat.* Série zool. **1**:126-270.

Sonneborn, T. M. 1930. "Genetic Studies on Stenostomum incaudatum. I. The Nature and Origin of Differences among Individuals Formed during Vegetative Reproduction." *J. Exp. Zool.* **57**:51-108.

Stolte, H. A. 1922. "Experimentelle Untersuchgen über die ungeschlechtliche Fortpflanzung der Naiden." *Zool. Jahrb. Abt. Allg. Zool. Physiol.* **39**:149-94.

Stone, R. G. 1932. "The Effects of X-rays on Regeneration in Tubifex tubifex." *J. Morphol.* **53**:389-420.

———. 1933a. "The Effects of X-rays on the Anterior Regeneration in Tubifex tubifex." *Ibid.* **54**:303-20.

———. 1933b. "Radium Irradiation Effects on Regeneration in Euratella chamberlin." *Carnegie Inst. Washington.* Pub. 435. Pp. 157-66.

Tsui, T. F., J. Chu, and S. Pai. 1946. "Fission and Disintegration in Stylaria fissularis at Different Temperatures." *Physiol. Zool.* **19**:339-45.

Turner, C. D. 1934. "The Effects of X-rays on Posterior Regeneration in Lumbriculus inconstans." *J. Exp. Zool.* **68**:95-115.

Van Cleave, C. D. 1929. "An Experimental Study of Fission and Reconstitution in Stenosto-
num." *Physiol. Zool.* **2**:18-53.
————. 1937. "A Study of the Process of Fission in the Naid Pristina longiseta. *Physiol.
Zool.* **10**:299-314.
Viguier, C. 1902. "Sur las valeur morphologique de la tête des Annélides." *Ann. sci. nat.*
Série zool. **25**:281-93.
Von Wagner, F. 1900. "Beitrage zur Kenntnis der Reparationsporzesse bei *Lumbriculus
variegatus.*" *Zool. Jahrb. Abt. Anat. Ont.* **13**:603-82.

ARTHROPODS

Weisz, P. B. 1946. "The Space-Time Pattern of Segment Forming in Artemia salina." *Biol.
Bull.* **91**:119-40.
————. 1947. "The Histological Pattern of Metamerical Development in Artemia salina."
J. Morphol. **81**:45-96.

VERTEBRATES

Bitjel, J. H., and N. W. Woerdeman. 1928. "On the Development of the Tail in the
Amphibian Embryo." *Konink. ned. akad. Weten. Proc.* Series C. **31**:1030-40.
Ford, P. 1950. "The Origin of the Segmental Musculature of the Tail of the Axolotl (Amby-
stoma)." *Zool. Soc. London Proc.* **119**:609-32.
Nakamura, O. 1938. "Tail Formation in the Urodele." *Zool. Mag.* **50**:442-6.
Pasteels, J. 1942. "New Observations concerning the Maps of Presumptive Areas of the
Young Amphibian Gastrula (Ambystoma—Discoglossus)." *J. Exp. Zool.* **89**:255-81.
Smithberg, M. 1954. "The Origin and Development of the Tail in the Frog Rana pipiens."
J. Exp. Zool. **127**:397-425.
Vogt, W. 1929. "Gestaltungsanalyse am Amphibienkeim mit örtlicher Vitalfärbung. II.
Gastrulation und Mesodermbildung bei Urodelen und Anuren." *Arch. entwick. Org.*
120:385-702.
Waddington, C. H., and E. M. Deucher. 1953. "Studies on the Mechanism of Meristic
Segmentation. I. Dimensions of Somites." *J. Embryol. Exp. Morphol.* **1**:349-56.

TUNICATES AND SCYPHOMEDUSAE

Berrill, N. J. 1949. "Form and Growth in the Development of a Scyphomedusa." *Biol.
Bull.* **96**:283-92.
————. 1950. "Budding and Development in Salpa." *J. Morphol.* **87**:553-606.
Brien, P. 1925. "Contributions à l'étude de la blastogenèse des Tuniciers. Bourgeonnement
chez Aplidium zostericola (Giard)." *Arch. biol.* **35**:155-204.
Brooks, W. K. 1893. "The Genus Salpa." *Johns Hopkins Univ. Biol. Lab. Mem.* **2**:1-396.
Chuin, T. T. 1930. "Le cycle évolution de scyphistome de Chrysaora." *Trav. stat. biol.
Roscoff.* **5**:1-180.
Delap, M. J. 1905. "Notes on Rearing in an Aquarium of Cyanea lamarckii P. and L."
Report Sea and Inland Fisheries Ireland Sci. Invest. (*for 1902*). Pp. 20-2.
Johnson, M. E. 1910. "A Quantitative Study of the Development of the Salpa Chain in
Salpa fusiformis-runcinata." *Univ. Calif. Pub. Zool.* **6**:145-76.

Komai, T. 1935. "On Stephanoscyphus and Nausithoë." *Kyoto Univ. Coll. Sci. Mem.* **10**:289-317.

Lambert, F. J. 1935. "Jellyfish. The Difficulties of the Study of Their Life Cycle and Other Problems." *Essex Naturalist.* **25**:70-7.

Neumann, G. 1906. "Doliolum." *Deutsche Tiefsee-Exped.* **12**:93-243.

Percival, E. 1923. "On the Strobilation of Aurelia." *Quart. J. Micro. Sci.* **67**:85-100.

Thompson, D'Arcy W. 1942. *On Growth and Form.* New York: Cambridge Univ. Press.

Chapter 14

Regeneration and Total Development: Ascidian Buds

THE SOMATIC TISSUES of ascidian and thaliacean tunicates in general exhibit a greater diversity of developmental, regenerative, and reconstitutive phenomena than does any other group. Quite apart from the fact that ascidian eggs develop in typical chordate fashion (p. 483) only to metamorphose into mature organisms no different from those of purely somatic origin, the somatic reconstitutions also follow diverse courses to attain the same result.

All degrees of reconstitution occur in ascidians: from a fairly typical regeneration of all missing structure through various types of reorganization involving numerous combinations of both old and new tissue—including the utilization of original structure without significant outgrowth of new tissue—to a virtually complete development from unspecialized somatic tissue in a fragment much smaller than any egg (notably in *Distaplia* and *Botryllus*). Moreover, in thaliaceans, particularly in *Salpa,* as already described, histogenesis and morphogenesis become well advanced in two dimensions before the third dimension is ever apparent. The discussion in this chapter is organized around these degrees of reconstitution.

In the early days of developmental morphology a teleological outlook was generally acceptable, so that the concept of a unit developing toward a predetermined goal—in other words, of an outcome being foreshadowed from the beginning—seemed less metaphysical than it does today (at least until Driesch attempted to evade the problem by evoking entelechies). Similarly, during the last two decades of the nineteenth century especially, the apparent evolutionary aspect of the development of eggs seemed to give substance to the theory of recapitulation—a theory now almost equally rejected. Nevertheless, both the apparent development toward a prescribed goal, observed in all organisms, and the historical or evolutionary aspect of the development of eggs of, for example, frogs, are indisputable phenomena, although past interpretations of these phenomena have been largely discredited. Units do develop and reconstitute *as if* they had a set goal, just as the disarranged cortex of a stentor re-establishes its normal pattern; and many types of eggs develop *as if* the past were impinging upon the present, just as the frog egg recapitulates certain ancestral aspects during its development. In discarding the language used to describe and discuss these two aspects of development, modern biologists have tended to deny their very existence, rather than to inquire into their significance, which even a casual study of the developmental morphology of tunicates attests is real.

Regeneration and Morphogenesis in Tunicates

Regeneration in ascidians and morphogenesis in bud development in ascidians are merely two aspects of the same phenomenon; the difference between them relates to the means by which the isolation process leads to development or reconstitution.

Whether regeneration or reorganization follows experimental or natural isolation of a fragment or the development of buds in organic continuity with the parental zooids, the tissue involved is histologically mature. In other words, the outer layer of any fragment is mature epidermis, not embryonic ectoderm; thus it is a specialized tissue with, of course, morphogenetic limitations. The fact that there are limitations plays an important role in the determination of the general course and nature of events in the various reconstitutive processes.

The Epidermis

The ascidian epidermis is covered by a collagenous tunic, which usually contains mesenchymal cells and which consists of two parts; a proteid cuticle, and the cellulose tunicin proper. Based on studies of both *Clavelina* and *Ciona,* Pérès (1948a and b) concluded that glucides, elaborated in the haemocoele, are absorbed and polymerized by the epidermis and secreted externally as the tunic. Although structurally simple, the epidermis is chemically and physiologically specialized, and in all ascidian reconstitutive processes never gives rise to any tissue other than new epidermis. This restriction in morphogenetic participation does not necessarily prevent the epidermis from playing an outstanding part in reconstitution as a whole.

The immediate consequence of this limitation of the ascidian epidermis is that regeneration depends on the presence of other tissues or cells that have multipotent, possibly totipotent properties. Such tissues exist in ascidians in three histological forms: (1) the atrial epithelium, lining the atrial or peri-branchial chamber on either side of the body; (2) the lining of the epicardium or perivisceral cavity adjacent to the heart and to the loop of the intestine; and (3) a mesenchymal septum, separating the afferent and efferent flow of blood in a vascular epidermal outgrowth. Depending on the morphology of a particular ascidian type, there may be one or more of these sources of multipotent cells. *Clavelina,* for example, has all three; *Ciona,* only the first two; and *Botryllus* and its related genera, only the first. An atrial lining is present in all types, epicardial tissue in most, and a vascular septem in only the few types which have a postcardial extension of the body.

According to Brien and Brien-Gavage (1927), other tissues of more restricted potency like the epidermis, can participate in the reconstructive processes to form new tissue of their own general kind: pharynx can form new pharyngeal tissue; the intestine can produce new intestinal tissue; and the dorsal cord, which is a delicate strand extending from the posterior end of the neural gland down to the gonads, can form a new neural complex. George (1935) reports that amputated siphons of *Styela* are readily regenerated by direct growth from epidermis and atrial epithelium. When *Ciona* individuals are cut transversely or obliquely at any level through the thorax, anterior regeneration proceeds rapidly from the posterior fragment, but posterior regeneration does not occur from the anterior fragment. Epidermis and atrial epithelium together regenerate new mantle, epidermis, and siphons; and according to Hirschler (1914), pharyngeal branchial tissues regenerate missing branchial structure, so that a complete individual is reconstituted to the normal *Ciona* form. In the styelid *Polycarpa,* specimens kept alive in aquaria frequently undergo evisceration, which in this organism involves expulsion of not only the branchial thorax but also the digestive tube, heart, and gonads; all that remains is the thick mantle wall, which consists of a complex connective tissue between the epidermis and the atrial epithelium. The missing internal structures then regenerate from atrial epithelium, folds of which grow into the central cavity from the mid-ventral line, according to de Sélys-Longchamps (1915). The complexity of the entire regenerating structure is immense, the process strikingly direct, and the phenomenon as a whole almost completely baffling.

The picture of the role of the epidermis in morphogenesis broadens when the stolon of *Salpa* (p. 341) is considered. The stolon originates in *Salpa* during a distinctly embryonic stage, and it is clear that the epidermis of the stolon retains its unspecialized ectodermal character throughout later stages of parental development. The established stolon consists essentially of an outer, ectodermal tube of simple epithelium; an inner, equally simple epithelial tube derived from the ventroposterior endodermal lining of the pharynx; and a compact ventral cord of mesodermal tissue. At the stolonic base and for some distance along the stolon, the ectodermal and the endodermal tubes are widely separated from each other by only haemolymph and scattered cells.

One of the striking features in the development of the stolon is the manner of formation of the neural complex. This complex differentiates from a neural tube, which *invaginates from the ectoderm* near the base of the stolon, just as the neural tube complete with neuropore invaginates from a medullary plate in chordate embryos. The ectoderm has retained its initial competence, and performs just as though it were truly embryonic. On the other hand, although the inner, endodermal component forms pharynx, intestine, and peribranchial chambers, it does not produce a dorsal neural evagination. Its doing so would, of course, be redundant; yet, it is striking that despite the considerable space between the ectodermal and the endodermal component, ectodermal activity suppresses equivalent development in the inner layer. How such suppression can be induced—by either ectoderm or both layers—is obscure; for any structural, even microstructural, continuity between the two layers is hardly likely. Moreover, since fluid circulates in the space between the layers to sustain the growing tissues, persistent patterns of diffusing chemical agents are equally unlikely. It is just conceivable—though there is no direct evidence to support such a speculation—that there are not only high-polymer, paracrystalline symplasmic continuities in the opposing, though not adjoining surfaces of ectoderm and endoderm but also an interplay of the electromagnetic fields of these surfaces—on

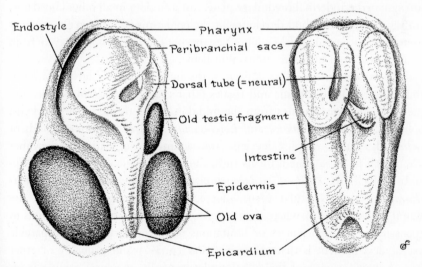

Figure 14-1. Left side and dorsal view of bud of *Glossophorum sabulosum,* showing complete reconstruction from epicardial component. (After Hjort.)

the assumption that there are such fields, as Szent-Györgyi contends exist in myofibrils (p. 5).

Other Tissue Components

The isolation of a fragment and its reconstitution to form a whole individual raises the question of the relative roles played by the various tissue components and their influence on one another.

In the great majority of ascidian reconstitutions, the isolating agent is unquestionably epidermal: in all merosomatous ascidians, fragments are isolated by epidermal constrictions of an elongating abdominal or postabdominal region; and in clavelinids and perophorids also, the epidermis seems to initiate the formative process. Not only does the epidermis isolate a fragment physically, but in organisms in which isolation is physiological rather than physical the epidermis defines the territory representing the new individual. However, in botryllids, including the polystyelids such as *Metandrocarpa,* it is debatable whether the epidermis or the atrial epithelium, which is an extension of it, is the primary activator. Nevertheless, the initial wholeness of the prospective new individal, i.e., the basic individuality, seems to be unmistakably epidermal, as, for example, in the development of form in the *Distaplia* bud.

On the other hand, morphogenesis, apart from epidermal form and integrity, relies heavily on the internal tissue of an isolated fragment or bud and therefore varies considerably according to the included tissue. The neural complex (neural ganglion, neural gland, and dorsal cord) develops in the ascidian embryo from the invaginated ectodermal medullary plate, just as it does in all other chordates; but in regeneration and bud development the epidermis is no longer competent to contribute to neural development. Yet the mature organisms derived from eggs and those from buds are indistinguishable.

In the polystyelids, including *Botryllus,* the inner vesicle, of atrial epithelial origin, gives rise to all organs and tissues other than the epidermis itself, and the neural complex evaginates from the mid-dorsal wall of the inner vesicle. In perophorids, such as *Perophora* and *Ecteinascidia,* and in the stolonic buds of *Clavelina,* the neural complex develops from a similar evagination of the inner vesicle, which in these genera forms from the mesenchyme of the stolonic septum. In polyclinids, such as *Amaroucium, Polyclinum, Distaplia,* and *Circinalium,* the inner vesicular tissue, of epicardial origin, similarly gives rise to neural structure. Thus whatever its origin, the inner tissue develops so as to compensate for the deficiencies or limitations of the epidermis. A comparable course of development is that of *Archiascidia* (p. 365) and probably of other genera, in which a piece of the dorsal cord (originally of neural origin) is included in the reconstituting piece: The distal end of this dorsal tissue gives rise to the new neural complex, so that the adjacent inner vesicular structure fails to produce the neural evagination.

What is the basis of such adjustments? What prompts the inner tissues to undertake to compensate for deficiencies or limitations of the epidermis, so that development will be completed? (The same problem arises in the discussion of morphogenesis in the stolon of *Salpa* on p. 346.) Certainly, the effects of inclusion of intestinal tissue in reconstituting pieces is also significant, as in species of *Eudistoma* (syn. *Archidistoma*), *Diazona,* and *Aplidium*.

Accordingly, just as dorsal-cord activity inhibits neural evagination of the adjacent vesicle, so the presence of intestinal fragments inhibits total reconstitution by the epicardium. Other tissues can compensate by supplying what is necessary—except that epidermal tissue cannot contribute to neural differentiation—but when the tissue histogenetically related to the missing structure is included in the fragment, it takes priority. Thus dorsal cord of neural origin forms the neural complex, and fragments of the old gut loop reconstitute a new intestine; but the epicardium, which can reconstitute any structure, does so only when necessary—i.e., when no related tissue is present. Timing may be a factor in epicardial participation; in other words, tissues already histologically determined may begin their own specific growth sooner than does a completely undifferentiated tissue such as epicardium.

Clavelina

Clavelina species have been favored for regenerative studies because the so-called budding stolon is primarily the greatly hypertrophied, posterior vascular extension of the epidermis and contains a mesenchymal septum separating the afferent and the efferent flow of blood. The large size of this stolon permits the entrance and therefore the accumulation of trophocytes, particularly toward the end of the life cycle of an individual zooid.

At the end of a breeding season the zooids regress and trophocytes multiply, become congested, and progressively accumulate toward the end of the stolon branches until only such richly stocked terminals remain. With the rise in temperature the following spring, each terminal mass develops into a new zooid, as do winter buds when brought into the warmer temperature of a laboratory.

Regeneration studies, chiefly by Brien, are mainly of reconstitution after section of the body below the thorax, both in *Clavelina* (1927a, 1930, and 1932) and in the clavelinid *Archiascidia* (1933). Besides reconstituting from fragments cut below the thorax, *Clavelina* can reconstitute from abdomen or from vascular stolon alone; buds, however, are formed only from the stolon.

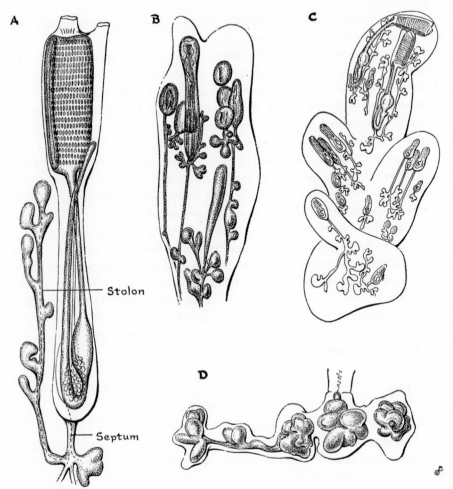

Figure 14-2.   Colonies in *Clavelina*.   **A.** Zooid of *C. picta* colony, showing general structure and hypertrophied epidermal vascular stolon containing mesenchymal septum.   **B.** Part of *C. picta* colony undergoing reduction within common tunic, showing reduction bodies of zooids and stolon ampullae congested with trophocytes.   **C.** Part of similar colony later, showing reconstitution of large zooids from reduction bodies and development of smaller zooids from isolated stolonic ampullary clusters.   **D.** Ampullary clusters of *C. lepadiformis* after complete reduction of zooids.

Reconstitution from Subthorax Fragments

1. In all such sections new epidermis forms from old epidermis as a continuation of growth following healing of the cut end. The rest of the elaborate thoracic structure grows as a double outgrowth from the epithelium that forms

the anterior end of the epicardium, which is probably homologous to the coelomic epithelium of vertebrates. The thoracic structure becomes fitted to the old esophagus, and if part of the esophagus had been excised, a new esophageal piece is also formed. According to Brien, the new neural complex in *Clavelina* seems to arise from the anterior end of the epicardiopharyngeal outgrowth; but in *Archiascidia* the same neural complex is unquestionably formed from the dorsal cord at the cut end, which proliferates in close contact with the mid-dorsal surface of the developing pharynx. (As Brien himself has suggested, further investigation of *Clavelina* may show that the neural complex forms in the same way.)

2. Brien (1932) found that the nature of regeneration posteriorly from short esophageal pieces of *Clavelina* depends on the time interval between the anterior and the posterior cut. In pieces cut anteriorly and posteriorly at the same time, thoracic regeneration occurs at both ends to produce typical bipolar structure—as in flatworms—and at both ends the epicardial tissue is the active agent, so that the process of reconstitution pursues the same course at both ends. When, however, the posterior cut is made 48 hours after the first, monopolar regeneration takes place: thoracic structure forms anteriorly, and abdominal structure posteriorly. In this type of isolation, posterior regeneration pursues a very different course from that at the anterior cut end (except that epidermis, as usual, gives rise to new epidermis): the epicardium elongates, but gives rise only to the pericardial heart; and the ends of the intestinal limbs proliferate, fuse, and reconstitute the posterior part of the intestinal loop.

Reconstitution from Stolon Fragments

1. During the first 24 hours after being cut, a piece of stolon shortens to about one-third its original length. According to Huxley (1926), long before the onset of morphogenesis, some fragments establish a kind of circulation, which is produced by changes in the epidermal epithelium alternating between a cuboidal and a columnar shape. The over-all shortening seems also to be the result of change in cell shape.

2. The stolon of *Clavelina* undergoes a definite evolution. In the newly functional zooid, i.e., until the attainment of sexual maturity, the stolon is a comparatively straight, simply branched extension of the postabdominal epidermis and contains a mesenchymal septum but virtually no blood cells. During the later part of the breeding season the terminal and subterminal parts of the stolonic branches expand to form clusters of small ampullae, each containing aggregations of trophocytes, which migrate from thoracic and abdominal tissue to eventually congest the ampullae. The evolutionary process culminates in the degeneration and sloughing of the greater part of the zooid.

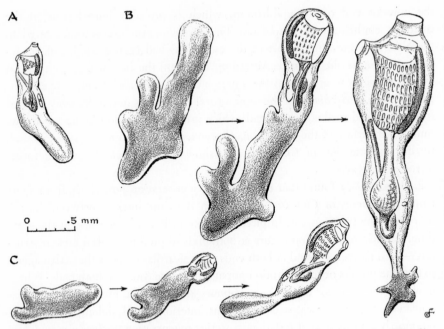

Figure 14-3. Reconstitution from cut ends of stolon fragments in *Clavelina lepadiformis.* **A.** Smallest example of complete development, showing 2 rows of gill slits. **B.** Typical example, showing 7 rows of gill slits and growth involving utilization of nutritive reserve cells in stolon. **C.** Smaller example, showing 4 rows of gill slits. (After Berrill and Cohen.)

Pieces of stolon cut during the earlier formative phase regenerate from the cut end; but as soon as the ampullary clusters appear, a cut end merely heals and a new zooid develops from one ampulla of a cluster. When the detached piece consists of several branches ending in ampullary clusters, a new zooid can develop from each cluster.

3. The mesenchymal septum may gain or lose cells, such as lymphocytes, from or to the haemocoele. In isolated stolon pieces the septal tissue withdraws into a comparatively compact mass which rapidly acquires an inner cavity. The vesicle thus formed subsequently gives rise to the whole of a new zooid with the exception of the epidermis, according to Brien (1930). The mesenchymal mass forms (at 20°C) within 24 hours of section; the vesicle, 48-72 hours; the epicardium, 96 hours; and the heart begins to beat after 130 hours, exhibiting its characteristic reversal of beat from this time. Development is complete about 200 hours after section.

The first sign of morphogenesis appears in the epidermis of the fragment as a local thickening, which defines the epidermal area. This area expands to form the new zooid's epidermis, the cells of which seem to be about twice as high as those of epidermis outside the area. Whether this columnar thickening results from contraction of the epidermal base or from local growth is uncertain, although growth undoubtedly occurs after this thickening (Berrill and Cohen, 1936).

4. The length of a stolon fragment is significant. A piece less than ½ mm long commonly does not exhibit any morphogenetic change; but one even slightly longer (about 0.6 mm) shows epidermal reaction either alone or accompanied by the formation of the mesenchymal vesicle, which develops to only an abortive extent. Pieces longer than 0.6 mm can complete development at a rate which increases in direct ratio to the size of the rudiment; e.g., pieces in the 0.6-1.0-mm range form 2-5 rows of gill slits; still-longer pieces form a maximum of 7 rows. With later growth many more rows are added in any piece more than 1.0 mm in length. In all instances of arrested development in long pieces, an abundance of food-laden cells remain, so that the cause of the arrest must lie elsewhere. The critical stage seems to be the onset of the heartbeat, which establishes a circulation whereby reserve cells become accessible to all parts of the zooid. When this stage is attained, development always proceeds to completion. Developmental arrest short of that stage seems to be related to the initial scale of organization.

In a stolon piece less than 0.6 mm in length, about one-quarter to one-third

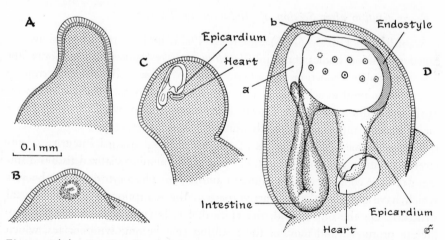

Figure 14-4. Small stolon fragments in *Clavelina lepadiformis,* showing arrested development at various stages depending on initial scale, all 9 days after isolation. **A.** Epidermal reaction only. **B.** Minute mesenchymal vesicle formed from mesenchymal septum beneath activated epidermis. **C.** Epidermal envelope enclosing inner vesicle exhibiting basic organization. **D.** Most-advanced stage at which arrest is possible, showing 2 rows of gill slits forming. (After Berrill and Cohen.)

of the piece, at the healed end, forms the epidermis of the presumptive zooid. In the smallest pieces this epidermal area is not only short but narrow. The mass of septal mesenchyme that accumulates centrally within this approximately hemispherical epidermal envelope is in proportion to the size of the envelope, so that the limits of growth and development seem to be determined either by the expansive potential of the epidermal cover or by the dimensions of the mesenchymal vesicle at the time of its formation. Nutrition of the growing tissues is probably the limiting factor. Without, or before the existence of, a circulation, only those trophocytes within the formative region yield their nutritive contents to the developing zooid; and once those contents are exhausted, development is arrested. The zooid at any stage of arrest seems to be perfectly normal in structure.

5. In summary: Response of a stolon fragment to isolation is immediate and consists of tangential contraction (with consequent thickening) and subsequent growth of the simple epidermal epithelium. The amount of area involved in condensation is proportional to the length—of a critical value—of the fragment, and must in turn be of a certain value in order to form a discernible mesenchymal mass. The establishment of this epidermal area, or disk, precedes the initiation of both the internal mesenchymal mass and the vesicle derived from the mesenchymal septum. The size of the septum is in turn proportional to that of the epidermal area.

Reconstitution after Section of Abdomen or Thorax

According to Huxley (1926) and Ries (1937), under certain conditions, so-called restitution bodies form in *Clavelina* after resorption of the thorax but before complete reduction to stolonic clusters. These bodies at first contain much of the postthoracic loop of the intestine and reproductive system, as well as the epicardium and heart. Except for the epicardium, which is an elongate sac of simple, unspecialized epithelium, all these structures are comparatively specialized tissues. With the loss of the individual's original integrity, a differential shift in cell population takes place. The more-specialized tissues disintegrate: their constituent cells undergo autolysis or phagocytosis by trophocytes which have already taken an active part in the resorption of the pharyngeal structure. On the other hand, the epicardial cells (unspecialized tissue) proliferate profusely, and most of the resulting cells become lymphocytes, which are no longer united as constituents of the epithelium.

At one time the food-laden trophocytes were thought to be the formative cells, but they have been shown to be nutritive only; it is now known that the formative cells are derived from either the septum or epicardium, depending on what is included in the fragment. It is significant, however, that the most-specialized cells degenerate, other specialized cells become actively phagocytic, and the least-specialized proliferate abundantly. The situation may be compared with the mixed cell populations in the reorganizing stumps of amputated amphibian limbs, with the competition between growth zones and differentiated tissues in hydroids or between developing medusa buds and supportive hydranths, and—according to Huxley (1921)—with the regression of mature zooids and ongoing development of buds in starving colonies of the ascidian *Perophora*.

Eudistoma

Produced in the long postthoracic region as the result of epidermal constrictions, buds are relatively narrow and sparse in trophocytes (p. 364). Each fragment thus isolated by constriction contains a section of the descending and ascending limbs of the intestine. Anteriorly the epidermis grows to form only epidermis for a new thorax, including the two siphons. A new pharynx, including the whole branchial and peribranchial structure, forms from epicardial tissue which extends forward to become attached to the anterior end of the esophageal fragments; this regenerative or reconstitutional process is essentially the same in all ascidians whose epicardium is of this general type.

Posteriorly the only comparable epidermal outgrowth is that which forms a small, branching, epidermal vascular process. The lower ends of the descending and ascending limbs of the intestine, brought close together by the epidermal constriction, fuse so that a gut loop becomes re-established anterior to the original stomach. Subsequently the lower section of the newly constituted gut loop differentiates *in situ* to form a typical stomach, poststomach, mid-intestine, and postintestine, all of which are recognizable not only in outward form but by the transformation of a low, cuboidal epithelium into a high, columnar epithelium (Berrill, 1947). The process is accompanied or preceded by a lengthening of the reconstituting unit as a whole; i.e., epidermis, epicardium, and intestinal canal extend in proportion to the whole. The striking event is the remodeling of the new digestive canal as a whole to a great extent within the confines of the old canal.

Several probably significant facts are noteworthy in this intestinal reorganization: Reorganization takes place in an endodermal tube that has undergone elongation; i.e., its originally regional character may have disappeared. Reorganization is apparent only after a new pharyngeal structure, of epicardial origin, has been added at the anterior end of the fragment and has become an

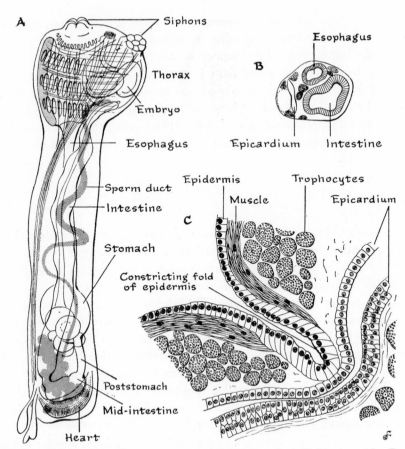

Figure 14-5. Zooid of *Eudistoma* (syn. *Archidistoma*). **A.** Individual zooid. **B.** Cross section of abdomen. **C.** Longitudinal section through abdominal constriction, showing activity of epidermis and passivity of internal tissues.

integral part of the new system. Thoracic or pharyngeal regeneration is the same no matter what the position of the original postthoracic fragment along the original anteroposterior axis; therefore the possible organizing influence of the regenerating anterior structure must not be overlooked. Besides these facts, the outstanding feature of intestinal reorganization is that it occurs within an epithelial tube that is not only reconstructed from a pair of old pieces but is in histological contact only with the surrounding haemolymph and the anterior regenerating region. In other words, apart from stimulus, or induction, from the anterior end, the reorganizing activity—and therefore the essential differentiation—takes place within the epithelial tube, and presumably has some substantial basis to which cells as individuals conform and in which a continuous and dynamic macromolecular pattern finds expression.

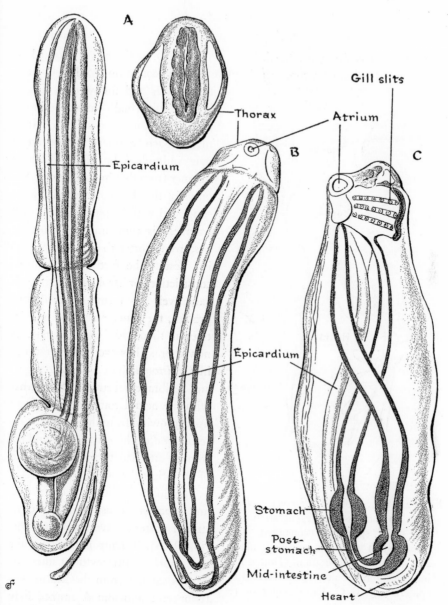

Figure 14-6. Subdivision of abdomen and subsequent reorganization of pieces in *Eudistoma*. **A.** Subdivision by epidermal constriction into thoracic and three abdominal pieces. **B–C.** Two stages in regeneration of new thorax and reorganization of gut loop within extended epidermal sheath.

Diazona

In *Diazona violacea,* a related, larger colonial genus, there is a seasonal regression of the thorax, accompanied by postthoracic congestion with yolk-laden trophocytes and followed by a strobilative epidermal constriction of the postthorax into a number of fragments. Each fragment reconstitutes a new individual but in a different manner from that in *Eudistoma:* Not only is the haemolymphatic space congested with trophocytes, but each fragment is comparatively short—at first even shorter than its diameter—i.e., despite the fact that the initial constrictive, or strobilate, activity is exclusively a form of epidermal growth, the inner epicardial and intestinal tissues do not elongate. A new thorax, however, regenerates as in *Eudistoma:* The new pharyngeal and peribranchial structures, together with the neural complex, form from tissue growing from the anterior transected end of the epicardium and becoming attached to the anterior end of the original esophageal tube.

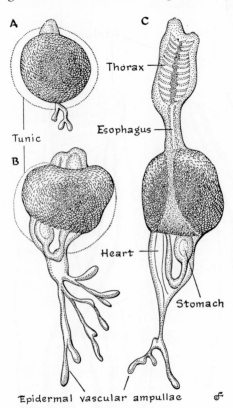

Figure 14-7. Isolated *Diazona violacea* buds in process of regeneration. **A.** Bud still spherical (congested with trophocytes) and just starting anterior and posterior regeneration. (Dotted line indicates zone of new tunic secreted by epidermis.) **B–C.** Later stages of regeneration.

The difference between *Eudistoma* and *Diazona* reconstitution is the nature of posterior reconstitution (Berrill, 1948b). In *Diazona* an extensive regenerative outgrowth of the epidermis extends from the healed posterior end and almost immediately exhibits its own typical pattern; namely, the outgrowth is an extension of the epidermal body wall which continues as an extensive, branching, epidermal vascular process. This pattern is apparently inherent in the epidermal layer, which is simple epithelium, for their is no inner tissue that possesses continuity enough to carry any imprint of pattern. Accompanying the outgrowth of the posterior epidermis, the newly conjoined intestinal loop also grows posteriorly to a comparable extent, and the several regional differentiations—stomach, poststomach, mid-intestine, and post-intestine—already noted in *Eudistoma* appear within the new tissue.

On the assumption that extensive growth takes place in gut, epidermis, and epicardium in both genera, the differences between the two are as follows: In *Eudistoma,* elongation or extension of tissues destroys (disorganizes) existing microstructure, so that new differentiation or pattern appears as a reorganization. In *Diazona,* the old intestine survives but does not elongate and retains some degree of its originally regional character, so that reconstruction is essentially a process of new growth extending from old tissue and the new differentiation appears in typical relation to the surviving part.

Aplidium

In *Aplidium zostericola* the zooids are somewhat similar in size and form to those of *Eudistoma,* but the postthoracic region (abdomen) undergoing elongation prior to strobilation includes the stomach, which elongates in proportion to the abdomen. Epidermis, epicardium, and both limbs of the intestine undergo extension, although as in *Archidistoma* and *Diazona* the epidermis alone strobilates and divides the internal tissues. Each piece thus contains a section of the elongate stomach and ascending intestine, a section of the epicardium, and a short length of dorsal cord. In subsequent reconstitution, the epidermis grows from the old epidermis the neural complex regenerates distally from the dorsal cord, and the new thorax regenerates from the epicardium. Posterior regeneration of the gut loop proceeds from the posterior ends of the stomach and intestinal fragments, although—as Brien (1925 and 1927b) pointed out—more by morphallaxis (as in *Eudistoma*) than by true regenerative outgrowth.

Other Polyclinid Ascidians

In most polyclinid ascidians, e.g., *Circinalium concrescens, Glossophorum sabulosum,* and *Amaroecium* species, the epidermal constrictions subdivide only the long postabdominal region (absent in *Aplidium zostericola*), so that the buds do not contain any section of the intestine. Thus, apart from the usual and limited contribution of the epidermis, morphogenesis is conducted by the fragment of epicardium.

Scott (1959), however, has subjected newly metamorphosed zooids of

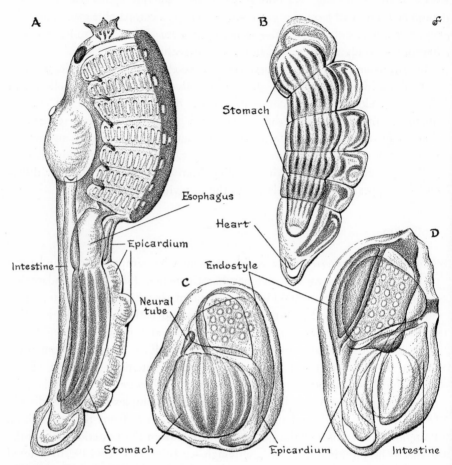

Figure 14-8. Budding in *Aplidium zostericola.* **A.** Individual zooid, showing elongating abdomen at onset of epidermal constrictions and elongating stomach ridges. **B.** Abdomen separated from thorax and undergoing segmental constriction. **C.** Bud at moment of isolation, showing fragment of elongated stomach together with regenerating thorax and epicardium from epicardial fragment. **D.** Later stage of same. (After Brien.)

Amaroecium constellatum to a process of fragmentation, with highly significant results. Fully formed but nonactive individuals, each derived from a single metamorphosed tadpole larva and each enclosed in its protective coat of tunicin, were fragmented by gentle pressure until the cavity of the tunic was a mass of floating bits of tissue. Reconstitution of new individuals from the tissue fragments occurred within 24 hours of fragmentation. If the epidermis was only slightly damaged during the experimental procedure, it spread until it became

whole again. In all such cases the fragmented tissues floating within the epidermal envelope reorganized into an essentially single digestive canal consisting
of branchial basket and typically differentiated U-shaped instenstine. Fragments
of all organs reassembled into their organ systems and established themselves
in the proper linear axis—behavior reminiscent of that of *Hydra* fragments
(p. 277). Reconstitution is extremely dynamic:

> Broken ends of oesophagus send streams of cells from their respective surfaces.
> They make their way through or about the caudal remnants, or other cellular
> debris, meet and unite. In some cases reunion necessitates movement of the
> pharyngeal epithelium over a relatively wide space. . . . The reconstituted stom-
> ach sends an oesophageal rudiment toward the pharynx, to join the original funnel-
> like opening there, and it sends a postgastric rudiment to connect with the intes-
> tine.[1]

The digestive tract completes whatever is lacking in its organization after
the fragments have been reunited in their proper respective positions. Epicardium and pericardium follow the movement of the digestive tract, and all three
structures orient themselves in their natural axial relationship.

Cell masses are commonly left over after the digestive tract and heart have
been reconstructed. These masses reorganize as structural units, according to
the special nature of their respective tissues, to form, for example, accessory
stomachs or long intestinal side loops. When this happens, each accessory part
lines up with its counterpart in the whole unit and joins at both ends with
that counterpart to form a double structure, unless the part is terminal; for
example, secondary branchial baskets produce an esophageal rudiment which
connects with the single stomach, resulting in a Siamese-twin effect. Cell masses,
therefore, exhibit type-specificity, and fragments of similar histogenetic character reaggregate and reassemble themselves in normal axial alignment by
ordered processes of spreading, sliding, and fusion. The epicardium responds
to the reorganizing behavior of the intestinal loop by forming a secondary
heart whenever a secondary digestive structure is formed.

In contrast, when experimental fragmentation ruptures the epidermis too
widely for this tissue to re-establish its continuity as a whole, the epidermis
forms two or three separate envelopes of smaller size. All masses contained
within each of these envelopes then reorganize within each into a complete
system of digestive tract, heart, etc., so that twins or triplets result. Therefore,
the property of wholeness and some of the information relating to the ordering
of the contained reconstituting cell mass apparently reside in the epidermis
(p. 321).

[1] Scott, *Acta Embryol. Morphol. Exp.* **2**:222 (1959).

Total Development in Buds

As mentioned early in this chapter, reconstitution by body fragments of ascidians ranges from the regeneration of structures through the reorganization and regeneration of substantial structures which serve as buds to the complete development of fragments extremely small relative to the size of the parental organism. Such development from minute fragments is most striking in the genera *Distaplia* and *Botryllus*.

Distaplia

Distaplia buds are formed by epidermal constrictions of an elongate probud, which is essentially an extension of the postabdominal region of either the tadpole larva or the adult zooid at the time of its regression. The total number of cells constituting a probud, which consists of an epidermal vesicle enclosing an epicardial vesicle, can be remarkably few.

During subsequent development the epidermis gives rise only to epidermis, and the epicardial vesicle forms all other systems and organs. At a very early stage, however, the epidermis exhibits the general form—including the long, postabdominal vascular process—of the presumptive zooid, but not until the epidermis invaginates locally to form the two siphons is it in actual contact with the inner vesicle. Also exhibiting a precocious and direct attainment of basic pattern or structure, the inner vesicle elongates and divides longitudinally into two components: one gives rise to the atrium and the gonads; and the other, by further elaboration as local evaginations or extensions, forms the pharynx, intestinal loop, epicardium, heart, and neural complex. In fact, the initial subdivision of the epicardial vesicle into two longitudinal components precedes the formation of buds by the probud—a phenomenon demonstrating that a morphogenetic process is active in two dimensions even before the third dimension has been defined, as in the *Salpa* stolon (p. 343). The main significance of the *Distaplia* type of development is that the fundamental organizational divisions of the organism are produced directly from a simple epithelial vesicle at as early a stage as seems mechanically possible and that from this early stage the general form of the future individual is exhibited in both inner and outer layers (Berrill, 1948a).

Two other phenomena in *Distaplia* are noteworthy. The first is that the several buds produced by the tadpole larva vary in size. The smallest buds never develop, but even the largest temporarily cease to develop after a very small increase in size, i.e., at a fairly early stage. Meanwhile, the oozooid completes its development, after which it feeds and grows as a functional individual for from 10 days to possibly several weeks. However, before it has attained sexual maturity, it resorbs within the tunic, or test, which envelopes both itself and its buds, thereby releasing into the common matrix the nutriment represented by its tissues. Simultaneously, the largest 1 or 2 buds proceed with their development to complete functional state. In other words, the development of partly

developed buds—those which have attained a certain stage—can be arrested without detriment to future continuation. Resorption of the adult organism occurs spontaneously after a certain period of development, and it is this resorption that allows bud development to proceed. Despite this sequential relationship between the two events, there is no organic tissue connection between parent and buds. Moreover, partly developed buds do not seem to be any threat to their parent. On the other hand, either inadequate amount or quality of food or some other poor environmental conditions adversely affect the histologically functional tissues of the zooid, but apparently do not affect those of the functionally immature bud. When buds remain in tissue continuity with the parent, the competitive situation is different (p. 217).

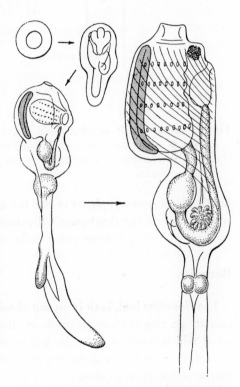

Figure 14-9. Four developmental stages in *Distaplia* bud, showing development of epidermal form.

The other phenomenon is that there are two periods in the course of development as a whole when a particular structure can be formed: the embryonic, primary developmental phase, or the postfunctional phase. Buds, however, are formed only during the late-embryonic period of an individual tadpole's development. The buds thus produced, however, and those of all later generations do not themselves form buds during their primary developmental period, but only at the very end of their lifetime—when the zooid has virtually regressed and only the postabdominal region remains. In both the tadpole and the buds produced, the process and the productive region of bud formation are the same,

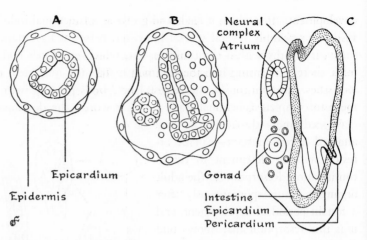

Figure 14-10. Early developmental stages in *Distaplia* bud. **A.** At time of formation as epidermal vesicle enclosing inner vesicle of epicardial tissue. **B.** Precocious segregation of prospective gonad. **C.** Evagination of primary organizational territories.

but the developmental phases of bud formation are far apart. This phenomenon suggests that in the developing individual there is a certain latent pattern that can find its full expression either early or late.

Botryllus

The *Botryllus* bud, both in its initial form and in its course of development, exhibits an almost classical simplicity. Because the life span of the individual zooid which develops from the bud is precisely limited in both the developmental phase and the mature functional phase, it is an outstanding example of precise temporal adaptation.

The bud first appears as a small disk of thickened atrial epithelium formed from the lateral body wall of an already well-established bud. At its maximum size the disk consists of about 20-150 cubical or columnar cells (5-14 cells across the diameter and one cell deep) and is surrounded by flattened epithelium. At this stage of bud development, the overlying epidermis, also a simple epithelium, shows a slight demarcation and no thickening. During subsequent development the epidermis forms only the epidermis of the new individual, and the atrial disk is chiefly responsible for morphogenesis.

When the disk attains its maximum size, it first arches to form a hemisphere and then proceeds to form a hollow sphere, or vesicle; at the same time, the overlying epidermis conforms to the disk's shape and size to serve as an enclosing envelope (but is never in contact with the atrial disk), so that a two-layered vesicle is produced.

As general growth proceeds and the inner vesicle, in particular, expands, two folds invaginate from its distolateral walls; the folds grow downward and divide the space within into a central and two lateral compartments, which represent the primary spacial divisions of the presumptive organism.

At about the same time, local evaginations from the posterior, posterolateral, and mid-dorsal wall of the central compartment form the presumptive intestine, heart, and neural complex, respectively, thus laying the foundation of the primary organ systems. The entire bud at this stage may be less than 0.1 mm in diameter and consists of only a few hundred cells.

Further bud development consists mainly of elaboration of the territories or organs already established. The heart evagination constricts and detaches itself from the side of the central compartment as a small vesicle which elongates and soon begins to beat. The neural evagination differentiates into brain and neural gland; the intestinal evagination lengthens and assumes the characteristic re-

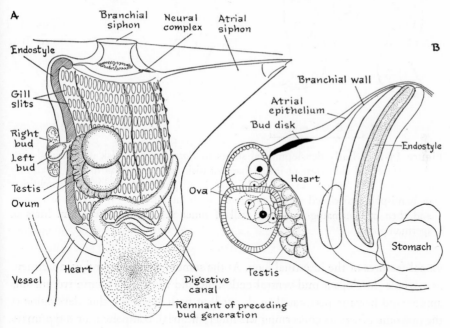

Figure 14-11. Development in *Botryllus*. **A.** Left side of zooid, showing mature testis and ovum, right- and left-side buds of next generation (stage 4), "ghost" or remnant of parental zooid of previous bud generation (stage 11), and epidermal vessel joining zooid to colonial vascular system (stage 10). **B.** Right half of young bud viewed ventrally (accordingly at left), showing developing testis, ova, and bud disk of next generation (stage 7 bearing stage 1).

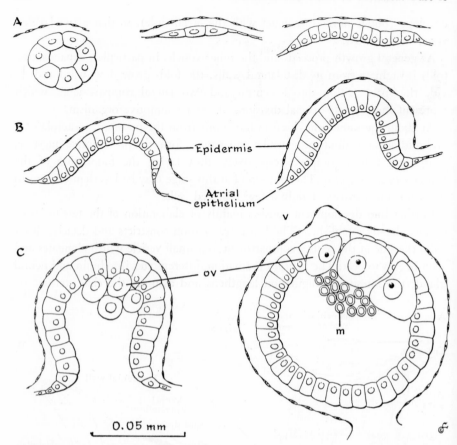

Figure 14-12. Early developmental stages in *Botryllus* bud disk of maximal size. **A.** Surface and side view of initial stage of disk and its growth to disk maximum (stage 1). **B.** Arching of disk to become hemisphere (stage 2). **C.** Conversion of hemisphere to closed sphere (stage 3) with lateral segregation first of oocytes and then of spermatogonia from wall of inner vesicle. **ov.** Ova. **m.** Mass of spermatogonia.

gional features of the intestinal loop. At the same time, the central compartment as a whole develops a mid-ventral endostyle, and rows of presumptive gill slits appear and become perforated in the wall at both sides. All this elaboration is the outcome of events concerning the inner, or atrial, component of the primary vesicle. The outer, or epidermal, layer forms, besides epidermis for the new individual, only the distal parts of the two siphons and, at a very early stage, an ampullary outgrowth, which fuses with the nearest adjacent blood vessel of the colony; at this time, the bud severs its connection to the parent.

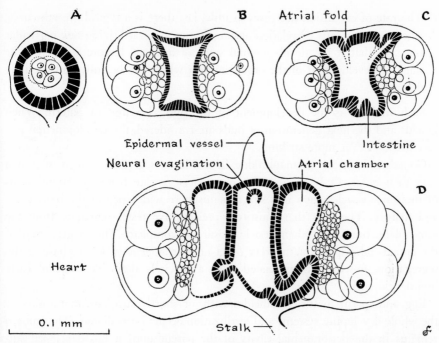

Figure 14-13. Later development in *Botryllus* bud disk of maximal size. **A–B.** Two stages in growth of segregated hermaphrodite gonads on each side. **C.** Subdivision of inner vesicle into central and lateral compartments. **D.** Evaginative expansions of central chamber to form intestine, heart, and neural complex.

Several aspects of the development of the inner component deserve particular emphasis:

1. The process as a whole can be described as a direct, progressive expansion of an area of simple epithelium which becomes the three-dimensional organism by foldings, contractions, and expansions.

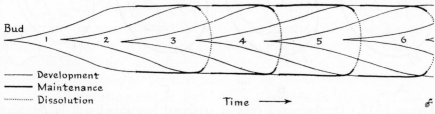

Figure 14-14. Growth curves for 5 successive bud generations, showing overlap between generations and three phases: development (stages 1-8), maintenance (stages 9-10), and dissolution (stage 11).

2. The rate of expansion decreases steadily, i.e., there is a typical growth curve.
3. There is a rigorous correlation among features characterizing any particular stage, so that if the developmental stage of any one feature is known, the stage of development of the organism as a whole can be fully and precisely defined.

Such in brief is the development of the bud as a unit of patern. When gonads and the next generation of buds are considered, the developmental picture becomes ever more striking.

Gonads, which are hermaphrodite in ascidians, segregate step by step at a very early stage. Only presumptive ova that segregate from the lateral walls of the inner vesicle when it is still a hemisphere wide open at the base, grow and mature. The cells that constitute presumptive testes segregate from the same place, but later, when vesicle closure is complete. Thus segregation of the gonads takes place at the very onset of morphogenesis. Each phase of the segregation, however, is associated with a precisely definable stage of early bud development.

Egg and sperm maturity coincides with, or shortly succeeds, maturation of the whole developing system; and fertilization occurs immediately. The embryo remains in the peribranchial cavity of the parent until it has developed sufficiently for liberation as a free-swimming tadpole. The parent undergoes dissolution a few hours before the next generation of buds mature. The duration of embryonic development is thus adjusted to the interval—usually a matter

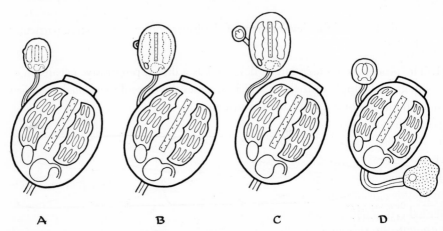

A   B   C   D

Figure 14-15. Developmental cycle of combined blastozooid unit in *Botryllus* colony. (After Oka and Watanabe.)

of a few days—between maturation of a bud and the bud's subsequent dissolution. Furthermore, temporal correlations hold for intergeneration relationships, which are definable in terms of the normal stages of *Botryllus* bud development:[2]

Stage 1. Disk of thickened atrial epithelium.

Stage 2. Disk evagination to form a hemisphere, together with segregation of gonads.

Stage 3. Two-layered vesicle: the inner formed by closure of hemisphere; the outer, by growth and evagination of epidermis.

Stage 4. Inner vesicle with intestinal evagination and beginning of atrial folds.

Stage 5. Inner vesicle divided into branchial and peribranchial compartments and digestive tube.

Stage 6. Heart distinct.

Stage 7. Bud disk of next generation.

Stage 8. Heartbeats.

Stage 9. Functional maturity: opening of siphons.

Stage 10. Maturation of gonads.

Stage 11. Dissolution.

Nature of Bud Rudiment

Since the cells of the atrial disk grow large and columnar before any change can be detected in the overlying epidermis, they can be regarded as the initiators of bud formation, either as individual cells or as constituents of a particular area of atrial epithelium. The increase in cell size within this area and the increase in basic-staining properties of the cytoplasm indicate a marked increase in local growth rate, especially since the region of locally increased growth rate is part of an epithelium which is already growing at a significantly rapid rate. A disk can be recognized—with difficulty—when it consists of no more than 8 cells (3 in cross section) and when it is only slightly thicker than the adjoining epithelium. With continued growth the cells become columnar and more numerous, and the disk grows, as a disk, until it is many times larger than when first recognizable. During this phase of growth the growth rate seems to be at the biological maximum.

The period of growth of the disk from its initiation to its final size in disk form coincides with the developmental period of the parental bud from the appearance of branchial ridges to just before the gill rudiments perforate. These periods comprise a definite phase and the correlation between them is precise, although the simultaneous conclusion of both is more readily recognized than their commencement. Thus there is a prominent period of growth of the disk, during which it seems to be as definitely and precisely a part of the developing parental pattern as is any other parental feature.

[2] Adapted from Berrill, *Biol. Bull.*, **80**:172 (1941a); and Sabbadin, *Boll. zool.*, **22**:249 (1955).

Onset of Morphogenesis

After a disk grows for a certain period to become either a circular or an approximately ovoid area of columnar atrial epithelium, it begins to arch. Whether this area forms 1 or 2 hemispheres and whether, if 2 are formed, they are of equal or unequal diameter depend on the shape of the disk: 1 hemisphere if the disk is circular, 2 hemispheres of equal diameter if the disk is twice as long as it is wide, and 2 hemispheres of unequal diameter if the disk is longer one way than the other but less than twice as long. Accordingly, it can be concluded with some assurance that growth without organization continues until the disk attains its maximum size and that the first sign of the subsequent arching to form 1 or 2 hemispheres is the onset of morphogenesis. In other words, the visible demarcation of a circular area or areas is the first step toward individuality; for, after the arching, the process of expansion, curving, and folding ensues and a single new organism develops from each such convexity. At the same time, the polarities of the parental organism are imposed on, or persist in, the substance of the disk and its derivatives, with regard to both the anteroposterior and the right-left axes.

A disk grows as a unit from its inception, but the quality of wholeness, which is the presumptive individual, is manifest at a definite time during the development of the parental organism. At that time (stage 1), the disk has attained its maximum dimensions, and arching is about to begin. The question arises: What is responsible for the transformation of a virtually flat circular area of the disk into a convex dome and then into a hemisphere or sphere? The arching or evagination is not merely the result of peripheral constriction of a local region of rapid growth by the adjoining atrial epithelium of the parent, although such constriction is possibly a contributing agent. The arching is primarily inherent in the disk, as it is in the invaginative plate of echinoderm embryos (p. 141), and the problem of the relationship between the whole and its parts once more arises: Do the properties of wholeness reside in either the basal or the free surface layer of the arching unit, or are the constituent cells coordinated in some other, as yet unknown manner? If the molding force resides in either of the two surface layers, there is no doubt that the basal layer is the active one, as evidenced by the formation of vesicles of different sizes, described below.

Whatever the developmental destinies of the two layers, a matter of geometry emerges at the outset, for only if the evaginating layer had no appreciable thickness would the problem of tissue mechanics fail to arise. As it is, the disk consists of large, dense, columnar cells which persist in the form throughout the process of arching and closing to form a spherical vesicle, and the behavior of the inner surface relative to that of the outer surface of the disk in forming such a hollow sphere is enlightening. In terms of cell number the degree of growth which accompanies the transformation from circular disk to closed, hollow sphere is virtually the same regardless of disk size. Disks at their maximum size (end of stage 1)—with about 20, 50, or 150 cells—transform into vesicles with about 33, 75, or 210 cells, respectively, which represent a proliferative increase of roughly 1½ times. The degree of expansion of the basal surface to form the outer, or basal, surface of the vesicle is almost exactly fourfold, again regardless of disk size. On the other hand, the free atrial surface, which becomes the inner surface of the vesicle, expands about 3½ times in the largest (about 150 cells) disks, but undergoes about a 100 per cent reduction in the smallest (about 20 cells) disks.

Apparently the behavior of the basal surface is consistent regardless of disk size; whereas that of the free surface is not. The conjoined surface layer, however, is not confined to the free surface of the cells, and it is possible that the continuity concerned in arching lies inward from the actual basal surface, as it does in the intercellular bridges of the *Volvox* embryo (p. 129). As long as the tension inducing arching is closer to the basal surface than to the other, the same differential behavior of disks of various sizes could be accounted for.

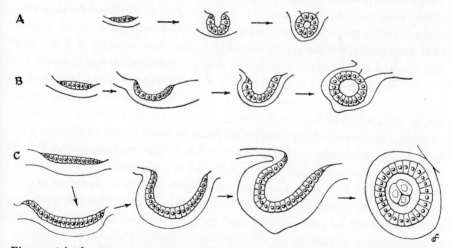

Figure 14-16. Early development of 3 bud disks of different maximum size. **A–B.** Minimal sizes from left and right sides, respectively, of first-generation blastozooids. **C.** Maximal right-side disk from later generation. Note determination of scale of organization and uniform degree of expansion of external surface of inner vesicle, but not of internal surface.

Rate of Growth and Expansion

From inception to maturation, a bud grows and develops in a unitary manner. Whatever the growth rate at the time the disk begins to arch, that rate declines from this initial maximal or near-maximal value until, by the time histological maturation is complete, growth ceases entirely. Linear measurements, which are an index of the expanding surfaces as a whole, show that the decline of the growth rate follows a typical sigmoid curve and that a particular stage of morphogenesis is associated with a particular point on the curve. The course of growth and time of maturation of oocytes, which segregate at the onset of development, and spermatogonia, which segregate soon after—despite their isolation—are similar to those of the somatic tissues. Development consists of cell division, enlargement, and maturation in various combinations in the various tissues; but all such combinations seem to be subordinate to, or conditioned by, the steadily declining growth rate. In other words, the disk has an initially high, specific growth rate which undergoes an equally specific decrement; therefore a disk of a certain size at stage 1 will expand n times before its growth or expansion rate declines to zero, and the value of n is constant for all disk sizes except for a disk so small (i.e., less than 20 cells at maximum state) that its development is inhibited.

Furthermore, since specific morphological stages of development are correlated with specific points along the growth curve, all intermediate stages between a disk's maximum size (stage 1) and its functional maturity (stages 9-10) are equally specific expansions of disk size. For example, if $d =$ initial disk size and a-n are expansion factors, $d \times a =$ two-layered vesicle (stage 3); $d \times b =$ subdivision of cavity (stage 4); $d \times c =$ segregation of heart vesicle (stage 6); . . . $n =$ functional maturity (stage 9-10). Such expansion values may be increasingly difficult to determine as the structural foldings become more complex, but the early stages can be analyzed in terms of cell number as well as by linear measurement of the disk as a whole. For example, for all values of d, a is approximately 1.6; b, about 6.0; and n, roughly about 500. Thus the area of the branchial wall at a given stage equals a specific expansion of d—regardless of the value of d—and a given morphological event commences after a certain amount of time—as expressed by the degree of expansion—so long as no inhibiting circumstances arise during or after the period of expansion.

Branchial differentiation, for instance, depends on (1) completion of the double wall, or fold, which separates the branchial from the peribranchial chambers and (2) local expansion of the inner wall of the partition to form transverse folds; both (1) and (2) are accomplished during a specific period from the time development begins. The amount of area available for branchial

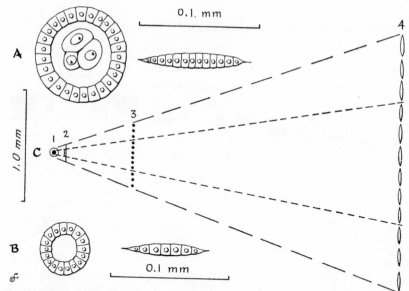

Figure 14-17. Disk size and gill-slit formation in *Botryllus*. **A.** Disk and closed-sphere stage of large size. **B.** Disk and closed sphere of small size. **C.** Sphere size and development of gill slits, at $\frac{1}{10}$ scale. **1.** Sphere size. **2.** Relative length of presumptive branchial wall. **3.** Relative lengths of branchial wall and number of gill slits at initial perforating stage. **4.** Relative length of emergent zooids when fully developed, showing size and number of rows of gill slits in each branchial wall of one side.

differentiation therefore varies only according to the initial disk size at the end of stage 1, since the duration and degree of expansion are constant. At the critical period, however, the number of transverse folds formed by the inner layer of the branchial wall is a function of the length of the wall; and subsequently, the number of gill rudiments per row is a function of the length of a transverse fold.

Evidently, there are limits to numerical, or quantitative, aspects of morphogenesis—limits set by the amount of available tissue (number of cells). Pattern emerges progressively with each step of an expansion process and is a function of the duration required for specific development, and the absolute area or the number of cells available at any particular stage can restrict the pattern quantitatively. Although both aspects of development are of major significance, the quantitative limitation is more readily analyzed.

Segregation of Gonads

Gonads either segregate during stage 2-3 or not at all. In disks consisting of more than 120 cells, presumptive mature ova segregate from the lateral walls

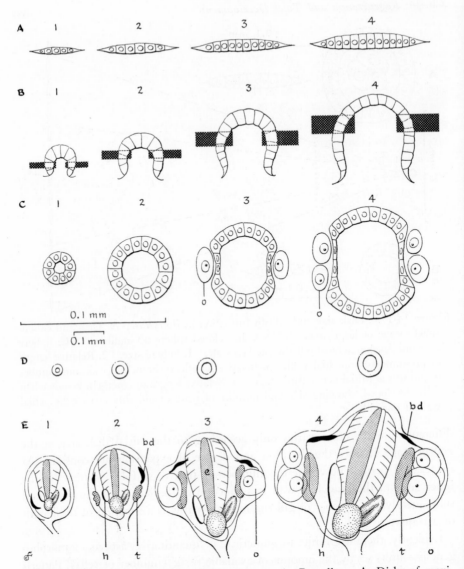

Figure 14-18. Disk size and gonad segregation in *Botryllus*. **A.** Disks of maximum sizes 1–4. **B.** Closing spheres developing from disks 1–4, showing extent of presumptive gonad area (black area). **C.** Continuation of same series 1–4 to closed-sphere stage, showing failure of ova to segregate in 1 and 2, single and double segregation in 3 and 4, respectively, corresponding to number of cells included in presumptive areas in preceding stage. **D.** Sphere stages of 1–4 reduced to ⅕ scale. **E.** Continuation of series 1–4 on reduced scale as far as bud-disk stigmata stage, showing subsequent development of gonads in the four sizes. **bd.** Bud disk. **h.** Heart. **t.** Testis. **e.** Endostyle. **o.** Ovum.

of the atrial vesicle at stage 2. Cells segregated from the same region after stage 2 but before stage 3 grow into recognizable oocytes, but do not grow after stage 3. Cells segregated from the same general region at stage 3 or a little later become testicular follicles enclosing spermatogonia. In disks consisting of less than 60 cells, reproductive cells do not segregate. In disks of intermediate size, cells may segregate to give rise to testicular components either alone or together with nongrowing oocytes.

The limiting factor in the segregation of ova seems to be the absolute number of cells constituting the disk at the end of stage 1. Obviously, the segregation of presumptive gonads from the lateral walls of the closing vesicle can only be the segregation of distinct cells. At the critical stage for the segregation of presumptive mature ova, however, the lateral walls of the vesicle consist of so few cells that only in the largest disks can as many as 3 or 4 cells be segregated. If the vesicle is two-thirds the maximum size possible, only 1 such cell can be extruded from the same relative presumptive area; and if the vesicle is even smaller, no segregation occurs, since the presumptive area consists of less than the surface of 1 cell.

Inhibition of the female components, however, does not necessarily imply inhibition of the male components, since the latter segregate later, when the total cell number has doubled. At this time, segregation of cells is mechanically possible; i.e., more cells comprise the lateral walls of the vesicle, so that a number of cells can be segregated while another number remain as a persisting lateral layer (Fig. 14-18). Consequently, buds developing from small disks not only are small but may be totally devoid of gonads. Somewhat larger disks develop into buds with male gonads alone; and depending on the specific size of still-larger disks, buds may form 1-4 mature ova.

In sum, presumptive mature ova segregate at the beginning of stage 2, when the growth rate of the whole is presumably at, or close to, its highest value (see also p. 208). Since the rate decreases from this time, the cells which segregate a little later (after stage 2 but before stage 3) to remain as nongrowing oocytes coincide with a slower growth rate. Cells destined to become male gonads segregate still later (stage 3), when the general growth rate is even slower.

Competition between Presumptive Areas

Both gonads and bud rudiments form from the lateral atrial walls of the parental bud; moreover, their presumptive areas adjoin and commonly overlap. The segregation of gonads is accomplished long before the first sign of a bud rudiment appears.

When disks are small, as in early generations of buds constituting a colony or as the result of starvation, no gonads are segregated. The presumptive gonadal area of the lateral walls remains undisturbed; and later, the shape of the presumptive bud area delimited is typical, for no tissue has been withdrawn from its prospective territory. Under these circumstances double buds of nearly equal size usually arise, at least on the right side.

When disks are somewhat larger, the gonads formed are substandard in size and apparently are segregated from all available tissue, including the posterior region of the presumptive bud area. The linear axis of the presumptive bud area is thus reduced, and only a single bud can be formed. This bud, however, can be relatively large compared with those developing from the ovoid, but smaller territories of earlier generations.

In the development of the largest disks, the full gonadal complement, which includes 3 or 4 fully potent ova, is segregated and apparently does not encroach on the presumptive bud area, for double buds arise.

This conclusion can be drawn: When development is full scale, adjoining presumptive areas do not compete for the available space; but when the scale is submaximal, an early-appearing presumptive area extends over a proportionately larger territory, so that on the basis of "first come, first served" a later-appearing region may have a correspondingly smaller territory. On the other hand, when the cellular state at an early stage is inadequate for the expression of the earlier presumptive area, the tissue of that area remains uncommitted, and the later-appearing area has full scope for development.

Size of a Disk

Whereas the shape of a presumptive bud area determines whether 1 or 2 circular disks are delimited and whether, if 2 are formed, they are of equal or unequal diameter, the size of a disk determines whether it will become a large or small zooid, whether it will segregate gonads, how many gill slits will comprise its branchial walls, and to a very great extent the size of the next-generation bud disks that form during bud development.

This matter of later-generation bud size raises problems of a rather subtle nature. How is the common, fairly steady increase in the size of zooids produced from successive generations of buds effected, since zooid size must at least partly depend on the size of the disk at its maximum state (end of stage 1)? What is the special nature of the atrial epithelium comprising the ovoid or circular area which precedes the actual demarcation of the presumptive individual? Is the growth of this area influenced differentially relative to growth of the parent?

An experimental approach—involving extirpation of buds at various stages— to these problems has been made by Sabbadin (1955 and 1958) and Watkins (1958). To determine whether the dissolution of adult zooids is induced by the final phase of rapid growth of the buds or vice versa, Watkins removed late-stage buds in an attempt to prolong the life of the parental zooids. Not one zooid persisted more than 24 hours beyond the time of dissolution of the controls; therefore Watkins concluded that dissolution of adults is innate and independent of the development of buds, just as it is in *Distaplia*.

Sabbadin found that induced premature dissolution of adults at functional maturity (stage 9) provokes an equally premature progression of the new bud generation from the appearance of heartbeats (stage 8) to functional maturity. In those adult zooids which persisted for longer than usual at stage 11, the next-generation buds also persisted at stage 8 for a longer time. Employing fine tungsten wire dipped in potassium hydroxide, Sabbadin extirpated buds of one side or the other, generation after generation, and observed that the surviving

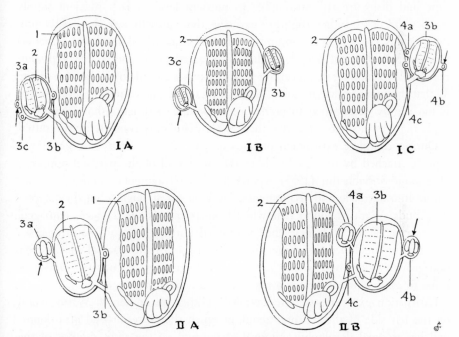

Figure 14-19. Ventral aspects of *Botryllus* blastozooids, i.e., right sides seen on left and vice versa. (Arrows indicate extirpations.) **Experiment I.** Combination stage 9-8-3. **IA.** Right anterior bud 3a extirpated. **IB.** Bud 2 of A matures, and buds 3b and 3c reach stage 7. Bud 3c is then extirpated. **IC.** Bud 3b produces buds 4a and 4c on right and bud 4b on left; i.e., smaller left-side bud grows larger and bears double right-side bud plus single left-side bud. **Experiment II.** Combination stage 9-8-5. **IIA.** Right bud of intermediate generation 2 bears right bud 3a of stage 5 and left, atrophied bud 3b. Bud 3a is extirpated. **IIB.** Bud 2 of A has matured, and bud 3b has resumed development and has produced buds 4a, 4b, and 4c. (After Sabbadin.)

buds attained larger dimensions, since relatively more nutrient per developing bud was supplied at the times of parental dissolution, and also lived longer in the absence of competition from other buds.

How developing buds can be differentially affected by the nutritive supply is indicated by the formulae of coexistent stages (Fig. 14-19). Stage 9 is an individual at functional maturity; it carries stage-7 buds, which in turn bear disks approaching their maximum state at the end of stage 1. The formula is thus 9.7.1. However, the parental zooids that produced the stage-9 individuals are about to undergo dissolution, so that for a brief period the formula is 11.9.7.1. This correlation seems to be virtually constant, signifying that the dissolution of the postmature generation coincides with the late phase of growth of bud disks. In other words, at the time of parental dissolution—which is also the time bud disks are still susceptible to nutrient level—a rich nutrient supply reaches the disks, either slightly increasing their growth rate or slightly prolonging their period of maximum growth. Whether rate is increased or period is prolonged, the final size of the disk is increased. Accordingly, it would seem reasonable to suppose that a certain plasticity persists until about the time (stage 3) that the inner vesicle is closed, but that thereafter the course of development is too rigidly determined to permit regulation. Experiments show that there is plasticity until stage 3, but that there is also regulatory competence thereafter.

During complete extirpation of buds at stage 2 or 3, fragments sometimes remain attached by their peduncle to the atrial wall of the parental bud even after stage 3. Sabbadin (1955) reports that such fragments commonly reconstitute into a vesicle and recommence development at stage 3 but do not produce buds. Watkins found that when hemispheres (stage 2) were damaged, 40 per cent of the buds nevertheless attained normal size and maturity. This investigator concluded that either the parental zooid retains actual control over bud growth or the bud is somehow induced to regenerate lost tissue and continue to grow until it attains a typical, mature size.

Extirpation experiments have revealed another phenomenon. When a bud on the left side of the parental zooid, or when the posterior bud of a double bud has stopped developing, removal of the bud on the right side, or of the anterior bud, respectively, causes the arrested buds to resume development. Somewhat conversely, when the bud on one side is extirpated during stage 2, but before stage 3, the bud of the opposite side divides in two, or if already so dividing, divides again to form three—presumably as a result of increased growth —which then develop. Also, when the bud of an oozooid which normally produces a bud only on the right side is extirpated, a bud usually appears on the left side and proceeds to develop. Some of these buds undergoing belated development commonly exhibit an abnormal condition known as *situs inversus:*

the digestive loop is on the right side, and the heart is on the left. This condition occurred in only 1 out of 1,500 controls, but in 30 per cent of the experimental zooids. Once established, the mirror-image symmetry persisted through successive generations. Such reversal occurs only in buds that recommence development when the parental influence is waning or gone.

It is very significant that extirpation of the relatively minute bud disk on one side of the parental organism is immediately compensated for by a surge of growth and consequent developmental activity of the presumptive bud area on the opposite side. In such extirpation it seems highly unlikely that either circulating fluids or neural transmissions are the mediating influences. A much more direct influence, involving structural continuity, seems to be operating; perhaps it is a shift or displacement effect associated with microstructural continuity, extending through the whole epithelial organization. Whatever the cause, the phenomenon seems to be essentially the same as that in compensatory regeneration or growth in bilaterally asymmetrical organs such as the chelae of certain crustaceans or the operculum of serpulid worms. When the large or functional organ is removed from these organisms, the small or abortive organ of the other side develops and replaces it, and a reversal of the asymmetry results. In the serpulid Hydroidea, neither circulating hormones nor the nervous system are involved, according to Okada (1933); and the phenomenon of both dominance and release from dominance associated with the presence and absence of the original one-sided organ seems to be a matter of an organizational balance that is much subtler than any system of chemical diffusion. This problem arises also in connection with vascular budding in botryllids (p. 398).

Symplegma

Bud formation in other botryllids, including polystyelid ascidians, is accomplished by essentially the same means as in *Botryllus:* The lateral body wall gives rise to an evagination consisting primarily of outer epidermis and inner atrial epithelium; the epidermis plays the same apparently passive role as it does in *Botryllus,* and the atrial epithelial vesicle is responsible for most of the morphogenetic processes. However, the time and place of bud formation, as well as the pattern of gonads and the size and extent of differentiation of segregated gonadal units, vary greatly among various species of adult botryllids.

Symplegma viridis offers the closest comparison with *Botryllus* in that the colonies look much like those of *Botryllus,* but differ mainly in that the individual zooids do not form systems around common cloacal apertures.

The individual zooid is about the same size as that of *Botryllus;* the gill-slit pattern is similar; and the hermaphrodite gonads are basically the same in

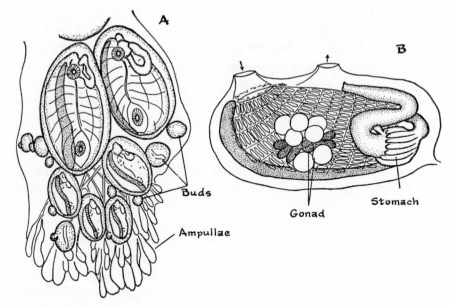

Figure 14-20. *Symplegma viride.* **A.** Margin of colony. **B.** Isolated, sexually mature blastozooid.

location, size, and composition. On the other hand, buds form only on the right side, though from approximately the same location on the atrial wall as the right-side *Botryllus* bud, and only in a temporal sequence of 4 per parental zooid, never as double buds as in *Botryllus.* Thus *Symplegma* is a typical botryllid except that there is only a right-side bud, which, when a rudiment, grows somewhat differently from a *Botryllus* bud rudiment.

The *Symplegma* bud rudiment, which develops from the wall of a bud also in process of development, as in *Botryllus,* appears earlier and persists longer, relative to the stages of parental development. Accordingly, it occupies a relatively greater part of the parental wall when it first appears, and it is potentially larger.

The atrial budding rudiment is approximately pear-shaped at all times. A circular area is delimited from the broad end of the pear and everts to form a hemisphere which becomes a closed vesicle, much as in *Botryllus;* but the narrow posterior residue does not evert at the same time—even to form an abortive vesicle. The residual area continues to grow until the shape and size of the original rudiment have been restored. Then an anterior circular area again everts to form a hemisphere, which becomes a vesicle.

This entire process takes place four times. The first bud begins to form in a parent bud at roughly the same time (stage 5) that the inner vesicle of *Botryl-*

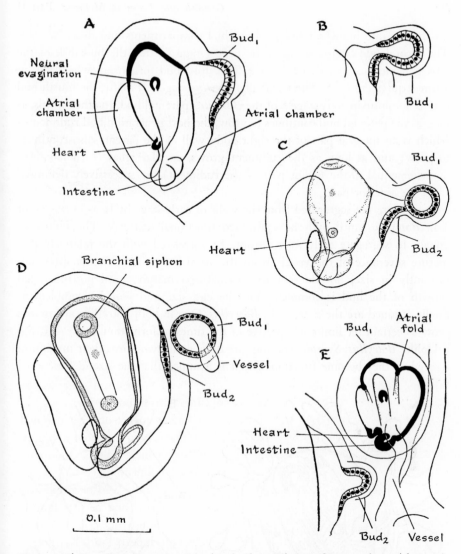

Figure 14-21. Development of buds of advanced *Symplegma* colony (shown in Fig. 14-20). **A.** Bud bearing in turn bud of succeeding generation, arising as convex thickening disk of right anterior atrial wall. **B–C.** Later stages in formation of bud shown appearing in **A**, showing formation of closed vesicle and residual area of disk. **D.** Three stages: development of larger zooid shown in **A**, its first bud now a closed vesicle with developing ampullary vessel, and residual disk representing first stage in bud development. **E.** More-advanced stage, showing residual disk forming second vesicle and further residual area; first bud has become attached to colonial circulatory system by its ampullary vessel and is completely separated from its parental bud.

lus divides into branchial and peribranchial compartments and digestive tube. The second bud begins at about the time (stage 7) that the bud disk of the next generation appears in *Botryllus*. The third bud forms somewhat later, at a time equivalent to the latter part of *Botryllus* stage 7. Finally, the fourth bud begins at approximately stage 8 of the *Botryllus* parental bud. In other words, as long as the parental bud continues to grow, the next-generation bud rudiment— which is an integral part of the right body wall of the parent—also continues to grow; and as long as the rudiment grows, circular areas (each having a diameter equal to the widest part of the rudiment) are successively delimited from the anterior part.

Gonads are not segregated from the walls of early-stage buds, as in stage 2 or 3 in *Botryllus,* until buds have attained functional maturity. The failure of gonads to segregate early may possibly be correlated with the relatively premature severance of the vesicle from the atrial lining. Ova do not grow significantly in size, nor do they or spermatozoa mature, until postfunctional growth of the bud is complete. Yet the egg size and gonadal complement finally attained are the same as in *Botryllus*. Ova in both botryllids apparently grow to certain set limits irrespective of the times of formation.

Unlike *Botryllus, Symplegma* buds are only about two-thirds their maximum size when they become functional. If it is assumed that the size at functional

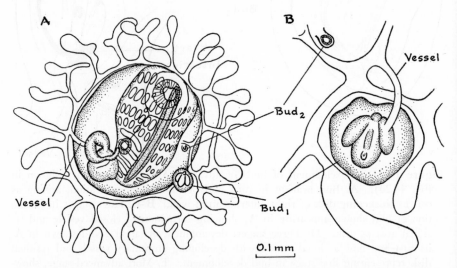

Figure 14-22. Functional oozooid of *Symplegma viride*. **A.** Ring of ampullary vessels, gill slits transforming from protostigmata to rows of definitive stigmata, and first and second buds forming from atrial wall. **B.** Later development of buds. The first bud has already become detached at point of origin and is now attached to ampullary vessels by its own epidermal vessel.

maturity (equivalent to *Botryllus* stage 9) is related to initial scale, either the initial size is too small to permit expansion to maximum size, or the initial rate of expansion or growth is too low. Whichever the cause, the failure to segregate gonads early becomes understandable.

The functional zooid as it first develops from the egg, i.e., the oozooid, is a little smaller than that of *Botryllus* and lacks buds. Unlike the *Botryllus* zooid, it continues to grow, and a succession of 4 buds are produced from the right anterior wall, i.e., from the equivalent site as buds formed during bud development. However, the oozooid buds, which are somewhat smaller than those formed earlier, are formed from the tissue of a functional zooid rather than from that of a nonfunctional, developing system.

In brief, therefore, a comparison between the two genera *Botryllus* and *Symplegma* reveals the following major points of difference: Oozooids and blostozooids of *Symplegma* continue to grow after reaching functional maturity; whereas those of *Botryllus*—no matter what their size or degree of sexual maturity—apparently do not. In *Symplegma,* buds and gonads, which fail to form during the early, prefunctional development of the oozooid and the bud, respectively, do so later. As long as growth continues, an opportunity lost in the early phase arises at a later one, so that the four-dimensional pattern becomes fully expressed.

Budding in Other Botryllids

Other botryllids are also of interest. Whereas in *Botryllus, Symplegma,* and *Kükenthalia,* the budding areas are precisely located on one or both sides of the body wall, or mantle; in others such as *Distomus, Stolonica,* and *Polyandrocarpa,* buds can arise from almost anywhere around the basal part of the mantle wall (Sélys-Longchamps, 1917; Berrill, 1949 and 1951). In *Metandrocarpa* the budding areas are commonly within the anterolateral areas, where, according to Abbott (1953), as many as 4 buds form during one short period of the postfunctional phase of bud growth. In all four of these last-mentioned genera, formation of the bud outgrowth, consisting of epidermis and atrial epithelium with some mesenchymal tissue between them, takes place after a profuse formation of ampullary growths from the epidermal vascular vessels of the test. If ampullary clusters indicate or establish local regions of high respiratory or other metabolic activities, they may thereby evoke local outgrowth of the body wall. Equally plausible, of course, is the converse: an incipient local outgrowth may stimulate regional growth of ampullae. Whichever the sequence, if adjacent zooids are too close to one another, both bud formation and ampullary growth are inhibited. The budding process as a whole is thus both a spatial progression, from diffuse or apparently indeterminate blastogenic territories of the mantle wall to precisely located, defined areas, and a temporal progression of the defined areas, from juvenile adults to early morphogenetic stages.

Vascular Budding in Botryllids

Oka and Watanabe (1959) demonstrated that in *Botryllus primigenus,* in addition to the atrial (palleal or peribranchial) budding of the type described above, new buds are formed also from aggregations of blood cells at the base of vascular ampullae within the colony. They have called this newly discovered process "vascular budding." In *B. primigenus,* vascular budding occurs in actively growing colonies concomitantly with atrial budding. In *Botrylloides violaceum,* the same authors (1959) report that vascular budding is never seen under normal conditions; however, in pieces of colonies devoid of zooids but containing portions of the epidermal vascular system, buds form and are irregularly distributed along the walls of the vessels. In both species, apart from the location of the vascular buds and their distinctive histological origin, the vascular buds follow the same developmental course as do atrial buds, commencing with the formation of an inner sphere enveloped by epidermis, with space between. Blastogenesis, according to Oka and Watanabe, is an exact replica of that in atrial buds, except that no gonadal tissue ever segregates. The special interest concerns the manner, place, and circumstances of vascular-bud initiation.

The outstanding feature of the initiation process is the mode of origin and

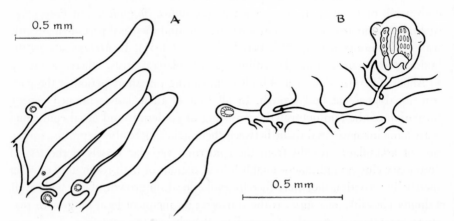

Figure 14-23. Vascular buds at base of epidermal ampullae in *Botryllus primigenus.* **A.** At end of budding period. **B.** Two buds formed consecutively on same ampulla. (After Oka and Watanabe.)

establishment of the inner vesicle of stage 3. The inner vesicle, instead of forming from a disk of atrial epithelium, as in palleal budding is constituted from an aggregation of small, round, unspecialized lymphocytic cells normally circulating in the haemolymph of the blood vessels and epidermal vascular outgrowths. An aggregation first appears as a small cluster of such cells closely opposed to the inner surface of the epidermis, at the base of an epidermal ampulla in *Botryllus primigenus* and of the epidermal vessels and elsewhere in *Botrylloides violaceum*. Once formed, a cell cluster grows as a compact mass, presumably by cell multiplication rather than by continuing aggregation, until it separates from the adjoining epidermis and develops an internal cavity, by direct cavitation, thereby becoming a vesicle typical of stage 3. This short-cut approach to stage 3 may well be responsible for the absence of ova in all vascular buds, in spite of adequate size, since ova segregate at stage 2.

Obviously, the cell mass that gives rise directly to the stage-3 vesicle in vascular buds is created *in situ* from previously loose or disorganized cells, and the mass cannot be regarded as the expression of inherent organization in a pre-existing tissue. No tissue corresponding to the atrial epithelium exists within the epidermal vascular outgrowths. The role of the epidermis therefore requires further scrutiny, for there is little doubt that a cell cluster originates and grows in response to an inductive influence which only the epidermis can supply.

The epidermal ampullae of a botryllid colony are actively growing formations. Although their shape remains fairly constant, each ampulla grows at its distal end, while its base progressively transforms into relatively narrow stalk; i.e., its size and shape reflect a process of active growth and epithelial transformation. In *Botryllus primigenus,* vascular buds form at an ampulla's base, where the epidermal transformation from wide ampulla to narrow vessel is imminent. After a time, as the result of this continual conversion, a particular bud which first formed at the ampullary base arises some distance along the narrow vessel, and new buds arise successively at the ampullary base as ampullary growth and transformation proceed. Buds are initiated at a certain critical distance from the growing tip and after certain intervals, in a manner directly comparable with successive bud initiations in the ascidian *Perophora* and in the hydroids *Obelia* and *Campanularia*. The difference is that in *Botryllus* the only organized tissue initially present is a single-layered epidermis; whereas an inner tissue is always present in the other genera. In a comparison between the initiation of vascular buds of *Botryllus* and bud initiation along the stolon of Campularian hydroids (p. 217), it is also significant that, in both, the common site of initiation is very close to the region where axial stolonic growth finally ceases.

The unmistakable fact in vascular-bud origination is that the epidermis is solely responsible for initiation; i.e., the initial change lies within the epidermal epithelium. Polarized growth at the ampullary tip determines that initiation will be a certain distance from the tip. In *Botrylloides violaceum,* the presence of colony zooids in continuity with the epidermal vessels inhibits bud initiation from the opposite direction, as in *Clavelina,* so that initiation occurs only after ampullary vessels have been severed from their associated zooids. Accordingly, when zooids are removed either by cutting or by disintegration, a change in the nature of some component of the epidermis, e.g., the macromolecular structure of a continuous basal surface, or an electromagnetic field, is transmitted a considerable distance to an ampullary base.

The residual question is whether in palleal budding the epidermis is responsible for initiating disk formation in the underlying atrial epithelium, even though there is no contact between the two tissues, or whether the atrial epithelium initiates disk formation without epidermal induction, and subsequently influences the overlying epidermis to develop accordingly. Atrial epithelium is continuous, as a simple epithelium, with the external epidermis, and it can legitimately be regarded as part of the external enveloping epithelium of the organism as a whole, equivalent to the epidermis itself. In view of the linkage between the two epithelia, this important question cannot be resolved at this time.

REFERENCES FOR
Regeneration and Total Development

ASCIDIANS (excluding Botryllids)

Berrill, N. J. 1947a. "The Structure, Development and Budding of the Ascidian Eudistoma." *J. Morphol.* **81**:269-82.
——. 1948a. "Budding and the Reproductive Cycle in Distaplia." *Quart. J. Micro. Sci.* **89**:253-89.
——. 1948b. "The Development, Morphology and Budding of the Ascidian Diazona." *Marine Biol. Assoc. U.K. J.* **27**:389-99.
——. 1948c. "Structure, Tadpole and Bud Formation in the Ascidian Archidistoma." *Ibid.* Pp. 380-8.
——. 1951. "Regeneration and Budding in Tunicates." *Biol. Rev.* **26**:456-75.
——, and A. Cohen. 1936. "Regeneration in Clavelina lepadiformis." *J. Exp. Biol.* **13**:352-62.
Brien, P. 1925. "Contribution à l'étude de la blastogénèse des Tuniciers. Bourgeonnement chez Aplidium zostericola (Giard)." *Arch. biol.* **35**:155-204.
——. 1927a. "Contribution à l'étude de la blastogénèse des Tuniciers. IV. Bourgeonnement de Clavelina lepadiformis (Müller)." *Inst. zool. Torley-Rousseau Rec.* **1**:31-81.
——. 1927b. "Contribution à l'étude de la blastogénèse des Tuniciers. Formation du

système nerveux et des glandes genitales dans les blastozooides d'Aplidium zostericola (Giard)." *Arch. biol.* **37**:1-45.

———. 1930. "Contribution à l'étude de la régénération naturelle et expérimentale chez les Clavelinides." *Soc. roy. zool. Belgique Ann.* **60**:33-46.

———. 1932. "L'hétéromorphose chez les Tuniciers. La régénération bithoracique et mono-thoracique des fragments oesophagiens de Clavelina lepadiformis (Müller)." *Acad. roy. Belgique* (Classe sci.) *Bull.* **18**:975-1005.

———. 1933. "Régénération thoracique chez Archiascidia neapolitana (Julin). Structure du système nerveux central." *Bull. Biol. France-Belgique.* **67**:100-24.

———, and E. Brien-Gavage. 1927. "Contribution à l'étude de la blastogenèse des Tuniciers. IV. Recherches sur le bourgeonnement de Perophora listeri Weigm. (Origine mesoblas-tique du septum stolonial.)" *Inst. zool. Torley-Rousseau Rec.* **1**:123-52.

Deviney, E. M. 1934. "The Behavior of Isolated Pieces of Ascidian (Perophora viridis) As Compared with Ordinary Budding." *Elisha Mitchell Sci. Soc. J.* **49**:185-224.

George, W. C. 1935. "The Formation of New Siphon Openings in the Tunicate Styela plicata." *Elisha Mitchell Sci. Soc. J.* **53**:87-91.

Hirschler, H. 1914. "Ueber die Restitutions -und Involutions-vorgänge bei operierten Exemplaren von Ciona intestinalis Flem." *Zeitschr. Zellforsch. mikro. Anat.* **85**:205-27.

Huxley, J. S. 1921. "Studies in Dedifferentiation. II. Dedifferentiation and Resorption in Perophora." *Quart. J. Micro. Sci.* **65**:643-97.

———. 1926. "Studies in Dedifferentiation. VI. Reduction Phenomena in Clavelina lepa-diformis." *Staz. zool. Pubb.* **7**:1-36.

Pérès, J. M. 1948a. "Recherches sur la genèse et la régénération de la tunique chez Ciona intestinalis L." *Inst. Océanogr.* (Monaco) *Bull.* No. 936. Pp. 1-12.

———. 1948b. "Recherches sur la genèse et la régénération de la tunique chez Clavelina lepadiformis Müller." *Arch. d'anat. micro. morphol. exp.* **37**:230-60.

Ries, E. 1937. "Untersuchungen über den Zelltod. II. Das Verhalten differenzierter und undifferenzierter Zellen bei der Regeneration, Reduktion und Knospung von Clavellina lepadiformis." *Arch. entwick. Org.* **137**:327-62.

Sélys-Longchamps, M. de. 1915. "Autotomie et régénération des viscères chez Polycarpa tenera Lacaze et Delage." *Acad. sci. Comptes rendus.* **160**:566-9.

BOTRYLLIDS

Abbott, D. P. 1953. "Asexual Reproduction in the Colonial Ascidian Metandrocarpa taylori Huntsman." *Univ. Calif. Pub. Zool.* **61**:1-78.

Berrill, N. J. 1940. "The Development of a Colonial Organism: Symplegma viride." *Biol. Bull.* **79**:272-81.

———. 1941a. "The Development of the Bud in Botryllus." *Ibid.* **80**:169-84.

———. 1941b. "Size and Morphogenesis in the Bud of Botryllus." *Ibid.* Pp. 185-93.

———. 1941c. "Spatial and Temporal Patterns in Colonial Organisms." *Growth* Supp. (Symp. 3) **5**:89-111.

———. 1947b. "The Developmental Cycle of Botrylloides." *Quart. J. Micro. Sci.* **88**:393-407.

———. 1949. "The Gonads, Larvae and Budding of the Polystyelid Ascidians Stolonica and Distomus." *Marine Biol. Assoc. U.K. J.* **28**:633-50.

———. 1950. "The Tunicata." *Roy. Soc. London.* Pp. 203-31.

Oka, H., and Y. Watanabe 1959. "Vascular Budding in Botrylloides." *Biol. Bull.* **17**:340-6.

Sabbadin, A. 1955. "Osservazioni sullo Sviluppo, l'accrescimento e la riproduzione di Botryllus schlosseri (Pallas), in conditioni di laboratorio." *Boll. zool.* **22**:244-63.

——. 1958. "Analisi sperimentale dello Svilluppo delle Colonie de Botryllus schlosseri." *Arch. ital. anat. embriol.* **63**:178-221.

Scott, F. M. 1959. "Tissue Affinity in *Amaroecium*. I. Aggregation of Dissociated Fragments and Their Integration into One Organism." *Acta Embryol. Morphol. Exp.* **2**:209-26.

Sélys-Longchamps, M. de. 1917. "Sur le Bourgeonnement des Polystyelinés *Stolonica* et *Heterocarpa*, avec quelques notes sur l'anatomie de ces deux genres." *Bull. Sci. France et Belge.* **50**:170-276.

Watkins, M. J. 1958. "Regeneration of Buds in Botryllus." *Biol. Bull.* **115**:147-52.

Watterson, R. L. 1945. "Asexual Reproduction in the Colonial Tunicate, Botryllus scholosseri (Pallas) Savigny, with Special Reference to the Developmental History of Intersiphonal Bands of Pigment Cells." *Biol. Bull.* **88**:71-103.

ANNELIDS

Okada, Y. K. 1933. "Remarks on the Reversible Asymmetry in the Opercula of the Polychaete *Hydroides*." *Marine Biol. Assoc. U.K.* / **18**:655-70.

Chapter 15

Development and Pattern:
Summation and Hypothesis

FROM THE foregoing chapters of Part II, it is possible to construct a comparative statement of relationships. Such a statement can hardly be called a general theory, but can at least serve as a framework of correlations from which a theory may eventually be elaborated. Since the account thus far has mainly concerned the various forms of development from somatic sources in coelenterates, sponges, worms, and tunicates and has also concentrated on morphogenesis rather than histogenesis, the summation of this chapter pertains primarily to these topics, especially with regard to the tunicates and hydroids. Nevertheless, the conclusions drawn are of general application—e.g., to vertebrate embryos—and in Part III are applied interpretatively to the developmental phenomena of the plant meristem.

Rates of Growth

1. Under standard environmental conditions, such as temperature, every specific tissue has a definite maximum rate of growth. When that maximum rate is attained, only growth occurs; i.e., no differentiation or establishment of individualized morphogenetic territories takes place. In other words, growth at the biological maximum rate for a particular tissue is incompatible with the development or maintenance of the fine microstructure associated with tissue organization and differentiation.

Examples:

a. The rapid, indeterminate growth of hydroid stolons and terminals, such as those in *Tubularia, Obelia,* and *Bougainvillia,* when the temperature rises above a critical value.

b. The first growth phase of bud disks, such as those of *Botryllus,* before the evagination begins to form 1 or 2 vesicles; and the initial, labile, indeterminate growth phase of regeneration blastemata.

c. The marginal territory of tissue explants cultured in vitro and almost every territory of malignant tissues, in vivo or in vitro.

d. The early growth phase of polyembryonic eggs, such as those of the bryozoan *Crisia.*

2. The initiation of the maximum growth rate within a certain region of an established organism or tissue characterizes that region as either a prospective bud or a fundamental disruption of the structural continuity of the organism.

Examples:

a. Bud disks.

b. Fission zones.

3. Definite morphogenetic territories become established only as the growth rate falls below the maximum, in somewhat the same manner that a colloidal gel forms only in the absence of excessive thermal agitation.

Examples:

a. The initiation by stolonic terminals, stem or otherwise, of emergent hydranth pattern in hydroids (e.g., *Tubularia*) only when the temperature falls, thereby inhibiting linear terminal growth.

b. The demarcation of prospective bud areas in the bud disks of *Botryllus* and *Symplegma* only after considerable growth has occurred and the growth rate is below the maximum.

c. The subdivision of the probuds in *Distaplia* and *Doliolum* to form definite buds only after the linear growth rate of the probuds has become much retarded.

d. The subdivision, or strobilation, of the abdomen in *Aplidium,* the post-abdomen in other polyclinid ascidians, and the stolon of the hydroid *Corymorpha* to form definitive units after extensive linear growth has reached a maximum and has slowed considerably.

e. The successive demarcation of medusa buds from the growing manubrium of a budding medusa just below the region of maximum growth.

f. The successive segregation of new segments in annelids from that side of the zone of growth associated with lower growth rates.

Shape of Territory

1. Whatever the shape of the activated territory that gives rise to prospective buds, the area segregated as a prospective individual is approximately circular. When the initiating structure is tubular, spheres are commonly segregated. Irregular shapes are never encountered.

Examples:
a. The production of large and small bud disks, or spheres, from noncircular activated territories in ascidians, such as *Botryllus, Distaplia,* and *Colella.*
b. The circular shape of the presumptive area of medusa buds in both hydroids and medusae.
c. The shape of most blastemata.

Availability of Space

1. When individual units are produced from a fairly flat surface, the circular area segregated is the largest portion (of that shape within the activated territory) that has a *uniform* submaximal growth rate. If part of the territory is still growing at the maximal rate, that part—irrespective of its shape—remains outside the organizing territory.

Examples:
a. The *simultaneous* segregation of *Botryllus* bud disks that are longer in one axis than in the other into two morphogenetic fields whose size relative to each other conforms to the amount of available territory.

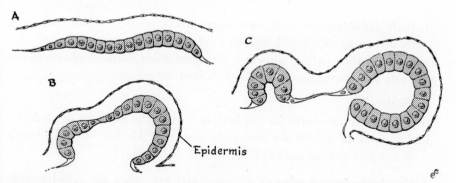

Figure 15-1. Three stages in subdivision of *Botryllus* bud disk of which one axis was longer than the other. **A.** Incipient arching indicative of two fields of unequal size. **B–C.** Separation and independent conversion, respectively, to vesicles. Note conformative growth of epidermis without contact.

b. In *Symplegma* the growth gradient of the activated territory (although longer in one axis than in the other) with one end persisting at the original maximal growth rate and remaining as unorganized meristematic tissue, so that only the territory growing at a submaximal growth rate organizes.

2. When a succession of prospective morphogenetic territories are segregated, the available area for each is delimited by the distance between the activated tissue of maximal growth rate and the tissue already organized or differentiating; i.e., the available area consists of activated, uncommitted territory growing at a submaximal rate. Within such a territory, the scale of organization—as indicated above—depends on the diameter of the largest circular area that can be encompassed.

Examples:
a. Bud primordia in *Hydra.*
b. Bud primordia in hydranths.
c. Bud primordia on the manubria of medusae.

3. When meristematic tissue is formed as a strip, helical or otherwise, primordia are successively produced, each with a diameter equal to the width of the activated strip at the level of segregation. Each succeeding primordium forms only after the addition of as much new material as a circular area of the same diameter can compass.

Examples: The same as for (2) above.

Growth Curves

1. Each developmental unit, once segregated, continues to grow, but at a steadily decreasing rate, which follows a typical curve from submaximal value to zero value; the extent of growth is thus predictable from calculation of the initial rate and the magnitude of the decrement.

Examples:
a. The growth rate curve for any bud, such as that of ascidians or medusae.
b. The curve of growth for blastemata generally, whether head blastemata of annelids or limb and tail blastemata of amphibians.

2. For developmental units of various initial size, when the initial growth rate and the decrement value are the same, the final size at the time of histological maturity is directly proportional to the initial size at the time of formation.

Example: The direct proportion of the final size of *Botryllus* buds to the size of the area of the prospective bud when first delimited.

3. The growth curve, and therefore the growth decrement, holds for the developing unit as a whole, whether growth takes place mainly by cell proliferation, mainly by cell enlargement, or by some combination of both.

Examples:
a. The growth of various tissues, e.g., epithelium, muscle, gonads, in the developing bud of *Botryllus*.
b. The growth of gourds, in which cell proliferation predominates in early phases and ceases in later phases.
c. The growth of oligochaetes, in which cell proliferation, segment formation, and cell and segment enlargement occur in varying ratios as growth proceeds.

The growth curve represents the total surface, or volume, expansion of the initial unit, and both cell division and cell enlargement conform to that expansion; i.e., the expansion is primary and unitary, and all other growth events are responsive. The expanding material—whose nature presents the basic problem in understanding the growth process—may be at least comparable to an expanding fibrous net, which is a close-packed, small mesh at the outset and a loose maximal mesh at the end of the growth process. The process of loosening, or expansion, becomes progressively slower as the mesh approaches its terminal state. (The expansion of the silver-line systems of ciliates is an example of such a process.) On the assumption that the expanding material is netlike, the growth decrement would be the progressive slowing down of the rate of net expansion. What underlies the packing and subsequent expansion of such a net is an as-yet unfathomed problem.

Course of Development

1. As development or expansion proceeds, territorial patterns progressively emerge, the largest or the most fundamental territories appearing first. Differential regional expansion combined with epithelial folding and segregation of local cell masses seems to be the basic process involved, rather than differential regional rates of cell proliferation. The process remains unitary, and cell number and cell shape conform to expansion and condensation of the epithelial sheets.

Examples:
a. The early phases of bud development in *Botryllus* and *Distaplia*.

b. Entocodal origination in the development of Hydromedusae.

c. Invaginative gastrulation as in starfish.

d. Development in general in *Amphioxus, Balanoglossus,* and related genera.

2. Besides the major organizational territories, subterritories that are prospectively histologically unique are delimited precociously either by expansive folding and subsequent constrictive segregation or by tissue segregation directly.

Examples:
a. The neural complex in ascidian buds.
b. The pineal and pituitary organs in vertebrate embryos.
c. The heart as an evagination in ascidian buds.
d. The gonads in hydroids and botryllid buds.

Because the segregation of tissues from a parental source suggests mutual incompatibility, such a segregation may be comparable to that which occurs in aggregation masses of mature or differentiating tissues, as in sponges. Therefore, the essential quality associated with gonadal or neural-complex differentiation may well be present before segregation and in fact induce segregation. On the assumption that the basic qualitative differentiation predates segregation, it is a property of the whole developing unit, as it is in eggs, and not of individual cells. In every such instance of segregation, differentiation is an integral and proportional part of the tissue surface long before there is any visible distinction among structures.

Each developmental territory appears at a prescribed time in the life cycle of an organism and at a precise location in the body; both time and location differ among territories. Apart from the major or primary subdivisions, each unit or morphogenetic field seems to consist of a center of activity surrounded by a more vaguely defined and less-active peripheral territory of the same general character. As growth proceeds, the center of each territory expands outward, so that the center commonly retains a relatively high growth rate and remains labile, and the peripheral region undergoes a rate decrement and exhibits differentiation from the margin toward the center.

Examples:
a. Hydranth development in *Obelia* and related Hydromedusae.
b. The regeneration blastemata of vertebrate limbs and heads of worms.

4. The number of cells in a particular prospective area at a given time is a controlling factor, in that an inadequate number may inhibit either direct or evaginative segregation. Such inhibitions may result from too small an initial scale of the developmental unit as a whole, so that the proportionate expansion of the territory in question is inadequate at the critical stage of development,

from too low a growth rate, so that the appropriate histological reaction (e.g., gonadal differentiation) does not take place, or from competition between presumptive areas, so that the extent or number of territories is diminished.

Examples:
a. The absence of gonads in early bud generations of a *Botryllus* colony.
b. The absence of buds in a newly functional oozooid of *Symplegma*.
c. The absence of entocodon in the earliest stage of hydranth development.
d. The anterior blastemata at posterior levels in many worms.

5. Adjoining morphogenetic fields commonly overlap, with a mutually re-strictive influence. In the absence of one, the other can increase its scale of organization. When the segregation of two adjacent territories occurs at dif-ferent stages in development, the earlier presumptive area organizes on a maxi-mum scale and may involve part, possibly all, of the territory of the later one. When the earlier is inhibited, the later one organizes on a maximum scale. The over-all scale of organization, however, may be so large that otherwise-competi-tive territories are each able to organize to the maximum extent.

Example: The formation of bud disks and gonads in developing *Botryllus* buds of various initial size.

Even without competition and although of maximum size relative to the size of the parental unit, a developing *Botryllus* bud disk may still be too small to yield territory for, e.g., gonadal segregation, as in the zooids of the first few generations in young colony.

6. Expansion of a developmental unit proceeds to a certain stopping point, which is determined by the initial size and the growth potential (potential $=$ rate $\times$ factor for decrement). This may be called phase 1.

At the end of phase 1, when growth has ceased and tissues or organs are functionally mature, an obvious transition to a new organismal or histological status initiates phase 2.

Growth is resumed from zero value in phase 2, and the rate of growth de-pends on environmental conditions, such as temperature and nutrition. Unless nutritionally or otherwise restrained, such growth proceeds to a second stopping point, which is unrelated to initial size and terminates about the time the full gonadal complement has been segregated.

Example: The growth pattern of any developmental unit.

7. Morphogenetic territories inhibited during growth phase 1 can appear during phase 2, and once formed, such territories persist and develop.

Example: Buds and gonads in developing zooids of *Symplegma,*

8. Growth, or expansion, of the whole continues in phase 2 until all morphogenetic territories that were undeveloped during phase 1 have been expressed. For example, if at the beginning of phase 2 a prospective territory must expand ten or a hundred times in area to express its potential structure, the organism as a whole expands correspondingly. When gonadal segregation, which seems to be the critical delimitation, occurs during phase 1, there is no growth in phase 2 at all. When segregation begins during phase 2, growth of the whole continues until segregation and maturation are complete. Thus growth as a whole continues until all presumptive organizational territories are fully expanded, i.e., expressed or developed.

Example: The growth pattern as it affects any morphogenetic territory.

9. Although the initial scale of organization determines the final size of a developing unit (end of phase 1), whether bud primordium or blastema, the final size of a particular organ at functional maturity is, in terms of mass or of cell number, a function of the stage of its initiation. In terms of functional value, the end in a sense determines the beginning; i.e., natural selection of terminal functional efficiency of an organ has determined that initiation occurs at the stage which allows the proper time for its development.

Examples:
a. The necessity for eggs of genetically determined size to segregate at a particular stage in bud development if they are to be able to attain that specific size by the end of bud development.
b. The correlation of the four phases of a colonial unit in *Botryllus*.

10. When one growing tissue approaches or comes into contact with another of the same organism, the tissue with the higher rate of growth as a rule becomes the pacemaker; and at least for a while thereafter, the two grow in conformity to each other.

Examples:
a. Adjacent bud layers of ascidians.
b. Constituent tissues of thaliacean stolons.
c. Entocodal and adjoining tissues in hydroids.
d. Optic vesicle and lens rudiment.
e. Mesoderm and epidermis during segmentation.

11. Organization may be fully established, and morphogenesis and histogenesis proceed, when only two of the three spatial dimensions of the prospective individuals has been determined.

Example: Development of the stolon in *Salpa*.

12. When the scale of one or two dimensions of a primordium has been established, growth proceeds until the remaining one or two dimensions attain a size proportionate to that of the established dimensions.

Example: The establishment of the scale of a salp by the dimensions of the cross section of the stolon; and the consequent growth of the third dimension —initially the thickness of one cell—until it is the same general size as that of the other dimensions.

13. Whenever possible, the initial inherent organization expands until every undeveloped territory or dimension attains its full expression. The whole pattern is represented in the cortical layer of the developmental unit at the time the unit is delimited, whether the unit is a circular bud disk consisting of a comparatively small number of cells, a sphere segregated from tubular tissue, or even a single somatic cell.

Example: Gonad and bud territories in hydroids and ascidians.

14. The property of wholeness, i.e., the individuality of the organism, apparently resides in the epidermis.

Examples:
a. Epidermal subdivisions in *Salpa* and *Pyrosoma*.
b. Epidermal fragmentation in *Amaroecium*.
c. Fission in oligochaetes.

15. Structural units, particularly of protein constitution, manifest like-to-like affinities at all levels of organization, as in the following sequence or organizational ascent:

1. Aggregation of protein molecules to form macromolecular fibers.
2. Aggregation of macromolecular fibers to form periodic fiber pattern.
3. Pairing of homologous chromosomes.
4. Orderly reorientation of scrambled cortical bands in cell cortex (e.g., *Stentor*).
5. Reaggregation of dissociated cells into tissue types according to original type-species (e.g., dissociated sponge tissue and vertebrate embryonic tissues).
6. Realignment and fusion of tissue fragments according to original type and location (e.g., hydroids and ascidians).
7. Mutual orientation of whole organisms (e.g., zooid clusters in developing ascidian colonies).

PART III

MORPHOGENESIS

IN PLANTS

The phenomena of morphogenesis and polymorphism in the development of the shoot and floral apex of vascular plants clearly have much in common with, for example, the successive formation of manubrial buds in a medusa and the processes of rhythmic growth as exhibited in segmentation and strobilation. There are also significant parallels between the developmental polymorphism of hydranth and medusa on the one hand and of vegetative and floral development on the other. It is equally obvious that the conclusions (summarized in the preceding chapter) concerning the initial size of rudiments, the permissive role of cell number, the competition between adjacent presumptive areas, and the general effects of growth rates, though drawn from animal morphogenesis, are pertinent to the apical meristem.

The morphogenetic study of processes in the plant meristem, however, is still in its infancy, so that this Part, on Morphogenesis in Plants, is of necessity limited to the descriptive information and experimental data now available. Accordingly, the following two chapters may serve as a basis for comparison with, or as an extension of, the preceding discussion of animal morphogenesis.

Chapter 16

The Vegetative Shoot

MORPHOGENESIS in plants has engaged the attention of botanists since the early nineteenth century, when Goethe formulated his semiphilosophical concept of plant form—a concept fully discussed in recent times by Engard (1944) and Arber (1950). For the most part, modern morphogenetic studies have been confined to the shoot meristem of vascular plants: ferns, studied primarily by Wardlaw, cycads by Foster, and flowering plants by Esau and Engard. These investigators, among others, have described the growth changes in meticulous detail, thus laying the essential foundation for experimental studies now under way. As Wardlaw has stated, however,

> the starting point lies in the work of Hofmeister, for he was not merely concerned with preparing descriptive accounts of changes in form during development, but also asked himself such questions as: How does the observed form come to be? To what processes of growth can the observed structural developments be related? What internal and external factors determine specific structural organization? [1]

The publication of Darwin's theory of phyletic descent long diverted botanists from this point of view, and the experimental approach to morphogenesis in plants is relatively recent and limited in scope compared with experimental animal morphology. Moreover, various hypotheses originally derived from zoological investigations have been utilized in attempts to explain morphogenetic events in the plant meristem. As Wardlaw (1955) pointed out, the concepts of diffusion gradients, metabolic gradients, and relative growth rates have been vigorously employed. However, the interpretation of such concepts as they pertain to plant morphogenesis remains a challenge.

In the present discussion an attempt is made to apply to the meristem, particularly of flowering plants, the various concepts based on animal morphogenesis and summarized in the preceding chapter, and from such applications,

[1] Wardlaw, *Nature,* **153**:127 (1944).

to interpret the morphogenetic events in plants in terms of shape, size, relative growth axes, and relative growth rates of initial morphogenetic units. At the same time, in this and in the succeeding chapter, detailed descriptions of the meristematic processes connected with shoot and flower development are described in detail insofar as they are known.

The Shoot Meristem

The shoot apex is the primary formative region in vascular plants. Once established, it exhibits a general organization fairly common to ferns, cycads, and flowering plants alike, though it varies greatly in size, shape, relative growth rates, and tissue layers. Problems of immediate interest that relate to the meristem proper—as distinct from foliar and floral appendages that may arise from it—concern the degree of plasticity of the apical meristematic tissue and that tissue's role in forming procambial tissue. The first problem involves the botanist's meaning of the term "regeneration" as distinct from the zoologist's meaning, and the other anticipates a discussion of the growth[2] of vascular tissue in relation to leaf-primordium determination.

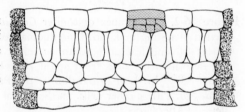

Figure 16-1. Initiation of new shoot after isolating piece of leaf of *Drosera*. (Dead cells at margin are crosshatched; activated cell is stippled). Note subdivision commences at inner level. (After Bünning.)

As employed by zoologists, "regeneration" denotes replacement or reconstitution of some structure from the surviving portion of an organism; for example, of an amputated hand from the stump of a vertebrate limb or of the amputated head region from the remainder of a flatworm body. Regeneration of this nature does not occur in vascular plants: mutilated leaves or floral units do not recover or replace their damaged or lost parts; instead, regeneration in plants denotes the establishment or re-establishment of a new shoot, or root, meristem, which usually reconstitutes the entire organization rather than a fragment thereof. However, the shoot meristem exhibits a plasticity which is strictly comparable to that of early blastemata in regenerating animal tissues. Thus Ball (1950) found that after excision of the central part of the shoot apex in *Lupinus,* a new

---

[2] In the present context, the term "growth" refers primarily to protein synthesis and cell proliferation, rather than cell enlargement; and the term "polarity" may refer either to growth activity in a given direction or to protoplasm in the sense of polarized microstructure and growth potential.

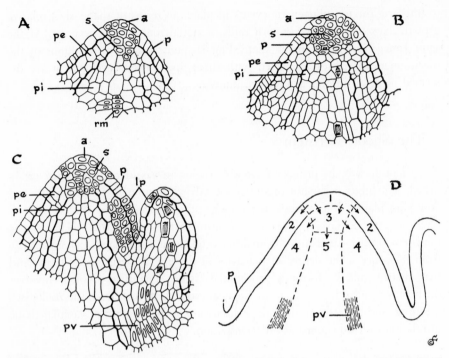

Figure 16-2. Shoot apex of conifer *Cunninghamia lanceolata.* **A.** Longitudinal section of resting bud. **B.** Longitudinal section of apex of expanding shoot. **C.** Apex, showing leaf primordium at initial and later stage. **D.** Diagram, showing derivation of zones from apical zone 1 and subapical initials 2. **a.** Apical initial cells. **lp.** Leaf primordium. **pv.** Provascular cells. **pe.** Peripheral meristem. **pi.** Pith mother cells. **s.** Subapical initials. **p.** protode. **rm.** rib meristem. (After Cross.)

apex, frequently one on each side of the wound, regenerated, as in a divided blastema. In each new shoot a ring of provascular tissue formed in the subapical region, above the base of the incisions, and subsequently united laterally with the original vascular tissue. The data suggest that the new vascular tissue forms in relation to the apical meristem rather than to the older tissue below, so that, according to Ball, the shoot apex is the organizer, which controls development by sending out basipetal stimuli.

Wardlaw (1957) describes the shoot apex as a whole as a dynamic, geometric system of the following interrelated zones:

1. A distal region, or summit, which is marked by a single apical cell in most ferns and by a group of embryonic initial cells in one or more layers in seed plants and which represents the meristem's center or focal point on which the integrity and sustained development of the primary axis depend.

2. A subdistal region, which consists of the superficial layer or layers of meristematic cells and in which growth centers are initiated.

3. An organogenic region, which is subjacent to the subdistal region and in which leaf primordia form and tissue differentiation begins.

4. A subapical region, which is subjacent to the organogenic region and in which the shoot widens and elongates considerably, the leaf primordia enlarge conspicuously, the parenchymatous cortex and pith form, and vascular tissue undergoes further differentiation.

5. A region of tissue maturation, at the base of the meristem.

Clearly there are two growth centers. One is the subdistal region, from which leaf primordia arise, either successively as single individuals in spiral sequence around the meristem or simultaneously as pairs on opposite sides of the meristem. Whatever the manner of formation, a primordium represents a local region of increased growth rate relative to the rate in its immediate surroundings. Whatever the sequential pattern, the process of initiation is rhythmic, and the time interval between the *initiation* of one leaf primordium and that of the next is known as a plastochron. The other growth center is the distal meristem from which growth of the shoot apex as a whole proceeds. This region, too, seems to exhibit rhythmic growth, which is seen as a regular change in shape correlated with each plastochron, during which the apex passes from a minimum to a maximum volume. The period of active shoot-apex growth falls during a plastochron, at which time, according to Gifford (1954), the site of the most active cell division may not be in the uppermost initial cells but in the subapical derivative cells. Whatever the precise site, there is evidence of a

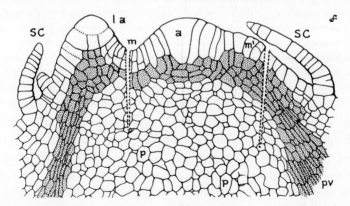

Figure 16-3. Longitudinal section of fern *Dryopteris aristata,* showing details of isolated meristem, with vertical incisions (broken lines) at sides. **m–m′.** Apical meristem, consisting of distinctive prism-shaped cells and large apical cell **a.** la. Leaf apex. **pv.** Provascular tissue. **p.** Pith. **SC.** Scale. (After Wardlaw.)

rhythm in the frequency of mitosis in the meristematic zone immediately distal to the zone of initiation of leaf primordia.

Although the remains of an apex regenerate new apices in a genus such as *Lupinus* (which typically has a group of apical initials), the punctured single initial cell of the fern *Dryopteris* is not replaced, and leaf primordia continue to appear in normal sequence until all the meristem is used up. Accordingly, the apical cell or the group of apical initials can be considered responsible for the continuing growth—or at least the integrity—of the apical meristematic tissue itself, though not directly for the secondary growth center, from which leaf primordia arise. Where a single apical cell bears the entire burden, its destruction terminates apical growth, but when a group of cells is responsible, partial destruction permits reconstruction of the apex from the remaining cells. Thus the apical meristem as a whole may be regarded as one or more apical initial cells which, through growth and proliferation, give rise to the subapical meristem. It is not known whether this growth is steady or rhythmic, but it is evident that there is a regular increase and decrease in the volume of the subapical zone and that the decline coincides with the period of leaf initiation in the zone immediately below.

The subdistal region is the region of possibly maximal growth rate and seems to exhibit rhythmic growth; the derivative tissue or zone basal to the subdistal region grows at a decreased rate. That a leaf primordium is initiated in this derivative zone is to be expected, since a local increase in growth rate can be expressed only within a field that is growing at a submaximal rate.

An aspect of shoot apices that has puzzled botanists is that

> whereas certain major functional activities are common to the apices in all classes of vascular plants, e.g., their regulated organogenic activities, their maintenance of the embryonic state, etc., apices may nevertheless be demonstrably very different in such matters as their size, shape, histological constitution and pattern of tissue differentiation. Apices may be conical, paraboloid or almost flat; their leaf primordia may be of small or large size relative to the apical meristem, and they may arise close to the apical cell or group of apical initials, i.e., close to the summit or centre of the meristem, or in a position on the flanks more or less remote from it.[3]

In a large, conical apex, such as that of *Dryopteris,* the rate of growth decreases from the apex steadily; in various paraboloidal apices the decrease is considerably more rapid; and in flat meristems the decrease is most rapid.

In the shoot apex of flax (*Linum*), the precursor of vascular tissue, the procambium, is directly below the apical meristem. According to Esau (1942), procambial cell divisions are evident in the initiation of the first leaf primordia, and there is no meristem between the apex and the procambium that could be designated as a meristematic ring.

[3] Wardlaw, *Am. J. Bot.,* **44**:176 (1957).

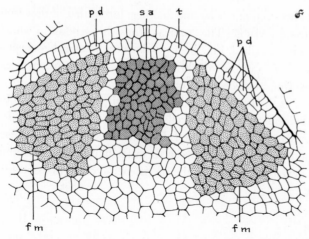

Figure 16-4. Vegetative apex of flowering plant *Succisia pratensis*, showing various zones. **pd.** Periclinal divisions. **sa.** Subapical initials. **t.** Tunica. **f.m.** Flank meristem. (After Philipson.)

The vascular region begins its existence at the apex not as a homogeneous meristem, not as a remainder of the apical meristem, but as a region in which certain groups of cells are in process of being differentiated as procambial cells. . . . The procambium strands differentiate in continuity with the procambium of the older traces within the axis and progress acropetally from the axis into the developing leaves . . . [Thus] the procambium may differentiate within the axis as a continuous system.[4]

Similarly in *Sequoia sempervirens*, according to Sterling (1945), the procambial strands differentiate acropetally in the shoot apex in continuity with older strands below, and are present in the apex before their respective leaf primordia emerge. Yet, as mentioned above, there is evidence that procambial differentiation is initiated from the apex—that new tissue is continuously added to that already present below the apex. Accordingly, the question arises: Are the leaf primordia themselves in some way responsible for the initial procambial differentiation within the apex? This question was answered by Wardlaw (1944), at least for the fern *Dryopteris*. Several weeks after he had removed all the leaves and all leaf primordia forming subsequently from the apex of the fern rhizomes, Wardlaw investigated the histology of the apex and found no leaf gaps in the vascular cylinder. In other words, the apex by itself forms a complete cylinder, and leaf primordia disrupt rather than initiate procambial differentiation.

In the short apex of lower vascular plants, it is difficult to recognize discrete

[4] Esau, *Am. J. Bot.*, **29**:741, 746 (1942).

layers or to define limits between one layer and another; but in that of flowering plants, such layers are usually visible. The "tunica-corpus" hypothesis, by which these layers are commonly described, may be a purely morphological concept—as is the concept of germ layers in the vertebrate embryo—but it is nevertheless useful. The apex in flowering plants consists of one or more peripheral layers of cells—collectively known as the tunica—which are primarily engaged in surface growth by anticlinal divisions; and of the cells of the interior—the corpus—which divide in various planes. Whatever the full significance of the two different manners of division, the distinction between them facilitates the description of both leaf and floral development and of the meristematic changes associated with such development.

Initiation of Leaf Primordia

Leaf primordia arise in succession around the base of the meristem and are initiated in the tunica, in the plants in which this is recognizable. When the tunica consists of only one layer, as in monocotyledons, a leaf primordium usually originates in this layer by periclinal divisions. In most dicotyledons, primordia usually arise in the second layer; then they undergo periclinal divisions in the third layer and often in the outermost corpus cells. In some dicotyledons, primordia arise only in the third layer. In *Spiraea* and other genera of the Rosaceae, leaf primordia are initiated on the flanks of the apex 2-5 cell layers below the surface, but usually in the second, according to Rouffa and Gunckel (1951). These investigators also found that during one plastochron the apex passes from a minimum to a maximum volume without a significant fluctuation in tunica stratification, although layers in the flank region are less stable during the minimum-volume phase. Thus leaf initiation seems at first associated with one tunica layer, which may be the surface layer, but is more commonly one or two layers beneath; and almost immediately thereafter involves the inner tunica and even corpus tissue at a deeper level. There is evidence, however, of periclinal divisions appearing first in corpus tissue immediately subjacent to the tunica.

Primordia arise on the apex in a definite and precise pattern, i.e., with phyllotaxis. As Wardlaw has emphasized, the mode of origin in ferns is different from that in angiosperms. In the fern not only does the tissue of the shoot apex derive from the divisions of a single, large, apical initial cell, but so do the leaf primordia. A single, superficial leaf-initiate cell is the first sign of a primordium. In ferns such as *Dryopteris,* primordia are circular in surface view and are widely spaced on the broad apex. In angiosperms, on the other hand,

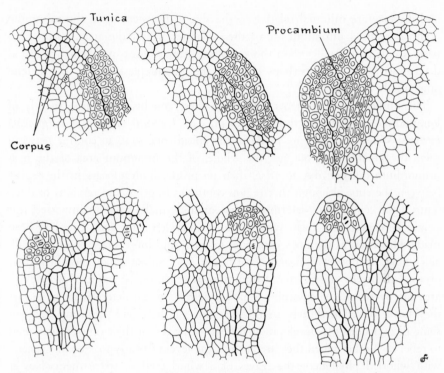

Figure 16-5. Origin and development of foliar organ (phyllode) in *Acacia*. (Heavy lines indicate delimitation of tunica and its derivatives from corpus and its derivatives. Nuclei are shown in cells most actively involved.) (After Boke.)

both the shoot apex and leaf primordium are represented by a group of cells rather than a single initial cell, and the leaf primordia are relatively closely packed on the flank.

Two interpretations—each bearing evidence—of primordia origins have been put forward: (1) The "field" theory, developed particularly by Wardlaw, is that leaf primordia are growth centers from which morphogenetic fields become organized and that adjoining fields exert mutual inhibition. (2) The "available-space," or "packing," theory, proposed by the Snows, states that a leaf primordium forms as a whole from the beginning and acquires position according to available space. Whether one or the other or a reconciliation of the two theories ultimately proves correct, a comparison between leaf initiation and bud initiation in hydras and medusae may well be profitable.

According to the Snows (1950 and 1952), a leaf is determined on any part of the apical cone of an angiosperm as soon as that part attains a critical width and distance below the extreme summit of the growing apex. Since this width and distance are minimal relative to the size of the stem apex, the minimum width is really a minimum arc of the circumference of the apex. Visible leaf bases are at first smaller than the supposed primary area over which a leaf is determined; but soon, perhaps early in the second plastochron, they extend tangentially beyond it.

The numerous experiments leading to these conclusions consist mainly of longitudinal cuts on the shoot apices of, e.g., *Lupinus albus* and *Euphorbia lathyris.* When two cuts are made in the apical cone so as to confine between them the greater part, but not the whole, of the minimum area of the next primordium but one due to arise, then no primordium appears in the region between the cuts, although that region continues to grow and remains healthy. Thus if the center were determined first, an abnormally small or imperfect leaf would be expected to form within the territory left after such cuts had been made. However, as the apex continues to grow and more space becomes available, new primordia appear. Accordingly, the Snows have modified their original theory by stating that a minimum width must be acquired before this area can overcome the summit's inhibitory effect on leaf determination. In fact they have moved definitely in the direction of the "field" concept: ". . . in some dicotyledons the developmental fields of leaves, or their centres, may tend to repel each other when they are at the same level or nearly so. For in a species with whorled phyllotaxis the leaves of a whorl tend to space themselves at equal distances after a disturbance." [5]

Leaf primordia arise relatively close together near the summit in most flowering plants, but form as centers relatively widely separated much further down the flank in ferns. A young fern primordium, situated above and between two older ones, occupies only a small portion of the cone. While the primordium is still part of the cone, its growth is relatively slow; but once in the subapical region, growth and morphological development are rapid. Wardlaw suggests, in line with Child's theory of metabolic gradients, that a leaf primordium is formed as a growth center of special metabolism and that the position of the center is apparently controlled by diffusion of substances from the apical cell or cell group at the shoot apex and by the two adjacent, youngest primordia. The growth center is described as a circular group of 7-9 meristematic cells, which contain a substance that determines the outgrowth of the lateral organ. This approach seems to be in keeping with the general tendency among students of plant development—e.g., Richards (1949)—to conceive of morphogenetic fields as diffusion fields and of diffusing substances, such as auxins, as determinants of pattern.

A different, though related, interpretation, based on the phenomena and con-

[5] Snow, M. and R., *Roy. Soc. London Philos. Trans.,* **139**:565 (1952).

cepts presented in earlier chapters (mainly in connection with *Hydra* and *Botryl-lus*) may prove more accurate. Whether the region of greatest growth in the shoot apex is the apical initial region of the summit or the subdistal tissue sub-jacent to it is of little consequence in this context of growth rate. Either may be regarded as the region of maximal biological growth, at least during each midplastochron growth phase. The growth rate decreases basipetally down the growth cone, and the tissue growing within this gradient must be significantly polarized in relation to the axis of growth. Leaf primordia are initiated only at a certain relative distance from the zone of maximal growth, at first as approxi-mately circular rudiments representing areas of tangential growth around a geometric center. The over-all process is therefore strictly comparable to the processes of column growth and bud formation in a hydra (p. 228).

The initiation of primordia in the region of maximal growth near the summit of the shoot apex is inhibited by two factors: (1) the high rate of growth, which seems to be incompatible with stable cytoplasmic fine structure or organi-zation and which makes impossible any local growth acceleration, since to be discernible such local growth would obviously have to exceed the maximal growth; and (2) the differential growth rate, with its consequent tissue polariza-tion, which is likely to hinder the establishment of local growth centers of tissue growing radially and tangentially to the axis of the shoot apex. As in *Hydra*, such centers can be initiated only where the rate of growth has dropped to minimal values. Only at such sites is it possible for a growth center to appear with planes of growth in opposition to the original axial polarization and for such growth to be sufficiently greater than that of adjoining tissue to appear as a local protuberance. In *Chrysanthemum* the region of maximum growth rate, according to Popham and Chan (1950), is a ringlike zone at some distance below the summit and immediately above the zone from which leaf primordia initiate.

The high growth rate and the tissue polarity of the apex also account for the sequence of primordia, since these two factors inhibit initiation until the tissue has become sufficiently displaced from the summit for the growth rate to be low and the polarity weak. On the other side of the formative zone, the boundary of available tissue is set by the limits of the most recently estab-lished primordium—again as in the successive formation of buds in *Hydra* and in the manubrial budding in medusae. The new tissue, forming continually, is in a plastic state conducive to the formation of primordia, which are bounded by tissue of too-active growth on one side and by pre-empted territory on the other. If the production of new, uncommitted tissue were continuous, rather than continual, there might be some difficulty similar to that encountered in strobilation (p. 352). However, the apparent rhythm of apical growth, as evi-denced by plastochrons alternating with visible primordium initiation, could indicate that tissue in the proper state for primordium initiation is formed intermittently.

In further comparison with the bud rudiments of *Hydra* and *Botryllus,* the initial primordium can be regarded neither as a true growth center nor as a whole from the outset, but as a combination of both—i.e., as a substantial area (of from one to many cells in pteridophytes and angiosperms, respectively) which expands to a certain extent before the primordium unit is delimited. Whatever the nature of this area, if the condition in *Hydra* is a criterion, available space for primordium initiation involves more than an angle subtended to the apex; both curvature and organizational scale must be considered. In *Hydra* the bud rudiment, or primordium, must be a relatively flat area of tissue of certain minimum dimensions; and just as a hydra column may be too narrow to permit initiation of such a unit, so also may a plant cone. Thus, besides the area or mass of tissue produced during each successive surge of apical growth, both slope and girth become determining factors.

If apical growth is symmetrical relative to the axis, each growth surge must result in a ring of available tissue above the primordia already established and below the region of pulsative growth. When the cone is narrow and presumptive primordia are relatively large, 2 primordia may appear simultaneously, one on each side. When the cone is somewhat wider, 3 or 4 primordia may arise simultaneously within the same latitudinal band, so that whorls of leaves are successively produced.

The majority of plants, however, exhibit spiral phyllotaxis, which can be assumed to result from an eccentric apical center of growth revolving about the axis as in the growing zone of *Phycomyces* (p. 28) or oscillating from side to side as in the growth terminals of the hydroid *Sertularella* (p. 212). In either course of growth, new material is formed in either spiral or alternate succession; and any particular phyllotaxis results from this eccentricity, the growth rate, and the dimensions of the primordium relative to those of the cone.

There seems to be competition between adjoining primordia to some degree in some plants. Such competition is probably comparable to that in *Botryllus* (p. 390), in which primordium fields overlap, so that in cases when two adjacent primordia are initiated simultaneously, there may be a mutual restriction in the marginal area between them.

Development of Leaves and Buds

Leaf and bud primordia form from meristematic tissue derived from the shoot apex, although buds usually become evident after the formative tissue has become separated from the apical region as a so-called detached meristem. The earliest stages of the two primordial types have been studied for *Sequoia semper-virens* by Sterling (1945), who contrasts them as follows: The leaf primordium develops a large buttress and, when only a few cells high, begins to grow upward; whereas the emergent bud primordium develops laterally and assumes a hemispherical form when about five cells high, at which height the leaf primordium is already starting to develop a pointed apex. The dual question arises: What determines whether a primordium develops as a leaf or as a bud, and what is distincitive about the initial development of each?

Determination of Primordium Type

In ferns the leaf primordia always appear, near the base of the apical meristem, as circular or elliptical groups of typical prism-shaped meristematic cells which grow out as a low mound of tissue surmounting deeper, small-celled tissue that is also in a state of active division. On further development, an approximately central, superficial cell of the primordium becomes much enlarged and at the same time acquires a two-sided shape. Divisions of the apical cell produce tissue in the tangential plane. According to Wardlaw, once the two-sided apical cell has been established, the foliar nature of the primordium is also established and irreversibly so. In flowering plants the leaf primordium forms in much the same manner, except that, as in the shoot apex of flowering plants, there is a group of centrally located initial cells rather than a large single cell.

In both types the problem arises concerning the two-sidedness of the developing leaf. Whereas responsibility might be assigned to the regularly alternating divisions of the single leaf initial in ferns, it cannot be assigned to the group of initials in flowering plants. The difference between the three-sided apical cell of the fern shoot apex and the two-sided apical cell of the leaf primordium is therefore not so much a cause of the two-sidedness of the mature leaf as it is a consequence of the two-sided quality of primordial growth. Wardlaw (1957) assumes that there is a real, not merely apparent, increasing rate of growth from the shoot apex to the subapical region where the primordia form; that the growth relationship, and consequent rate of differentiation, between the adaxial and abaxial sides of the primordium is different from the outset; and that this differential is responsible for, or related to, leaf symmetry and orientation.

Some of Wardlaw's experiments (1955) throw some light on the "two-sided" problem. In the fern, when the territory of the next presumptive primordium is isolated from the shoot apex by a deep tangential cut, a bud and not a leaf forms from the primordium subsequently produced. Similarly, when the two youngest primordia are isolated before the initial two-sided cell appears in each, buds are formed instead of leaves. Wardlaw's suggestion that the provascular tissue underlying the prism-shaped cells of the meristem is important to these morphogenetic developments is in line with observations by Esau (1942) that the divisions initiating the procambium in the shoot apex of flax are localized beneath the areas where leaf primordia emerge and are evident before the primordium forms a bulge on the surface of the apical cone. Furthermore, according to Sussex (1951) isolation of the apical meristem in the potato, by means of vertical incision, and subsequent induction of a subsidiary meristem indicate that the dorsoventrality of the leaf results from influences proceeding from the apical meristem.

When a presumptive leaf primordium is experimentally isolated from the shoot apex, so that a bud forms instead of a leaf, the bud primordium induces a complete, symmetrical provascular ring. The polarizing influence of the primary apex is effectively eliminated by the intervening incision, so that the meristematic tissue on the abapical side of the incision can develop freely, i.e., as a bud or shoot. The inevitable conclusion seems to be that the tissue which becomes the provascular tissue of the growing leaf primordium is obtained from the prospective provascular ring of the stem apex as a whole *after* such tissue has become histogenetically determined, even though such histogenesis is nonobservable; otherwise, leaf gaps in the ring would not appear. On the assumption that this interpretation is valid, the leaf primordium incorporates, as its own prospective provascular tissue, cellular material that is already incipiently organized as part of a ring growing primarily in the direction of the apicobasal axis of the shoot. In other words, in a bud or a shoot apex, provascular tissue

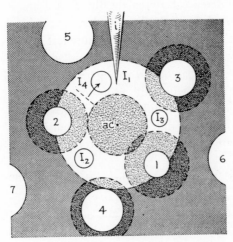

Figure 16-6. Operation on apex of *Dryopteris,* showing expected result when I_1 position of presumptive leaf primordium is incised, i.e., before its field has developed. I_2 and I_3 will occupy normal positions for these primorda, but I_4 will arise closer than normal to I_1 position. (Numbers 1–7 are established leaf primordia.) (After Wardlaw.)

differentiates as an axial ring located symmetrically about the axis; whereas in a leaf primordium, the prospective provascular tissue, being already incipiently differentiated, is a sector of a ring with an established tendency to grow and elongate parallel to the stem axis of which it is at first a part, rather than directly outward toward the primordium center as it would if it were entirely free to undergo complete self-differentiation.

Accordingly, a leaf primordium grows *as if* an arc of provascular tissue were extending into it from a region farther down the shoot and were impinging across its horizontal diameter, although the manner of growth is not by intrusion of tissue, but by incorporation of characteristic tissue shape and growth orientation. Thus the primordium grows, not as a cone growing outward symmetrically about its own polar axis, but rather as a disk folding across a growing

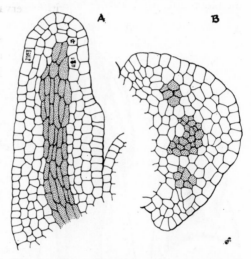

Figure 16-7. Developing leaf primordium in *Rubus*, showing development of provascular tissue (stippled) from basal arc. **A.** Longitudinal section. **B.** Transverse section through base. (After Engard.)

edge that approximately coincides with its horizontal diameter and forming a two-sided mass. In a bud the only difference is that the deeper, prospective procambial tissue organizes exclusively in relation to the primordium center, and the outgrowth as a whole is radially symmetrical about the center.

Development of the Leaf Embryo

According to Engard's monograph (1944) on the vegetative and floral morphogenesis in *Rubus* (raspberry), the leaflets and stipules in three-foliate leaves arise simultaneously. (The manner of origin in seven-foliate leaves is basically the same.) The leaflets arise as two papillae diverging from the central mass of the primordium; the stipules, as similar papillae at the two adaxial, basal corners of the somewhat adaxially flattened primordium. The leaflet primordia originate in the second layer of the primordium, although deeper cells are subsequently involved.

In brief, the main events in the development of the leaf embryo are:

1. Early in the ontogeny of the leaf, the leaflet undergoes by far its greatest relative growth.

2. Through rapid elongation, the stipules become a very prominent feature of the older embryonic leaf.

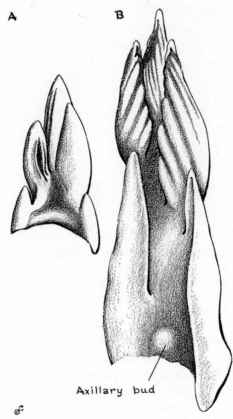

Axillary bud

Figure 16-8. Embryonic leaves of raspberry *Rubus idaeus*. **A.** Leaf base, stipule rudiments, and three-lobed leaf embryo. (×120.) **B.** Later stage, showing axillary bud within leaf base, extensive growth of stipule rudiments, and differentiation of three-lobed leaf embryo. (×27.) (After Engard.)

3. During most of the ontogenetic development, the leaflet continues to grow and attains its greatest length, while the petiolar region grows the least.

4. In later maturation stages, when all structures have appeared, the petiolar region undergoes rapid extension.

5. Still-later stages are characterized by lamina (surface-area) development and further petiolar elongation, which carries the stipules upward.

Thus the prominent features of the embryonic leaf are the vertically developing leaflets and stipules. An axillary bud primordium, close to the leaf base, is primarily a proliferation of leaf base tissue, rather than of axial stem tissue.

The older concept of the growth of leaf or leaflet lamina, according to Foster (1936), involved three distinct histogenetic processes: marginal growth (known in greater detail mainly because of the relative ease of study); surface growth and regulation of form; and differentiation and maturation of tissues. The production of the mesophyll and the vascular system is traceable to marginal and submarginal dividing cells, and the blade is essentially a lateral, winglike outgrowth of the primordial mass; the mesophyll and the vascular system basically constitute the midrib, and the blade is a product of a vertical tier of marginal initials and a vertical tier of submarginal initials. Engard, however, found that the concept as a whole does not explain lamina growth in *Rubus* and also that the initiation of the primary lateral veins is a function of the primordial midrib. Even apart from Engard's findings, it is difficult to conceive of marginal growth resulting in nearly vertical acropetal vein development. According to Engard, there is a general meristematic activity of the entire marginal layer of cells, which derive from the original development of the margin of the primordium.

Thus the growing leaf primordium can be visualized as a circular or ellipsoid mound of tunica tissue on the flank of a shoot apex—tissue which grows as if it were being pushed somewhat parallel to the shoot axis by deeper tissue arising from a somewhat lower level down the stem, as an arc with its convex side outermost. The general form of the leaf, in this concept, is determined primarily by the pattern and the growth orientation of the inner component, i.e., the procambial or presumptive provascular tissue. The outer tissue, essentially of tunica origin, although clearly responsive to growth stimulation from the inner tissue, in all probability has innate growth limitations; i.e., the outer tissue cannot produce an unlimited supply of intercostal tissue and may be outreached, so to speak, by elongative growth of vascular tissues. In other words, a basis for relative growth effects is present from the beginning.

In the development of the three-foliate, stipulate leaf of *Rubus,* as the example, the leaf pattern seems to be determined at the onset as a set of papillae corresponding topographically to the arc, or crescent, of presumptive provascular tissue contained in the base. A question fundamental to the problem of emergent pattern arises: What initiates or is responsible for the primordium subdivision which clearly foreshadows the essential structure of the leaf? In the three-foliate leaf the stipules form as a pair of papillae from the horns of the crescent; the lateral leaflets, as a pair of originally larger papillae from between the area of stipules and the center of the crescent; and the terminal leaflet, as an outgrowth of the central region. Such formation of a set of five circular areas symmetrically marked out of a crescent area suggests comparison with the delimitation of leaf primordia from the shoot apex. The similarity becomes even more striking when the order of growth and maturation is considered, for at a certain stage the stipules are larger than the embryonic lateral leaflets, which in turn are larger than the terminal leaflets. (Eventually, the stipules are the smallest units, and the terminal leaflet outgrows the lateral leaflets.)

Thus it may be concluded—tentatively perhaps—that the determinative process proceeds from the margins of the primordium toward its center. Support for this conclusion derives from the fact that a leaf primordium separated from its presumptive vascular tissue develops as a shoot. This long-standing concept of the leaf as a modified shoot is further supported by the original appearance of the axial bud as an integral part of the base of the embryonic leaf.

If the leaf primordium is regarded as initially a bud or a potential shoot—as Wardlaw's experiments indicate—which develops as a leaf owing to the polarized growth of the crescent of subjacent, presumptive vascular tissue, it is obvious—on territorial and mechanical grounds alone—that primordial tissue which is off-center from the concave side of the arc will not be involved in leaf formation. This concave area becomes the axillary bud; therefore, it seems more reasonable to regard it as an integral part of the primordium proper than as a "detached meristem" figuratively torn from its mooring as part of the interprimordial meristem of the stem apex. In effect, then, a leaf is a shoot with an extremely off-center apex (the axillary bud) and one large foliar appendage. However, the fallacy in such a definition is to limit pattern to a purely spatial quality, since pattern is either actually or potentially an open system with a temporal quality as well.

Variation in Leaf Shape and Size

If, as suggested, the embryonic leaf is produced by a combination of the directional growth tendencies of the superficial layers of the primordium and of the presumptive provascular tissue and if the two components should be differentially influenced by factors of the internal or external environment, then leaf pattern should vary according to circumstances. This it does, as observed in the sequential change in the shape of the leaf as it proceeds from the base to the tip of a shoot of, for example, young box elder (*Acer negundo*), whose leaf may change from an entire to a pinnate shape. Variation in leaf pattern has also been observed in the sequence of increasing leaf size. Ashby (1948 and 1950), analyzing *Ipomoea caerula* in terms of leaf area, cell number, and cell size, found that there are significant gradients in all these features from leaf to leaf up the stem. In the development of all leaves there is an initial phase of cell division and then a phase of cell enlargement virtually unaccompanied by further cell division. The division phase lasts longer in leaves from the upper part of the stem than in lower leaves, before cell enlargement commences.

In addition, whereas the water supply to the shoot influences leaf area and cell number (but not cell size), disturbance of the water supply by removal of all adult leaves from a stem has no effect on the morphology of new leaves developing at the apex. The greater the supply of water, with its contained solutes, to the shoot apex, the larger become the primordia and their subsequent growth. Thus the supply must be obtained from the stem axis and not from leaves already formed. In ferns, for example, the apical meristem grows at the normal rate even when completely separated from vascular stands and continuous only with the pith tissue on which it is seated, so that it must get water, salts, and nutrients through the pith. That the supply derives from the stem axis can account for the relatively large size of leaves produced, for instance, on adventitious shoots growing from the cut stumps of healthy trees whose root systems remain intact. In other words, either the rate or the duration of the initial growth phase in the undetermined primordium is influenced by the nutrient supply, just as in the initial bud disk of *Botryllus* (p. 392).

Ashby and Wangerman (1950) point out, however, that gradients of cell size and cell number in leaves from successive nodes, although affected by the environment, are basically a response to the position of the leaves on the shoot and are symptomatic of an aging process in the apical meristem. That "aging" is the proper term in this context is doubtful. Yet the apical meristem does undergo a progressive change as reflected in the size and shape of leaves at successive levels of the stem—changes to be analyzed in terms of the apex rather than solely of the interaction of the leaf primordial tissues.

As already indicated, a shoot grows not only longer but wider. As growth of the whole continues, the apical base and the stem grow wider, so that although the first leaves formed develop from small primordia on a small apex and are attached to a slender stem, those leaves are in what eventually becomes the thickest region of the stem. At the time of initiation, the small basal leaves therefore reflect the size of the apex meristem; but since they are associated with the stem region that has the longest period of growth in width, they are ultimately relatively as well as absolutely the smallest, for they themselves do not share that growth. Furthermore, since the size of the axillary bud in its dormant state is proportionate to the size of the leaf at whose base it forms, the size of axillary buds is in general inversely proportional to the width of the stem.

At the same time, as a shoot grows, its rate or extent of linear growth changes: a plot of the length of successive internodes along a stem shows that length increases from the base to the middle region of a stem and then decreases from the middle region to the apex. Once more the time of formation is significant, for the basal internodes are produced when the shoot apex is small; i.e., relative to apical dimensions, internodes are longer at the time of formation than at any later time. Accordingly, it can be concluded that the length of the stem internodes relative to the size of the apex decreases progressively as apical growth proceeds. Particularly noticeable in shoot growth from the twigs of trees, the internodes between successive leaves finally become so short that they are virtually nonexistent, so that leaves and their axillary buds end in a whorl.

Two questions of interest arise: What underlies the progressive diminution in length of internode and in the duration and, possibly, the intensity of cell proliferation in the leaf rudiment as the shoot grows? How does stem elongation and width affect the shape or size of axillary buds? (This second question is discussed on p. 433.)

To the first question, there are two possible answers, neither excluding the other and neither wholly satisfactory:

1. The sap supply, with its contained nutrients, wanes, either because the root system fails to maintain a crucial level or pressure as developing shoots increase their demands, or, in the case of trees especially, because the water table drops, as in late spring, with essentially the same result. (It is significant that in deciduous trees the ascent of sap occurs before leafage, and therefore transpiration cannot be evoked as a sap-lifting force. Root pressure may well be the sustaining force, according to Priestly (1929) and White (1938).

2. The other possibility is that each stem apex has an inherently limited growth potential, in common with virtually all other developing units, and that this potential is progressively exhausted, so that a typical growth curve or decrement is to be expected. Since the decrement concerns primarily linear extension of the axis rather than mass accretion of primordia, a waning auxin supply might be the immediate causal factor. Each surge, or unit quantity, of growth at the apex, however—a quantity corresponding to one plastochron, during which one primordium forms—is responsible for, or expressed by, the amount of internodal stem tissue associated with the leaf or leaves formed. Is internodal length, therefore, a function of the amount of auxin or other substance available during that unit of biological time under consideration? Whaley (1959), studying the changes in the volume of meristem, in cell size, and in nuclear size in the apex of *Lycopersicon* found that as growth of the whole proceeds, the apical meristem increases to a maximum volume late in the grand period of growth and then decreases sharply to a constant size. The cell and nuclear sizes also decrease, and when minimal, ontogeny begins.

Development of Axillary Buds and Stipules

The problem of whether stipules and axillary buds are parts of the leaf or of the stem has been subject to some debate. Engard (1944), for example, regards stipules as integral parts of the leaf base which develop essentially as do leaflets. The problem, however, is directed to their relation to the leaf primordium, before the leaf is fully determined.

The considerable variability in shape of leaf primordia, from species to species, seems significant to this problem. Stipules originate at the very beginning of leaf development and are characteristic especially of primordia whose bases extend close to 180 degrees around the subapical base of the shoot meristem. When the stipular area of such a primordium is removed, the leaf develops without the stipule having formed, according to Snow and Snow (1933). Stipules are not known to form from primordia that are approximately circular. Somewhat similarly, an axillary bud forms in the adapical central part of a primordium whether the primordium is elliptic or approximately circular, but if markedly elliptic, such as those that produce stipules, three axillary buds may be formed: a relatively large median one, and a pair of smaller lateral buds.

This variable pattern in shape may be at least provisionally explained by the phenomena and conclusions of determination in the bud disk of *Botryllus* (p. 390), in which the area actually organized as a presumptive individual is circular. When the length of the *Botryllus* primordium is about twice its width, two circular areas organize within it. When the primordium is approximately pear-shaped, one large and one small circular area organize within it.

In the leaf primordium the same general organizing phenomenon seems to operate within the tunica layers, but it is modified by the directional growth and form of the subjacent presumptive provascular tissue. Diagrammatically, in an approximately circular primordium, a simple arc-shaped leaf rudiment and a single axillary bud rudiment are delimited within the territory. In a primordium whose base extends close to half the circumference of the shoot apex—i.e., the primordium is relatively long in the horizontal axis and tapers laterally—the two end regions are not incorporated in the primary leaf organization. Instead, these end regions somewhat independently organize as developing units similar to the leaf rudiment, but on a much smaller scale, comparable to the small residual bud unit of a pear-shaped *Botryllus* primordium. To what degree stipular rudiments are parts of the leaf primordium and to what degree they are independent units are mainly a matter of definition. Similarly, in such a leaf primordium, the residual region available for axillary bud formation is much longer in the horizontal than in the vertical axis; consequently, it subdivides symmetrically into three units all approximately circular: a relatively large central area, and one smaller area at each side. However, it must not be supposed that all stipule-producing primordia have typical axillary buds or that triple buds can appear only in conjunction with stipule formation. The lateral buds are commonly so small as to be abortive, just as in extremely small residual bud rudiments in *Botryllus*.

Chapter 17

The Floral Apex

THE TRANSFORMATION of a shoot apex into a flowering apex has long been the subject of intensive experimentation, although mainly to determine the nature of the controlling agencies in the internal and external environment. As valuable as is knowledge of the nutrients, "flowering hormones," temperature, and light which participate in this transformation, such information may be no more explanatory than that of the thyroid hormone to the understanding of amphibian metamorphosis. As Heslop-Harrison has said: "It is no doubt true that the details of floral organogenesis are accomplished by processes the nature of which still entirely elude us," [1] though he does propose the hypothesis that floral organogenesis is regulated but not determined by auxin levels at the apex. Basically, metamorphic change is the outcome of changes in growth of tissues.

The comparatively few investigations of ontogenetic events, however, have brought to light some few significant correlations. In *Rubus,* according to Engard (1944), the transformation of the apical meristem from the vegetative to the floral condition is marked by a virtual suppression of plastochrons and by meristematic activity of the corpus, the latter event leading to a reshaping of the apical dome. In *Amygdalus,* Brooks (1940) describes the apex as becoming larger and more domed as the result of cell division within the corpus while the tunica, which is 4 cells deep in the vegetative shoot, becomes reduced to a 2- and finally 1-cell layer in the flower rudiment. Gregoire's theory (1938) that the conversion process is a new construction of floral tissue on the vegetative apex has not been widely accepted, and the general consensus has been summarized by Philipson on the basis of his investigations on, e.g., *Bellis* and *Succisia:*

[1] Heslop-Harrison, *Biol. Rev.,* **32**:81 (1957).

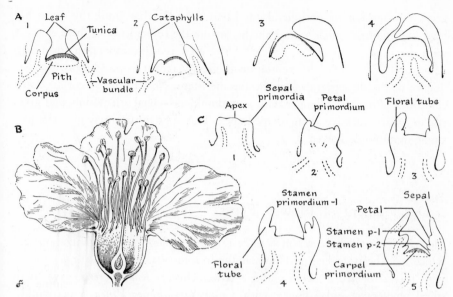

Figure 17-1. Initiation and development of flower in *Amygdalus*. **A1–4.** Transformation from vegetative to floral apex, showing change in shape of apex and progressive thinning of tunica layer. **B.** Fully developed flower. **C1–5.** Development of flower from floral apex. (After Brooks.)

. . . the peripheral meristem of the inflorescence apex can be seen to be derived from the peripheral meristem of the vegetative apex by its extension down the flanks of the apex. As the vertical growth of the inflorescence ceases the central zone of the corpus disappears, being replaced by the peripheral meristem, which then extends continuously over the apex.[2]

Thus the change from vegetative to floral apex is a progressive transformation, and the inflorescence rudiment is not constructed on the vegetative apex as an entirely new organization, as Gregoire proposed. The tunica of the vegetative apex persists and extends as the tunica of the floral apex. The transformation, whether slow or sudden, represents a progressive developmental change in apical growth and constitution. According to Boke (1947), *Vinca rosea,* from its earliest stages to flowering, forms three types of apex, all with a two-layered tunica and an active corpus: a juvenile apex, which forms only leaf primordia; an adult apex, which bears both leaf and flower primordia; and a floral apex, which forms only flower primordia.

Within the incipient floral apex, two changes seem to be paramount: (1) the cessation of growth of the central initial zone, or apical mother cells, with a corresponding disappearance of the underlying file meristem; and (2) the pro-

[2] Philipson, *Biol. Rev.,* **24**:41 (1949).

gressive extension of peripheral meristem toward what until this time has been the tip of the shoot apex. Whether or not the coincidence of these changes with the decrease in auxin content of the apex is significant, pith formation ceases and provascular differentiation is greatly reduced—possibly as a consequence of the disappearance of the file meristem. Growth of the corpus tissue continues, although no longer with a predominantly axial orientation, and gives rise to a fairly massive, dome-shaped apex. This apex in turn may give rise to a single or multiple flower.

Development of Flower Types

Flower types which represent a wide range of inflorescence organization and which have been studied developmentally in some detail are *Frasera,* a gentian, with leaves formed as whorls and a flower cluster; *Rubus* (raspberry), with spiral phyllotaxis and a flower with a carpellary dome; *Vinca,* a periwinkle, with a succession of leaves and flowers produced by the adult apex; and *Bellis,* a daisy, with a composite flower head.

Frasera

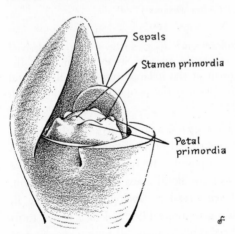

Of the various types of flower ontogeny, the type described by McCoy (1940) for *Frasera carolinensis* is the simplest. Leaves are formed as whorls separated by long internodes; i.e., there is no indication of spiral growth. The vegetative apex apparently produces a succession of meristematic rings, each of which forms several primordia simultaneously, as in plants that produce opposite leaves. A flower cluster forms from the upper part of the flowering shaft, and an individual flower is 2-3 cm broad, sympetalous, with a deeply parted corolla, and 4 epipetalous stamens.

Figure 17-2. Young flower in *Frasera carolinensis* with 2 sepals removed. (After McCoy.)

The flower primordium first appears as a conical axillary outgrowth, which upon more rapid growth, especially in width, of the distal portion differentiates into a receptacle and pedicel. The receptacle then differentiates successively into the following structures:

Sepals. A calyx primordium develops as an undulating collar, or zone, of meristematic tissue encircling the receptacle. First one pair and then a subsequently formed pair of opposite protuberances diverge from the edge of the collar. The edges of the first pair overlap those of the second, which formed slightly higher on the floral axis, presumably during a plastochron that closely followed the first. These protuberances become the 4 sepals.

Petals. After the calyx arises, the outer margin of the residual, small, rounded mass of apical meristem becomes raised to form a second collarlike ring, which is the beginning of the common petal-stamen primordium. The rim of the ring soon becomes noticeably elevated at 4 loci, from which 4 lobes arise simultaneously on the lateral face of the ring. These 4 petals alternate in position with the 4 previously formed sepals.

Stamens. Stamen primordia arise from the same ring of meristematic tissue and at about the same time as the petals, but at four distinctly higher apical loci, which alternate with the petal primordia on the upper and inner margin of the ring. At first each stamen is a conical papilla, but it soon differentiates into a basal region, which grows in length, and a distal region, which expands both vertically and laterally as the anther. The common petal-stamen zone finally develops en masse to form a short tube upon which the stamens are inserted a short distance from the receptacle.

Carpels. After the petal and stamen primordia have become well established, two carpels form at the tip of the apex or floral axis as transverse ridges whose ends are not united at first. The small cuplike depression between the ridges is the end of the floral axis and forms the floor of the ovarial cavity. The two meristematic cushions, during their vertical and lateral growth, become somewhat pointed, gradually approach each other, and finally unite at their edges to form a syncarpous gynoecium.

Sepals, petals, stamens, and carpels all arise on the side of the apex by periclinal cell divisions within the inner tunica and by supplementary irregular divisions within the corpus. Although the floral apex is reduced after each successive differentiation, a certain amount of lateral and vertical growth does occur after each—a phenomenon which suggests that there perists a succession of plastochrons. There are two other very significant aspects of such differentiation: (1) The succession of differentiations is from the periphery toward the center of the meristematic area. (2) The differentiations consist of vertical growths from a progressively decreasing slope; furthermore, adjacent organs or tissues, formed at the same time and initial growth rate, tend to fuse.

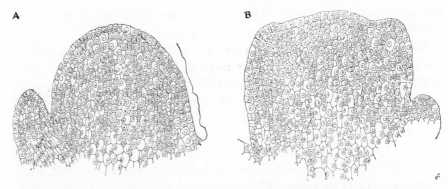

Figure 17-3. Transition from vegetative to reproductive apex in *Rubus*. **A.** Vegetative apex with last foliage leaf primordium at left. **B.** Reproductive-apex origin of sepal primordia with virtual elimination of plastochron. Acropetally developing provascular trace sepal) discernible at left. Corpus cells are beginning meristematic activity which will initiate raising of carpellary dome. (After Engard.)

Rubus

In *Rubus,* studied mainly by Engard (1944), the first obvious step in the transition to the floral apex is the formation by the apex of a whorl of primordia at approximately the same time, i.e., a series of brief plastochrons closely follow one another. Originating in the second tunica layer, as do leaf primordia, these primordia constitute the sepal primordia and are continuous where they join the axis; i.e., they are not separated by internodes. The next observable change is that the central corpus cells become meristematic and raised, so that the apex enlarges into a dome. Provascular traces differentiate in relation to the sepal primordia, much as in leaf primordia except that each trace divides: one branch within the sepal, and the other within the dome.

Then, in somewhat similar manner to sepal formation, the petal primordia arise, one between each pair of sepals. There is an initial asymmetry, which indicates that the petal primordia are initiated one after the other, although in close succession; but when all the petals are full grown, symmetry is attained. Each petal primordium forms a single provascular trace, which connects with a sepal trace.

Stamen primordia arise successively from the territory between the petal ring and the central dome, each with a single trace. Each of the first 10 formed is median to the overlapping edges of a petal and a sepal; the later stamens, which are numerous, develop centripetally between the petals and the dome. Stamens and carpels both originate in the second tunica layer and, like the stamens, each carpel has only one provascular trace. The carpels originate close to one another and successively from the base of the dome toward the apex in a spiral course.

Thus, as in *Frasera,* the delimiting of primordia is from the peripheral to the central area of the apex; but unlike *Frasera,* development is spiral—a phenomenon implying that growth of the carpellary dome is spiral also. There is also an implication that for any given species the primordial and final size of stamen and carpel is predetermined and that the number of stamens and carpels produced is a function of the available area of suitable meristematic tissue—an example of the applicability of the Snows' available-space concept (p. 421). It is also significant that although carpel primordia form on the sides as well as on the upper surface of the central dome, their axis of growth is at right angles to the meristematic surface and is therefore equivalent to vertical growth from the upper central region of the dome.

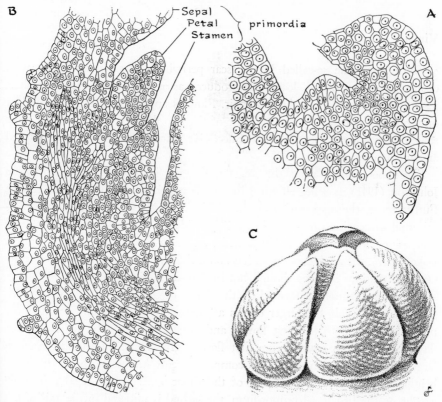

Figure 17-4. Development of flower in *Rubus.* **A.** Origin of petal primordium on hypanthium; young sepal shown on right. **B.** Relationship of sepal, petal, and stamen primordia and provascular traces, in left portion of hypanthium. **C.** Floral bud, showing asymmetrical development of sepals and view of carpellary dome. (After Engard.)

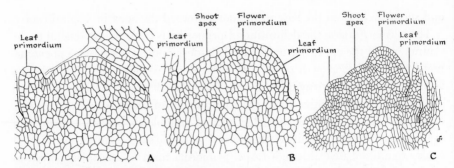

Figure 17-5. Development of flower primordia in *Vinca*. **A.** Adult shoot apex. **B.** Apex, showing initiation of leaves and first flower of a pair. **C.** Meristematic complex evolved from **B.** (After Boke.)

Vinca

Vinca rosea, the so-called Madagascar periwinkle, has oblong leaves with decussate phyllotaxis, i.e., leaves are produced as opposite pairs, each pair alternating at right angles to the next pair. Flowers appear in pairs in the axil of only one of the paired leaves, and one of the two flowers is never in the same stage of development as the other. Successive pairs of flowers appear in spiral arrangement on the axis.

The shoot apex during the vegetative stage is described by Cross and Johnston (1941) as being elliptical in outline and having a two-layered tunica. Furthermore, the apex varies in size during the course of a plastochron: its minimal dimensions are 70 μ long, 45 μ wide, and 3-5 μ high; its maximal dimensions are 160 μ long, 110 μ wide, and nearly 50 μ high. According to Boke (1947), the adult shoot apex has about the same dimensions and about the same fluctuations in size—though not so flat—as the juvenile apex.

Paired leaf primordia, formed on the flanks of the adult apex, are very different in size. Only the larger one is a "common" primordium; i.e., it is the primordium of both the first flower and the leaf which subtends from that flower. The primordium of the first flower, growing rapidly to become the largest part of the meristematic complex, seems almost terminal since its height becomes greater than that of the shoot apex. Before sepal primordia appear on the first floral primordium, the second is initiated in the axil of the subtending leaf primordium and at the base of the first floral primordium. Once initiated, all floral primordia become hemispherical first, then widen at the apex, and finally, just before sepal initiation, become rather flat at the summit.

Bellis

Studied primarily by Philipson (1946), the transformation in the English daisy, *Bellis perennis,* is somewhat similar to that in *Vinca,* except that it is more sudden. The primordium of the uppermost leaf is exactly the same as the primordia formed earlier, but before this primordium has developed beyond the very earliest stages, the apex of the axis transforms.

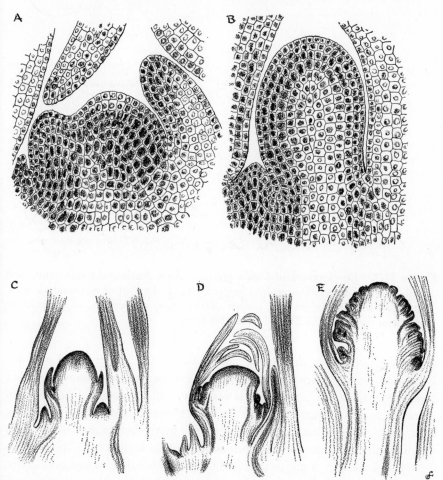

Figure 17-6. Floral development in *Bellis perennis.* **A.** Apex doming toward inflorescent start; leaf primordium shown at left. **B.** Inflorescence rudiment, with meristem extending down flanks. **C.** Stage when involucral bracts are forming. **D.** Stage when flower primordia are beginning to form. **E.** All floral primordia have been formed, and floral organs are differentiating. (After Philipson.)

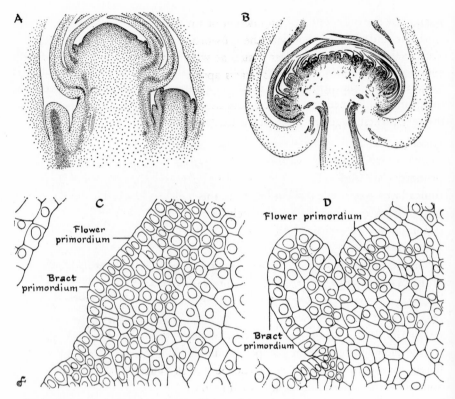

Figure 17-7. Development in *Dahlia gracilis.* **A–B.** Sections through young and well-developed terminal capitulum, respectively. **C–D.** Part of meristematic mantle of **A**, showing origin and further development, respectively, of receptacular bract and flower primordium. (After Philipson.)

The vegetative apex during the phase of maximal area is broad and flat. Even after the appearance of leaf primordia, at the time of minimal area but greatest curvature, the apical arch is very low, and a young leaf primordium immediately grows faster than does the apex. However, the earliest sign of the onset of flowering is immediately recognizable by the higher arch of the apex, which at this time towers over the primordium of the highest leaf. This increase in height of the apex is due to increase in depth, rather than in number of cells, so that extension of the apex is the result of distortion, or transformation of internal shape.

Once such extension has occurred, however, the shallow dome grows by cell proliferation into a high dome, the capitulum. Within this dome, the direction of growth is as much lateral as vertical, so that the inflorescence rudiment becomes covered by a hemispherical meristem. This peripheral meristem is concerned not only with the production of involucral bracts, which are similar to sepals, and of florets, but with the growth in breadth of the central mass and with the growth of provascular tissues as well. During the enlargement of the inflorescence primordium, the 13 bracts appear successively in a compact zone within the second tunica layer midway up the inflorescence rudiment to form a double circle which divides the peduncle from the receptacle.

As the hemispherical receptacle enlarges, the cells of the central tissue increase in size and vacuolate, but the peripheral layers remain actively meristematic, dividing both parallel, and at right angles, to the surface. Flower primordia are initiated from this peripheral region, primarily from the second tunica layer, but with contributions from the subjacent layer. The hemispherical primordium, once formed, immediately flattens until it is stud-shaped; then its circumference grows upward to form a cup. The primordia appear acropetally in a spiral whorl.

The outermost flowers, however, consisting of about two whorls, are soon outstripped in stature by the younger primordia; and it is the outer, so-called backward flowers that form the ray florets. According to Philipson, the stuntedness of these flowers is partly due to the lack of stamen rudiments and partly to the capitulum, whose active growth perhaps robs these florets of space for full development. Philipson also points out that the period of anthesis of these florets may even precede that of the outer-disk florets, so that the order of anthesis is the same as the order of initiation.

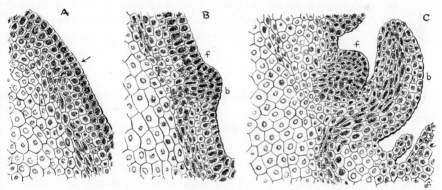

Figure 17-8. Development of floral bracts and flowers in *Dipsacus fullorum*. **A.** Initiation of bract (indicated by arrow) above which peripheral meristem is uniform; below which it is differentiated into provascular meristem and peripheral meristem. **B.** Single tunica layer encloses highly meristematic bract primordium **b** to which provascular strand is connected; peripheral meristem above bract is patch of dense cells which constitute axil of bract **f**. **C.** Same patch in axil is forming flower primordium. (After Philipson.)

Polymorphic Development

The transformation from the vegetative to the floral apex has both positive and negative correlations. In consequence of the diminishing activity of the corpus initial cells, less and less file, or rib, meristem is produced in a given time. Auxin content also decreases, although whether the relationship between this decrement and apical elongation is causal is not at all clear. Whatever the basic cause of the generally diminishing activity, its effects raise questions, such as: How are the various floral units produced? What role does procambial tissue play in the initiation and development of floral units?

In the initiation of leaf primordia, as discussed earlier, the corpus material incorporated is already incipiently provascular, possesses an ingrained growth polarity, which is derived from the primary polar orientation of the shoot apex, and exhibits a spectacular capacity for growth. Floral primordia seem to differ from leaf primordia in two ways: (1) the former grow at varying, but different angles relative to the same axis or to its functional equivalent; and (2) the corpus material incorporated is either nonprocambial or is procambial but much less determined than at the primordial level of the vegetative axis. To what extent the different floral appendages can be legitimately interpreted as foliar structures and to what extent the floral apex as a whole can be regarded as a modified shoot apex have been the subject of long-standing debate—much of which is semantic: Can a petal or a stamen be properly considered a modified leaf? The answer in great part depends on the definition of a leaf. The point of view of the ensuing discussion is that such semantic questions may obscure the search for a general interpretation of polymorphic development in plants.

Certain postulates must be made before any interpretation of events may be attempted:

1. A primordium of any kind, vegetative or floral, is an approximately circular area, or disk, of tissue, which forms in the first or second subdermal layer of the apical meristem and exhibits a relatively high rate of growth.

2. A primordium may be large or small, both relative to the dimensions of the apex and absolute to the degree that the scale of the apex is itself a variable.

3. A variable amount of subjacent corpus tissue is incorporated as part of the primordium, as though the high growth rate of the primordium initial gradually influences the subsequent growth activity of the deeper tissue.

4. The corpus material involved in primordium growth may or may not be already determined as procambial or provascular tissue. The greater the degree of determination, however, the greater the tendency of the primordium to grow parallel to the stem axis, whatever its placement on the apex.

5. A primordium is not initially determined as a developing structure of any particular foliar or floral type, but is an uncommitted "morphogenetic field" with dynamic and unified qualities.

6. A portion of a primordial field, if not otherwise involved in morphogenesis, may remain a presumptive or prospective whole; i.e., a part of the original whole may exhibit the properties of the whole under certain circumstances of isolation, which is a general property of fields.

7. "Detached meristems," which give rise to axillary buds, are essentially adaxial districts of primordia originally formed at the apex and are not to be regarded as meristematic fragments from interprimordial sites—fragments accidentally displaced by the growth displacement of primordia.

8. A flowering plant cannot be defined in terms of any one stage or structure; rather, it is the expression of the continuing growth activity of a self-multiplicative primordium. Moreover, final structures cannot properly be interpreted in terms of one another but only as the products of primordia developing under variable circumstances.

On the basis of these postulates, it is possible to present a consistent account of the polymorphism in vegetative-floral development.

Vegetative Phase

1. During the vegetative phase of apical growth, the primordia formed incorporate a relatively large quantity of corpus tissue from a level where provascular determination is already taking place in a ring of meristematic tissue. Such primordia grow as though they were thrust by a thick arc of provascular tissue in the direction in which this tissue was originally differentiating; i.e., tissue of a predetermined shape and growth polarity is incorporated within the primordium. The result is a leaf embryo with large vascular bundles well developed mainly on its abaxial side and with a relatively small, residual, area, which remains as the presumptive axillary bud, on its adaxial side.

2. As the shoot apex grows, primordia increase in size proportionately— for a time without any other significant change—so that leaves and their respective axillary buds become progressively larger.

3. With continuing apical growth, apical circumference increases, but vascular differentiation and elongation decrease somewhat. Primordia then incorporate proportionately less provascular meristem, partly because of the slackening rate of provascular determination and differentiation within the corpus tissue and partly because of a concomitant decrease in the primordium's vigor of growth, so that relatively less provascular tissue is influenced by the primordium's growth rate. Moreover, the thinner arc, operating only in the

more abaxial territory of the primordium, may fail to affect the primordium center and thus work with tissue of a lower growth potential. The effect is to reduce the thickness of the arc of provascular tissue differentiating in the primordium, so that proportionately more adaxial territory remains to form the presumptive axillary bud. If both factors are involved, development is affected differentially: the leaf component grows to a less extent, but the presumptive bud remains apparently unaffected at this time.

4. The nature of an axillary bud depends, in part at least, on the circumstances of its formation. Primordia which are formed during the earlier phases of shoot growth and which, separated by long internodes, become associated with the lower and middle stem regions, grow more directly in line with the stem axis than do those forming later. Either because of the direction of growth in the primordium or because of the elongation of the leaf base in conformity with stem and leaf elongation, the axillary buds of this region are high relative to their base width; i.e., their shape is typical of vegetative shoots.

5. As an apex matures, the primordia produced seem to incorporate so little provascular arc tissue that leaves become small and simple—some become only bracts—while the bud territory becomes compensatorily large. In such primordia and as a direct result of the reduced provascular growth activity, buds are low domes whose base width is greater than their height. Conversely, when primordia grow as leafllike structures that even converge toward the central axis, so little territory remains for bud formation, and that which does remain is so elevated on the leaf base, that buds are virtually suppressed.

Floral Phase

Floral apices, which may be single flowers or composite inflorescences of various types, may arise indirectly by conversion of a vegetative shoot or directly from subterminal buds.

Formation of a Single Flower

6. The circumferential territory—the presumptive perianth—of the floral apex usually gives rise to two whorls of primordia which commonly differentiate as sepals and petals, respectively. However, in some floral apices—e.g., that of the onion *Allium cepa*—the common distinctions of color and form are not developed. Both whorls of primordia grow in much the same manner as do those of the vegetative shoot that become leaves. This similarity does not imply that these primordia are modified leaves any more than that leaves are modified sepals or petals, merely that the pattern of growth is similar. In both leaf and floral primordia, however, the provascular tissue incorporated is approximately

arc-shaped at the base. After the foliar form is produced from such incorporated material, the area remaining at the adaxial side of the foliar base is the presumptive axillary bud territory. The outer set of primordia—the sepals—however, develop distinct provascular traces; whereas in the inner set—the petals—there is no subdivision of the trace, so that sepals are in general more leaflike than are petals. (For a discussion of the special differentiations of sepals and petals, see p. 437).

7. Axillary buds from sepals and petals develop into stamens, at least in the simpler flower types such as onion and *Frasera,* although such origination does not necessarily preclude the origin of similar axillary buds of essentially similar size and final location from the meristematic tissue immediately adaxial to the petal-stamen primordia and therefore independent of a foliar appendage.

8. Stamens develop from primordia which, in their early growth stages, grow upward in the general direction of the primordial axis of symmetry, i.e., at approximately right angles to the surface of the initiating tissue, and not obviously obliquely as do all foliar appendages. In other words, a stamen pri-

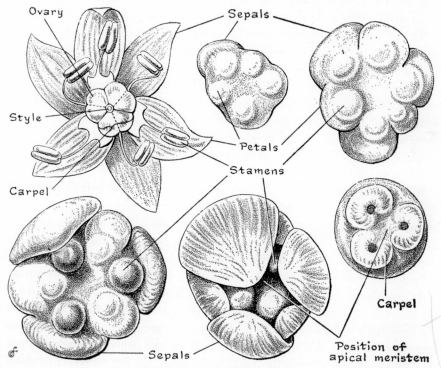

Figure 17-9. Development of flower of onion *Allium.* Diagram of opened-up flower. Each sepal and petal has stamen growing from its base, axils of petals being relatively wide. At center are 3 fused carpels. (After Jones and Emsweller.)

mordium grows in a manner comparable to that of a whole shoot or floral apex: symmetrically about an axis, and with a single provascular trace centrally located. Although the degree to which foliar form may develop is somewhat variable within an abortive range, the absence of completed foliar form is readily attributable to the precocious histological differentiation of the stamen. The specific differentiation of stamen structure may be due partly to the small initial primordium size and partly to factors correlated with the position of the stamen on the developing apex.

9. Carpels form from primordia on the more central apical meristem. As do stamen primordia, each carpellary primordium develops a single provascular trace and grows outward, not obliquely, from its base. Carpel primordia differ from stamen primordia not only in relative position on the floral apex but also in initial size. Since the scale of organization is probably innate for any given species, size may also be determined. Small primordia can develop into stamens, but not into carpels. Among other factors, availability of space on the floral apex permits the formation of primordia which are large enough to develop into carpels. Even carpel primordia can originate from active meristem, i.e., from tissue growing in a progressive, though relatively subdued manner.

The various developmental units described above can therefore be regarded as variations in the development of a morphogenetic field, namely, the undetermined primordium. These variations are the result of variation in the initial size of the primordium, in the relationship of the primordium's axis of symmetry to the axis of provascular differentiation, and in the chemical or nutritive condition of the tissue from which the primordium is formed. However, all these variations apply to the developmental capacities of the primordium as a single unit, even though one of the products of its development may be another primordium.

Wherever spiral phyllotaxis is evident in leaf production, it is also found in floral development. Thus even in as simple a flower as the onion, the double whorl of perianth structures exhibits a spiral sequence in initiation although not in final size. The same spiral sequence is evident in the growth of the carpellary apex, though hardly discernible when only 2 or 3 carpels are produced.

In *Rubus,* however, and other plants—e.g., the cocklebur—with a similar carpellary dome, carpellary primordia are initiated in a genetic spiral from low on the sides of the dome to the apex. In other words, the floral apex retains the same basic pattern of growth as that of the vegetative apex, but the growth rates and growth gradients are sharply reduced: the general growth rate progressively decreases to zero, the apical center being the last area to cease growing. During this combination of dome expansion and growth-rate decline, carpellary primordia successively form along the spiral course of growth. The determination of size and number is based on the same underlying features of meristematic growth as in the vegetative shoot; i.e., there is a band of meristematic tissue—of a certain width—immediately above tissue which is already determined or differentiating and below more apical tissue which is still growing at too high a rate for participation in primordium formation. Within this band and in line with the genetic growth spiral, circular areas are successively delimited as carpellary primordia; the diameter of the primordia is determined by the width of the band of physiologically available tissue, and the number of primordia is determined by the final over-all length of the spiral.

This interpretation, which implies that the size and the sequence of carpellary primordia are determined by the nature of growth in the carpellary dome as a whole, seems to be borne out by the presence of a single spiral provascular trace which develops acropetally within the floral apex.

Formation of a Composite Inflorescence

Between the inflorescence of the daisy and the single flower of the raspberry, there are both a general similarity and particular differences. In general, the transformation from the vegetative to the floral apex is much the same both in shape and in meristematic organization. Primordia form in spiral sequence in much the same way and are comparable in size. Whorls of bracts, more numerous and somewhat smaller, simulate the whorl of sepals: each bract primordium develops in a leafllike manner and has vascular traces of a foliar pattern, though the bract seems more like an abortive leaf than does the sepal.

The main difference between a flower and an inflorescence concerns the floral apex. The primordia which form on the inflorescent dome, or spike, give rise to more or less complete flowers, i.e., florets, in place of carpels alone. In fact, the floret can be regarded as a single flower which is constructed on an extremely reduced scale and which exhibits at that scale most, if not all, of the developmental phenomena of flowers of far greater magnitude. However, such an interpretation must be balanced by a recognition that reduction in scale of organization imposes modifications, since the size of the constituent cells is virtually unaffected and histological limitations become operative. Thus both scale of organization and histogenetic induction are involved.

Initiation of a Flower

The general problem of the meristem exists at two different levels of inquiry: the correlation between the dynamic, histological constitution of the meristem, together with its emergent creations, and the nature of external and internal environmental factors affecting that constitution. Any factor that affects the size, shape, and growth pattern of the meristem indirectly controls the nature of the developmental units produced by it. Thus although the study of hormonal effects in both plants and animals may throw light on processes of growth and development, such a study is at least one step removed from the actual developmental event, so that influential *effects* must not be confused with either fundamental *causes* or *courses.*

The shape of an apical meristem reflects, generally, the nature and conditions of its growth. An overabundant supply of sap, with its contained water and nutrients, commonly produces and maintains apical meristems in a vegetative form. Sap pressure may be a factor, as evidenced by the fact that certain meristematic apices receive a more abundant or rich nutrient supply than do others. Whether or not pressure is the crucial cause, it is, according to White (1938), a force that is generally underrated.

Besides pressure, direction may be a factor, since there is evidence that the effective supply of sap, by whatever means it is transported, is greater when passing in a direct line than after being deflected; for example, terminal shoots grow more vigorously than lateral shoots, and the growth-promoting supply seems to decrease with every successive branching structure, which represents a departure from the main axis. An entirely comparable situation is that in the branching systems of many hydroids, in which the streaming hydrocoelic fluid represents the sap, and the terminal or subterminal growth zones the apical meristems (p. 181).

Some indication of the importance of the quantity of nutrient supply as a factor in floral initiation is readily seen in the succession of axillary buds along a stem, as during the seasonal growth of a twig in many trees. In the box elder (*Acer negundo*), for example, each node bears three axillary buds: one relatively large central bud, and two smaller lateral buds. Each set of three buds increases in mass progressively along the stem, in proportion to the size of the leaf with which the set was originally associated; the scale of both the leaf and its buds was determined by the width or girth of the stem at the time of their initiation, i.e., at the time that part of the stem was the base of the apical meristem (p. 433). Leaf and bud sizes are roughly proportional therefore to the distance along the stem from the beginning of the stem's seasonal growth to its tip. Since the length of internodes along a stem also varies, increasing to a maximum in the middle region and finally decreasing almost to zero at the tip, it may influence the shape of a bud. Thus a leaf's position along the stem may have several effects on its buds: elongation

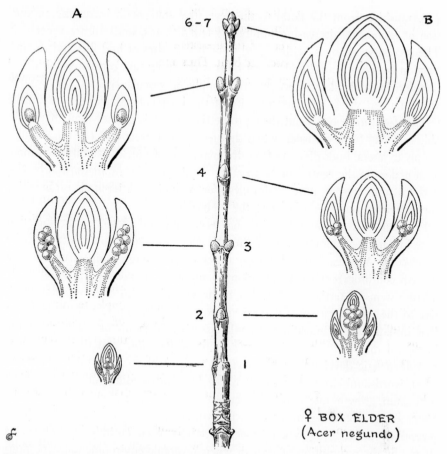

Figure 17-10. Foliar and floral condition of succession of triple buds along terminal shoot in female box elder *Acer negundo*. **A.** Progression from leaves to flowers to leaves in lateral buds. **B.** Progression from flower rudiments to leaves in central buds.

size, supply of nutrient, and relative position and size of lateral to central buds in each set.

In a typical terminal shoot (Fig. 17.10), the smallest, proximal bud set consists centrally of a bud with embryonic leaves surrounding three female units and laterally of very much smaller buds with minute leaves. The buds in the second, more-distal set are similar but larger. In the third, the smaller lateral buds are like those of the central bud of the first-two sets, i.e., with female units, but the central bud is vegetative. In the remaining more-distal sets, the central buds are vegetative, but the lateral buds from set to set show a progressive transition from the floral to the vegetative state, with increasing production of leaves and decreasing number of female units, until all are vegetative. The successive changes, with regard to both the series of bud sets and the distinction between central and lateral buds within a set, are correlated with, and therefore possibly determined by, the supply of vascular vessels. The supply varies in two ways: (1) the quantity of the incipient vascular ring available to later-forming leaves (and their prospective buds) is greater than that available to leaves formed earlier, when the stem is narrower; and (2) the vascular supply to lateral buds diverges from the main axis of nutrient supply even more than it does to the central buds. In other words, there is an anatomical basis—perhaps a matter of pure hydraulics—for assuming differences in nutritive supply within bud sets and along the stem axis.

Pertinent to this context of internal environmental factors is the fact that once initiated, a hydranth primordium develops to completion under virtually all circumstances except high temperatures (even at high temperatures simple growth continues); whereas initiation is commonly suppressed in the relatively narrow stems of tertiary and higher branches and is also comparatively sensitive to the concentration of certain amino acids. The conclusion reached was that initiation of primordium growth essentially depends on establishing the biological maximal rate as in the bud disk of *Botryllus,* and that subsequent morphogenetic development is correlated with a submaximal growth rate. In other words, initiation and maintenance of a primordium at the biological maximum growth rate demand a critical nutritive level, which may not be sustained. It is also significant that, in hydroids at least, a primordium site is potentially active even when it seems to have been entirely suppressed, as shown by the resumption of growth when favorable environmental conditions are restored.

By analogy with animal tissue, it is reasonable to suppose that growth initials in plant meristems exhibit a similar sensitivity to nutrient supply. The growth zone of initials in flowering plants may be the zone of biological maximum growth rate, which may persist only as long as the critical level of critical nutritive substances is maintained, although even when the level is maintained, the growth system as a whole may be suppressed, as in arrested bud development when all growth differentials are in abeyance.

On the assumption that this relationship between maximal growth zone and nutritive level is valid, the morphogenetic transformation from a vegetative to a floral apex may be interpreted very simply. That the vegetative character of a shoot apex depends on the maintenance of the zone of growth of the group of initials is an experimentally established fact; and, as discussed above, this zone dominates the growth organization of the apex as a whole. Any factor that reduces the growth activity of the initial zone, without completely suppressing apical meristematic growth, inevitably changes the over-all pattern of growth in the direction of floral organization. A comparable change is the conversion of the continuing, but pulsative growth of a hydroid terminal to the establishment and development of a hydranth or gonangial rudiment in such organisms as *Obelia* and *Campanularia*. In both plant and hydroid, when the rate of terminal growth is reduced, a growing rudiment of a generally unitary nature and massive character appears. In the plant, any relative reduction of the growth rate in the initial zone has two simultaneous consequences: the zone of peripheral meristem progressively extends toward the apical center and finally covers the whole; and the differentiation of the ring of polarized, vascular, file tissue correspondingly diminishes. The combined effect of this extension of available meristematic area and of the progressive lessening of polar determination of provascular tissue is to permit various organizational events that promote floral development and to suppress, gradually or abruptly, the provascular pattern that induces foliar development. The two counterbalanced patterns of growth gradients are as decidedly alternate types of apical growth in plants—leading to vegetative and floral structure, respectively—as are the alternate types of terminal growth in hydroids—leading to hydranth and medusa structure, respectively.

The determination of flowering, accordingly, has both a negative and a positive aspect. Factors that reduce the apical center of growth initials without affecting the submaximal growth areas, at least to a comparable extent, permit or encourage meristematic growth of a floral type. And such reduction may be all that is necessary, for the situation seems comparable to that in various types of animal metamorphosis; i.e., at least partial elimination of the established organization is as necessary as is stimulation toward growth of residual tissues. Both elimination and stimulation are generally sensitive to a variety of internal and external environmental influences, as are the counterpart processes in the plant.

However influential environmental factors may be, there is still the innate growth limitation of the growth zone of initials. Such a limitation exists just as surely as a leaf primordium or an amphibian limb blastema has a prospective growth expansion of specific amount. The zone of growth initials undergoes a process of waning vigor, as though a limited stock of some essential growth constituent were becoming progressively depleted. When such an "aging" event comes to an end, terminal growth either ceases altogether or transforms to the floral state. However, the transformation from the vegetative to the floral state in the apical meristem or the initiation of the floral state in an incipient bud is regarded by many as the result of an accumulation of an agent generated elsewhere in the plant, i.e., as a positive induction of the floral state in some area other than the growth zone, rather than a dampening or suppression of the vegetative state within that zone. Neither concept excludes the other, for positive stimulation of adjacent meristematic tissue of tunica and corpus may well coincide with the negative weakening of the growth intensity and potential of the initial zone.

Heslop-Harrison (1957) suggests that such an inductive agent may be nucleo-proteins with limited powers of self-duplication and that the formation of effective quantities of this agent depends on the conditions—of which auxin balance is the most critical—prevailing for its self-duplication. Certainly, high auxin levels inhibit floral initiation, and low levels seem to promote it; but it is possible that the auxin may work indirectly by maintaining activity in the growth initial zone and thereby prohibiting changes which depend on the zone's weakening.

Development of a Flower

The flowers produced by the various floral apices are almost infinite in variety of form, size, degree of completeness, and number. Although all these features may be influenced by external or internal factors, such as hormonal and nutritive supplies, they are basically polymorphic expressions of a general form of meristematic growth that is comparable in many ways to the polymorphism expressed in hydromedusae. In other words, whatever the controlling or modi-fying agents, the meristematic tissue is the instrument and principal performer, just as the embryonic ectoderm of the amphibian has the innate capacity to develop the pattern and differentiation of the brain and sense organs even without mesodermal and archenteric inductors, as Holtfreter (1945) has pointed out. In both floral and hydromedusan polymorphism, the most difficult and significant problem of development arises: the capacity of unspecialized tissue to undergo self-organization.

Variation among floral types in scale of organization is enlightening, since scale differences are correlated with cell numbers, whereas cell sizes are virtually constant no matter what the scale. Thus when a primordium's dimensions are below a certain limit, the cell number becomes so reduced that restrictions are thereby imposed that would not otherwise be felt. On this basis alone, a primordium could conceivably develop a stamen, but not a carpel.

Growth of the Floral Apex

On the other hand, the growth pattern of the floral apex as a whole takes precedence over the developmental units produced by it. This apical growth pattern can be properly understood only as a space-time continuum. A floral apex has both a shape and a course of growth. Whether apical growth is spiral or otherwise, the course of growth is from the periphery toward the apex center and is essentially similar in pattern to that of the vegetative apex but very different in intensity. The condition is in fact very much like that produced by Wardlaw experimentally in the fern: extirpation of the apical initial resulted in the cessation of formation of material at the apex and in the continued formation of leaf primordia successively toward the apex until all the subapical territory was utilized. In the floral apex much the same event occurs, except that the general apical growth, which persists even though the growth initial zone has disappeared, continues at a waning rate. Accordingly, territory available for organization continues to form from the marginal area inward toward the center; in other words, growth continues in the central area, and available territory is in effect shunted continuously outward from the center toward the margin (as if the whole were a flat disk). This territory available for organization is delimited in the same way as were the leaf primordia; i.e., an outer limit to the whole is defined by structure or tissue previously determined, and an inner limit is set by tissue still growing at a rate too high to permit the crystallization of protoplasmic pattern.

The width of the band of tissue thus bounded, within which organizational units become established, may be narrow or wide, depending on the growth rate at the center, the rate of decline at the center, and the rate of decline between the center and the periphery. There is, moreover, clear evidence that the available territory becomes organized in a successive manner, just as in leaf determination, either in spiral or alternate sequence. Each primordium produced seems at first to be approximately circular—the shape observed also in medusae, hydroids, and ascidians—so that the width of the band of available territory must represent one diameter of a primordium and therefore of the initial scale of organization. Accordingly, either serially in a spiral course or simultaneously in sets from successive concentric territorial circles, primordia form and eventually give rise to floral or inflorescent structural units. Consequently, when a territorial band is very narrow, correspondingly small but numerous primordia appear; whereas when a band is broad, and its outer circumference is the same length as before, a small number of large primordia segregate. Such a system permits an extraordinary degree of variation, and such variations underlie the more obvious genetic distinctions of floral pattern visible in flowers.

Formation of Structures

Another variable, which affects the general appearance of a flower, though in an accessory capacity, is related to the formation of sepals and petals in the flower and of bracts and ray florets in the inflorescence. These are all developmental creations which are neither leaves nor fully reproductive units, although collectively they are intermediate. All four develop from primordia which are established on the outer slopes of the transforming apex at a time and in such a way that provascular tissue is incorporated to a greater extent than in the reproductive dome proper, although to a lesser extent than in the earlier-formed leaf primordia. Whether these somewhat leaflike appendages form at all and the extent to which they do form depend on the nature of the transformation from vegetative to floral apex. A sharp transition commonly produces a purely reproductive head, such as in tree flowers of dwarf type; whereas a gradual transition can yield variations of the accessory appendages according to circumstance. Thus a genetically determined form of transition from the vegetative to the floral apex or, perhaps more significantly, the developing form of the floral apex independently of any previous vegetative state determines the floral pattern as a whole.

Determination of Sexuality

Although the seat of floral-pattern determination is at present more a matter for investigation than for exposition, some of the scope in pattern, particularly of sex expression, may indicate aspects for such investigation. Flowers may be hermaphrodite or unisexual. There is debate about which condition prevails in the primitive flower, but such debate is tempered by a growing belief that angiosperms are polyphyletic. From the point of view of developmental mechanics, the question is: How do separately sexed and even sterile or abortive flowers derive from the hermaphrodite state? Genes and hormones may be determinants but, in different degrees of remoteness, they operate *upon* a system whose response—the actual event—is the prime interest. Furthermore, in the complete flower, with accessories, the sequence of formation is acropetal, i.e., from the periphery toward the center, so that sepals, petals, stamens, and carpels form in concentric circles in the sequence in which they are listed and that the first three organogenetic units are more closely associated with the shoulders of the apex.

Such a setting—of sequential formation and of genetic and hormonal influence—is subject to modification in two ways: (1) The pattern of growth may be altered and consequently effect the nature, size, and number of primordia dependent on that pattern. (2) The tissue may be forced into lines of histogenesis by hormonal impact, although it is not known whether in plants hormones act directly in determining histogenesis or indirectly by modifying the growth pattern. That all plant hormones act similarly is as unwarranted a generalization at present as that in animals the sex hormones, embryonic inductors, and histogenetic öoplasms act alike. Both polycyclic hydrocarbons and nucleoproteins are involved and substances, chemically so different, may well operate in different ways, even if their effects are the same.

In this context of internal and external influences on the sexuality of flowers, the effect of auxins and that of temperature are pertinent. By applying auxins to leaves of male plants of hemp, Heslop-Harrison (1956) induced complete sex reversal: all auxin-treated genetically male plants at first formed female flowers, but later reverted to the production of normal male flowers. It is also significant that low temperatures during early growth promote femaleness and suppress maleness in monoecious plants, such as the acorn squash (*Cucurbita pepo*), according to Nitsch and his co-workers (1952); whereas high temperatures have the reverse effect. Moreover, the temperature prevailing during the dark period seems to be critical. In the acorn squash the male plants show a directional progression, from the lowest nodes upward, of undeveloped through normal to inhibited flowers; the female plants show a sequential progression from the first-formed flowers, which are normal, to later-formed flowers, which have excessively large ovaries. In female plants growing in short days at low temperatures the female condition is so extreme that the flowers form fruit parthenogenetically.

Whatever the agents modifying the full expression of the hermaphrodite organization, the process is primarily the suppression of pistillate structures in the male and staminate structures in the female. These suppressions are to be regarded as functions of the growth pattern that leads to the formation of a complete flower and as correlations with the initial scale of the floral apex.

Suppression of the pistillate structure is most readily observable in the flowers of plants that also produce complete flowers. Except that the exclusively male flowers are smaller than complete flowers, and develop from correspondingly smaller apices, floral development is apparently normal. Accessory appendages and stamens develop typically from the outer whorls of available tissue, but the residual central area associated with carpel formation is too small to permit pistillate organization. In other words, structures that normally form first do so at the expense of later-forming structures, whose presumptive territory has been usurped. In most simple scale reductions, the residual apical center has been most reduced, as though size reduction results primarily from a curtailment or relatively rapid waning of the apical growth rate.

Tree flowers commonly show size reduction. In the sugar maple (*Acer saccharum*), for instance, flowers terminate on long, slender stalks which emerge simultaneously from a flowering branch terminal. In the same flower cluster, all these flowers may be male, all may be complete, or some may be male and others complete. In the last-named possibility, the male flowers are invariably borne on stalks that are considerably narrower than those bearing complete flowers and that commonly are collateral stalks emerging from the sides of the wider ones (Fig. 17-11). In this species neither the male nor the complete flower carries petals, both have 8 stamens, but the central area of the male is smaller and undeveloped. In the Norway maple, the scale of organization is somewhat larger, the main stems of the flower cluster bear hermaphrodite flowers, and the thinner adjoined stems bear male flowers, as in the other species. The difference is that even in the male type a sterile pistil

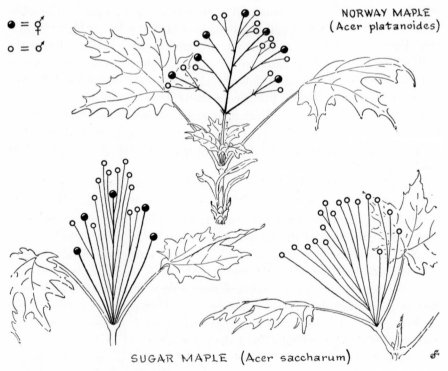

Figure 17-11.   Distribution of complete and male flowers in flower shoots of two maples.

base is visible; on the other hand, accessory structures are also present in both male and complete types in the form of 5 sepals and 5 petals in addition to the 8 stamens. Therefore, the two species together illustrate the entire range from the complete flower, which has both sepals and petals, to the extremely reduced flower, which has only stamens.

In other plants, such as hydrangea, there is, together with the fertile flowers in a flower head, an abortive, smaller flower, which has petals but neither staminate nor pistillate structure. Accordingly, the progressive reduction of the central area of the developing floral apex may be so great that only the outermost, if any, territory remains and produces only foliar organs. When only the reproductive territory of the apex remains, as when the peripheral slope associated with the development of foliar units is virtually absent, reduction in initial scale may result in purely unisexual flowers, without foliar appendages, even in monoecious plants such as hazel and oak. In fact, in all species that produce catkins, the unit male flowers of the inflorescence are initially of such a small scale that there isn't enough territory for the formation of even one pistillate unit, although such scale reduction by itself merely accounts for the absence of female structure and not for the presence of male. Female flowers, which form on a larger initial scale elsewhere on the same plants, consist of the central pistil-carpel structure only; i.e., the female flowers represent the central area only of the apex.

Since hormonal and nutritional conditions must be the same for both sexes in the same plant, the choice of sex formed must primarily result from differences in the initial scale of organization: When territory is adequate, female organization forms; when territory is too small to form the female pattern, it may form the male, as in ascidians. It is noteworthy, however, that the male flower, however small, never consists of only one central stamen but of at least two or three surrounding an unoccupied central point. Thus stamens form from a circular band of territory which may be in a somewhat different state of growth from that which gives rise to female structure; consequently, in exclusively female flowers this territory must be eliminated by modification of the pattern of apical growth.

In brief, then, the various differences between unisexual and bisexual flowers, with and without floral appendages, may be interpreted in relation to initial scale, to whether or not the floral apex forms shoulders, and to whether or not the apex forms the peripheral and the central zone associated with staminate and pistillate structures, respectively. These three factors depend on the variable extent of polarized tissue persisting from the vegetative phase and on the variable characteristics of the apical growth gradient during its course of existence. The size of rudiments, the shape of meristems, and the presence and effectiveness of hormonal and nutritional chemicals, however, are all subject to genetic control and to external circumstances of time and place.

Development of Single vs. Composite Flowers

Finally, the one outstanding problem remaining concerns the developmental difference between single flowers and the inflorescence of composites. The general pattern of apical growth seems to be essentially the same in both; the difference is in the developmental potentiality of the successively produced primordia. In the peripheral zones, there does not seem to be a very significant distinction between bracts and sepals or between petals with stamens and ray florets with pistils, except that the axillary bud associated with the foliar appendage in both single flowers and composite is relatively larger in the latter. The main difference between the two kinds of flowering concerns the reproductive area proper, in which the primordia develop peripherally as stamens and centrally as carpels in the single flower and as complete flowers on a minute scale in the composite flower.

Why does the carpellary dome of a raspberry or cocklebur, for instance, initiate carpellary primordia in a regular spatial and temporal sequence progressing toward the apex; whereas the capitulum of a daisy similarly produces primordia, but each develops in addition to a stalk a central carpel surrounded by stamens and marginal appendages? Primordial size is not the answer, for size differences are hardly significant. Scale of organization however may yet be significant for scale is relative.

As in reconstitutions from cell aggregates of hydroid tissue (p. 267), an essentially complete pattern may be expressed in a relatively small aggregated mass; whereas much larger masses may develop only part of the organization. Because of its genetic constitution, the complete floral unit organization, in the daisy and other composites, can be expressed within a minute space. On the other hand, there is little doubt that the genetically based pattern of scale prevents the initial pattern of a typically single flower such as raspberry or magnolia from being expressed in such a small space. Primordia develop into whatever is possible genetically, with or without coercion from histogenetic agents. For example, a stamen has a fairly definite size depending on the species; and for a stamen of a given species to be formed, a primordium of precisely related size must precede it. Primordial size is in turn species-specific; for example, a stamen primordium of a raspberry is much smaller than that of a tulip stamen, but a stamen primordium as small as that of a daisy floret could not possibly produce a staminal or any other structure in either raspberry or tulip.

Thus the scale of pattern imprint—whether of stamen, carpel, or complete flower—is a variable; and scale itself becomes relative to the fineness or mesh of the basic imprint, so to speak, and the dimensions of the primordium in which it is imposed. Clearly, cells as units of structure are subordinate to the scale, so that the question asked about floral initiation is the same as that about floral-type development: At what material level or in what substance does the seat of organization lie? And the answer seems to be the same: Organization lies in the continuity—tenuous or confluent—throughout the cortical cytoplasmic layer, which unites cells and overrides their individuality. The nature of this patterning is the primary and persistent problem, although the comparisons among species above indicate that the same pattern may be expressed in coarse or in fine texture—a phenomenon suggesting that a dynamic, expansive fibrillar organization pervades the cortical material of tissues and represents the essence of the organism. In this connection Whaley's electron-microscope study of the growing root apex in maize is highly significant, for although morphogenesis in the root apex is less spectacular than in the shoot apex, the intercellular relations in the growing tissue is probably the same in both. Successive stages of root development show an extensive, general vesicular reticulum of which the nuclear envelope of the interphase cell is a component; this reticulum provides a relatively great lipoprotein surface area between the nucleus and the cytoplasm, between the cytoplasm and the surface of the cell, and, through intercellular connections, between the reticula of neighboring cells. Even though the system fragments during the prophase of mitosis, the fragments remain in much the same location and the system re-forms.

Thus the intercellular integrity of a tissue may persist if the percentage of cells undergoing mitosis is not too high, but may be incompatible with proliferative growth of a maximum rate. In other words, the incipient and developing organization of a primordium—whether of a leaf, a carpel, or the bud of *Botryllus*—may depend on the establishment and maintenance of such an intercellular reticulum; whereas in regions where growth rate is maximal the reticulum may be too unstable to permit such expression or even the territorial identity of a primordium.

REFERENCES FOR
Morphogenesis in Plants

Arber, A. 1950. *The Natural Philosophy of Plant Form.* New York: Cambridge Univ. Press.
Ashby, E. 1948. "Studies on the Morphogenesis of Leaves. II. The Area, Cell Size and Cell Number of Leaves in *Ipomoea* in Relation to Their Position on the Shoot." *New Phytol.* **47**:176-92.

———. 1950. "Studies on the Morphogenesis of Leaves. VI. Some Effects of Length of Day upon Leaf Shape in *Ipomoea caerula.*" *Ibid.* **49**:375-87.

———, and E. Wangerman. 1950. "Studies on the Morphogenesis of Leaves. IV. Further Observations on Area, Cell Size and Cell Number of Leaves of *Ipomoea* in Relation to Their Position on the Shoot." *Ibid.* Pp. 23-35.

Ball, E. 1948. "Differentiation in the Primary Shoots of *Lupinus albus* L. and of *Tropaeolum majus* L." *Soc. Exp. Biol.* Symp. **2**:246-62.

———. 1950. "Isolation, Removal, and Attempted Transplants of the Central Portion of the Shoot Apex of *Lupinus albus* L." *Am. J. Bot.* **37**:117-36.

Boke, N. H. 1947. "Development of the Shoot Apex and Floral Initiation in *Vinca rosea* L." *Am. J. Bot.* **34**:433-9.

Brooks, R. M. 1940. "Comparative Histogenesis of Vegetative and Floral Apices in *Amygadlus communis,* with Special Reference to the Carpel." *Hilgardia.* **13**:249-306.

Cross, G. L. 1942. "Structure of the Apical Meristem and Development of the Foliage Leaves of *Cunningham lanceolata.*" *Am. J. Bot.* **29**:288-301.

———, and T. J. Johnson. 1941. "Structural Features of the Shoot Apices of Diploid and Colehicine-induced Tetraploid Strains of *Vinca rosea* L." *Torrey Bot. Club Bull.* **68**:618-35.

Engard, C. J. 1944. "Organogenesis in *Rubus.*" *Univ. Hawaii Res. Pub.* No. 21. Pp. 1-233.

Esau, K. 1942. "Vascular Differentiation in the Vegetative Shoot of Linum. I. The Procambium." *Am. J. Bot.* **29**:738-47.

———. 1943. "Origin and Development of the Primary Vascular Tissues in Seed Plants." *Bot. Rev.* **9**:125-206.

Foster, A. S. 1943. "Zonal Structure and Growth of the Shoot Apex in Microcycas calocoma (Miq.)." *Am. J. Bot.* **30**:56-73.

Gifford, E. M., Jr. 1954. "The Shoot Apex in Angiosperms." *Bot. Rev.* **20**:477-529.

Gregoire, V. 1938. "La morphogénèse el- l'Autonomie morphologique de l'Appareil floral. I. le Carpelle." *La Cellule.* **47**:285-452.

Heslop-Harrison, J. 1956. "Auxin and Sexuality in Cannabis sativa." *Plant Physiol.* **9**:5-88.

———. 1957. "The Experimental Modification of Sex Expression in Flowering Plants." *Biol. Rev.* **32**:38-90.

Hotlfreter, J. 1945. "Neurulization and Epidermization of Gastrula Ectoderm." *J. Exp. Zool.* **98**:161-209.

McCoy, R. W. 1940. "Floral Organogenesis in *Frasera carolinensis.*" *Am. J. Bot.* **27**:600-9.

Millington, W. F., and E. L. Fisk. 1956. "Shoot Development in *Xanthium pennsylvanicum.* I. The Vegetative Plant." *Am. J. Bot.* **43**:653-65.

Nitsch, J. P., *et al.* 1952. "The Development of Sex Expression in Cucurbit Flowers." *Am. J. Bot.* **39**:32-43.

Philipson, W. R. 1946. "Studies in the Development of the Inflorescence. I. The Capitulum of *Bellis perennis* L." *Ann. Bot.* **10**:257-70.

———. 1947. "Studies in the Development of the Inflorescence. II. The Capitula of *Succisia pratensis* Moench. and *Dipsacus fullonum* L." *Ibid.* **11**:285-97.

———. 1948. "Studies in the Development of the Inflorescence. IV. The Capitula of *Hieracum boreale* Fries and *Dahlia gracilis* Ortg." *Ibid.* **12**:65-75.

———. 1949. "The Ontogeny of the Shoot Apex in Dicotyledons." *Biol. Rev.* **24**:21-49.

Popham, R. A., and A. P. Chan. 1950. "Zonation in the Vegetative Stem Tip of *Chrysanthemum morifolium* Bailey." *Am. J. Bot.* **37**:476-83.

Priestly, J. H. 1929. "Cell Growth and Cell Division in the Shoot of the Flowering Plant." *New Phytol.* **28:**54-81.

Reeve, R. M. 1943. "The 'Tunica-corpus' Concept and Development of Shoot Apices in Certain Dicotyledons." *Am. J. Bot.* **35:**65-75.

Richards, F. J. 1949. "The Geometry of Phyllotaxis and Its Origins." *Soc. Exp. Biol.* Symp. **2:**217-45.

Rouffa, A. S., and J. E. Gunckel. 1951. "Leaf Initiation, Origin, and Pattern of Pith Development in the Rosaceae." *Am. J. Bot.* **38:**301-7.

Salisbury, F. B. 1955. "The Dual Role of Auxin in Flowering." *Plant Physiol.* **30:**327-34.

Snow, M. and R. 1933. "Experiments on Phyllotaxis. II. The effect of displacing a primordium." *Roy. Soc. London Philos. Trans.* B. **222:**353-74.

———. 1949. "On the Determination of Leaves." *Soc. Exp. Biol.* Symp. **2:**263-75.

———. 1952. "Minimum Areas and Leaf Determination." *Proc. Roy. Soc. London.* Series B. **139:** 545-66.

Sterling, C. 1945a. "Growth and Vascular Development in the Shoot Apex Sequoia sempervirens (Lamb). I. Structure and Growth of Shoot Apex." *Am. J. Bot.* **32:**118-26.

———. 1945b. "Growth and Vascular Development in the Shoot Apex Sequoia sempervirens (Lamb.) II. Vascular Development in Relation to Phyllotaxis." *Ibid.* Pp. 378-86.

Sussex, I. M. 1951. "Regeneration of the Potato Shoot Apex." *Nature* (London). **167:**651.

———. 1952. "Regeneration of the Potato Shoot Apex." *Ibid.* **170:**755.

Wardlaw, C. W. 1944. "Experimental and Analytical Studies of Pteridophytes. IV." *Ann. Bot.* **8:**387-99.

———. 1952. *Phylogeny and Morphogenesis: Contemporary Aspects of Botanical Science.* New York: St. Martin's Press, Inc.

———. 1957. "On the Organization and Reactivity of the Shoot Apex in Vascular Plants." *Am. J. Bot.* **44:**176-85.

Wetmore, R. H., and C. W. Wardlaw. 1951. "Experimental Morphogenesis in Vascular Plants." *Plant Physiol. Ann. Rev.* **2:**269-92.

Whaley, W. G. 1931. "Developmental Changes in Apical Meristems." *Nat. Acad. Sci. U. S. Proc.* **25:**445-8.

———. 1959. "Dynamics of Cell Ultrastructure in Development and Growth." *Sci.* **130:**1425-6.

White, P. 1938. " 'Root Pressure' an Unappreciated Force in Sap Movement." *Smithsonian Inst. Washington Ann. Report.* Pp. 489-97.

PART IV
DEVELOPMENT OF EGGS

The study of egg development has been so intensive for nearly a century that embryology has assumed the aspect of an independent science; clearly, though, it is but a part of developmental biology as a whole. It is such a large part, however, that selectivity of only those aspects that are especially pertinent to the present treatment as a whole is characteristic of this part of the book.

Chapter 18

Nature and Development of Eggs

EGGS, the typical means of development and reproduction in the animal kingdom, are clearly more than just developmental devices. In all probability, gametes, meiosis, and conjugation antedated the evolution of eggs as developmental units. Yet, unspecialized cells in general, given suitable circumstances, possess the capacity to grow and develop into a complete replica of the organism from which they originate. It is this capacity which has been "grafted," so to speak, onto the meiotic-gametic phase of egg evolution—a phase of primarily adaptive or evolutionary significance. In the context of this book, eggs are pertinent only as they relate to the phenomena of developmental potencies and processes.

The processes of maturation and fertilization may be regarded as sophisticated avenues leading to the developmental event—as "obligations and rituals" that are performed for purposes of pedigree before the protoplasmic potential can be expressed. From the standpoint of development, therefore, eggs are essentially large cells which are designed to be set free from the parental organism and which possess sufficient mass for the production of cells from which a functional organism is formed.

The primary feature of an egg, apart from the fact that it is a single cell, is its size, because large size implies a latent capacity for division into many small cells. And, depending on their number and degree of specialization, any group of unspecialized cells physically or physiologically isolated from the parental organism has the capacity to undergo morphogenesis and histogenesis.

In most organisms, growth of the individual egg cell is confined to the pre-maturation phase as an oocyte, and subsequent development is typically that of a closed system. However, growth is not always in this sequence and pattern: extensive growth may occur during the early stages of development and in various ways. Moreover, the egg at the onset of development may be just a large gametic cell whose cytoplasm is either completely unspecialized or is so differentiated that symmetries and histogenetic territories are already established. Such alternate possibilities underlie the following brief discussion of the various types of eggs considered.

Growth of Eggs

The simplest type of oocyte grows, from initiation to maturity, as a single cell without follicular contributions. Among the many organisms whose eggs are of this type are the hydroids *Clava* and *Hydractinia* and the echinoderm *Asterias*. In the hydroids *Tubularia* and *Hydra,* however, the full-grown oocyte is formed by the fusion of many smaller cells: the primary oocytes fuse together in plasmodial areas, each of which consists of a syncytium and many nuclei. In *Hydra* each area gives rise to a single ovum, all but one of the nuclei dissolving into the common cytoplasm. In *Tubularia,* whose egg is much larger than that of *Hydra,* the primary fusion masses fuse a second time to form a single, large amoeba-shaped oocyte which later rounds up into a typical egg shape; as in *Hydra,* only one nucleus survives—all others dissolve into the cytoplasm. Whether the fastest-growing oocyte virtually engulfs and absorbs the others is not known; but some such process must occur to explain the survival of only one nucleus.

In many ascidians the oocyte grows rather independently and becomes surrounded by a layer of follicle cells which are descendants of abortive oocytes. In some species, e.g., *Styela plicata,* the follicle cells make temporary contact with the surface of the growing oocyte, but eventually lie free in the perivitelline space. In others, e.g., *Botryllus,* the growing oocyte actually engulfs the follicular cells at a very early stage, but ejects them later. Whether the follicular cells are merely involved in growth or actively contribute to that growth is not known at present.

Oogenesis raises profound questions of cell physiology regarding both growth and the inhibition of cell division, but such questions are subsidiary to the main problem of organization. Growth—all proliferation, in fact—seems to be more a concomitant or condition of organizational expression than a positive determinant. Eggs do not necessarily have to be large even relative to somatic cells in order to develop, although as a rule a certain critical mass seems necessary. This critical diameter usually ranges from about 60 μ in some polychaetous annelids to the enormous eggs in reptiles and birds; yet, according to Bigelow (1909), in the jellyfish *Pegantha smaragdina,* each egg is associated with a nurse cell and is little larger than adjacent tissue cells—about 8 μ in diameter at the onset of cleavage. Much larger than Pegantha, though still small compared to typical egg size, is the 30-μ egg of the stalked jellyfish *Haliclystus*. In all such unusually minute eggs and in some larger eggs, growth as well as cleavage occurs during the subsequent development phase.

Nature of Eggs

What, then, is an egg? Apart from its gametic features, is it any more than a protoplasmic mass of sufficient size and/or growth potential to form a comparatively large number of cells? In other words, is it necessarily or fundamentally a developmental unit with a special built-in chemical or organizational imprint that determines its developmental course more specifically and elaborately than the course of any uncommitted somatic cell? In many organisms, such elaborateness is evident; but in others, there is no evidence that anything more than a quantitative preparation for expansive growth has occurred before development begins. This appears to be the case among coelenterates especially, and, as Whitaker (1933) found, is even more striking in an autotrophic egg—such as that of *Fucus*—in which even the polar axis may be determined by any one of several environmental agencies.

The Indeterminate Egg

The smaller coelenterate eggs, which are translucent cells with a diameter between 100 and 200 μ, do possess an initial polarity related to the polarity of the parental epithelium; but experimental separation of blastomeres of both hydroid and medusa eggs indicates that no organizational determination occurs during early development. Isolated blastomeres of 8-cell and even 16-cell stages seem to be indifferent and reasonably capable of forming larvae with normal proportions; only the qualitative reduction of material within the blastomere indicates that an inhibitory factor is operative. In the larger egg of *Tubularia*, the fertilized egg cleaves to produce a mass of cells pressed in cup-shaped form around the spadix of the gonophore in which it lies, so that a marginal area forms as an accident of position. This marginal area is the presumptive tentacle rim and also establishes the polar axis of the embryo. The number of tentacles formed depends on this area's circumference. All that is required for development in *Tubularia* is that the egg be sufficiently large and capable of cleaving, for subsequent development of pattern is direct and independent of special directive agencies. Moreover, a polyp developing from a planula is the same—except for its temporarily ciliated external surface—in size, shape, and tissue as that developing from an asexual planuloid (p. 167).

If eggs in general are primitively, or initially, unspecialized cleaving protoplasmic masses, resulting from the fusion of gametes, how many of the more-general developmental phenomena of the relatively specialized types of eggs are a consequence of special devices for reinforcing, accelerating, or diverting the innate course of development, and how many are inherent in the egg cell as a cell, rather than an egg?

The various organisms exhibiting polyembryony, relevant to this question, include *Crisia* among bryozoans, Harmer (1893); *Platygaster,* according to Leiby and Hill (1924), and *Litomastix,* according to Silvestri (1936), among parasitic insects; and certain scyphomedusae. Typically, in polyembryonic development, the egg is minute and develops within a nutrient medium supplied by parent or host. Rapid growth occurs simultaneously with cleavage, and a somewhat amorphously shaped mass of proliferating cells is produced which later gives rise to a succession of smaller, secondary masses, or clusters. The formation of the amorphous masses continues as long as growth and cell proliferation are maintained at their maximal rate, as in metastatic growth in tissue cultures and tumors, and morphogenesis commences only when the rate of growth subsides and each isolated mass represents a presumptive individual. The species of *Platygaster* eggs may thereby give rise to 1-8 definitive embryos; that of *Litomastix,* to as many as 2,000.

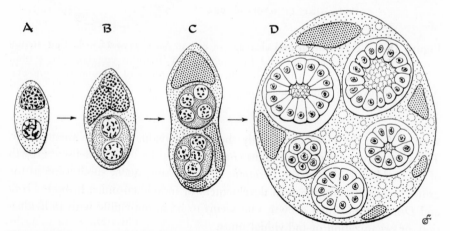

Figure 18-1. Polyembryony in *Platygaster.* **A.** Fertilized egg, showing polar nucleus above and segmentation nucleus below. **B.** Fertilized egg, showing both polar nucleus and segmentation nucleus divided. **C.** Embryonic region divided to form twin germinal masses. **D.** Polyblastular stage with 5 embryos at early blastular stage. (After Leiby and Hill.)

   With such a disorderly process, especially as seen in *Crisia,* no initial organization of the egg can be presumed to have any subsequent influence; for each
secondary mass, which develops into an individual organism, seems to be as
self-determined as, for instance, a *Distaplia* bud growing and developing within
the nutritive matrix of the ascidian colony. In fact, this polyembryonic process
is comparable to the whole budding process in *Distaplia* and other budding
organisms. The *Crisia* egg undoubtedly grows at its maximal rate at first, at
which rate and time it gives rise to a large multicellular mass; somewhat
later, when the growth rate is almost certainly decreasing, secondary masses

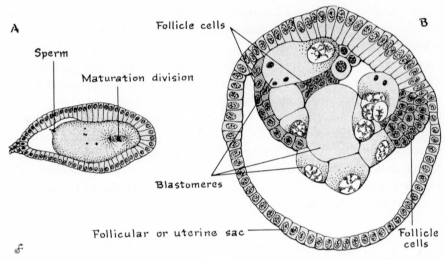

Figure 18-2. Fertilization and cleavage of egg in *Salpa pinnata.* **A.** Egg at moment of sperm penetration and first maturation division; egg enclosed within
follicular sac. **B.** Section through follicular sac at phase of blastomere segmentation and invasion among blastomeres by follicle cells (calymnocytes). (After
Stier.)

separate from the large mass. Only then do individuation phenomena occur.
The whole sequence of events thus corresponds to the initial extensive growth
of the budding region of the *Distaplia* larva, the subsequent subdivision as the
growth rate falls, and the final development of individual units. In both *Crisia*
and *Distaplia,* maximal growth rate seems to be incompatible with individuation, or organization of individual units.
   In both bryozoans and insects the rapidly proliferating and growing
blastomeres resulting from the single ovum are enveloped by nurse cells, or
follicle cells. A similar envelopment occurs in all species of the tunicate *Salpa,*
in which the small ovum lies within an ovoid follicular sac near the mid-dorsal

plane at the posterior end of the parental organism. Cleavage of the *Salpa* ovum is total, unequal, rapid, and very irregular; it has been described as a process of blastomeric budding. According to Brien (1928), the whole process is very quickly masked by a migration of follicle cells (calymnocytes) from the lining of both the oviduct and the follicular chamber into the so-called egg chamber; there these follicle cells distribute themselves between the blastomeres, thereby prohibiting any possibility of an orderly process of cleavage. This combination of follicular migration and distribution, in relation to the polarities and dimensions of parental origin, leads to one of the most amazing developmental performances in the animal kingdom.

External Determination in Salpa

According to Brooks' classical description of this genus,

the most remarkable peculiarity of the Salpa embryo is this. *It is blocked out in follicle cells* which form layers and undergo foldings and other changes which result in an outline or model of all the general features in the organization of the embryo. While this process is going on the development of the blastomeres is retarded, so that they are carried into their final positions in the embryo while still in a very rudimentary condition. Finally, when they have reached the places which they are to occupy, they undergo rapid multiplication and growth, while the scaffolding of follicle cells is torn down and used as food for the true embryonic cells.[1]

On the other hand, Brien demonstrated that the follicular structure is not the actual structure of a presumptive salp and that the blastomeres are definitely independently active. Nevertheless, there remains enough of Brooks' interpretation to suggest a unique process.

The complex extraembryonic structures produced in *Salpa* derive partly from the follicle cells and partly from adjacent parental tissue. Blastomeres are rather closely packed within an enveloping mass of follicle cells, and adjacent cloacal epithelium of the parent thickens to form an epithelial cone around the follicular mass. At a later stage cloacal folds meet above the cone and completely enclose it. The complex structure ultimately formed is bilaterally symmetrical: its long axis corresponds to that of the parent, which has a dorsal and ventral aspect; and its specific dimensions are proportionate to those of the parent. Within a central tongue of tissue in this structure, the blastomeres lie as three isolated groups of cells separated from one another. Thus formation of an organized structure which has definite polarities, symmetries, and scale takes place before the blastomeres show any indication of morphogenesis.

The three groups of blastomeres are (a) two pharyngeal masses in the

[1] Brooks, *Johns Hopkins Univ. Biol. Lab. Mem.*, **2**:83 (1893).

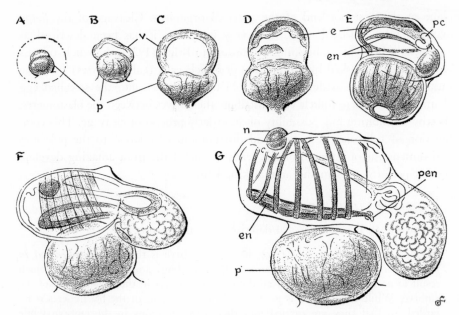

Figure 18-3. Development in *Salpa democratica.* **A–C.** Three stages, showing growth of placenta **p** and vesicular dome **v** of follicular tissue. **D–E.** Embryo **e** forming within vesicle. **en.** Endostyle. **pc.** Pericardium. **F–G.** Later development. **n.** Neural complex. **pen.** Posterior end of endostyle, foreshadowing budding stolon.

form of lateroventral vesicles; (b) ectoblastic masses, consisting of an independent, anterior, vesicular neural mass; a cloacal mass; and ectoderm along the mediodorsal line which progressively envelops the embryo like a kind of retarded epiboly; and (c) a mesoblastic mass, which forms the pericardium, genital mass, eleoblast, and general mesenchyme. These three groups are merely presumptive; until they have grown together to form a continuous tissue, their constituent blastomeres exhibit only growth and proliferation. Only when continuity has been established, does morphogenesis begin. In fact, before the groups are continuous, the follicular tissue represents a sort of crude modeling of the organism; blastomeric growth slowly catches up quantitatively and, after the blastomeres have been separated by follicle cells, they become united. The scale and polarities of the presumptive organism are set when blastomeric tissue first fuses to form a coherent whole. From this time forward, the follicular tissue progressively disintegrates while morphogenesis proceeds.

Two distinct phases are thus evident: In the first phase the growth of follicular tissue is dominant, forming the follicular structure, and blastomeres undergo rapid growth as well as division. In the second phase the follicular tissue disappears, and the blastomeric tissue, having attained continuity and lacking direct follicular nourishment, grows at a slower rate. Morphogenesis commences with the onset of this second phase, and the initial scale of the presumptive area set by the follicular structure is an index of the size finally attained by the new individual.

Comparative studies fully substantiate the relationship between initial and mature size. In *Salpa,* however, size determination is a circular relationship: in *S. democratica* the sexually mature parental organism is small and produces a correspondingly small follicular dome, which houses a comparably small initial embryo whose ultimate size is definitely a multiple of the initial size. The same internal size relationships hold for the large *Salpa maxima* and for the intermediate *Salpa fusiformis.* In each species the initial scale of the embryonic tissue at the time of fusion determines the final dimensions, and the extent of expansion of the blastomeric tissue present at the time of fusion is the same.

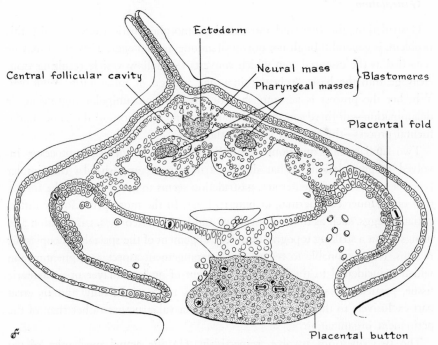

Figure 18-4. Transverse section of *Salpa maxima,* showing setting and scale of salp pattern in temporary follicular tissue, with blastomeres contained as three separate masses: a median presumptive ectodermal and neural mass, and a left and right presumptive pharyngeal mass—all lying above placenta and placental folds. (After Brien.)

The Determinate Egg

The foregoing examples as a whole show that eggs may be no more than masses of unspecialized cellular material which is somehow able to produce many cells, from which one or more individual organisms form. These masses are no more initially determined or organzed to pursue a particular developmental path than are comparable masses of unspecialized somatic tissue. On the other hand, most animal eggs exhibit either some degree of preliminary organization or the rudiments of a progressive determinative process, which either reinforces or diverts the developmental course along or from its natural path. These eggs, however, must be regarded as secondarily specialized tissue; moreover, such eggs raise the problem of distinguishing between the special devices introduced to ensure or modify the developmental procedure and the innate characteristics of all cells that have not become too specialized.

Gastrulation

Gastrulation, the first and most obvious morphogenetic event to raise this problem, is general though not universal among animal eggs. This event may be regarded as the essential step which converts the hollow vesicle resulting from cleavage into the basic "tube-within-a-tube" design of most animal embryos. Whether the process is accomplished by invagination, unipolar ingression, or delamination—all three commonly regarded as modifications of the same fundamental process—the result is virtually the same.

Two fundamental questions arise: What are the protoplasmic processes involved in gastrulation? What does gastrulation accomplish? In the simpler types of eggs, as in coelenterates, gastrulation seems to result only in the formation of an inner endodermal, or enteric, layer. In the more precociously differentiating eggs, such as those of ascidians and lower vertebrates, gastrulation also brings about a striking topographical rearrangement of the specialized ooplasmic regions. It is reasonable to regard this rearrangement aspect as a phenomenon which is incidental to the basic establishment of an inner layer of archenteric tissue, particularly since such rearrangements relate specifically and in great part exclusively to the formation of transient larval structure rather than of the permanent organization.

The two questions involve, respectively, (1) the actual mechanics of the gastrulation process, which inevitably encompasses the invaginative and delaminative processes as a whole; and (2) the nature of precocious specializations and their relationship to the gastrulation process. (Invaginative gastrulation as a process is discussed on pp. 141; the nature and significance of the ooplasmic specializations are of more immediate concern.)

Ooplasmic Differentiation in Ascidians

It is evident, from the foregoing account, that egg cells or groups of cells derived from cleaving eggs can develop the adult-type organization directly, i.e., without passing through a distinctive larval form and without having any discernible initial organization (other than the polarity common to all cells). Many invertebrate eggs, however, although at the lower end of the 50-200-μ range of diameter, show marked ooplasmic differentiation and form larval organisms that are very different from the adult organisms. Of these, the eggs and larvae of ascidians are a striking example and afford a direct comparison with somatic buds.

Pelagic eggs of ascidians are 0.1-0.2 mm in diameter. The development of the pelagic eggs of *Styela partita* is typical of ascidians. With the rupture of the membrane of the germinal vesicle, the nuclear sap mixes with the surrounding cytoplasm, thereby lowering cytoplasmic viscosity, and a polar cap of clear ectoplasmic substance flows across the equator of the egg. In *Phallusia mamillata* and *Ascidia malaca,* Ortolani (1955) attached colored chalk granules to the egg surface to demonstrate that at the time of fertilization there is a general cortical shift from the animal pole toward the equator and a subsequent convergence from the equator to the vegetal pole. After rupture of the germinal vesicle, at the time the polar cap is formed, and whether or not an egg has been fertilized, a yellow crescent appears in the equatorial zone and extends halfway around the egg, according to Conklin (1905); the color is that of contained mitochondria. A similar crescent is orange in *Boltenia echinata* (Berrill, 1948), reddish in *Pyura squamulosa* (Millar, 1951), and colorless in other species. According to Dalcq (1938), the crescent lies immediately subjacent to the egg cortex.

Cleavage patterns are rigidly determined by the ooplasmic organization: the first cleavage plane coincides with the bilateral axis of the egg; and subsequent divisions eventually segregate the several ooplasmic territories from one another. Thus, at the onset of gastrulation, after 6 cleavages, the so-called animal half of the cleaving egg consists of 26 ectodermal and 6 neural plate cells; the vegetal half, of 4 neural-plate, 4 chordal, 4 muscle, and 10 endodermal cells.

Before gastrulation, by invagination, which takes place during the interval between the sixth and seventh cleavage, the yellow crescent substance becomes limited to a crescent of cells symmetrically arranged as two arms around the presumptive posterior side of the organism. The chordoneural plate cells are symmetrically arranged as two arms around the anterior side. During gastrula-

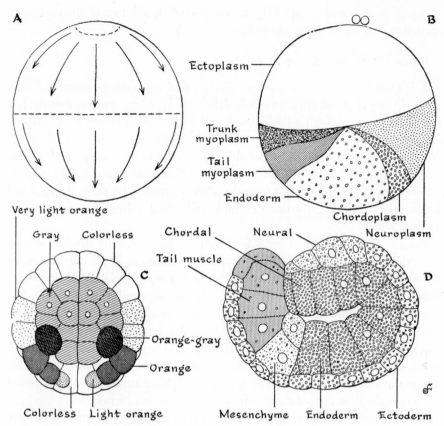

Figure 18-5.   Organization of ascidian egg.   **A.** Flow of peripheral cytoplasm from animal pole to equator and from equator to vegetal pole at time of fertilization, as shown by surface markers on egg of *Phallusia.* (After Ortolani.)   **B.** Side view of fertilized egg in *Styela,* showing pattern and presumptive nature of ooplasms visible in living egg. (After Conklin.)   **C.** Vegetal aspect of cleaving egg in *Boltenia* just before onset of gastrulation, showing segregation of different ooplasms within distinct cell groups, with colors and pattern as seen in living embryo.   **D.** Section of *Ciona* embryo at close of gastrulation, showing chordate organization and ooplasmic segregation. (After Conklin.)

tion the two arms of chordal cells are brought together in the middle line of the embryo as a short column 4 cells long and 4 cells wide and are fitted between the two arms of the posterior yellow crescent. In other words, two crescent regions, originally widely separated in the egg and blastula, are brought together by gastrular movement to form a single dorsoposterior mass. The central core of chordal cells gives rise to a notochord, consisting of about 42 large vacuolated cells in single column. The cells of the yellow crescent give rise to two lateral bands of tail-muscle tissue, each band consisting of about 18 cells arranged roughly in three equal rows.

Tail formation consists of the following events:

1. Localized histogenetic ooplasmic localizations in the egg prior to cleavage.
2. Spatial convergence during gastrulation of previously separated presumptive muscle and notochordal territories.
3. Rapid histodifferentiation of muscle and notochordal territories and consequent cessation of cell division preceding final differentiation, so that each tissue consists of small numbers of comparatively large cells.
4. Outgrowth of the tail as the result of swelling and interdigitation of the notochordal cells.
5. Histodifferentiation of the muscle bands as they become stretched because of contact with the extending notochord.

At the close of gastrulation, when the blastopore has virtually disappeared and its ventral margin has brought the two arms of presumptive muscle tissue alongside the chordal cells of the dorsal region, the dorsal ectoderm assumes the typical shape and character of a neural plate. A neural groove appears, and the lateral margins of the plate converge toward the mid-dorsal region, where they fuse to form a neural tube. The dorsal tube thus formed gives rise to three parts: an anterior vesicle from whose inner wall an ocellus and otolith differentiate; a residual region (of the posterior part of the vesicle), which later gives rise to the adult neural complex; and a narrow neural tube, which extends mid-dorsally throughout the tail.

Investigating the development of centrifuged *Styela* eggs, Conklin (1931) found that the mitochondria (responsible for the yellow color of the crescent) which had been considered to be related to the myogenic functions of the crescent material, could be displaced entirely without affecting the development of the crescent region into tail-muscle tissue. Furthermore, at least four visibly different regions of the egg were found to have specific developmental capabilities as early as the first cleavage: the posterior crescent (which contains the colored mitochondria); the anterior grey crescent, which gives rise to chordal tissue; the transparent hyaloplasm of the ventral hemisphere, which is destined to become ectodermal and neural tissue; and the yolk-filled dorsal hemisphere, the future endoderm. These four regions cannot be displaced without causing serious dislocations of tissues and organs in the larva.

Conklin considered the different capabilities as deriving not from any visible inclusions, but from the cytoplasmic ground substance whose physical and chemical make-up differ from area to area. Whatever their cytoplasmic basis, the locations and types of histogenetic agents are already established prior to the first cleavage of the egg, although both Dalcq (1932 and 1938) and Reverberi (1947) have shown that egg fragments obtained before or immediately after fertilization can develop into almost complete dwarf larvae. Yet the significant fact is that determination of the pattern of ooplasmic specializations takes place between the final maturation of the egg, when the egg nucleus ruptures, and the onset of the first cleavage after fertilization.

Apparently, none of the specialized regions acts as the organizer of another region, for a visibly normal nervous system develops even after the chordal cells or the presumptive mesoderm has been removed from the pregastrular embryo. In fact, a map of presumptive areas of the fertilized but as-yet uncleaved egg is more than a projection of prospective differentiations; it actually represents a distribution, or pattern, of histogenetic agents whose destruction or disarrangement is followed by a corresponding disturbance of the tissue and structure of the tadpole larva. In other words, the organization of the tadpole is essentially portrayed in the cortical layers of the undivided egg, but in such a spatial arrangement that associated parts converge only as the result of the gastrulation process. That gastrulation has this effect does not imply that the process is purposive in this respect, but that the agents responsible for the precocious differentiation of neural, chordal, and myogenic tissue have been localized in the egg in the relation to the position of each tissue in the postgastrular embryo.

It is highly significant that all the precociously differentiating tissues are features of the transient structure in the chordate tadpole larva and not of the permanent organization and that these larval structures of locomotion and orientation appear, function, and disintegrate long before the cleavage process has ended and before the permanent structures, such as the digestive and circulatory systems, have become functional. A similar statement of equal validity pertains to the precocious differentiation and functioning of the distinctive locomotive and orientative structures of the polychaete trochosphere larva. In the development of the egg of both organisms, there is a duality which leads to the formation, at different times, of two very different types of organization.

General Significance of Ooplasmic Differentiation

Development of an ascidian egg, and of specialized eggs in general, may be regarded as a triple event: cleavage, the formation of the permanent organization, and precocious differentiations. The first event, cleavage, is typical of all holoblastic eggs, which may be considered overlarge cells that are in a peculiar metabolic state until fertilization or activation, when, according to Whitaker (1933), more-normal metabolic conditions are restored. With the restoration of normality, the egg undergoes consecutive divisions, which progressively lead to the establishment of cell sizes characteristic of the somatic tissues of the species. Thus the course of cleavage is rapid at first and becomes progressively slower as the later-forming cells approach normal size.

The cleavage process as a whole is essentially the production of as many standard-size cells as can be derived from the initial egg mass. The number produced varies according to the final cell size, which in turn depends on the species and mass—volume and density—of the egg. Thus the small ascidian and echinoderm eggs, both of whose diameter is about 130 μ, can form several thousand cells during the course of cleavage; the number produced among ascidians may vary considerably within the several-thousand range, depending on a particular egg's size and yolk content. In the developing ascidian egg, as in the ascidian bud, the nature of the functional, feeding organism at the close of the first phase of proliferation varies according to the final size, which in the egg depends on the initial egg mass, rather than the size of a maximal bud disk. The small (0.13 mm in diameter) egg of *Styela, Ascidia,* and *Ciona,* for example, yields about 3,500 cells and forms branchial walls perforated by only 3 or 4 single gill slits; whereas the larger (0.7 mm in diameter) egg of *Ecteinascidia* yields about 175,000 cells and forms branchial walls with about 200 gill slits arranged in 12 rows. In other words, unit structures of innately determined size form in numbers proportional to the area of available tissue at the critical developmental time.

Toward the end of the cleavage period, two somewhat competitive developmental activities proceed: the development of the permanent organization, and the precocious differentiation of certain regions to form the special larval structure. The development of the permanent organization of the ascidian adult is in many ways similar to that of the "temporary" organization of a developing ascidian bud. The significant differences are: A bud consists initially of a considerable number of cells, these cells are already in the form of a vesicle with an inner and an outer layer, and the outer layer is a unipotent epidermis—i.e., forms only epidermis. By contrast, the egg divides a number of times before the number of cells is comparable to that in the initial bud, the vesicle (blastula) produced invaginates to form the two layers, and the outer layer is a multipotent ectoderm—i.e., forms other tissues in addition to epidermis. Thus in the embryo the outer layer (ectoderm) invaginates to give rise to the neural complex and the pair of peribranchial (atrial) sacs; whereas in the bud the inner layer infolds to form the neural complex and the sacs. Thus, apart from the limited formative ability of the mature epidermis and apart from the initial difference between the two-layered cell cluster in the bud and the single large egg cell in the embryo, the development of permanent organization is fundamentally similar in bud and egg. Therefore the egg cannot be regarded as having been specifically designed to form a permanent organization. It may thus be concluded that the ability to develop the adult structural organization is latent in any histologically unspecialized somatic cell, so long as conditions conducive to continuing growth and expansion prevail—conditions which are supplied in advance in eggs, in the form of ovocytal growth, and supplied by various means in buds.

Dissociability of Cleavage and Differentiation

It is against this backdrop of proliferation and emergent permanent organization—a setting strictly comparable to that in the development of, for example, a *Botryllus* bud—that the phenomena of specialized ooplasms and precocious differentiation and development of larval tissue and structure are considered in the following discussion. As Needham pointed out, in what he suggested might be called "prolegomena to any future theory of the integration of the developing organism," [2] the various processes of ontogenetic development, though normally in gear with one another, are fundamentally dissociable, as evidenced by a comparative analysis of the development of ascidian eggs whose individual size and yolk content differ.

First, despite the size of a particular egg, gastrulation occurs between the 64-cell and 128-cell stage (i.e., between the sixth and seventh cleavage), and the number of cells constituting the notochord is about 40. Thus, although the pre-

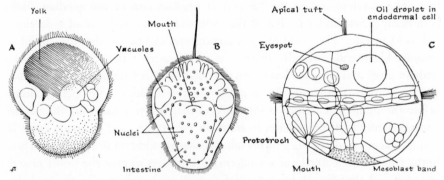

Figure 18-6. Differentiation without cleavage in *Chaetopterus* egg. **A.** Activated but unfertilized, undivided egg after 24 hours. (After Lillie.) **B.** Normal 24-hour larva. (After Lillie.) **C.** Normal *Nereis* trochophore, showing large size of cells of larval tissue. (After Costello.)

[2] Needham, *Biol. Rev.,* **8**:220 (1933).

sumptive chordal cytoplasm undergoes 5-6 divisions regardless of egg size, the presumptive tail-muscle tissue subdivides 5-6 times in the smallest eggs, but 8-10 in the largest; evidently, histodifferentiation of tail-muscle tissue is less rigidly determined than is chordal tissue.

Second, a comparison based on quantity of yolk—which is related to size—reveals a differential effect on the relative rate of development of larval and permanent tissues: With small eggs of minimal yolk content, the tadpole larva is functionally differentiated when only one-fifth of the developmental period has elapsed—long before the organism is capable of feeding. With the larger eggs of the stylid type, larval development is complete about halfway through the developmental period. The slower rate of development in the larger stylid eggs is due to the relatively heavy load of yolk granules incorporated by the chordal cells; as a result, the rate of chordal differentiation, and therefore of larval-structure differentiation as a whole, is retarded, since the notochord seems to be the pacemaker. In the eggs of perophorid, calvelinid, and distomid ascidians, in which the yolk is relatively more condensed in the vegetal half, the yolk content of presumptive chordal tissue is proportionately even higher, so that the rate of cleavage and differentiation of chordal tissue (and tadpole larval structure in general) is extremely retarded. In fact, attainment of histological and functional maturity of larval tissue closely coincides with that of the tissues destined to be part of the permanent adult organization. Thus, because the developmental period of larval tissue can continue into that of adult tissue, the essential duality—larva and adult development—of the total developing system is unmistakable (Berrill, 1935 and 1941). From such observation, it is evident that all the special features of the ascidian egg pertain only to the development of the precociously differentiating transient larval structure. Apart from the convergence of associated structures effected by the gastrulation process, the basic (permanent) developmental process leading to the permanent organization of the adult ascidian has had a secondary (transient) developmental system superimposed on it. This secondary system is active from the very onset of development and is competitive for a territory or material with the basic process. Significantly, independently in several species of molgulid ascidians, whose eggs are minimal in size and in yolk content, the whole larval complex has been genetically eliminated without in any way disturbing the sequence or character of developmental stages in the remaining organism (Berrill, 1931).

In sum, the determinate egg is a specialized cell not only because of its large size and its peculiar metabolic state but because of the specific arrangement of its specialized cytoplasmic territories. Thus specific differentiation of the egg cell is essentially the same as that of, e.g., the sensory retinal cell; the difference is the egg's ability to undergo progressive subdivision and formative development, with the incorporated special ooplasm playing a deflective, accelerative, or reinforcing role. It is equally evident that the pattern of cell division is not determinative, but is determined by the cytoplasmic pattern, and therefore has no influence on the adult organism. (The cleavage pattern is not even discernible after the first few cleavages.) The bilateral symmetry of the crescent ooplasms of the ascidian egg defines the embryonic axis and determines the orientation of the first cleavage spindle so that the cleavage plane and the axis coincide. Similarly, the presence and shape of the myogenic and chordal crescents determine the cellular segregation of these territories, not vice versa; and, according to Fauré-Fremiet and Mugard (1948), in the bivalve *Teredo,* the cleavage pattern conforms to, but does not determine, the regional differentiation of the cortex.

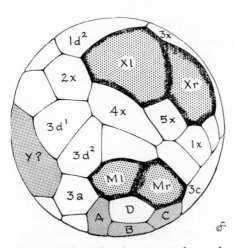

Figure 18-7. Silver-impregnated morula in *Teredo norvegica,* showing correspondence of local cortical differentiation and cleavage pattern. (Letters and numbers refer to cell lineage.) (After Fauré-Fremiet and Mugard.)

It is tempting to postulate a structural basis (presumably of protein character) associated with the cortical layer, not only for the processes concerned in gastrulation but for the animal-vegetal axis as well. In other words, the conclusion would be that the various manipulative and chemical agencies which modify the normal developmental course are all playing upon a highly refined, dynamic, and labile structural instrument. In view of electron-microscope studies on the presence and nature of periodic structure within and outside cells and tissues, it is quite plausible that the explanation of such phenomena will be found in knowledge of ultimate structure.

Chapter 19

Nature and Development
of Eggs (continued)

The Chordate Egg

Some insight into the ooplasmic specialization of the primitive chordate egg may be gained by a comparison among the embryonic developments in the ascidian *Styela,* the larvacean tunicate *Oikopleura,* and *Amphioxus.* Typically, all three types of organisms have small, transparent eggs and develop the basic chordate organization including notochord, hollow dorsal nerve cord, lateral tail muscle, and gill slits. In all three, at or before fertilization, cytoplasm flows to certain zones where it constitutes posterior and anterior crescents which impose or reflect the primary bilaterality of the egg, as shown for *Styela* by Conklin (1905 and 1931), *Oikopleura* by Delsman (1912), and *Amphioxus* by Conklin (1932).

In each of these three organisms, the crescent plasmas, which are histogenetically the prospective chordal, muscle, and neural tissues of the chordate embryo, are initially separated from one another and later brought into mutual alignment by the gastrulation process. The differences among the three lie in the time or stage, relative to the process of cleavage, at which gastrulation and histodifferentiation occur.

Histogenesis and Cleavage

Of the three, the histogenesis within the chordal and the mesodermal crescent is most rapid in *Oikopleura,* less rapid in *Styela,* and slowest in *Amphioxus.* In each, cleavage or subdivision in these zones is gradually inhibited as histological differentiation becomes visible. The most striking features are the close correlation between the rate of histodifferentiation of chordal tissue and that of mesodermal tissue; the correlation between this correlated rate and the time of inception, relative to the cleavage period, of the process of gastrulation; and the marked, unitary manner in which these processes and correlations vary among the three types. In *Amphioxus,* gastrulation occurs between the ninth and tenth cleavage; in *Styela* and all other ascidians, between the sixth and seventh cleavage; and in *Oikopleura,* between the fifth and sixth. In *Amphioxus* the number of both chordal and lateral muscle cells formed from the egg is about 300, which represents between 8 and 9 subdivisions of the original zones. In *Styela* the number of chordal and tail-muscle cells is about 40 each, which represents between 5 and 6 divisions; and in *Oikopleura* the number is 20 each, which represents between 4 and 5 divisions. Since the volume of the original territories subdivided is about the same in all three organisms, the size of the cells produced in each is inversely proportional to their number.

Cleavage and differentiation in chordate eggs

GENUS	DIAMETER OF EGG IN MM	GASTRULATION		NOTOCHORD		TAIL MUSCLE	
		CLEAVAGE NUMBER	APPROX. CELL NUMBER	DIVISIONS OF PRESUMPTIVE MATERIAL	APPROX. CELL NUMBER	DIVISIONS OF PRESUMPTIVE MATERIAL	APPROX. CELL NUMBER
Oikopleura	0.09	5-6	38	4-5	20	4-5	20
Styela	0.13	6-7	76	5-6	40	5-6	36
Amphioxus	0.12	9-10	780	8-9	330	8-9	400
Petromyzon	1.00	11	2,200	9	500	—	—
Trituris	2.60	14	16,000	11-12	1,200	—	—

The conclusion drawn from these facts is that the precocious histogenetic processes and the process culminating in the visible topographical rearrangements effected by gastrulation represent a single, coordinated mechanism which is established in these eggs at or before fertilization and is responsible primarily —and possibly exclusively—for the development of chordate structure and organization. The rate at which this mechanism operates varies as a whole among the three, and it is reasonable to assume that this variation relative to the cleavage rate is due to variations in the initial concentration either of the reacting substances or at least of an enzymatic pacemaker. Thus, compared with the ascidian type of organism, the precocity of differentiation in the larvacean (*Oikopleura*) type represents an acceleration of about 20 per cent relative to cleavage rate; whereas, compared with the larvacean type, in which gastrulation and histodifferentiation occur one division sooner than in the ascidian, the two processes occur three divisions later in *Amphioxus*—a retardation of at least 50 per cent relative to cleavage rate.

Although quantitatively such variations—which can be readily attributed to a 20 per cent acceleration and a 50 per cent retardation of an essentially chemical reaction relative to the general metabolic rate which controls the cleavage rate—are comparatively small, the total effects are striking. In the larvaceans such as *Oikopleura*, the onset of both gastrulation and histodifferentiation seems to represent the extreme in precocity, compared with the ascidian type as a standard; for it would seem to be mechanically impossible for either gastrulation or histodifferentiation to occur in a blastula earlier than the 32-cell stage. On the other hand, in *Amphioxus* the relative retardation which amounts to a postponement of gastrulation and histodifferentiation by only three cleavages, has striking consequences. In both the ascidian and the larvacean tunicate, the cell numbers in the precociously differentiating chordal and mesodermal tissues are so small and the cell sizes so large that direct segregation is the only possible method of separating these tissues from adjacent tissues; i.e., the precociously differentiating tissue simply slides out of its position of immediate contact with adjacent tissues. By contrast the three additional divisions in *Amphioxus* increase cell numbers and decrease cell sizes by a factor of 8, so that evaginative separation of histogenetically related tissue is possible. The contrast can be visualized by examining the effect upon the development of the ascidian egg of retarding the determinative process relative to cleavage to the same degree.

Thus if the gearing ratio of cleavage to gastrulation and histodifferentiation of the ascidian egg is shifted to that of *Amphioxus*, segmentation in the ascidian egg will proceed beyond the ninth cleavage (three divisions later than normally) before gastrulation begins. Consequently, a blastula similar to that of *Amphioxus* is produced, and invagination into a blastocoele, or large segmentation cavity, occurs. The formation of the large, hollow blastula does not involve any process of determination; its developmental existence is merely cut short by the onset of gastrulation. Rather, it results from the segmentation of the egg, the adhesion of cells to form an epithelium, and the innate attempt of each cell to assume as spherical a shape as possible. With every successive cleavage, the wall of the blastula gets thinner, its surface area more extensive, and its blastocoele larger. If the onset of gastrulation were postponed by one more division, the blastocoele might survive gastrulation as it does in echinoderms. Furthermore, differentiation of chordaplasm and myoplasm would be retarded correspondingly, so that each would give rise to relatively large numbers of comparatively small cells.

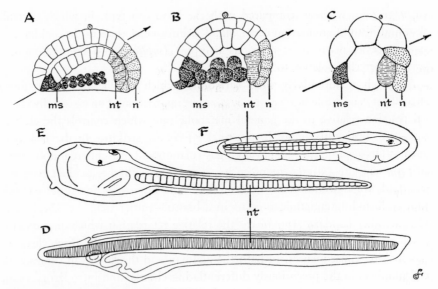

Figure 19-1. Development in *Amphioxus, Ascidia,* and *Oikopleura,* showing general similarity of segregation and relationship between cleavage stage at gastrulation and nature of notochord and larva. **ms.** Tail mesoderm. **nt.** Notochord. **n.** Neural. **A** and **D.** *Amphioxus.* **B** and **E.** *Ascidia.* **C** and **F.** *Oikopleura.* (After Conklin and Delsman.)

Postponement of the onset of gastrulation from the sixth to the ninth cleavage would nearly double the period from fertilization, for intercleavage intervals become successively longer, especially after the seventh cleavage. Similarly, the interval between gastrulation and the actual establishment of the notochord and tail-muscle bands would be longer, so that differentiation of these tissues would be completed during the late- and postgastrular period; at this time, there are both sufficient space and sufficient cells of small enough size for the formation of notochord and mesodermal units by evagination, as in *Amphioxus*.

From the analysis by Conklin (1932) of *Amphioxus* embryology, it is evident that the origin of the notochord is essentially the same as in ascidians: subdivision of a chordal crescent which undergoes the same relative translocation, but differentiates more slowly than in ascidians and continues through three more divisions. The plate of chordal tissue produced then separates from the adjoining endoderm by folding off dorsally, i.e., evaginating, from the archenteron. Although the *mode* of separation of histogenetically distinct tissue is specifically prescribed in each of the three types of organisms, it seems unlikely that the actual process of direct segregation, in the more-condensed larvacean and ascidian types, is fundamentally different from the process of evaginative separation in *Amphioxus,* whose tissue cells are smaller, more numerous, and united as an epithelium. Rather, the same primary activities seem to be involved, and the somewhat different methods seem to result from the different mechanical situations. In other words, evagination and direct segregation have essentially the same physical and chemical basis—the nature of which remains an outstanding problem in biology.

It is noteworthy that the development of isolated blastomeres of *Amphioxus* depends on whether they represent either the left or right half of the originally bilaterally symmetrical egg—i.e., whether they contain an adequate amount of both the chordal and the mesodermal crescent—or whether they represent either the anterior or the posterior half, both of which inevitably lack either the chordal or the mesodermal component. When a blastomere contains both crescents complete, though, half-size larvae develop; but when it does not, only partly developed larvae result, according to Conklin (1933). Despite this dependence on the presence and topographical symmetries of the initial ooplasms, the developmental process in the *Amphioxus* embryo is quite reminiscent of that in the ascidian bud: the process as a whole seems to consist essentially of the extension, infoldings, and outfoldings of a simple epithelial tissue. The final organizational and structural pattern is, of course, different; but the general means of attaining the individualistic patterns seems to be basically the same, despite the presence of specialized, but perhaps diluted, ooplasms in the *Amphioxus* egg.

The comparison among larvacean, ascidian, and *Amphioxus* development is also interesting in another way. As indicated above, the ascidian tadpole larva typically matures as a nonfeeding larva functioning to secure a suitable settling site for permanent attachments; well after the larva is permanently attached—in fact, when its tissues have disintegrated—the developmental terminal of the egg proper is attained.

If the following interpretation (Berrill, 1955)—that the simple, functionally mature ascidian organization is fundamentally primitive and that the chordate tadpole larva has evolved within the ascidians for the explicit function of selecting a suitable settlement site—is valid, other chordates are essentially evolutionary neoteinic derivatives of this kind of larva. Whether or not this interpretation is accepted, it is significant that both in the larvaceans such as *Oikopleura* and in *Amphioxus,* metamorphosis does not involve the destruction of the basic chordate features. In *Oikopleura* there is virtually no doubt that the larval character is a permanent feature; for the tail persists throughout the life of the organism, and all postlarval growth of the 20 notochord cells and of the two bands of 9 tail-muscle cells, which are formed early in embryonic development, consists only of cell enlargement, and not of proliferation. In other words, histological stability, possibly associated with the intensity of ooplasmic determination, is acquired early and is maintained throughout life.

Although the type of metamorphosis in *Amphioxus* is different from that in ascidians, the manner of feeding and general behavior of the adult in both are quite similar. In *Amphioxus*, the comparatively weakly determined ooplasmic pattern in the uncleaved egg leads not only to a larval type similar to that of the ascidian—except for rate of histogenesis—but to the specifically determined larval tissues and to tissues comparable to the permanent tissues of the ascidian. However, both the larval and the presumptive permanent tissues grow simultaneously, but in different ways, in *Amphioxus*. Apparently, not only is metamorphosis virtually suppressed, but the larval pattern initiated by the distinctive ooplasms becomes part of the permanent organization. (For a general discussion of the nature of metamorphosis, see p. 525.)

The Vertebrate Egg

The type of primitive vertebrate egg typical of most amphibians, the more-primitive bony fishes, and lampreys, is clearly the same type in all three. In contrast to the echinoderms and the lower-chordate egg, whose diameter range is 0.1-0.2 mm, the fresh-water vertebrate egg has a diameter range of 1.0-2.5 mm, which represents a thousandfold increase in volume or mass. By means of differential dye reduction and other methods, Child (1928) has shown that the early development of the amphibian egg is characterized by a gradient whose high point is at the animal pole. Dalcq and Pasteels (1937) postulate two gradients: one corresponds to the obvious vitelline gradient, as evidenced by the progressively decreasing size of yolk platelets from the animal to the vegetal pole; and the other, a cortical, sulfhydryl-ribonucleoprotein gradient whose high point is at the dorsal side of the egg. Whatever the specific nature of such a gradient, the primary question is whether there is a balanced system at all.

Experiments on early dorsal-lip stages of the embryos of *Rana pipiens* by Paterson (1957) indicate that such a system is necessary for normal development. In these experiments the animal third, the vegetal third, or both, were removed from the embryo (the vegetal cut was consistently made above the dorsal lip). The vegetal two-thirds differentiated into a new embryo; the animal two-thirds did not—in fact, the ectoderm and endoderm remained completely undifferentiated; but the middle third alone developed into a normal embryo, regenerating a new dorsal lip, in about 50 per cent of the experiments. An animal-vegetal balance is further indicated by the results of experiments in which only the vegetal third was removed: the middle third did not develop into a swim-

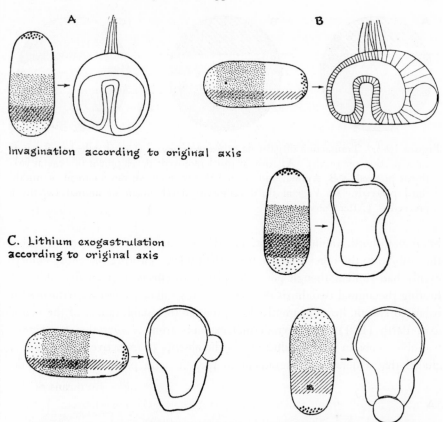

Figure 19-2. Sea-urchin eggs stretched by centrifugal force, showing stratification of light, clear, and heavy material and also unchanged position of cortical pigment band (crosshatched), followed by nearly normal larval development relative to original egg axis. **A.** Stretched perpendicularly. **B.** Stretched in line with egg axis. **C.** Irrespective of axis of stratification in relation to egg axis, eggs are equally sensitive to exogastrulative effect of lithium ions. (After Hörstadius.)

ming tadpole. Apparently, the presence of the animal third inhibits such development.

The general conformity of Paterson's result to Hörstadius' (1949) findings with isolated and recombined layers of the sea-urchin embryo is unmistakable, except that treatment with lithium chloride in *Rana pipiens* does not restore the balance in the animal two-thirds, whereas animal halves of sea-urchin blastulae treated with lithium may gastrulate (Hörstadius, 1953 and 1955). However, differential heat treatment of an intact embryo is most effective, Paterson found. Heating the animal two-thirds had no effect, i.e., there was no further differentiation; but heating the vegetal third stimulated typical development of

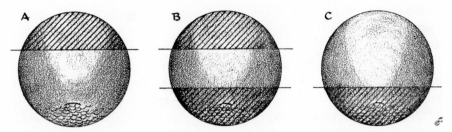

Figure 19-3. Transaction of gastrula of *Rana pipiens*. (Crosshatched region indicates portion removed.) **A.** Animal third removed; development of vegetal two-thirds proceeds. **B.** Animal and vegetal thirds removed; development of middle third proceeds. **C.** Vegetal third removed; development of animal two-thirds proceeds. (After Paterson.)

brain, notochord, somites, and gut. The temperature difference from one end of the embryo to the other was not more than 0.1°C. Heating the entire embryo evenly had similar, though not so pronounced, effects. Incidentally, although heating the animal two-thirds did not have any salutary effect on restoring the balanced system, heat apparently does promote the establishment of the neural axis. Barth (1941), using ectodermal explants from *Ambystoma punctatum,* was able to obtain neural inductions in the absence of inductive tissue; he concluded that a dorsoventral gradient in the ectoderm plays a role in the de-

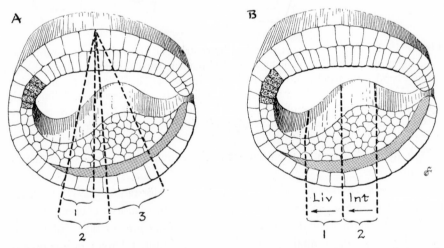

Figure 19-4. Early neurula of *Hyla*. **A1, 2,** and **3.** Experiments consisting of removal of anterior quarter, anterior third, and middle third of mesenteron, respectively. **B1** and **2.** Experiments consisting of rotating anterior third of midgut and middle third, involving prospective liver **Liv** and intestine **Int**, respectively. (After Kemp.)

termination of the nervous system. Since, in all these experiments the neural axis was parallel to the groove along which the temperature was graded, a spatial control of the axis seems certain.

A somewhat similar process of spatial regulation is observed in the archenteron of the neurula stage of *Hyla regilla*. Experiments by Kemp (1946) consisted of (1) cutting the neurula longitudinally and rejoining it so that one side is in the reversed position; (2) extirpating a wedge-shaped sector, including the middle-third of the archenteron, from the ventral side; (3) introducing a similar wedge into the middle region; (4) removing a middle section and reinserting it in reversed position; and (5) leaving the dorsal layers of the embryo intact. The results were: regional epithelial characteristics are retained, but a single digestive tube forms from two histologically different primordia; normal form is re-established after extirpation of the central third or the anterior quarter; a normal structure is obtained from the fusion of two primordia after the addition of histologically similar material; and normal polarity of liver and pancreas is attained despite the reversal of tissue position. Apparently, as long as the greater part of the endodermal tube remains intact, the tube is able to impose its polarity upon the whole embryo, so that the whole differentiates regionally according to position, much as in the case of the reorganizing gut in *Eudistoma* (p. 369).

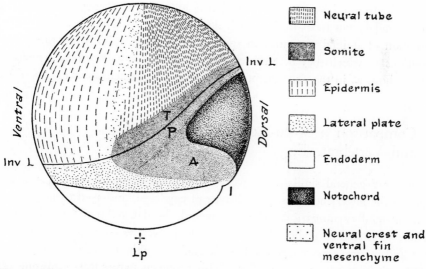

Figure 19-5. Map of presumptive regions of early gastrula of urodele *Triturus pyrrhogaster*. **A.** Somites 1–9 in anterior region of trunk. **P.** Somites in posterior region of trunk. **T.** Somites in tail region. **1.** Site of first invagination. **Inv L.** Invagination line. **Lp.** Lowermost pole at this stage. (After Nakamura.)

On the other hand, the presumptive endoderm of gastrulae and early neurulae of *Triturus pyrrhogaster* can differentiate into tissues and organs so long as the endoderm is surrounded by mesenchyme. Okada (1955) combined pieces of anterior and posterior endoderm with pieces of mesoderm from various parts of the marginal area. He found that no matter where the mesoderm came from, the anterior endoderm differentiated into pharynx—and sometimes into stomach and intestine as well—and the posterior into intestine—and sometimes into pharynx and stomach as well. This pattern of differentiation is similar to that in the regeneration of a gonophoral zone in both parts of a transected *Tubularia* regenerate (p. 264). So long as a large amount of mesenchymatous tissue was included in each endodermal explant, nearly all parts of the *Triturus* digestive tract differentiated from each small explant.

Thus, in the neurula stage, each area of presumptive endoderm is neither irrevocably determined nor self-differentiating. Moreover, since the mesodermal tissue is effective, regardless of its original site in the marginal area, it cannot impose pattern. Rather, mesoderm apparently stimulates a nonspecific induction or general growth in order to cause a small piece of endoderm to assume the general character of the whole. The seat of organization becomes even more obscure in those experiments of Oppenheimer (1955) involving the implantation of shield grafts into the embryo of the teleost *Fundulus*. The gut-forming cells induced in the shield become located at some distance from the graft and from other, secondary structures; and tissue movements which carry the secondary gut tubes to their final destination not only result in the differentiation of unified structures in specific locations but also bring them into positions in which they become selectively continuous with related structures of the host, as in the experiments of Scott on fragmented ascidians (p. 375).

Comparison between Lower-Chordate and Vertebrate Egg

There is a general similarity between the holoblastic vertebrate egg and the eggs of the lower chordates. In each type of egg, crescent-shaped regions represent bilaterally symmetrical prospective neural, chordal, and mesodermal tissues. There is an unmistakable likeness between the yellow crescent of the *Styela* egg and the grey crescent of the frog or toad egg, although in the anuran egg ooplasmic bilaterality is not established either soon enough or intensively enough to induce the same rigidity of correspondence between mitotic orientation and cytoplasmic symmetry as in the ascidian or *Amphioxus* egg.

The most obvious initial difference, that of size, is reflected in the final number of cells resulting from cleavage. The typical ascidian egg divides to produce about 4,000 cells, and the similar-size egg of *Amphioxus* produces about 8,000 somewhat smaller cells. The egg of a frog, on the other hand, subdivides to produce almost a million cells whose individual dimensions are comparatively larger than those of either the ascidian or *Amphioxus* egg. The sheer increase in mass and in potential cell number at the end of embryonic development, which coincides with the onset of feeding state, is clearly pertinent to the morphogenetic and histogenetic processes in such an egg.

Another less-obvious difference is the rate of cleavage and of development. The small, transparent ascidian egg of, e.g., *Ascidia, Ciona,* and *Styela,* cleaves at about the same rate as the same-size egg of *Amphioxus,* provided the temperature is the same. At a temperature of about 18°C, the egg of *Ascidia,* for instance, undergoes subdivision first about once an hour, commences gastrulation 7 hours after fertilization (i.e., between the sixth and the seventh cleavage), and begins neurulation 10 hours after fertilization. The larger eggs of other ascidians—such as *Perophora* (0.25 mm in diameter), Botryllus (0.45 mm), and *Ecteinascidia* (0.72 mm)—divide more slowly in direct relation to the volume:surface ratio of the egg; for example, in *Ectinascidia* the intercleavage interval is at first almost 10 hours (at 18°C), and gastrulation occurs about 110 hours after fertilization. By extrapolation, an ascidian egg 1.0 mm in diameter would require several weeks to complete its development. Accordingly, it is rather surprising that the developmental rate of primitive vertebrate eggs, whose diameter range is 1.0-2.0 mm, is not comparably retarded. According to Weisz (1945), the egg of the so-called African toad *Xenopus laevis* (1.0 mm), for example, though relatively less yolky than most other amphibian eggs, undergoes cleavage about once an hour at first, completes gastrulation between the seventh and thirteenth hour after fertilization, and commences neurulation 15 hours after fertilization. Thus, despite the tenfold difference in diameter between the average-size *Xenopus* egg and the smallest ascidian egg, the cleavage and developmental rate in the former is almost as rapid as that in the latter. The larger eggs of some species of *Rana* also divide roughly every hour; but the interval between fertilization and the onset of gastrulation is relatively long, and the subsequent period of development is of somewhat longer duration. Even in *Rana,* however, the rates are comparable to those of *Xenopus* and not to the slow pace of the larger ascidian eggs. It may be significant that although the egg of the lamprey *Petromyzon* is about the same size as that of the amphibian and its development generally resembles that of a urodele, gastrulation apparently occurs at a relatively early cleavage stage and the number of embryonic chordal cells is correspondingly small. This phenomenon suggests that there is a quantitative difference in the specialized ooplasms compared with those of the amphibian—a difference comparable to that between the larvacean and ascidian eggs among tunicates (see the table on p. 484).

Apart from the yolk gradient in the holoblastic vertebrate egg, resulting in the contrast in cell size and cell number between the animal and the vegetal hemisphere and resulting in the displacement of the blastocoele and invagination center toward the animal pole, possibly the most significant difference between the primitive chordate and the vertebrate egg concerns the cellular constitution of the blastula wall. Instead of being a continuous simple epithelium, as in *Amphioxus,* which transforms entirely by invaginations and evaginations, the blastula wall in the vertebrate egg is at least two cell-layers thick throughout and even thicker in the animal hemisphere, where the more-important morphogenetic events occur. There is, however, considerable variation even among amphibians, not only between anurans and urodeles as a whole but between one anuran and another, with regard to both the cellular make-up of the pregastrular wall and the relative role played by the superficial coat.

According to Holtfreter (1943), the surface layer of the amphibian egg is formed before, not after, fertilization, and is not hyaline, but contains small yolk grains and in some regions is characterized by a dense accumulation of spherical granules of black pigment. In contrast to the echinoderm egg, in which the hyaline layer apparently does not participate in the cleavage process, the coating of the amphibian egg is superficially divided by the dividing blastomeres. It is a more-integral and indispensable part of the living cell; for when torn off experimentally, the cytoplasm, having no other means of support, dis-

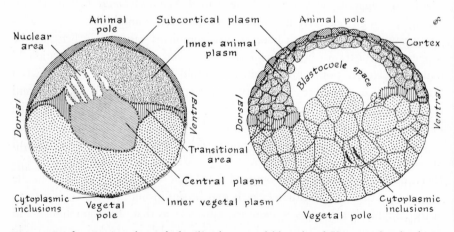

Figure 19-6. Section through fertilized egg and blastula of *Xenopus laevis,* showing same distribution of plasms in two stages and segregation of outer and inner cell layer separating subcortical plasm from inner animal plasm. (After Nieuwkoop.)

integrates. If the surface of a fertilized or unfertilized egg or of a blastomere is even slightly injured with a glass needle, the wound edges retract to several times the original width of the surface—a phenomenon indicating that the coat is normally in a state of tangential stress.

In *Xenopus laevis,* the nature and role of the surface coat are somewhat clearer than in other anurans studied. During cleavage and formation of the early blastula, the entire surface becomes covered by a superficial cellular layer which is fused to the cortical film and sharply differentiated from the deeper layer. This superficial layer consists of small, cubical cells in the animal hemisphere and larger cells, richer in yolk granules, in the vegetal hemisphere, with apparently no zone of intermediate-size cells between the two hemispheres. In the course of development, the animal part of this layer forms only the superficial layer of the ectoderm and proctodeal ectoderm; the vegetal part invaginates entirely to become the endodermal epithelium of the archenteric tube; and the deeper layer alone gives rise to both the neural ectoderm and the mesoderm. Thus morphogenesis is already greatly conditioned by the simple process of cortical delamination instead of by the foldings characteristic of the *Amphioxus* blastula. The delamination process in the amphibian egg may well result from the surface-depth differentiation of the relatively massive cortical region.

Gastrulation

The first visible sign of gastrulation is the transverse, crescent inrolling at the center of the grey crescent area of the anuran embryo and of the corresponding area of the urodele embryo. In the urodele, according to Schechtman (1934), this invaginative process is preceded by a unipolar ingression, during which the surface layer of the vegetal pole expands into the deeper layer. Schechtman believes that this unipolar ingrowth is effected by an even earlier ingrowth of an underlying gelated cortex material which drags the pigmented surface along with it. Somewhat similar and more-extensive movements of both the vegetal and animal hemispheres of the pregastrular embryo of *Xenopus* have been described by Pasteels (1949).

At the time of invagination, the blastoporal tissue already possesses the innate capacity for organized development; for, according to Holtfreter (1944), when the blastoporal lips of *Ambystoma punctatum* are excised, exposed to alkali for 10 minutes (so that the lip tissues disintegrate into individual cells), subsequently intermingled by a glass needle, and finally neutralized, the cells reaggregate and reorganize into an axial system with two cellular layers.

The actual process of amphibian gastrulation has been described for the anuran *Hyla regilla* by Schechtman (1942) in terms of tissue mechanics. The presumptive region of the dorsal blastoporal lip has an autonomous capacity for extension under diverse experimental conditions; however, the invaginative process requires continuity between the presumptive chordal tissue and the lateral regions of the marginal zone and is therefore correlative. On the other hand, the circumblastoporal region of the dorsal lip—i.e., the presumptive pharyngeal endoderm and head mesoderm—has an inherent capacity for invagination, as evidenced by explanation, transplantation, and isolation experiments. Although the lateral regions of the marginal zone, including the presumptive somites and tail mesoderm, have inherent capacities for stretching, involution, and invagination, these regions are incapable of dorsal convergence and of constriction over the yolk mass. The latter activities take place only if there is continuity between the lateral marginal zones and the dorsal-lip tissues; however, the capacity to converge and constrict is not limited to specific regions of the marginal zone, for the ventral marginal zone of the early gastrula can perform the same functions of the dorsal or dorsolateral marginal zones when experimentally placed in the same position.

It seems, therefore, that the presumptive chorda is invaginated by the inward-directed tension, or pull, exerted by the invagination and involution of

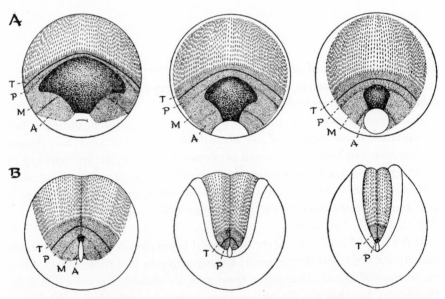

Figure 19-7. Gastrulation and neurulation in *Triturus pyrrhogaster*. **A.** Early, middle, and late gastrular stages from posterior aspect. **B.** Late gastrular (or early neurular), middle neurular, and later neurular stages. **A, P, T,** and shading as in **Fig. 19-3. M.** Mesoderm. (After Nakamura.)

the lateral marginal zones and that the lateromarginal zones are then pulled downward and inward in the dorsal position by the autonomous stretching and simultaneous narrowing of the presumptive chorda. Finally, constriction of the blastoporal lips over the yolk mass results from the progressive withdrawal of marginal-zone tissue by dorsal convergence, i.e., by a purse-string effect.

The stretching process leading to gastrulation involves the surface material of the egg and embryo as a whole, and not merely the region around the invagination center. Since the tissue movements are clearly of a supracellular nature, the fundamental question is: What is the nature of the directing force or substance? Even in aging, unfertilized frog eggs, according to Holtfreter (1943), a simulation of the stretching and convergent movements of the cortical material takes place without any cell division taking place. Diffuse streaks of darker pigment radiate from the animal zone across the dorsal grey crescent toward the imaginary blastopore; and other cortical patterns recall the cellular subdivision and convergent stretching of the dorsal mesoderm of a developing embryo. The cortical expansion in the aging, unfertilized egg consists of a centrifugal movement of particles along lines of flow toward the blastoporal region. Thus "the segmentation of the egg substance seems to be a superimposed process that interferes with the tendency of expansion which is of a submicroscopic nature." [1] As Holtfreter points out, this concept is not new: Differentiation without cleavage has been demonstrated in *Chaetopterus* by Lillie (1902 and 1906); for fertilized and unfertilized eggs of this polychaete can develop into ciliated, trochophorelike structures without undergoing cell division. In such development, the ectoplasmic layer stretches toward the vegetal pole while the endoplasm moves further into the interior, where it assumes a bilaterally symmetrical arrangement.

That epibolic spreading is the result of a submicroscopic process of fibrilization has been suggested by a number of investigators, who recognize that such spreading or extension requires a definite surface (such as yolk, endoderm, or even glass) over which to spread. In epibolic extension in *Fundulus,* according to Trinkaus (1951), the factors primarily responsible for blastodermal extension are intrinsic; the blastoderm expands after it has been provided with a specific substratum to which it can adhere; and the periblast, which has the capacity to expand over the yolk independently of the blastoderm, is that substratum. The data support Morgan's earlier suggestion (1895) that epiboly of the blastoderm and the periblast entails expansion of material already in these areas at the onset of gastrulation and, by projection, probably much earlier—possibly soon after fertilization. Similarly, Trifonowa (1934), after parthenogenetically stimu-

---

[1] Holtfreter, *J. Exp. Zool.,* **94**:287 (1943).

lating the eggs of various fresh-water teleosts, observed that the blastoderm commonly spreads over the yolk in an epibolic movement even when no cell or nuclear division takes place. There is little doubt that subdivision into cells, in this connection, is a secondary process, although it is a mechanical necessity for the extension of such movements in the form of invaginative processes.

The question remains: When the cellular state does exist, are the epibolic movements of the whole sheet those of individual cells acting in unison, or are the cells caught up in, or subordinate to, some supracellular entity. Holtfreter observed that isolated ectodermal cells of the amphibian embryo expand in much the same way as when they are incorporated in an epithelial sheet—

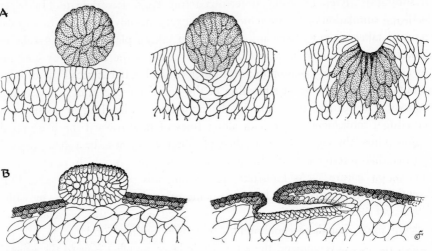

Figure 19-8. Blastopores in *Ambystoma*. **A.** Graft of blastoporal cells, partly covered by surface coat, sinks into endodermal substratum and forms blastoporal groove. **B.** Coated piece of marginal zone of blastopore invaginates under host ectoderm and forms archenteron. (After Holtfreter.)

a phenomenon which indicates their common inheritance of the expansive material or property, but does not directly answer the question. He suggests that the significance of the substratum for the spreading of cells seems to be that it orients the polar molecules of the protoplasm in definite patterns, and that the syncitial surface coat of the amphibian egg and of all epithelia derived from it is primarily instrumental in lowering the surface tension of the peripheral layers so that the layers can spread. He also considers the morphogenetic movements that are associated with gastrulation as being significantly related to specific changes in the shape and arrangement of the cells involved.

As endodermal invagination proceeds, the coated surface layer extends into the interior to form the superficial surface of the archenteron. The continuing expansion or spreading movement, according to Holtfreter (1944), is manifested in the former blastoporal area—which at this time occupies the mid-ventral archenteron—by a folding up of the lateral edges of the hind gut to form the dorsal roof of epithelium and by a general widening of the pharyngeal cavity.

> The endodermal spreading process is in principle identical with the epiboly of the ectoderm; in fact, both germ layers are potential competitors in their tendency to cover surface areas. . . . The driving power in both cases resides mainly within the coated surface layer.[2]

This tendency to cover is shown in the experimental removal of the anterior part of both the neural plate and the thin mesodermal layer of an *Ambystoma* neurula; the endoderm was left intact, except for two perforations in the dorsal epithelium. The uncoated surface of the exposed endoderm was rapidly covered by the remaining epidermis and by the coated epithelium of the archenteron: the epidermis simply moved forward; the epithelium spread over the denuded surface by passing through the dorsal holes. Close examination showed that the advancing fringe of the coated endoderm was not separated into solitary cells, that it had a well-marked boundary line, and that the spreading cells were flattened and strongly elongated in the direction of movement (cf. wound healing, p. 115); whereas during the covering process the cells of the uncoated endoderm decreased in individual surface area to become cylindrical and even bottle-shaped.

> These antagonistic potencies—expansion or contraction of the cell layer—are always present within the endoderm. . . . It is the non-adhesiveness of the coat that prevents a mutual overgrowth of the different epithelia. . . . Invagination must be due to the specific location of the endoderm within a system that suppresses the membranous expansion and compels the cells to reduce their surface area. . . . We have to look for factors additional to the competitive ectodermal spreading to account for the invagination movement.[3]

Other experiments showed that the cells of the blastoporal lip are endowed with an inherent tendency to stretch in a uniform direction and that this innate tendency, rather than external mechanical forces, is the principal cause of their directed movement within the framework of gastrulation—a phenomenon recalling the activity of the mesenchymal cells associated with gastrulation in the sea urchin (p. 142).

Further experiments by Holtfreter on the mechanics of invagination showed that (1) contracted endodermal cells of the blastopore can penetrate into an adequate cellular substratum and assume a bottle shape; (2) the migrating

---

[2] Holtfreter, *op. cit.*, **95**:171 (1944).
[3] Holtfreter, *op. cit.*, p. 175.

cells, by dragging along in their wake an elastic, tapering neck portion together with part of the surface coat, are responsible for the formation and deepening of the blastopore; and (3) adjacent cells become stretched in the direction of the invaginating cells. He concludes that invagination is mainly a matter of adsorption and cell polarity. According to this interpretation, cellular invagination is a special case of spreading, and whether a graft will infiltrate another cell layer or spread over its surface depends both on the graft's intercellular cohesiveness and on the gradient of surface tension between the graft and the other layer. This gradient and the integrative action of the syncitial surface-coat layer are considered to be the supracellular factors which integrate the single cells into collective mass movements. That the gradient and the surface coat are important is indubitable; but, for a complete explanation of mass tissue movements, they are inadequate—particularly in view of the complex morphogenetic foldings of simple epithelia in the absence of any firm substratum.

In *Xenopus,* according to Nieuwkoop and Florschütz (1950), the active role of the superficial coat in both gastrulation and neurulation is very pronounced, but the role of the cortical film is a temporary phenomenon limited to precocious development in Anura. Furthermore, these investigators suggest that in urodeles the superficial marginal tissue plays an active role in gastrulation and in the induction of the neural plate; whereas in anurans this tissue probably does not participate in neural induction, but only directs the invagination of the archenteron, so that the neural plate is induced by internal tissue alone.

Thus, in *Xenopus* the phenomenon of inrollment is accompanied by strong epiboly of the internal animal material.

Xenopus, in fact, shows a somewhat aberrant, but relatively simple type of anuran development. From an early stage onward, the developing egg is clearly double-layered. The outer, superficial layer represents both the presumptive epithelial layer of the ectoderm and the lining of the entodermal epithelium of the archenteron. The inner layer, however, consists of three presumptive areas: an animal cap, which represents the sensorial layer of the ectoderm; an equatorial ring, which represents the inner marginal zone or future mesodermal mantle; and a thick vegetative layer, which represents the nutritive yolk mass. During early development the presumptive sensorial layer of the ectoderm and the presumptive marginal-zone material—which forms, among other things, the dorsal archenteron roof—are constantly separated from the cortical layer by the subcortical plasm. According to Nieuwkoop (1956), the inner marginal zone, from which the craniocaudal axis develops, seems to develop from a material which at no time has any direct topographical relationship to the cortical layer of the egg.

Physiological relationships, if present, must therefore be mediated by the layer of subcortical plasm which separates both. These relationships can only be established during the developmental period previous to the middle blastula stage, at which intercellular spaces begin to appear between both layers, separating them physiologically (physiological interaction is generally characterized by an intimate cellular contact).[4]

Neurulation and Induction

The phenomenon of neurulation, particularly in developing amphibian embryos, has been the subject of numerous investigations and much speculation. In the selective discussion that follows, the primary problems in neurulation—the actual nature or mechanics of the transformation of the medullary plate into the neural tube and its derivative, and the inductive and self-differentiating properties of the invaginating archenteron and chordomesodermal sheet—are briefly surveyed.

In general the conversion process is comparable to the invaginative process of gastrulation and raises all the problems of mass movement of tissues. Townes and Holtfreter (1955) mixed up cells from various parts of a neurula-stage amphibian embryo and found that such cells perform individually the same kinds of directed movements as do corresponding tissue fragments of the same kind of cell. For example, neural-plate or mesodermal cells that have been combined with either epidermal or endodermal cells move inward and unite with one another while the epidermal cells move outward and undergo spreading at the periphery of the plate (p. 105). Tissue segregation becomes complete because of the emergence of a selectivity of cell adhesion. However, the surface coat of a neural-plate fragment, as in the whole plate, remains intact and retains its normal property of distal nonadhesiveness. During neural invagination the coated surface contracts; and though such contraction is not the cause of the invagination, the coated surface becomes significantly engaged in the formation and maintenance of the central lumen. Moreover, a neurocoele can form not only as a result of the infolding of any portion of the coated neural plate but also through secondary cavitation of a compact mass formed by individual neurogenic cells. This secondary process is a more complex version of the basic problem of mass tissue movement.

During the neurulation process the number of cells clearly increases in both the plate and the adjoining epidermis, and the neural-plate cells especially undergo marked changes in shape and possibly size. In *Ambystoma maculatum,*

[4] Nieuwkoop, *Staz. zool. Pubb.*, **28**:247 (1956).

Gillette (1944) has shown that despite the change in cell shape, the size of both neural and epidermal cells decreases slightly during neurulation, and that cell proliferation, although detectable, is insufficient to account for the relative changes in total areas of the tissues concerned and is not significantly different from the proliferative rate generally. Gillette concludes that the characteristic thickening of the neural plate is accomplished by a tangential contraction of the plate toward the dorsal mid-line and by a corresponding expansion and thinning of the nonneural epidermis.

Folding of the neural plate seems to be a continuation of this contraction process; and since contraction proceeds at a greater rate at the external surface than in the depths of the plate, Gillette considers the surface coat responsible for the folding of the neural plate. It seems more likely, however—despite the indisputable fact that the coated layer does contract at this time—that several other factors participate in the actual formative event: the increased cohesion of neural cells, the tendency of distinct tissues to segregate from one another, and an *expansion* of the deeper (or basal) material of the neural tissue. The evident contraction of the external, coated surface may be no more than a response to a decrease of, or release from, a state of tangential tension (see the discussion on p. 385 of the factors concerned in the formation of the inner vesicle of a *Botryllus* bud).

Once again the problematical nature of an invaginative or evaginative movement arises. In amphibian neurulae the simple, columnar neural plate, as convergence toward the mid-dorsal line proceeds, transforms into an infolding mass several cells thick, and the two surfaces of the mass can be regarded as no more than *contributory* to the event. It is possible, even likely, that symplasmic union between cells takes place and that the "cohesion" involved is comparable to the ectoplasmic fusion, which seems to be responsible for the invaginative process, in developing *Volvox* colonies (p. 130). In other words, although direct evidence in amphibian embryos is lacking, the entire invaginative process involved in the condensation and infolding of the neural plate may be just a contraction of symplasmic fibrillar matter (at the ectoplasmic level) extending not only through the superficial and basal surfaces of the plate tissue but, more significantly, through the conjoined vertical cell walls and passing from the neural margin toward the mid-dorsal line. Furthermore, as the contraction process continues, the individual cells slip among themselves in a manner that suggests a process of flow in the ectoplasmic matter.

An ectoplasmic flow may be indicated by experiments of Roach (1945), in which the anteroposterior axis of neural-plate grafts, with and without underlying mesoderm, was reversed during early neurular and preneurular stages in *Ambystoma*. In both types of experiments, i.e., with and without mesoderm,

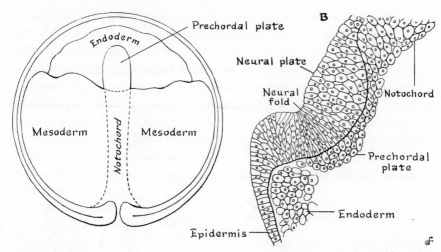

Figure 19-9. Early neurular stages. **A.** Diagram of early neurula of *Triton,* showing extent of underlying endoderm, mesoderm, notochord, and prechordal plate. (After Vogt.) **B.** Median section through cranial part of young neurula of *Xenopus,* showing neural fold and plate in relation to same underlying tissue. (After Pasteels.)

the neural pattern that developed was reversed—a phenomenon demonstrating that neural pattern was already established in the medullary ectoderm of the preneurular stage. On the other hand, after reversal of the lateromedial axis, with and without substratum, the grafts developed according to their new position; evidently this axis—the axis of flow toward the mid-dorsal line— remains labile. Fibrillar organization, of a liquid crystal quality, may extend throughout the neural plate at all stages; and formation of the neural tube may be primarily a combination of flow, condensation, extension, and orientation of the symplasmic substance. Presumably, individual cells can withdraw from or become part of such a system, or can unite to form one, without losing their innate individuality or seriously disturbing the system itself. The phenomena of wound healing and of coherence in slime molds present basically the same problems.

The stimulus for neural differentiation and the determination of limits of the neural plate derive from the underlying invaginated archenteron and dorsal mesoderm, as Spemann and his students demonstrated. The nature of this stimulus and of its transmission is still a subject of conflicting opinion. Thus, Waddington (1956) describes the essential process as the liberation of an "evocator" at the blastoporal lip, its transport cranially by the mesoderm, its diffusion laterally within the mesoderm and dorsally into the overlying ectoderm; and he attributes many of the properties of the so-called neural field to the spatiotemporal relations of these sequential aspects of the process.

Nieuwkoop (1952), however, regarding this concept inadequate, summarizes the development of pattern in the central nervous system as a very close interaction of four factors: the spatially structured inductive influence of the archenteron roof, the competence of the overlying ectoderm, the inherent property in the ectoderm of autonomous self-organization, and time—on which the other three aspects depend. According to this interpretation, there are no qualitative regional differences in the primary induction of the whole central nervous system; rather, a single, general induction from the underlying archenteron roof evokes the formation of a single, homogeneous neural field which spreads

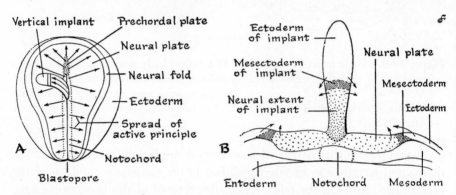

Figure 19-10. Spread of activating principle in early urodele neurula. **A.** Diagram of spread of activating principle through prospective neural and mesectodermal areas of host and vertical implant. **B.** Diagram of cross section through normal neural plate of host and longitudinal section through elongated implant, showing equivalent extension of activating principle in horizontal direction through neural plate of host and in vertical direction through implant. (Arrows indicate directions of morphogenetic movements which lead to closure of neural tube and to later shape of activated part of implant.) (After Nieuwkoop.)

in the ectodermal layer in all directions and determines the extension and shape of the whole neural formation. This concept is based mainly on experiments consisting of the implantation of folded strips of ectoderm vertically into the nervous system of a host neurula; the experimental results indicate that an activating agent spreads into the perpendicular fold for a distance equivalent to the lateral extension of the neural plate of the host at the level of implantation. The spreading, activating agent, liberates the capacity for prosencephalic differentiation throughout the ectoderm, in which activation is brought up to threshold.

Development of the various cranio-caudal structures of the central nervous system is based upon various intermediate degrees of counteraction between prosencephalic neural differentiation tendencies and mesodermal influences from the substrate. The later finer segregation of various areas of the neural field into a number of subfields for special organ systems must, however, depend on intrinsic properties of the neural field itself.[5]

This concept was developed further by Nieuwkoop and Nigtevecht (1954) from experiments combining explants of competent urodele ectoderm with fragments of anterior notochord; the latter demonstrated a strong activating, but a weak transforming influence. Both the activating and the transforming influences, manifested in fieldlike behavior, spread with decreasing effect in the overlying ectoderm; and both influences, especially activation, can spread over an area much larger than that in which a direct contact between inductor and ectoderm occurs.

The dynamic aspects of neural induction have been analyzed experimentally by Eyal-Giladi (1954) in several preneurular stages of *Ambystoma mexicanum* and *Pleurodeles waltii*. At each of five stages, this investigator excised strips from the whole area above that part of the archenteron roof already invaginated and folded and grafted them into the ventral side of the same embryo in order to test for the ability to differentiate autonomously. The invaginative gastrulation on which the experiments are based is briefly summarized spatiotemporally as follows: The invaginating, anterior part of the archenteron roof glides under the superficial layer and passes gradually first under the more-posterior prospective mesoderm still situated on the surface of the egg, then under the presumptive tail mesoderm which later forms the posterior part of the neural plate, next under the presumptive spinal cord, and finally under the presumptive hind- and forebrain areas—its ultimate destination.

Eyal-Giladi's experimental results, in support of Nieuwkoop's general concept of activation, and transformation are:

1. The induction process starts from the first contact between the superficial (or primary ectodermal) layer and the anterior part of the archenteron roof; as a result of this contact, activation of the superficial layer is very rapidly brought up to the level necessary for neural differentiation. This differentiation then proceeds autonomously, in a prosencephalic direction.

2. After a certain portion of the superficial layer has established permanent contact with the anterior part of the archenteron roof, that portion of the superficial layer gives rise only to the prosencephalon, regardless of the duration of this contact. On the other hand, when, because of cranial shifting of the

[5] Nieuwkoop, *J. Exp. Zool.*, **120**:105 (1952).

archenteron roof, the same activated area establishes contact with a chordo-mesodermal portion of the archenteron roof, its direction of differentiation is transformed and the material gives rise to caudal structures.

3. The transformation, like the induction process, takes place very quickly. The degree of transformation of any activated area of the superficial layers is independent of the duration of influence, but depends on the craniocaudal position of the chordomesodermal material with which it establishes contact.

4. The presumptive tail mesoderm is probably an integral part of the neuro-ectodermal area. During the process of normal induction, this material, after being induced to form neural tissue, is transformed to form tail mesoderm, which represents the highest degree of transformation.

In other words, when the anterior portion of the archenteron roof—the portion in which the inducing factor plays the chief role—establishes contact with the neuroectoderm, a wave of activation at a certain threshold enables neural competence to become manifest. From its onset, this wave of activation is in the form of a neural field with very definite spatial configurations and ultimately leads to the formation of a prosencephalon. This field, of a very dynamic character, is constantly moving in an anterior direction during the period of invagination. The relative movements are of paramount significance; i.e., the archenteron roof shifts anteriorly at the same time that the neuroecto-derm shifts posteriorly. Thus a wound which causes a relative posterior shifting of the caudal ectoderm toward the wound results in caudalization. Conse-quently, the lowest level of neural differentiation is always neural crest material.

From other experiments by Ter Horst (1948) and by Sala (1955), involving regular explants and implants of archenteron sandwiched between two layers of competent ectoderm, it is evident that the size of induced neural structures increases rapidly toward a maximum in the anterior notochordal area and then progressively decreases in the more-posterior areas. The regional differences in inductive capacity of the archenteron seem to be unmistakably quantitative for the most part.

As a whole, the fairly detailed experimental analyses of the processes and associated phenomena of gastrulation and neurulation in amphibian embryos demonstrate that a reactive system involving a responsive superficial layer and an inductive invaginating layer is a basic feature of the total event. Thus Nieuwkoop's distinction between activation and transformation is important to such an interpretation. Equally significant is the fact that responsive, or competent, tissue reacts in a manner typical of its histogenetic type e.g., the prosencephalic response of the dorsal neuroectoderm to the archenteric influ-ence, the lens-vesicle response to contact with the optic vesicle—although isolation experiments by McKeehan (1954) on lens rudiments indicate that the capacity for self-differentiation develops slowly and is not merely triggered—and the response of the anterior sensory plate to the anterior end of the archenteron to form olfactory invaginations. It is particularly striking that in the last-mentioned response, the experimental broadening of the advancing end of the archenteron results in an increase in the number of olfactory invaginations, according to Holtfreter (1935); this result indicates that the nature and dimensions of such developing structures are innate in the tissue concerned and that, as in lens induction from competent ectoderm of abnormal location, the actual developmental event is a response to a nonspecific activation at a certain developmental time. The importance of the time factor in all structural differentiations can hardly be overemphasized.

Apart from this over-all description of the dynamic developmental event, the primary problems remain (except that they are more clearly defined). For example, the material basis of involution—whether in gastrulation, neural formation or lens invagination—remains unidentified. The chemical nature—assuming it is chemical rather than physical—of the activating and transforming influences is still quite obscure. It is possible that the nonspecific activator is a general growth stimulus similar or comparable to that operating between cooperatively growing tissues in tunicate stolons and medusa buds; whereas transforming influences probably are primarily histogenic chemical substances, such as amino acids, vitamin A, and perhaps complex nucleic-acid derivatives, which act on a dynamic substratum. The most fundamental phenomenon in the developmental event as a whole is the autonomous capacity of tissues to undergo self-differentiation, as evidenced, for example, by a simple epithelium evolving the essential pattern of the whole without coercion (strikingly observed in the *Botryllus* bud) and by the regional, cephalic differentiation of the activated neural plate.

Lest the living developmental entity be submerged in a close examination of specific induction agents, a quotation from Harrison may be pertinent here:

Let us now return to the development of the ear, in consideration of which stress has been laid on the consequences of reorientation of the rudiment by grafting. The importance of this rests upon the fact that in rotation experiments only a simple vectorial change is imposed, from the results of which a simple conclusion may be drawn. . . . The ear rudiment is at first isotropic about an axis perpendicular to the surface ectoderm, following which at a certain stage evidence of definite antero-posterior polarization is found. The whole ectodermal region is, in fact, so plastic during the stages of gastrulation and early neurulation that neither local nor vectorial differences in its qualities can readily be detected. Just as the neural folds are closing, something occurs which orients in an antero-posterior direction some constituents of the material destined to form the ear. At about the same time the direction of the ciliary beat of the ectodermal cells and the direction of outgrowth of the lateral line become fixed, all of which shows that a fundamental change in protoplasmic structure then takes place in the ectodermal layer.[6]

[6] Harrison, *Conn. Acad. Arts Sci. Trans.,* **36:**297 (1945).

Chapter 20

Tissue Interactions

INDUCTIVE influences of one tissue on another are of widespread occurrence. Certain basic facts stand out: Whenever possible, protoplasm moves toward a center—no matter how distant—of chemical attraction. Pseudopodia of an amoeba flow toward a nearby prospective food source; myxamoebae, toward an acrasin center. A spermatozoan orientates and moves toward an egg surface while still some distance away (relative to its own dimensions); and at a close, but still considerable distance, a fertilization cone of cortical cytoplasm projects from the egg surface toward the sperm. Such egg-sperm attraction is probably comparable to that of the matching protuberances of conjugating filamentous algae such as *Spirogyra*.

Tissue Interactions in Tunicates

Induction of growth in multicellular epithelia at some distance from the inductive source is also well established in certain organisms, perhaps most clearly in the epidermal outgrowths of ascidians. The developing bud of *Botryllus* exhibits this phenomenon in two ways: The bud's inner vesicle, derived from atrial epithelium, is separated from the adjacent epidermis by a space through which the body fluid circulates; yet, as the atrial vesicle enlarges, the adjacent epidermis does so correspondingly. Shortly after such enlargement, the central region of the involved epidermis protrudes as a long ampulla which eventually unites with the epidermal vascular network of the *Botryllus* colony. As the growing tip of the ampulla approaches a nearby vessel of the common system, a similar outgrowth appears on the vessel and advances to meet and fuse with the tip of the approaching ampulla. There is no doubt that the first outgrowth induces the growth of the second and that the first must be within a certain distance before a response is evoked. Growth is both induced and oriented across a gap filled with highly viscous, collagenous tunicin. It is possible that substances liberated as a result of growth activity of the inductor tissue diffuse from the terminal zone of growth and that within a certain distance of another epidermal vessel the concentration exceeds a certain value and evokes a response. However, if a chemical gradient is the cause, then as the distance between the inductor and responder shortens and the concentration becomes higher, it might be expected that the response would be more extensive—perhaps involving more than one vessel; this does not happen. Hence, a chemical gradient with its critical concentration value seems inadequate explanation. It is conceivable, for instance, that electromagnetic fields associated with rapidly growing tissue may induce a similar activity in adjacent tissue within a certain distance.

Induction of growth at a distance also occurs in those ascidians in which elongation of the abdomen or postabdomen precedes subdivision or strobilation by the epidermis. In *Eudistoma* and *Aplidium,* strobilation of the abdomen involves three components: epidermis, epicardium, and the digestive canal. In other genera, the strobilative process concerns only the postabdomen and involves only the epidermis and the epicardium. In both types of strobilation, the component tissues either elongate simultaneously and independently of one another, or one tissue elongates and induces a comparable elongation in the other one or two tissues. The first possibility seems highly unlikely; even if the alternate is assumed to be true, several questions arise: What is the nature of the elongation? the nature of the inductive agent? the identity of the inducting tissue?

The question of identity narrows down to either epidermis or epicardium since these alone are involved in postabdominal extensions; accordingly, in abdominal extensions the digestive tube must elongate in conformity with one or the other; i.e., either epidermal or epicardial tissue elongates and induces a corresponding elongation of the limb of the digestive tube, though there is no epithelial contact between them. Whether epidermis or epicardium is the inducting pacemaker is more difficult to determine. The evidence favors the epicardium, particularly in bud production by postmature zooids of *Distaplia,* in which the proximal tip of the epicardium clearly induces a local growth of the adjacent wall of the epidermis. Thus it is most likely that epicardial extension or growth is the primary factor in related genera. In other words, it is most probable that abdominal or postabdominal extension occurs primarily in the epicardial epithelium and that all neighboring tissues, including the epidermis and the digestive tube, elongate responsively to a corresponding degree. Epidermal constriction or strobilation, however, is a response of the epidermis to its own growth activity; i.e., it is not directly induced. As in the chick limb bud described above, there is a maintenance factor in the inner (mesodermal) layer and a responsive but determinative field active in the outer (ectodermal) layer.

Among tunicates more-complex examples of tissue interaction are seen in the incipient budding stolons of the thaliaceans *Pyrosoma, Doliolum,* and *Salpa.* In all three genera the stolon outgrowth is at first a hemispherical evagination of the ectoderm (or epidermis); within this evagination is a blind tubular extension of the posterior end of the endostyle (which may well be homologous to the embryonic epicardial outgrowth) together with a mesodermal mass of

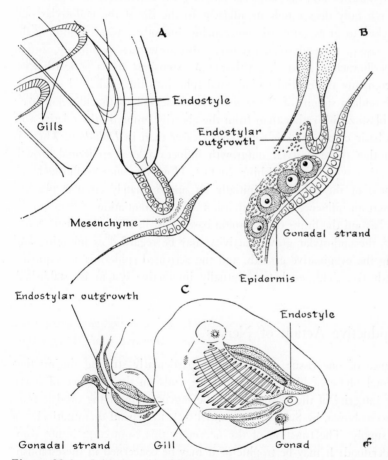

Figure 20-1. Initiation of budding stolon in thaliacean tunicate *Pyrosoma,* showing growth of blind endodermal tube from end of endostyle, and overlying cap of actively growing epidermis, with activated gonadal strand in older type. **A.** Primary blastozooid. **B.** Secondary blastozooid. **C.** Chain of three developing individuals.

presumptive genital tissue, and in *Doliolum* other tissues as well. Here again
the question is whether the ectodermal-epidermal layer is induced or inductive.
From studies on *Pyrosoma* and *Doliolum,* Godeaux (1957) considers that the
endostylar (endodermal) outgrowth is primarily responsible for stolon growth
as a whole. This interpretation seems most likely, for the extension of the
posterior end of the endostyle commences at a considerable distance from an
apparently inactive ectoderm or epidermis, and the latter becomes visibly active
only after the intervening space has been considerably reduced. Accordingly,
it can be assumed—although not with certainty—that the endostylar tube is
the primary inductor and that adjacent tissues grow extensively in response. In
Doliolum, not only does ectoderm adjacent to the tip of the endostylar outgrowth thicken to grow outward conformably, but other tissues adjoining the
flanks and base of the endostylar extension also grow correspondingly, so that
a pair of peribranchial (cloacal, perithoracic) extensions and a pericardial extension accompany the central endodermal tube into the epidermal envelope.

In the secondary stolons of *Pyrosoma*—i.e., those stolons that develop from
functional blastozooids rather than from the abortive embryo—the relative role
of the two basic components is somewhat clearer than in *Doliolum.* There is
little doubt that the endostylar outgrowth induces a local epidermal growth
and evagination without establishing contact. However, once established, the
activated area of the epidermis, initially an approximately circular placode,
seems to exert an influence on the extension and differentiation of genitomeso-
derm even beyond the limits of the stolon base (Berrill, 1950). In other words,
as in *Salpa,* the endostylar growth activity may be regarded as initiating and
maintaining the evaginative growth, and the activated epidermis as acquiring
its own inductive field, so that a mutually interactive system is established.

The Inductive Action of Nerves

The influence of one tissue on another is particularly striking during that
developmental phase of blastemata which depends on the presence of nerve
fibers. The integrity of the blastema and the regeneration of the urodele forelimb has been shown, by Singer (1946 and 1952), to depend quantitatively on
the nerve supply. The function of the nerves in this connection is not completely understood: It may be trophic or it may be concerned in the establishment of tissue polarities.

Singer found that the effective nerve supply can come from the motor,
sensory, or sympathetic nervous system. Similarly, in both flatworms, according
to Beyer and Child (1930), and nemertean worms, according to Coe (1930

and 1934), the regeneration blastema that leads to the formation of a head after decapitation is associated with the presence of nerves. The nemertean *Lineus socialis,* which has a pair of lateral, widely separated nerve cords, regenerates from all longitudinal strips only when they contain a piece of nerve cord. Coe considers that the cut nerves liberate some growth-stimulating substance which acts on the dormant cells of adjacent parenchymal cells, transforming them into active regenerative cells and which directs their movements anteriorly. In the sabellid polychaete *Myxicola aesthetica,* Okada (1934) found that a second head could be induced at any level merely by cutting the ventral nerve cord, without sectioning the body, and that a head regenerates from longitudinally split pieces deprived of anterior and posterior ends and even intestine but containing nerve cord, but does not regenerate in the absence of the nerve cord even though the intestine is present.

Okada with co-workers Kawakami (1943) and Tozawa (1943), in a series of implantation and excision experiments on the earthworm *Eisenia foetida,* noted different responses in two races from different localities—a situation demonstrating the difficulty of standardizing experimental material in earthworms. In both races, however, implantation or deflection of ventral nerve cord into the dorsal body wall induced new, segmented outgrowths containing brain

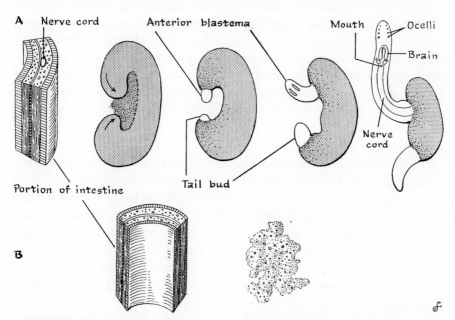

Figure 20-2. Influence of nerve on regeneration in nemertean *Lineus socialis.* **A.** Fragment with portion of nerve cord regenerates. **B.** Fragment without nerve cord disintegrates after 30 days. (After Coe.)

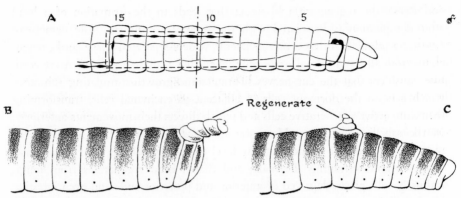

Figure 20-3. Influence of nerve on regeneration in earthworm. **A.** Diagram of method of deflecting nerve cord to dorsal side, consisting of some 10 ganglia so displaced (vertical broken line shows level of decapitation after healing of wound). **B.** Experimental result: formation of new segments because intestine extends from old segments into new. **C.** Induction of new segments by implanting anterior ventral nerve cord of 4 ganglia beneath dorsal wall of head-regenerating region. (After Okada and Kawakami.)

structure when implantation was in anterior regions, but induced new tails in posterior regions. The inductive capacity—in its effect, at least—is strongest at the anterior end and gradually decreases posteriorly. During the development of a new tail, the intestine first undergoes a thickening in response to the inductive stimulation of the nerve and then takes the form of folds before it evaginates to form the intestine of the new tail. Okada also recognizes the induction as a humoral effect that is due to the secretion of an active substance from the cut ends of the nerve, for new growth is not induced in the body wall when contact with the nerve is along its length.

Whether the "growth stimulus" supplied by nerves is specifically neural is another question. Overton (1950) found that implants of spinal-cord tissue into the base of the dorsal fin of *Ambystoma* larvae brought about a rapid local increase of cells which then became incorporated into the fin structure. Neural outgrowth can be greatly augmented by nonneural tissue; for example, mouse sarcoma tissue transplanted to the allantoic membrane of four-day-old embryos evokes extensive growth of the sympathetic system, though not of the somatic motor system or parasympathetic system. Levi-Montcalcini (1952) considers this effect as being induced by a protein.

From histological studies of the effects of delayed denervation on regenerative activity in limbs of urodele larvae, Butler (1949) concluded that neural influence is essential for the early mobilization of the cells concerned with the establishment of a blastema and for the initiation of morphogenetic activities within the blastema. On the other hand, neural influence is *nonessential* for growth, morphogenesis, and histogenesis of the regenerate. Regeneration of the tail in urodeles similarly depends on the activity of nerves, especially of the spinal cord; according to Holtzer (1956), nerve ganglia are not effective. On the other hand, Kamrin and Singer (1959) report that spinal ganglia transplanted to denervated forelimb regenerates, in the newt, may induce an abnormal amount of regenerated tissue—a phenomenon which supports a threshold theory of neuronal influence.

Two alternate possible explanations of the inductive capacity of nerves arise: The initiative effect of nerves may establish or maintain a necessary tissue polarity which is otherwise absent, or the nerves may induce the growth necessary for the first steps in blastema production. The first possibility is far from conclusive, for Butler (1955) has shown that tissue polarity can be reversed without permanently damaging normal regeneration. After inserting the carpel end of an amputated forelimb of a urodele larva into a pocket in the body wall immediately posterior to the shoulder and allowing the graft to heal, Butler reamputated the limb through the proximal part of the upper arm. After a period of regression, a blastema formed on the reversed limb near the elbow joint, and a fairly normal hand regenerated.

Quite apart from the initial trophic or polarizing influence of nerve tissue adjoining tissues, the pattern of nerve outgrowth in a developing limb blastema poses a problem. The factors determining pattern formation are known to be intrinsic to the limb, and it is supposed that those determining the specific arrangement of the limb—whether by means of chemical attraction, electric potentials, or mechanical guidance—reside in the various limb components. However, in the development of the frog hind limb, Taylor (1943) has shown that the factors determining the various branches in the innervation pattern are to be found in the already existent pattern in the mesostroma. He suggests that although a neuroplasmic reticulum is present within the mesenchyme of

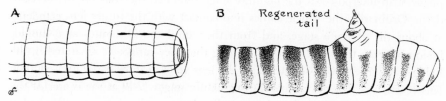

Figure 20-4. Influence of nerve on regeneration in earthworm. **A.** Diagram of method of deflecting ventral nerve cord to dorsal side; cord passes over anterior cut surface and is bent backward to become imbedded beneath dorsal wall. **B.** Experimental result: regenerated tail.

the early limb bud, determination and establishment of the pattern is not simultaneous, but progressive from the base to the tip of the bud. How the mesostroma acquires its pattern is not known, except that there is stress exerted by a shift of the limb axis relative to a specialized area of somatopleure. This process of pattern establishment may be comparable to that in the hydra bud, in which the wave of neural differentiation apparently spreads from the tip toward the base of the bud.

Okada and Kawakami emphasize that in the earthworm the posteriorly decreasing gradient in the ventral nerve cord's capacity for inducing regeneration is nearly tantamount to a definite orientation in the nerve's mode of action. However, the existence of such an orientation does not imply the existence of qualitative differences such as those between head and tail differentiation. Thus, when the body wall from a level posterior to the middle of the worm body is grafted onto the anterior end, a tail forms under the influence of the head end of the nerve cord; moreover, the induction is as effective on a cut surface at a posterior level as on one at an anterior level. The extent of the nerve's role in the qualitative determination of the regenerate is in determining the dorso-ventrality of the new segments: whether during the induction process or during regeneration, the side that is directly contiguous to the nerve cord always becomes ventral, and the opposite side—the one more remote from the nerve cord—becomes dorsal and acquires pigmentation before the ventral side does.

Tissue Interactions in Vertebrate Embryos

Lens Induction

The experimental approach to the inductive influences of tissue on tissue has been both various and extensive. One type of experiment is to interpose a barrier between two adjoining tissues. For example, in investigating lens induction by the optic vesicle in the chick, McKeehan (1951) found that a cellophane strip interposed at the 6-somite stage inhibited the response of the ectoderm. Ordinarily, the vesicle comes into contact with the future lens ectoderm at about the 9-somite stage, and from then until the 21-somite stage the adhesion between the two layers is so strong that they cannot be separated in the living state; after the 26-somite stage, the lens vesicle can be easily excised. The lens reaction is first discernible during somite stages 12-19 at the center of the adherent ectoderm, as a loss in vacuolization, a palisading of cells, and a doubling in cell number. The area of contact seems to be determinative, as Rotmann (1940) has shown in experiments on urodeles, in which the ectoderm from haploid donors was grafted onto normal optic vesicles. The initial size of the lens rudiment results from the activation—not of a specific number of ectodermal cells—but of an epidermal area of specific dimensions. Thus, when first discernible, the lens rudiments in haploid ectoderm contained about 70 per cent more cells than, but had the same total dimensions as, diploid lens rudiments of the same species—a situation reminiscent of Fankhauser's observations on cell size, number, and organ structure of heteroploid salamanders (p. 89).

Although the nature of the contact reaction is not known precisely, it is definitely not specific in the optic vesicle. Experiments by Jacobsen (1955) on *Triturus,* confirms such nonspecificity. In the early neurula of *Triturus,* the presumptive lens ectoderm is underlaid by the entodermal wall of the archenteron, and the anterior edge of the lateral plate mesoderm lies immediately

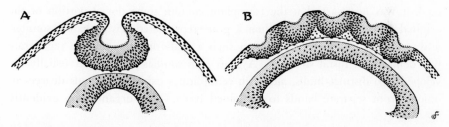

Figure 20-5. Induction of nasal sacs in ectoderm of amphibian embryo by abnormal widths of anterior end of forebrain. **A.** Single sac from narrow end. **B.** Multiple sacs from broad end. (After Holtfreter.)

posterior to the presumptive lens site. When Jacobsen combined isolates of the presumptive lens ectoderm with either entodermal-wall tissue or lateral plate mesoderm (or both), lenses formed in a substantial percentage of such combinations, but not in the isolated ectoderm alone. The interpretation of these results is that lens induction derives from synergistic action of either the optic vesicle or the underlying nonneural tissues and is therefore nonspecific. Such a composite cause in lens induction is comparable to that in the emergence of other invaginative structures, such as the otocyst, olfactory vesicle, and hypophysis—which are under the influence of more than one inductor according to Holtfreter (1951). Area of contact, or of near contact, however, between inductive and responsive tissues is apparently as important in the development of olfactory vesicles as it is in the lens, although the response is variable in terms of numbers of vesicles rather than of scale of organization of a single organ.

Limb-Bud Regeneration

Most morphogenetic activities are performed either by epithelial or other tissue alone or by two tissues adjoined to each other. Examples of the former are the invaginations and evaginations in the development of ascidian buds, embryos of *Amphioxus,* and many invertebrates; examples of the latter, in the limb buds of vetebrate embryos and the imaginal disks of larval insects. Experiments on both embryonic limb buds and regeneration blastemata are particularly relevant, since they concern the relative role of ectodermal and mesodermal layers and their interaction.

The sequence of growth of the urodele embryonic limb bud has been shown by Takaya (1938), who transplanted limb disks of varying size from the normal dorsal site to more-ventral locations. For any subsequent development to proceed, it was necessary that the transplant contain some dorsal portion of the original surrounding tissue. When a graft represented fewer than 4 somites, it was usually absorbed; even when such a graft did develop, the regenerate was deficient in distal structure. A graft of 6 somites in diameter usually developed two distinct limbs, although reduplicates exhibited various degrees of union: from separate bands to conjoined bases. Thus organization evidently

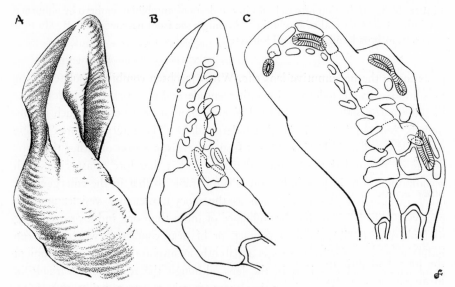

Figure 20-6. Tail skin and spinal cord transplanted to naked stumps of elbow and wrist result in taillike outgrowth. **A–B.** Elbow transplant, showing fin and skeleton. **C.** Wrist transplant, showing axial skeleton. (After Glade.)

proceeds from the basal ring toward the distal center, in conformity to the general process observed in limb regeneration.

In the regeneration of urodele limbs, the relative role of the tissue components in determining pattern is indicated in experiments by Ichikawa and Okada (1954), who interchanged one or more limb components. When hind-limb muscle or bone of adult *Triturus* was implanted into a forelimb from which either tissue had been removed and then the forelimb was amputated, typical hind-limb pattern developed in about 30 per cent of such experiments. When unirradiated hind-limb muscle and/or bone was implanted into an irradiated forelimb, a hind limb formed in 70 per cent of such experiments, i.e., from regenerating blastemata consisting mainly of implanted tissue. Evidently each of the component tissues of the limb possesses factors which determine the pattern of development, as further in-dicated by experiments in which tail-skin and spinal-cord tissues were transplanted to cover the naked stumps of forelimbs amputated at the wrist and elbow. In each such experi-ment, an outgrowth that was taillike both in form and skeletal struc-ture developed, according to Glade (1957). In other words, the regional character of differentiated tissue re-tains a determinative influence, which, however, may or may not be overridden by comparable qualities of the host tissues.

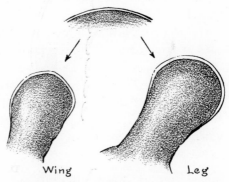

Figure 20-7. Diagram of generalized limb bud of chick embryo at early stage, when apical ectodermal ridge is uniform, from which asymmetries develop in wing and leg. (After Zwilling.)

The general interaction between limb ectoderm and mesoderm has been investigated more effectively in chick embryos by Zwilling (1955 and 1956), who employed a versenate-maceration technique for completely removing the ectoderm of limb buds, thereby exposing smooth, naked mesoderm to which other ectoderm could adhere. The mesodermal base alone or covered with nonlimb ectoderm fails to develop and gives rise only to girdle elements; apical-limb ectoderm seems to be necessary. The reactivity of the limb-bud mesoderm to the outgrowth stimulus of the ectodermal ridge was tested (1) by dividing the limb bud into an apical sliver and a basal section and placing the latter in an ectodermal jacket, and (2) by placing apical ectoderm on each lateral surface of the mesoderm of one limb bud. Every level of the limb-bud mesoderm responded to the outgrowth stimulus and formed distal structure: in (1), an entire limb formed from each part (i.e., from the apical sliver and from the basal section); and in (2), the limb formed varied in degree of duplication from one lateral surface to another. When the most active sections of the apical ectodermal ridges were taken from 2 or 3 buds and were placed in tandem along the free mesodermal surface of one bud, no duplication occurred —a result indicating that the ectoderm conforms to the mesodermal pattern. Thus the area of the outgrowth-inducing ectodermal surface, and not the mesodermal mass, determines whether or not limb duplication will occur.

When ectoderm and mesoderm from a polydactylous limb bud were inter-

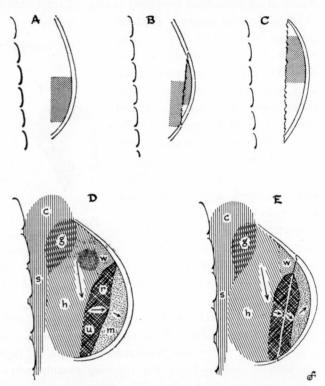

Figure 20-8. Diagram, showing distribution of ectodermal-ridge maintenance factor in mesoderm of normal right limb bud of chick. **A.** After reorientation of anteroposterior axis of apical zone, factor acts horizontally by spreading through intervening tissue to initiate inductive capacity of postaxial (originally preaxial) portion of apical ridge. **B.** After reorientation of anteroposterior axis of limb bud severed at level of body. **D–7.** Fate of wing bud, showing prospective wing parts before and after, respectively, reorientation of apical zone. **c.** Coracoid. **g.** Glenoid. **s.** Scapula. **h.** Humerus. **w.** Web. **r.** Radius. **u.** Ulna. **m.** Manus. (After Saunders, Gesseling, and Gfeller.)

changed with components from a genetically normal bud, the resulting limb was normal when the combination was polydactylous ectoderm and normal mesoderm, but was polydactylous when the combination was normal ectoderm and polydactylous mesoderm. Since more-extensive ectodermal ridges developed when normal ectoderm was combined with polydactylous mesoderm, it is evident that the mesoderm contains some factor which influences the development of the ridge. The pattern of the limb, therefore, results from a difference in the distribution of the maintenance factor which leads to a response in the ectodermal ridge, and this response in turn results in a more-extensive preaxial outgrowth, according to Zwilling and Ames (1958). In sum, the most likely seat of organization seems to be the interacting, adjoining surface layers of the two tissues; and whether this territory is called intercellular cement, interpenetrating matrices, or fused hyaloplasms is mainly a matter of emphasis or definition.

Feather-Papilla Induction

A comparable relationship exists between epidermis and subdermal tissue in the development of feather papillae. Of composite origin, a papilla consists of a dermal core and a thin covering of follicular epidermis—the so-called regeneration cells. These cells give rise to the papilla's epidermal coat, from which the collar of a regenerating feather is derived. According to Lillie (1942), the dermal component induces bilateral symmetry in an originally potent ectoderm, and this symmetry leads to the specific orientation of the feather. In continuation of this work, Wang (1943) devised a technique by which the epidermal coat of a papilla was destroyed and the denuded dermal core was transplanted autoplastically into a follicle of the reciprocal tract, the papilla of which had been previously removed. Local epidermis grew to cover the transplant, giving rise to a composite papilla with dermal portion from one tract and epidermal portion from the other. Such papillae invariably produced feathers characteristic of the host tract in growth rate, shape, form, structure, and pigmentation. The change in tract specificity of the denuded papillae must be ascribed to a recoating of the transplanted dermal core by the epidermis of the reciprocal tract.

Therefore tract-specificity is not a function of the dermal components of breast and saddle papillae—although, according to Lillie (1942), they undoubtedly have a nutrient function—but of the epidermis alone. However, since denuded papillae which have been transplanted in random orientation give rise to feathers of corresponding orientation, the factors determining feather symmetry reside in the dermal tissue. In fact, a feather cannot form in the absence of a dermal papilla; thus it seems that dermal tissue imparts a specific activation to the overlying epidermal cells.

In addition to activation, the dermal tissue induces or imposes on the epidermis a polarity corresponding to its own, which in turn initiates the formation of a series of parallel barbs dorsally and thus determines the position of the future rachis. With the establishment of feather symmetry, the epidermis becomes a self-differentiating system; and it is noteworthy that saddle epidermis, for instance, produces only saddle feathers, and breast epidermis produces only breast feathers, although no morphological distinction between the respective cells has been discerned. Nonfollicular epidermis does not react at all.

The problem of feather pigmentation is even more complex and has much in common with that of pigmentation pattern determination in urodeles: In both cases the phenomena of migration and aggregation of neural crest cells to specific sites and the formation of the specific pattern of melanophores are involved, according to Willier and Rawles (1940), Watterson (1942) and Twitty (1945 and 1949).

Urodele-Balancer Induction

The relative role of epidermal and mesodermal (or mesectodermal) components is a matter of general pertinence to the determination of pattern and structure, as shown, for example, in Harrison's classical analysis (1925) of the development of the urodele balancer. During gastrulation, the areas of the ectoderm that give rise to gills, balancers, nose, ear, other placodes, hypophysis, and lens are segregated as a pattern without any definite boundaries of its several constituents but as centers of differentiation. Within each such center, the capacity for differentiation is most intense at a certain location, from which it gradually diminishes peripherally. An intermediate area, e.g., between balancer and gills, may give rise to one or the other organ, depending on the influence to which they are subjected.

The factors evoking the development of the balancer are apparently localized in a certain ectodermal region which overlies the mandibular arch. This localization is shown by several experiments:

1. When transplanted to any other region of the head at any time during or after the neural-plate stage, this ectodermal area gives rise to a balancer at that site.

2. When the presumptive balancer ectoderm of *Ambystoma punctatum* is implanted in an embryo of *Ambystoma triginum,* which normally lacks a balancer, a balancer develops.

3. When balancer ectoderm is replaced by ectoderm from some other region of the body or from the corresponding balancer region of *Ambystoma triginum*, no balancer develops.
4. Removal of the mandibular-arch mesoderm, or of the neural-crest mesectoderm in the mandibular region, does not interfere with balancer formation.

Thus the capacity to form balancer structure definitely lies within the epidermis, but the stimulus for formation seems to be inherent in the underlying tissue. Moreover, whether or not a balancer forms depends on the age of the epidermal graft, and not of the host embryo which receives it. A balancer develops only when the young ectoderm is transplanted to the head—a phenomenon suggesting that the mesectoderm of the head region is necessary; but later, just before the balancer bud would appear, if the balancer ectoderm is grafted even to the trunk, a balancer develops. However, when young balancer ectoderm together with some mesectoderm is grafted into the trunk region, a balancer develops.

Anuran-Larva Adhesive-Disk Induction

The relative role of ectoderm and mesectoderm in the formation of a balancer in urodeles is closely related to that in the development of adhesive disks, or ventral suckers, in anuran larvae. According to Holtfreter (1936), ectoderm transplanted from an anuran gastrula to the head region of a urodele embryo develops typical anuran suckers; similarly ectoderm transplanted from a urodele gastrula to the head region of an anuran embryo develops a balancer. Schotté and Edds (1940) report that only two distinct regions of the urodele embryo release or stimulate formation of the adhesive sucker: the central substomodaeal area, and the two balancer sites. Thus, both in urodele and anuran embryos, there seems to be a balancer-sucker field which is associated with the underlying mesectoderm and to which competent ectoderm responds according to its genetic constitution. The ectoderm carries the histospecific factors; and the underlying tissue, the topographical inductive factors. When either component is deficient, a given structure fails to develop. Thus, according to Mangold (1931), transplantation between the two urodeles *Triturus taeniatus* (which has balancers) and the axolotl, *Ambystoma tigrinum*, (which does not) reveals that the balancer-inducing factor is present in the mesodermal tissue of both, but that the axolotl ectoderm has lost the capacity to respond to it. The inductive pattern is inherent in the mesodermal tissue, even though the mesendoderm as a whole is a self-differentiating system with its own distinctive pattern, as seen in exogastrulation. The regional inductors—their nature and their capacity to form a patterned system—remain an unsolved mystery.

Ciliary-Beat Induction

Another example of induction concerns the pattern of ciliary beat in developing amphibians. Tung and Chang (1949) report that impermeable mica sheets placed between mesoderm and ectoderm in frog embryos prevented determination of the direction of the epidermal ciliary beat; whereas permeable agar sheets permitted normal development of the ciliary beat pattern—a phenomenon indicating some mediation by diffusible substances. Similarly, Brahma (1958) sandwiched a membrane with pores 4 mμ in diameter between two layers of ectoderm and then placed the "sandwich" between axial mesoderm and presumptive neural plate. Induction occurred only on that side of the ectoderm that was in contact with the mesoderm—i.e., not on the side separated by the membrane—and no discernible cytoplasmic connections passed through the pores; therefore, Brahma concluded that the lack of induction on the side separated by the membrane was due to the failure of the active substance to diffuse or to attain a sufficiently high concentration on that side. Neither contact effects nor diffusion, however, rules out the possibility that the other is operative also.

Interpretation

The fundamental question concerning tissues interaction is: How are the changes in one tissue effected by the induction of another, contiguous tissue? Are histogenetic changes primary and morphogenetic developments resultant, or vice versa? Niu (1956) explanted an organizer tissue into a hanging drop of modified Holtfreter solution; 7-10 days later he introduced a small piece of reacting tissue (young ectoderm) some distance from the explant. The inductor tissues used were from regions of the dorsal area, the medullary plate, the notochord, somitic blocks, neural folds, and endoderm; the reacting tissue was the presumptive ectoderm of *Triturus* and *Ambystoma* species. With each type of inductor, there developed a culture of pigment cells whose nerve fibers radiated from the center mass. However, when the reacting ectoderm was introduced after 12-18 days, myoblasts instead of pigmented neuroblasts formed. Furthermore, the cell-free medium was just as effective after the original explant had been removed. Spectroscopic analysis indicated the presence of nucleoproteins in the cell-free medium; but whether the RNA or the protein component was the effective inductor was not shown.

Such induction is comparable to that underlying the histological changes resulting from the exposure of the epidermis of chick embryos to media enriched with vitamin A. Fell and Mellanby (1953) showed that in such a medium the epidermis undergoes metaplasia from a squamous, keratinizing epithelium to a cuboid, mucus-secreting one. According to Weiss and James (1955), the same reaction is obtained whether the exposure is intermittent or continuous for as brief a period as 15 minutes. The result is interpreted to mean that even a brief contact with the chemical diverts the cellular mechanism of differentiation (i.e., of specific synthesis) into an alternate pathway.

That chemicals are effective does not, however, rule out the need for substantial contact between reacting tissues for orderly induction. Experimenting with mouse-embryo rudiments in vitro, Grobstein (1954) found that the epithelium of both kidney and submandibular gland rudiments and of embryonic spinal cord can induce tubule formation in kidney mesenchyme. Direct contact seemed to be unnecessary, and even porous membranes did not prevent differentiation although the pores did become filled with long pseudopodia. He believes that mediation is normally effected through the so-called intercellular cement—i.e., by something intermediate between full cellular contact and free diffusion:

> Cellular boundaries are not always sharp and the distinction between hyaloplasm and matrix may be primarily conceptual. Matrix reactions would be assumed to be a governing relationship rather than an exclusive one . . . . The properties of the continuum, assumed to be produced by the cells, naturally would in part be dependent upon the cellular properties. But since the continuum provides the immediate environment of the cells, the properties of the cells would be in part dependent upon the continuum.[1]

Where matrices formed from two adjacent tissues are completely incompatible (nonpenetrable), an interface would form, and exchange would be limited to those materials capable of passing through both matrices; i.e., there would be no adhesion and little influence of one tissue on the other. Where matrices are compatible (interpenetrable), there would be adhesion and a new matrix with new properties that might affect each component cell layer. Interactions at the molecular level might lead to new components which might be a source of something new—e.g., a basement membrane—that forms continually and serves as a stabilizing force in development.

By use of trypsin and versene, King and Briggs (1955) dissociated the cells of late-stage frog gastrulae; and from these and other experiments Brachet

[1] Grobstein, in Rudnick (ed.), *Aspects of Synthesis and Order in Growth,* p. 251 (1954).

(1958) concludes that the matrix which holds cells together is a ribonucleo-protein-calcium complex. Moreover, Yamada and Takada (1956) report that ribonucleoproteins extracted from the liver and from the kidney of amphibian gastrulae can induce cephalic organs, such as eyes and brain, and spinocaudal differentiation, respectively. That proteolytic digestion of such inductors results in loss of activity suggests that the integrity of the protein is essential to the inductive effect. Weiss (1947) has presented evidence that substances in cell extracts promote the growth of homologous tissue either by being incorporated into corresponding cells or by neutralizing growth inhibitors present in the medium. Support for this interpretation is found in the numerous examples of histospecific types of tumors by respective cell-free extracts. However, the histospecific agent and the agent responsible for the unrestrained, malignant quality may not be the same, even though each may be a part of a liponucleo-protein complex.

Grobstein suggests that

> a labile intercellular continuum, locally alterable in penetrability and other prop-erties by physiological shifts, endowed with a high degree of specificity, closely responsive to the genotype, able through polymerization to condense to higher levels of order thus providing boundaries and interfaces, seems to be exactly the missing piece in many puzzles of development. Affinities and disaffinities in cel-lular aggregation and disaggregation, mass cellular movements such as stretchings and foldings, gradients and field phenomena, regulation, mutant developmental abnormalities, all might find a rationale at least in part in such a complex bio-physico-chemical matrix.[2]

This interpretation is essentially the same as that maintained throughout this book, as in discussions in preceding chapters concerning the symplasmic state and protoplasmic fine structure. Weiss (1933), too, called attention to the role of intercellular ground substance during development, although he did regard the colloidal ground substance as nothing more than the product of cell secretion.

Whether the so-called intercellular cement is a cellular secretion or a true extension of the hyaloplasm is to a great extent a matter of definition, and it seems clear that in many organisms there is no sharp distinction.

[2] Grobstein, *loc. cit.*

Chapter 21

Metamorphosis

THE PHENOMENON of metamorphosis is clearly related to the specialization of eggs. Specialized ooplasms are responsible for the development of specialized larval organs and tissues; i.e., larvae are equipped with precociously differentiated locomotory, sensory, and commonly attachment organs, whose development can usually be traced to the initial presence and pattern of the fertilized, as-yet uncleaved egg. Frequently such structures or tissues become functionally differentiated before the developing organism attains the feeding state. Furthermore, larval organization—at least as observed in the functional tissues—is markedly different in activity and general structure from adult organization. Obviously, a transformation or metamorphosis is necessary in order that the adult organism may function (except that the neoteinic type of organism functions and grows as a larva until it attains maturity). Thus the metamorphic process, which may be gradual or sudden, is primarily one of differential destruction and development of the various tissues comprising a larval organism.

Metamorphosis in Marine Larvae

Most invertebrate organisms produce larvae that eventually undergo metamorphosis. In general, marine larvae may be categorized as either undergoing a cataclysmic transformation, not only sudden but differentially destructive (e.g., the pilidium larvae of many nemerteans, the pluteus larvae of echinoids and ophiuroids, and the tadpole larvae of ascidians) or as metamorphosing less dramatically (e.g., the trochophorous larvae of polychaetes and the gastropod veliger)—even without loss or destruction of larval tissue (e.g., the coelenterate planula and sponge embryo). A planula, in fact, is considered a larva only because its ectodermal layer is externally ciliated and because it is mobile;

essentially the same may be said of sponge larvae. In both, the later transformation is mainly a process of continued development which involves some change of form—the change depending on contact with a substratum. Such a transformation is radically different from that of, for example, a free-swimming chordate tadpole larva into a sessile filter-feeder, as in ascidians.

The general significance of the more-specialized types of marine larvae has long been debated. An early, generally discarded interpretation—that they represent ontogenetic relics of ancient ancestral types—may not be as completely invalid as De Beer (1957) suggests. The generally accepted current belief is that echinoderm larvae (Fell, 1948), ascidian tadpoles (Berrill, 1955),

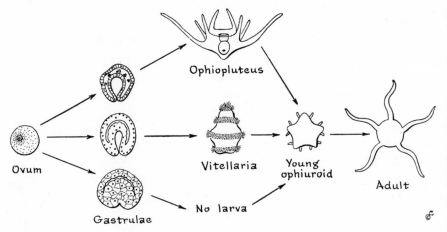

Figure 21-1. Divergent modes of development in ophiuroid echinoderms. (After Fell.)

and nemertean larvae (Smith, 1935) are evolutionary novelties "interpolated" into an originally more-direct developmental cycle.

Whatever their general significance, marine larvae undergo a metamorphic process that may be discussed in terms of the tissues and structures that are destroyed, the tissues that escape destruction, and the predisposing and triggering factors.

In those specialized marine larvae whose metamorphosis is profound and involves extensive destruction, the tissues destroyed are doubly significant. First, they are almost exclusively those tissues that differentiated precociously from the especialized cytoplasmic territories of the egg. The implications of this fact—somewhat beyond the scope of this book—are that if, as suggested above, the larval specializations are evolutionary elaborations superimposed on a primitively direct development, then the original ooplasmic territories which give rise to such specialized tissues are also "interpolations." In other words, there is further, if indirect, evidence that the egg is a specialized developmental device; and the more differentiated the egg, the more specialized (and the least primitive) it is. The second facet of the significance of the destroyed tissues is that they are highly differentiated or specialized, commonly incapable of continuing cell division, and comparatively histologically aged.

Metamorphosis is thus both retrogressive and progressive. To a considerable, but variable, extent the two aspects are related, for the retrogression of some tissues commonly accelerates the development of others. Any stimulus whose *apparent* result is the resorption or histolysis of precociously differentiated tissue is likely to induce or accelerate the progressive aspect of the metamorphic process; thus it is not surprising that the factors normally inducing metamorphosis in marine larvae seem diverse and debatable.

Factors Inducing Metamorphosis

The very act of larval attachment to a substratum may be sufficient to induce or permit certain changes in form. In polychaete larvae, attachment (or settlement) and metamorphosis are virtually synonymous, and the stimulus for the first seems to be the stimulus for the other, as shown in an extensive investigation by Wilson (1955) on *Ophelia* larvae. These larvae metamorphose after becoming attached to certain sand grains—those that bear living microorganisms, such as bacteria, of a fairly definite concentration; those that bear the films of dead organisms or other nonliving organic matter may even inhibit metamorphosis. For a long time, the traces of copper in the substratum of oyster larvae were considered the inductive factor in settlement and metamorphosis. Although this conclusion is now strongly questioned, traces of metallic elements do have a striking effect—as either enzymatic poisons or developmental accelerators—especially when present in histologically mature and fully differentiated tissue.

The factors inducing metamorphosis—investigated in ascidian larvae perhaps to a greater extent than in any group other than vertebrates and insects—have been systematically studied by Grave and Nicoll (1939) in the tadpole larvae of *Ascidia* and *Polyandrocarpa*. These investigators found that, besides iodine, nontoxic concentrations of iron, copper, and aluminum salts accelerated the metamorphosis of larval tissues, but that the chlorides of a number of other metals did not. They suggested that, in addition to an aging factor, copper is particularly active in effecting the destructive phase of metamorphosis. Experiments on *Styela* larvae by Glaser and Anslow (1949) support this hypothesis.

Increased acidity or carbon-dioxide tension of the water also induces metamorphosis in a variety of ascidian tadpole larvae; whereas increased alkalinity delays the onset of the destructive phase (Berrill, 1929). A histological analysis of tail resorption in the ascidian tadpole larva indicates that the so-called aging factor is probably the progressive exhaustion of the yolk reserves of the tail epidermis, which is the tissue affected first of the three tissues (epidermis, muscle, and notochord) involved in tail resorption. Nutritive exhaustion culminates in the centripetal contraction of the epidermal envelope—a contraction which seems to have a disruptive effect on the tissues within the envelope (Berrill, 1947). Since there is no circulatory system at this stage of development, the metabolites resulting from muscular activity of the tail accumulate; and this activity and accumulation are almost certainly contributory factors in inducing metamorphosis. High alkaline concentration in the external environment would accordingly tend to buffer the accumulated acidity within the larva.

Metallic traces and metabolite accumulation, either separately or together, are at most partial explanations of the metamorphic event. A full explanation would have to account for the phenomena observed in the large, red tadpole larvae of *Styelopsis grossularia* of northern Europe. Upon settling and metamorphosing, these larvae undergo two obvious changes apart from structural development: The color—apparently of all tissues—rapidly changes from red to yellow-orange and at the same time mesenchymal cells traverse the epidermis of both body and tail to occupy a position within the layer of tunicin. The red pigment is extractable and does not seem to be a pH indicator (Berrill, 1929). Nevertheless, the over-all metamorphic process in these larvae proceeds in a fairly normal way and responds to the same inductive stimuli even when the tail has been amputated.

Metamorphosis thus consists of a destructive phase, possibly induced by a combination of internal and external factors, and a constructive phase, during which juvenile tissues develop at the same time that differentiated larval tissue disintegrates. The profound difference in susceptibility between the immature adult tissue and the mature larval tissue is that the former is only indirectly affected by any agent capable of inducing metamorphosis in larval tissue.

Much of the metamorphic process as a whole can, in fact, be explained by this difference in suceptibility as well as by the limited nutritive supply in the larval tissue. Larval structures in the mature functional state are metabolically vulnerable, just as the head of a *Tubularia* or other hydroid is vulnerable compared to the stalk tissue. When there is competition for the nutritional supply, the young, vigorously growing tissue may actually induce resorption of differentiated structure, as the growth of a medusa bud from the base of a hydranth may cause hydranth resorption. Conversely, the sudden disintegration of mature tissue releases abundant nutrient, so that adjacent juvenile tissue, previously in

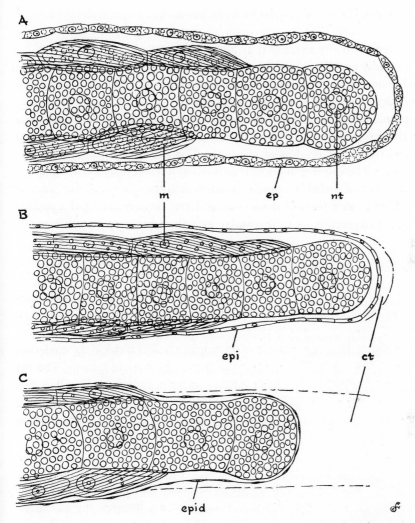

Figure 21-2. Three stages in tail absorption in ascidian *Stolonica socialis,* showing relative changes in epidermis, muscle, and notochord, respectively, and progressive starvation and contraction of epidermal envelope. **m.** Muscle cell. **ep.** Yolk-laden epidermis. **nt.** Yolk-laden notochord cell. **A.** Tip of tail at time of liberation of tadpole. **B.** Tail after several hours of active swimming; tip just starting to shrink from enveloping tail cuticle **ct; epi.** epidermis with yolk exhausted, but cell length changed only slightly. **C.** Tail of tadpole at end of free-swimming existence, with tip of tail well withdrawn from cuticle.

a state of suspended growth, undergoes a corresponding surge of developmental growth. The phenomenon is comparable to that in a *Distaplia* oozoid, whose first buds remain partly developed as long as the oozoid is functional and intact, but rapidly complete their development when the oozoid undergoes autolysis and phagocytosis. Similarly, in a reduction body of a *Clavelina* zooid and in many other organisms, the degenerative process does not affect the unspecialized coelomic epithelium of the epicardium, but includes all the specialized tissues; the result is that one tissue proliferates as the others disintegrate, and there is no significant net loss of substance. The situation is the same in the tadpole larva: The disintegration of the specifically larval tissues releases nutritive substances and probably growth stimuli within the system, so that there is both a release from a possible dominance of larval organization and a sudden supply of the material basis for expansive growth of presumptive adult tissue.

Suppression of Metamorphosis

Metamorphosis can fail to occur during a course of development which normally includes this phenomenon. There are two reasons for such suppression. Increase in egg size may obviate the necessity for a larval stage, or it may make difficult the attainment of that stage. In echinoderms, for example, increased egg size is correlated either with absence of the larval type characteristic of a class or with modification or partial suppression of that type. With the absence or degeneration of larval organization, the need and basis for metamorphosis disappear. How this happens is fairly clear. Particularly in an organism whose larva is primarily a floatative device that enables the organism to feed and grow as a planktonic constituent and to attain a size sufficient for the development of a permanent organization, rather than a device for selecting a suitable settlement site (as it is in all ascidians), increase in egg size, or content, eliminates the need for such growth and therefore of such survival value as the larva may have had. All that is necessary to suppress the formation of transient larval structures is reduction of the initial intracellular pattern or gradient systems of specialized ooplasms. Other ooplasmic differentiations may be substituted, as in certain echinoderms, or, more commonly, development of permanent organization may become secondarily but completely direct; i.e., metamorphosis is completely lacking from the cycle because there is nothing to change.

The second reason for the suppression of metamorphosis—namely, neoteny—is far more difficult to understand. Neoteny appears perhaps in its simplest state in the larvacean tunicates such as *Oikopleura,* whose larval structure develops even more precociously than that of the tadpole larvae of ascidians.

In the *Oikopleura* notochord, histological differentiation commences and cell division ceases when only 20 cells have been produced; and in each tail-muscle band, when only 9 or 10 cells have been formed. Despite this precocity and cellular reduction, the larval structure persists and becomes part of the sexually mature adult organism. All subsequent growth of the tail results from cellular enlargement, not proliferation. Furthermore, the tissues of the nonlarval structure remain fairly restricted in cell number—strikingly so in certain genera— so that development as a whole is essentially the growth of a somewhat condensed tadpole larva which attains sexual maturity without having undergone significant cell multiplication.

It seems likely that there is some causal relationship between the suppression of the first (disruptive) phase of metamorphosis and the failure of the second (progressive) phase to proceed much beyond whatever rudimentary structure can be formed from the cellular material available at the stage of tadpole organization. The question is: What restricts the proliferation of cells? Notochordal and tail-muscle cells may, of course, be too highly differentiated even during the tadpole stage to continue division; but there is no indication that the cells constituting the digestive tract and the heart in larvacean tunicate are any more specialized than those making up these structures in ascidians. Yet in the larvacean genus *Fritillaria*, for instance, the heart in the full-grown adult consists of only 2 rhythmically contractile cells.

The suppression of metamorphosis has bearing on the core of the problem of organization. Since the growth of larval forms, whether or not they lead to neoteinic adults, seems to be generally accompanied by increase in cell size rather than number, it seems reasonable to conclude that the two types of growth are not mere coincidence. The tissues of the neoteinic, perennibranciate urodeles consist of cells that are many times the size of their counterparts in tissues of other urodeles. It may be significant that the basal metabolic rate of neoteinic urodele tissues is low relative to that of other urodele tissues.

Pertinent to this context is the fact that the cells lining the mid-gut of the larvae and pupae of the mosquito *Culex* are large and polyploid and that with the onset of metamorphosis these cells undergo a succession of divisions which lead to the formation of the intestine of the imago. According to Berger,

> the great increase in size, unaccompanied by division, which these cells undergo during larval life, and the subsequent series of divisions accompanied by progressive reduction in size, give further evidence that growth of cells and cell or nuclear division are distinct processes, and that cell division is caused by something outside the dividing cell and not by a mere surface-volume ratio. Cell division is apparently neither a function of the cell's geometric dimensions nor of its chromosome number.[1]

[1] Berger, *Carnegie Inst. Washington Pub.*, No. 496, p. 229 (1938).

Once again there arises the concept of a supracellular basis for organization and pattern—a material continuity that in one state can expand (i.e., grow) without inducing proliferation of the cells involved (which grow nevertheless) and that in another state induces intense cellular proliferation without actual tissue growth. What it is that changes during a change in state and what the nature and procedure of the inducing agents seem to be questions that apply especially to metamorphosis in amphibians and insects. The most pertinent aspects of these problems are discussed below; for a general review, see Etkin (1955).

Metamorphosis in Amphibian Larvae

The agent inducing metamorphosis in amphibian larvae is the hormone from the thyroid gland, which, according to Allen (1938), seems to act directly, i.e., without the mediation of other organs. Experimental removal of the gland or spontaneous failure to synthesize normal thyroxin results in failure to metamorphose. Premature increase in the amount of thyroxin induces premature metamorphosis. Permanently neoteinic urodeles fail to metamorphose because their tissues do not respond to the thyroid hormone. The axolotl, however, according to Bytinsky-Salz (1935), normally fails to metamorphose, not because its tissues are insensitive to thyroid hormone or because the thyroid gland is insensitive to pituitary stimulation, but because the pituitary gland does not stimulate the thyroid. Metamorphosis, therefore, depends on the attainment of a certain concentration of thyroxin and on tissue sensitivity to this substance.

The primary effect of thyroxin in mammals is to raise the general level of oxidative metabolism in tissues and thereby to liberate heat energy. The impact of the hormone in the growing organism may well be differential, since the older, more highly differentiated cells and tissues succumb under the influence of thyroxin; whereas the younger, less-specialized cells and tissues survive. The initial phase may be fundamentally disruptive, as in the metamorphosis of ascidian tadpoles; and the destruction of a precociously induced organization may, in part at least, serve to permit other events. On the other hand, Etkin did not find conclusive evidence of a general increase in metabolic rate during amphibian metamorphosis, nor do compounds other than thyroxin that raise the metabolic rate induce metamorphosis; therefore thyroxin *per se* assumes a significance quite independent of its effect on the metabolism of the higher vertebrates.

In insects the corpus allatum, which produces the molting hormone, has a definite influence on the metabolic rate. Thomsen (1949) found that removal of the gland results in a 20 per cent decrease in metabolic rate and that reinsertion of the gland restores the normal rate. Metamorphosis, however, is inhibited by the activity of the corpus allatum. The metamorphic process is initiated by the brain, which stimulates the prothoracic glands to secrete a hormone which in turn induces transformation in the tissues. Thus the question concerning the nature of the agents inducing metamorphosis is only partly answered.

The primary problem—namely, the nature of the transforming system—is even more obscure, partly because the system responds both directly and indirectly to the initiating factor. Furthermore, there is the differential response of different areas of a given tissue to the stimulating agent. For example, Clausen (1930) reports that in the frog tadpole, tail muscle and skin respond by histolysis, but back muscle and skin do not and that there are regional differences even within the tail. Since there seems to be no significant difference between the functional epidermal or muscle tissue of one territory and that of another, the pattern of sensitivity is apparently based on something other than histological variance.

There is also a time factor in tissue sensitivity to the inducing agent: some tissues respond later than others, either because their sensitivity is weaker and their reaction slower or because they require a somewhat higher concentration of the hormone before reacting. Thus the low thyroxin concentration in the frog tadpole at the beginning of metamorphosis is sufficient to activate the very sensitive leg primordia, but not until the climax of metamorphosis is the concentration high enough to activate the least-sensitive tissues, such as the tail and the tympanic ring.

Quite apart from the factors of time and concentration, there is a pattern of sensitivity which relates to the fundamental difference between the larval and the adult pattern of organization, whether in the amphibian, insect, or ascidian. The metamorphic process as a whole is the destruction of a pattern of organization that resulted from the particular pattern of cortical cytoplasmic specializations in the egg—an organization which has been superimposed on a system that originally seems to have been directed toward a different developmental outcome. There is little doubt that a frog tadpole is merely the inevitable outcome of the development of a frog egg; that the organization of the tadpole depends on that of the egg shortly after fertilization; and that if the conditions for growth were satisfactory, an isolated, totipotent somatic cell would develop directly into an adult, i.e., without passing through the tadpole phase. This last may never be experimentally feasible, but the growth of a complete carrot plant from an isolated cell, reported by Steward (1958), and the development of an ascidian from a small group of somatic cells (p. 378) make it theoretically plausible. In eggs this direct development has been interfered with and complicated by the intrusion of "interpolated" structure and the distortion of pattern that are induced by the specialization inherent in the egg but not in other cells. It is significant that these egg-induced structures and tissues are primarily those which are sensitive to the disruptive influence of the metamorphic hormone. Many, if not most, of the subsequent developmental events may be no more than the continuation or re-establishment of the whole pattern inherent in every unspecialized cell.

Both in insects and in amphibians the positive event is the embryonic development of a special larval organization, which depends on an unobservable but definite organization of the egg cortex; and metamorphosis is primarily the disintegration of a differentiated organization that is limited in extent of development anyway. But only with the dissolution of these tissues and the consequent flood of nutrient into the system can the suppressed components of a permanent pattern develop to assert the adult type. It should be noted, however, that the comparatively early dissolution of the balancer in urodele larvae and of the external gills and ventral suckers in frog tadpoles is in the same category as the later resorption of the urodele external gills and of the frog tail. As to dissolution of the balancer, autotomy is accomplished by epithelial proliferation across the balancer base, so that it seems evident, according to Kollros (1940), that the life span of the balancer is essentially determined by factors intrinsic to it. It is also significant that larval structure can be readily regenerated in many organisms; e.g., the urodele balancer and the anuran tail can regenerate, even though they are ultimately destined to undergo destruction.

The constructive aspect of metamorphosis is more than a response to the release from dominance by larval structure and to the increased availability of growth nutrient. In ascidians, for instance, the progressive acquisition of adult structure proceeds fairly normally even when the larval complex of tail and adhesive disks has been amputated. In the frog tadpole, tissues may respond directly to a local increase in thyroxin level, i.e., independently of metamorphic change as a whole; this phenomenon has been observed experimentally: introduction of thyroid fragments into the hindbrain brings about a localized premature maturation of the eyelid-closure mechanism. In general, it seems evident that stimuli supplied by the thyroid gland and by other tissues have a dual, simultaneous effect: inducing the histolysis of certain highly differentiated larval tissues, and initiating the growth and differentiation of other tissues. Such a double effect may be comparable to the pattern of growth and development of hydroid terminals (p. 211), whose development either proceeds to completion or does not commence. Initiation of growth depends on the nutritive level; but once begun, development is independent of the nutrient supply

—i.e., the terminal grows at the expense of any adjacent histologically mature tissue. This distinction between the initial triggering process, which may be entirely a matter of raising growth metabolism to the biological maximum, and the subsequent drive to developmental completion represents an almost exact duplication of the situation in amphibian metamorphosis.

Metamorphosis in Insects

In insects, besides the hormone's control over the time of onset of metamorphosis, the profound developmental transformation associated with pupation in holometabolous insects is of great interest. Henson (1946) regards insect metamorphosis in general as the repetition of the same developmental processes that occur during embryogenesis, rather than as the activation of imaginal rudiments, and he considers the type of metamorphosis typical of holometabolous insects as resulting from the removal of checks on the continuation of early development. If it is assumed that removal of checks is equivalent to the destruction of a larval structure that was originally induced by ooplasmic differentiation, then Henson's hypothesis is essentially the same as the concept of metamorphosis as primarily an induced dissolution (histolysis) of precociously differentiated tissues and consequent elimination of a dominating organization and as a growth stimulation for the surviving nondifferentiated tissues. In support of this thesis are experimental data, obtained by Bodenstein (1939), which suggest that pupal development is mainly a response to a high nutrient level and oxygen supply.

Formation of the imago during pupal development is mainly the result of local expansive growth and differentiation of epidermal imaginal disks and intestinal rings of comparable nature. Consequently, much attention has been given to the nature and development of the disks, particularly in *Drosophila* by Chun (1929). During the pupation of this and other similar types of insects, the epidermis of the imago derives from the ectoderm, which spreads from the imaginal disks to cover the whole surface of the individual. Each disk gives rise to one of the clearly defined areas of the imago: antennae originate from antennal disks; eyes, from optic disks; legs and wings, from specific leg and wing disks. Data on the origin of the less clearly delimited epidermal areas of the imago is as yet inconclusive. A comparison with *Euplotes* (p. 60) may be instructive.

Stern (1940) raises the question: Do the various imaginal anlagen in the larva constitute a completely predetermined mosaic of the later adult; or is the full extent of the imaginal surface to be formed by each disk not fixed at an early stage, but subject to mutual adjustments among various adjacent disks? From transplantation and marking experiments, he concluded that the prospective significance and potency of mesothoracic disks is definitely fixed before metamorphosis. In his later study (1954) of *Drosophila* bristles, each of which is a complex organ consisting of a bristle cell, a socket cell, and a nerve cell, Stern asks these pertinent questions: What causes certain cells at specific locations in the imaginal disks to differentiate into bristles instead of remaining epidermal cells? How do genes participate in this determination? He concludes that in a population of genetically identical somatic cells, the differentiation of those at certain locations must be due to a superimposed differential organization. Stern suggests that one or more systems of "prepatterns" precede the establish-

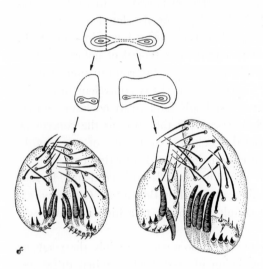

Figure 21-3. Parts of imaginal (genital) disk of *Drosophila*, showing differentiation of male anal plates. The position of the bicentric field-district for anal plates and its reorganization after cutting is indicated above. Each separated part differentiates a pair of anal plates. Large plate at right is normal size and presumably arose from intact right field center; other three plates, all harmoniously reduced in size and number of bristles, presumably arose from blastema of divided left field center. (After Hadorn.)

ment of the differentiation pattern; but what the nature of such systems may be is unknown. Quantitative or spatial differences may be factors; for example, it may be significant that the development of the mutant Bar-eye in *Drosophila* is determined by differences in the original size and cell number of the optic disks since the degree of expansion, or growth, of the disks is the same in the wild and the mutant forms (see the discussion of initial disk size in *Botryllus* on p. 390).

Imaginal disks, however, do seem to constitute a potential developmental mosaic. Hadorn (1953) cut the genital imaginal disk of the third instar larva of *Drosophila* into medial, paramedial, and transverse fragments and implanted the fragments into a host's body cavity, where they were well supplied with nutrients and hormones. At the onset of metamorphosis, when the fragments differentiated, Hadorn observed that each fragment develops only its district-specific elements; thus he concluded that the different parts of the genital disk had been localized within the disk at the time of operation. On the other hand, each such district has the properties of a field, so that individual regions of the field-district can restore the normal organization of the field. It is noteworthy that such regulation is always accompanied by cell proliferation until normal cell number, shape, and size of the district are restored.

Accordingly, in the larval state, the organism consists of a pattern within a pattern: one is functional and histologically mature, and the other is apparently discrete and disparate. Since each is probably based in a material continuum, it is entirely likely that coexistence entails some mutual interference. For instance, what are the relative roles played by epidermis and mesoderm? With this question in mind, Shatoury (1955) has compared the development of the imaginal buds in normal wild-type and the lethal no-differentiation mutant of *Drosophila*. In normal development the imaginal disks are simple epithelial invaginations during the first and second instars and exhibit no local differentiation in the form of folds; but as soon as mesodermal tissue reaches these invaginations, the folding begins and the first sign of differentiation appears— a characteristic pattern of folds which correspond to the segments of the organ which the buds will form. In these folded buds the groups of mesodermal cells fit snugly into the epithelial folds; such fitting must be the result of mutual interactions. In the lethal no-differentiation type, some or all of the imaginal buds fail to continue differentiation during the third instar because the imaginal-disk mesoderm undergoes some abnormal proliferation; some, such as the leg disks, are strongly affected, and others, such as the genital disks, are only slightly disturbed. In other words, abnormal behavior of the mesoderm upsets the whole system; however, since the mesodermal cells arrive in place when they are still undetermined, they probably supply no more than a general stimulus, so that whatever orderly form they develop is probably induced by the epidermis to which they have migrated. The situation is comparable to that in the limb bud of chick embryos (p. 519).

From the studies by Wigglesworth on *Rhodnius*—studies relating to both hormonal control and the nature of the actual metamorphic change in insects— it is known that each bristle (a complex structure comparable to that in *Drosophila*) arises from a plaque consisting of a group of specialized epidermal cells. At each molt the number of plaques on an individual segment of the abdomen increases, and by the comparison of new and old cuticles it is evident that most of the plaques appear at the old sites. Between these old sites, formerly occupied by simple epidermal cells, there arise new plaques. These plaques are obviously not predetermined in number, since the number can be varied experimentally. That, in the normal insect, the new plaques appear between the most widely separated existent plaques may be a phenomenon of general significance, for it is not unique with plaques or even with insects: a similar interpolation of new structural units is typical of the growth of the branchial wall in many ascidians, in which new stigmatic perforations appear between adjacent gill slits wherever space permits; similarly, in a hydroid such as *Syncoryne* number of gonophores between tentacles progressively increases.

In *Rhodnius,* when a wide area of the epidermis of the second-stage nymph is destroyed by excessive heat, the epidermal cells at the margin of the burned area divide repeatedly and spread inward, below the cuticle covering the burned area, to give rise after the next molt to epidermis devoid of plaques and bristles. However, by the third stage, plaques and bristles have reappeared to cover much of the healed area. In analyzing this process, Wigglesworth (1954) reports that during molting, the epidermal cells begin to divide profusely —far beyond requirements—and a large number subsequently die. Yet the final spacing of nuclei in the new epithelium is very orderly, so that the plaques—which are evenly distributed throughout the epithelium—must be separated from one another by approximately constant numbers of cells. This even distribution was found even when the integument had been greatly extended by experimental blocking of the anus, so that the plaques were correspondingly more widely separated. Thus the number of cells, rather than the amount of space between plaques, is the important factor. Such regular histological spacing recalls the pattern of single and double cones in the retina of vertebrates (p. 139).

By feeding fifth-stage nymphs, and at various periods after feeding, decapitating them and transfusing with blood from the fourth-stage nymph whose corpus allatum is intact, the fifth-stage nymphs commenced metamorphosis, but after a varying interval resumed nymphal development. Even the adult could be caused to molt by transfusion with blood from several molting fifth-stage nymphs. Thus it is possible, by producing a *Rhodnius* whose form is intermediate between nymph and adult, to study the transformation of cuticle from nymph to adult and vice versa. According to such experiments, the different elements in the cuticular structure of the *Rhodnius* abdomen become determined at different times, and the resumption of the nymphal cuticle when the adult is hormonally induced to molt again are evidence that the system which can produce this type of cuticle is latent in the adult cell community.

When a cylindrical segment from a limb is implanted in the abdomen of *Rhodnius,* the epidermal cells spread outward and backward from the two cut margins until they unite to form a continuous inner coat (cyst) which can be induced to molt simultaneously with its host. Similarly, when two insects are decapitated and united neck to neck, the epidermis of one unites with that of the other, so that it becomes impossible to tell where one insect ends and the other begins. When united in this way, they molt simultaneously; but when joined by a capillary tube (so that only blood is shared in common), they molt independently.

What is the nature of this continuity? Wigglesworth suggests that it is a chemical linkage and that "the epidermis is a chemical continuum upon which the cells are strung at intervals like beads upon a net. In other words, the organism is a giant molecule—for the form of the organism is determined by the epidermis, and we should probably not be far from the truth in regarding the epidermis *as* the organism." [2] He envisions the organism as a giant molecule with potentially active centers which require specific hormones in order to act, with individual groups of cells, each group maintaining the integrity of one fragment of the continuum, and with the whole being an expansion of the cortical plasm of the oocyte.

> If metamorphosis is produced experimentally in the 1st-stage nymph soon after hatching from the egg, the epidermis of the immature adult that results is made up of less than one-hundredth part of the number of cells which compose the normal adult. This simple observation alone shows how limited is the importance of the cells in defining the form of the body. The cells do not "cooperate to mould the body form"; they merely carry and care for a small segment of the continuum which is the organism and of which they are the servants. The movements and secretions which bring about the visible process of growth and metamorphosis are, of course, the products of cells. But the cells are only the agents of the pervading web which is the organism itself. [3]

The experimental control of the time and size of metamorphosis in anuran amphibians could well lead to a similar statement—which is the main theme of this book—and the fundamental nature of the continuum itself remains the problem of overwhelming importance to any understanding of organism and organization. From all sources, the evidence indicates that the basal continuity of the epidermis carries the primary imprint of the organism; that interruption of this continuity affects the integrity of the organism; that the pattern inherent in this layer, in its initial state, is activated and maintained, but not determined, by more-general influences from inner tissues; and that the emergence of pattern during development is, apart from such activation, the outcome of the material expansion of the continuum. It is probable that the continuum includes the basal continuity of the digestive tube, since there is obvious continuity of surface between this tube and the epidermal-ectodermal layer. Admittedly, this is an oversimplification; but it is a note to end on which may well be but the beginning.

[2] Wigglesworth, in *Essays on Growth and Form*, p. 37 (1945).
[3] Wigglesworth, *op. cit.*, p. 39.

REFERENCES FOR
Development of Eggs

NATURE AND DEVELOPMENT OF EGGS

Barth, L. G. 1941. "Neural Differentiation without an Organizer." *J. Exp. Zool.* **87**:371-82.

Berrill, N. J. 1931. "Studies in Tunicate Development. II. Abbreviation of Development in the Molgulidae." *Roy. Soc. London Philos. Trans.* Series B. **219**:281-346.

———. 1935. "Studies in Tunicate Development. III. Differential Retardation and Acceleration." *Ibid.* **225**:253-336.

———. 1941. "Spatial and Temporal Growth Patterns in Colonial Organisms." *Growth* (Supp.). **5**:89-111.

———. 1948. "The Nature of the Ascidian Tadpole, with Reference to Boltenia echinata." *J. Morphol.* **82**:269-85.

———. 1950. "Budding and Development in Salpa." *Ibid.* **87**:553-606.

———. 1955. *The Origin of Vertebrates.* Fairlawn, N. J.: Oxford Univ. Press.

Bigelow, H. B. 1909. "The Medusae." *Mus. Comp. Zool. Harvard Bull.* **37**:9-245.

Brien, P. 1928. "Contribution à l'étude de l'embryogénèse et de la blastogenèse des Salpes." *Inst. zool. Torley-Rousseau Rec.* **2**:5-116.

Brooks, W. K. 1893. "The Genus Salpa." *Johns Hopkins Univ. Biol. Lab. Mem.* **2**:1-396.

Child, C. M. 1928. "The Physiological Gradients." *Protoplasma.* **5**:447-76.

———. 1936. "A Contribution to the Physiology of Exogastrulation in Echinoderms." *Arch. entwick. Org.* **135**:457-93.

Conklin, E. G. 1905. "The Organization and Cell-Lineage of the Ascidian Egg." *Acad. Nat. Sci. Phila. J.* **13**:1-119.

———. 1917. "Effects of Centrifugal Force on the Structure and Development of the Eggs of Crepidula." *J. Exp. Zool.* **22**:311-419.

———. 1931. "The Development of Centrifuged Eggs of Ascidians." *Ibid.* **60**:1-119.

———. 1932. "The Embryology of Amphioxus." *J. Morphol.* **54**:69-151.

———. 1933. "The Development of Isolated and Partially Separated Blastomeres of Amphioxus." *J. Exp. Zool.* **64**:303-75.

Costello, D. P. 1945. "Segregation of Öoplasmic Constituents." *Elisha Mitchell Sci. Soc. J.* **61**:277-89.

Dalcq, A. 1932. "Études des localisations germinales dans l'oeuf vierge d'ascidie par des expériences de Mérogonie." *Arch. d'anat. micro. morphol. exp.* **28**:223-333.

———. 1938. "Étude micrographique et quantitative de la Mérogonie double chez Ascidiella sdarbra." *Arch. biol.* **49**:397-568.

———, and J. Pasteels. 1937. "Une conception nouvelle des bases physiologiques de la morphogénèse." *Ibid.* **48**:669-710.

Delsman, H. C. 1912. "Weitere Beobachtungen über die Entwicklung von Oikopleura dioica." *Nederl. dierk. ver. Tijdschr.* **12**:197-205.

Eyal-Giladi, H. 1954. "Dynamics of Neural Induction in Amphibia." *Arch. biol.* **65**:180-259.

Frankhauser, G. 1948. "The Organization of the Amphibian Egg during Fertilization and Cleavage." *N. Y. Acad. Sci. Ann.* **49**:684-708.

Fauré-Fremiet, F., and H. Mugard. 1948. "Ségrégation d'un matériel cortical au cours de la segmentation chez l'oeuf de Teredo norvegica." *Acad. sci. Comptes* rendus. **227**:1409-11.

Gillette, R. 1944. "Cell Number and Cell Size in the Ectoderm during Neurulation (Ambystoma maculatum)." *J. Exp. Zool.* **96**:201-22.

Harmer, S. F. 1893. "On the Occurrence of Embryonic Fission in Cyclostomatous Polyzoa." *Quart. J. Micro. Sci.* **135**:199-242.

Harrison, R. G. 1945. "Relations of Symmetry in the Developing Embryo." *Conn. Acad. Arts Sci. Trans.* **36**:277-330.

Holtfreter, J. 1943. "A Study of the Mechanics of Gastrulation. I." *J. Exp. Zool.* **94**:261-318.

——. 1944. "A Study of the Mechanics of Gastrulation. II." *Ibid.* **95**:171-212.

Hörstadius, S. 1953. "The Effect of Lithium Ions on Centrifuged Eggs of Paracentrotus lividus." *Staz. zool. Pubb.* **24**:45-60.

——. 1955. "Reduction Gradients in Animalized and Vegetalized Sea Urchin Eggs." *J. Exp. Zool.* **129**:249-56.

Horst, J. ter. 1948. "Differenzierungs- und Induktionsleis-tungen verscheidener Abschnitte der Medullarplatte und des Urdarmdaches von Triton im Kombinat." *Arch. entwick. Org.* **143**:275-303.

Kemp, N. E. 1946. "Regulation in the Entoderm of the Tree Frog Hyla regilla." *Univ. Calif. Pub. Zool. 1942-47.* **51**:159-84.

Leiby, R. W., and C. C. Hill. 1924. "The Polyembryonic Development of Platygaster verualis." *J. Agric. Research.* **28**:829-40.

Lillie, F. R. 1902. "Differentiation without Cleavage in the Egg of the Annelid Chaetopterus pergamentaceus." *Arch. entwick. Org.* **14**:477-99.

——. 1906. "Observations and Experiments concerning the Elementary Phenomena of Embryonic Development in Chaetopterus." *J. Exp. Zool.* **3**:153-268.

Millar, R. H. 1951. "The Development and Early Stages of the Ascidian Pyura squamulosa (Alder)." *Marine Biol. Assoc. U. K. J.* **30**:27-31.

Morgan, T. H. 1895. "The Formation of the Fish Embryo." *J. Morphol.* **10**:419-72.

Needham, J. 1933. "On the Dissociability of the Fundamental Process in Ontogenesis." *Biol. Rev.* **8**:180-223.

Nieuwkoop, P. D. 1952. "Activation and Organization of the Central Nervous System in Amphibians." *J. Exp. Zool.* **120**:1-108.

——. 1956. "Are there direct relationships between the cortical layer of the fertilized egg and the development of the future axial system in Xenopus laevis embryos? *Staz. zool. Pubb.* **28**:241-9.

——, and P. A. Florschütz. 1950. "Quelques caractères spéciaux de la gastrulation et de la neurulation de l'oeuf de Xenopus laevis, Dand., et de quelques autres Anoures. I. Descriptive." *Arch. biol.* **61**:113-50.

Nieuwkoop, P. D., and G. V. Nigtevecht. 1954. "Neural Activation and Transformation in Explants of Competent Ectoderm under the Influence of Fragments of Anterior Notochord in Urodeles." *J. Embryol. Exp. Morphol.* **2**:175-93.

Okada, T. S. 1955. "Experimental Studies on the Differentiation of the Endodermal Organs in Amphibia. I, II, III." *Kyoto Univ. Coll. Sci. Mem.* Series B. **21**:1-22.

Oppenheimer, J. 1955. "The Differentiation of Derivatives of the Lower Germ Layers in Fundulus following Implantation of Shield Groups." *J. Exp. Zool.* **128**:525-60.

Ortolani, G. 1955. "I movimenti corticali dell'uova di Ascidie alla fecondazione." *Riv. biol.* 47:171-8.

Pasteels, J. 1949. "Observations sur la localisation de la plaque prechordale presumptive au cours de la gastrulation chez Xenopus laevis." *Arch. biol.* 60:235-50.

Paterson, M. C. 1957. "Animal-Vegetal Balance in Amphibian Development." *J. Exp. Zool.* 134:183-205.

Reverberi, G., and A. Miganti. 1947. "La distributione delle potenze nel germe di Ascidie allostadio di otto blastomeri, annalyizzata mediante le combinazioni e i transplanti di blastomeri." *Staz. zool. Pubb.* 21:1-35.

Roach, F. C. 1945. "Differentiation of the Central Nervous System after Axial Reversals of the Medullary Plate of Amblystoma." *J. Exp. Zool.* 99:53-77.

Runnström, J. 1929. "Über Selbstdifferenzierung und Induktion bei dem Seeigelkeim." *Arch. entwick. Org.* 117:123-45.

Sala, M. 1955. "Distribution of Activating and Transforming Influences in the Archenteron Roof during the Induction of the Nervous System in Amphibians. I. Distribution in Cranio-caudal Direction." *Konink. ned. akad. Weten. Proc.* Series C.

Schechtman, A. M. 1934. "Unipolar Ingression in Triturus torosis." *Univ. Calif. Pub. Zool.* 36:303-10.

————. 1942. "The Mechanism of Amphibian Gastrulation. I. Gastrulation-promoting Interactions between Various Regions of an Anuran Egg (Hyla regilla)." *Univ. Calif. Pub. Zool.* 1942-47. 51:1-40.

Silvestri, F. 1936. "Insect Polyembryony and Its General Biological Aspects." *Mus. Comp. Zool. Harvard Bull.* 81:469-98.

Townes, P. L., and J. Holtfreter. 1955. "Directed Movements and Selective Adhesion of Embryonic Amphibian Cells." *J. Exp. Zool.* 128:53-118.

Trifonowa, A. 1934. "Parthenogenese der Fische." *Acta Zool.* 15:183-213.

Trinkaus, J. P. 1951. "A Study of the Mechanism of Epiboly in the Egg of Fundulus heteroclitus." *J. Exp. Zool.* 118:269-320.

Von Ubisch, L. 1933a. "Untersuchungen über Formbildung. IV." *Arch. entwick. Org.* 129:45-67.

————. 1933b. "Untersuchungen über Formbildung. V." *Ibid.* Pp. 68-87.

————. 1936. "Untersuchungen über Formbildung. VII." *Ibid.* 134:644-67.

Waddington, C. H., 1956. *The Principles of Embryology.* London: George Allen and Unwin.

Weisz, P. B. 1945. "The Development and Morphology of the Larva of the South African Clawed Toad, Xenopus laevis." *J. Morphol.* 77:163-217.

Whitaker, D. M. 1933. "On the Rate of Oxygen Consumption by Fertilized and Unfertilized Eggs. V. Comparisons and Interpretations." *J. Gen. Physiol.* 16:497-528.

TISSUE INTERACTIONS

Berrill, N. J. 1950. "Budding in Pyrosoma." *J. Morphol.* 87:537-52.

Beyer, K., and C. M. Child. 1930. "Reconstitution of Lateral Pieces in Planaria dorotocephala and P. maculata." *Physiol. Zool.* 3:342-65.

Brachet, J. 1958. *Biochemical Cytology.* New York: Academic Press, Inc.

Brahma, S. K. 1958. "Experiments on the Diffusibility of the Amphibian Organizer." *J. Embryol. Exp. Morphol.* 6:418-23.

Butler, E. G. 1949. "Effects of Delayed Denervation on Regenerative Activity in Limbs of Urodele Larvae." *J. Exp. Zool.* **112**:361-92.

———. 1955. "Regeneration in the Urodele Fore-Limb after Reversal of Its Proximo-distal Axis." *J. Morphol.* **96**:265-82.

Coe, W. R. 1930. "Regeneration in Nemerteans. II. Regeneration of Small Sections of the Body Split or Partially Split Longitudinally." *J. Exp. Zool.* **57**:109-44.

———. 1934. "Analysis of the Regenerative Processes in Nemerteans." *Biol. Bull.* **66**:304-15.

Fankhauser, G. 1945. "Maintenance of Normal Structure in Heteroploid Salamander Larvae, through Compensation of Changes in Cell Size by Adjustment of Cell Number and Cell Shape." *J. Exp. Zool.* **100**:445-555.

Fell, H. B., and E. Mellanby. 1953. "Metaplasia Produced in Cultures of Chick Ectoderm by High Vitamin A." *J. Physiol.* **119**:470-88.

Glade, R. W. 1957. "The Effect of Tail Tissue on Limb Regeneration in Triturus irridescens." *J. Morphol.* **101**:477-522.

Godeaux, J. 1957. "Contribution à la connaissance des Thaliaces." *Soc. roy. zool. Belgique Ann.* **88**:1-285.

Grobstein, C. 1954. "Tissue Interaction in the Morphogenesis of Mouse Embryonic Rudiments in Vitro," in D. Rudnick (ed.). *Aspects of Synthesis and Order in Growth.* Soc. Study Dev. and Growth. Symp. 13. Princeton, N. J.: Princeton Univ. Press. Pp. 233-56.

Harrison, R. G. 1921. "On Relations of Symmetry in Transplanted Limbs." *J. Exp. Zool.* **32**:1-136.

———. 1925. "The Development of the Balancer in Amblystoma, Studied by the Method of Transplantation and in Connection with the Connective Tissue Problem." *Ibid.* **41**:349-427.

Holtfreter, J. 1935. "Morphologisch beeinflussung von Urodelen ektoderm bei xenoplastischer Transplantation." *Arch. entwick. Org.* **133**:367-494.

———. 1936. "Regionale Induktion in xenoplastisch zusammengesetztsen Explantation." *Ibid.* **134**:466-550.

———. 1951. "Some Aspects of Embryonic Induction. Soc. Study Dev. and Growth. Symp. 10. Pp. 152-171.

Holtzer, S. 1956. "The Inductive Activity of the Spinal Cord in Urodele Tail Regeneration." *J. Morphol.* **99**:1-39.

Ichikawa, M., and Y. K. Okada. 1954. "Studies on the Determination Factors in Amphibian Limb Regeneration." *Kyoto Univ. Coll. Sci. Mem.* Series B. **21**:23-8.

Jacobson, A. G. 1955. "The Roles of the Optic Vesicle and Head Tissue in Lens Induction." *Nat. Acad. Sci. U. S. Proc.* **41**:422-5.

Kamrin, A. A., and M. Singer. 1959. "The Growth Influence of Spinal Ganglia Implanted into the Denervated Forelimb Regenerate of the Newt, Triturus." *J. Morphol.* **104**:405-40.

King, T. J., and R. Briggs. 1955. "Changes in the Nuclei of Differentiating Gastrula Cells As Demonstrated by Nuclear Transplantation." *Nar. Acad. Sci. U. S. Proc.* **41**:321-5.

Levi-Montcalcini, R. 1952. "Effects of Mouse Tumor Transplantation on the nervous System." *N. Y. Acad. Sci. Ann.* **55**:330-43.

Lillie, F. R. 1942. "Recent Work on the Development of Feathers." *Biol. Rev.* **17**:247-66.

Mangold, O. 1931. "Das Determinations problem III." *Ergebn. Biol.* **7**:193-404.

McKeehan, M. S. 1951. "Cytological Aspects of Embryonic Lens Induction in the Chick." *J. Exp. Zool.* **117**:31-64.

Niu, M. C. 1956. "New Approaches to the Problem of Embryonic Induction," in D. Rudnick (ed.). *Cellular Mechanisms in Differentiation and Growth.* Soc. Study Dev. and Growth. Symp. 14. Princeton, N. J.: Princeton Univ. Press. Pp. 155-72.

Okada, Y. K. 1934. "Régénération de la tête de *Myxicola aesthetica* (Clap.)." *Bull. Biol. France-Belgique.* **68**:340-81.

———, and T. Kawakami. 1943. "Transplantation Experiments in the Earthworm, *Eisenia foetida* (Savigny), with Special Remarks on the Inductive Effect of the Nerve Cord on the Differentiation of the Body Wall." *Tokyo Univ. Fac. Sci. J.* Series 4. **6**:25-96.

Okada, Y. K., and H. Tozawa. 1943. "Supplementary Experiments of Transplantation in the Earthworm. The Induction of a Tail by the Transplanted Nerve Cord." *Ibid.* Pp. 635-47.

Overton, J. 1950. "Mitotic Stimulation of Amphibian Epidermis by Underlying Grafts of Central Nervous Tissue." *J. Exp. Zool.* **115**:521-60.

Rotmann, E. 1940. "Die Bedeutung der Zellgrosse fur die Entwicklung der amphibien-lense." *Arch. entwick. Org.* **140**:124-56.

Saunders, J. W., M. T. Gasseling, and M. D. Gfellar. 1958. "Interactions of Ectoderm and Mesoderm in the Origin of Axial Relationships in the Wing of the Fowl." *J. Exp. Zool.* **137**:39-74.

Sehotté, O., and Mac. V. Edds. 1940. "Xenoplastic Induction of Rana pipiens Adhesive Discs on Balancer Site of Amblystoma punctatum." *J. Exp. Zool.* **84**:199-221.

Singer, M. 1946. "The Nervous System and Regeneration of the Forelimb of Adult Triturus. V. The Influence of the Number of Nerve Fibers, Including a Quantitative Study of Limb Innervation." *J. Exp. Zool.* **101**:299-338.

———. 1952. "The Influence of the Nerve in Regeneration of the Amphibian Extremity." *Quart. Rev. Biol.* **27**:169-200.

Takaya, H. 1938. "Transplantation of Limb Discs of Varying Size." *Kyoto Univ. Coll. Sci. Mem.* Series B. **4**:321-33.

Taylor, A. C. 1943. "Development of the Innervation Pattern in the Limb of the Frog." *Anatomical Record.* **87**:379-413.

Thornton, C. S. 1958. "Inhibition of Limb Regeneration in Urodele Larvae by Localized Irradiation with Ultra-violet Light." *J. Exp. Zool.* **137**:153-79.

Tung, Y. F. Y., and C. Y. Chang. 1949. "Studies on the Induction of Ciliary Polarity in Amphibia." *Zool. Soc. London Proc.* **118**:1134-79.

Twitty, V. C. 1945. "The Developmental Analysis of Specific Pigment Patterns." *J. Exp. Zool.* **100**:141-78.

———. 1949. "Developmental Analysis of Amphibian Pigmentation." *Growth* Supp. (Symp. 9) **13**:3-61.

———. 1955. "The Eye," in B. H. Willier *et al.* (eds.). *Analysis of Development.* Phila., Pa.: W. B. Saunders Co. Pp. 402-14.

Wang, H. 1943. "Development of the Innervation Pattern in the Limb Bud of the Frog." *Anatomical Record.* **87**:379-413.

Watterson, R. L. 1942. "The Morphogenesis of Down Feathers with Special Reference to the Developmental History of Melanophores." *Physiol. Zool.* **15**:234-60.

Weiss, P. 1933. "Functional Adaptation and the Role of Ground Substances in Development." *Am. Nat.* **67**:322-40.

———. 1947. "The Problem of Specificity in Growth and Development." *Yale J. Biol. Med.* **19**:255-78.

———, and R. James. 1955. "Skin Metaplasia in Vitro Induced by Brief Exposure to Vitamin A." *Exp. Cell Research* Supp. **3**:381-94.

Willier, B. H., and M. E. Rawles. 1940. "The Control of Feather Color Pattern by Melanophores Grafted from One Embryo to Another of a Different Fowl Breed." *Physiol. Zool.* **13**:127-200.

Yamada, S., and S. Takada. 1955. "An Analysis of Spinal-Caudal Induction by the Guinea Pig Kidney in the Isolated Ectoderm of Triturus Gastrula." *J. Exp. Zool.* **128**:291-331.

Zwilling, E. 1955. "Ectoderm-Mesoderm Relationship in the Development of the Chick Embryo Limb Bud." *J. Exp. Zool.* **128**:423-41.

———. 1956. "Interaction between Limb Ectoderm and Mesoderm in the Chick Embryo." *Ibid.* **132**:157-253.

———, and J. F. Ames. 1958. "Polydactyly, Related Defects and Axial Shifts—A Critique." *Am. Nat.* **92**:257-66.

METAMORPHOSIS

Allen, B. M. 1938. "The Endocrine Control of Amphibian Metamorphosis." *Biol. Rev.* **13**:1-19.

Berger, C. A. 1938. "Multiplication and Reduction of Somatic Chromosome Groups as a Regular Developmental Process in the Mosquito, Culex pipiens." *Carnegie Inst. Washington* Pub. 496. Pp. 209-32.

Berrill, N. J. 1929. "Studies in Tunicate Development. I. General Physiology of Development of Simple Ascidians." *Roy. Soc. London Philos. Trans.* Series B. **218**:37-78.

———. 1947. "Metamorphosis in Ascidians." *J. Morphol.* **81**:249-68.

Bodenstein, D. 1939. "Investigations on the Problem of Metamorphosis. VI." *J. Exp. Zool.* **82**:329-56.

Bytinsky-Salz, H. 1935. "Heteroplastic Transplantation of the Hypophysis in Ambystoma." *J. Exp. Zool.* **72**:81-93.

Chun, T.-Y. 1929. "On the Development of Imaginal Buds in Normal and Mutant Drosophila melanogaster." *J. Morphol.* **47**:135-99.

Clausen, H. 1930. "Rate of Histolysis of Anuran Tail Skin and Muscle during Metamorphosis." *Biol. Bull.* **59**:199-210.

De Beer, G. R. 1957. *Embryos and Ancestors.* Oxford: Clarendon Press. 159 p.

Etkin, W. 1955. "Metamorphosis," in B. H. Willier *et al.* (eds.). *Analysis of Development.* Phila., Pa.: W. B. Saunders Co. Pp. 631-63.

Fell, H. B. 1948. "Echinoderm Embryology and the Origin of Chordates." *Biol. Rev.* **23**:81-107.

Glaser, O., and G. A. Anslow. 1949. "Copper and Ascidian Metamorphosis." *J. Exp. Zool.* **111**:117-40.

Grave, C., and P. A. Nicoll. 1939. "Studies of Larval Life and Metamorphosis in Ascidia nigra and Species of Polyandrocarpa." *Carnegie Inst. Washington* Pub. 517. Pp. 1-46.

Hadorn, E. 1953. "Regulation and Differentiation within Field Districts in Imaginal Discs of Drosophila." *J. Embryol. Exp. Morphol.* **1**:213-6.

Henson, H. 1946. "The Theoretical Aspect of Insect Metamorphosis." *Biol. Rev.* **21**:1-14.

Kollros, J. J. 1940. "The Disappearance of the Balancer in Amblystoma Larvae." *J. Exp. Zool.* **85**:33-52.

Lynn, W. G., and H. E. Wachowski. 1951. "The Thyroid Gland and Its Functions in Cold-blooded Vertebrates." *Quart. Rev. Biol.* **26**:123-68.

Shatoury, H. H. 1955. "Lethal No-D'fferentiation and the Development of the Imaginal Discs during the Larval Stage in Drosophila." *Arch. entwick. Org.* **147**:523-38.

Smith, J. E. 1935. "The Early Development of the Nemertean Cephalotrix rufifrons." *Quart. J. Micro. Sci.* **77**:337-81.

Stern, C. 1940. "The Prospective Significance of Imaginal Discs in Drosophila." *J. Morphol.* **67**:107-22.

———. 1954. "Genes and Developmental Pattern." *Caryologia* Supp. **6**:355-69.

Steward, F. C. 1958. "Growth and Organized Development of Cultured Cells." *Am. J. Bot.* **45**:709-14.

Thomsen, E. 1949. "Influence of Removal of Corpus Allatum on the Oxygen Consumption of Adult Calliphora erythrocephala." *J. Exp. Biol.* **26**:137-49.

Wigglesworth, V. 1940. "The Determination of the Characters at Metamorphosis in Rhodnius prolixus (Hemiptera)." *J. Exp. Biol.* **17**:201-22.

———. 1945. *Essays on Growth and Form.* Oxford: Clarendon Press.

———. 1954. *The Physiology of Insect Metamorphosis.* New York: Cambridge Univ. Press.

Wilson, D. P. 1955. "The Role of Micro-organisms in the Settlement of Ophelia bicornis Savigny." *Marine Biol. Assoc. U.K. J.* **34**:531-43.

Index